Factor Categories with Applications to Direct Decomposition of Modules

PURE AND APPLIED MATHEMATICS

A Program of Monographs, Textbooks, and Lecture Notes

Contributions to *Lecture Notes in Pure and Applied Mathematics* are reproduced by direct photography of the author's typewritten manuscript. Potential authors are advised to submit preliminary manuscripts for review purposes. After acceptance, the author is responsible for preparing the final manuscript in camera-ready form, suitable for direct reproduction. Marcel Dekker, Inc. will furnish instructions to authors and special typing paper. Sample pages are reviewed and returned with our suggestions to assure quality control and the most attractive rendering of your manuscript. The publisher will also be happy to supervise and assist in all stages of the preparation of your camera-ready manuscript.

LECTURE NOTES
IN PURE AND APPLIED MATHEMATICS

1. *N. Jacobson,* Exceptional Lie Algebras
2. *L.-Å. Lindahl and F. Poulsen,* Thin Sets in Harmonic Analysis
3. *I. Satake,* Classification Theory of Semi-Simple Algebraic Groups
4. *F. Hirzebruch, W. D. Newmann, and S. S. Koh,* Differentiable Manifolds and Quadratic Forms (out of print)
5. *I. Chavel,* Riemannian Symmetric Spaces of Rank One (out of print)
6. *R. B. Burckel,* Characterization of C(X) Among Its Subalgebras
7. *B. R. McDonald, A. R. Magid, and K. C. Smith,* Ring Theory: Proceedings of the Oklahoma Conference
8. *Y.-T. Siu,* Techniques of Extension of Analytic Objects
9. *S. R. Caradus, W. E. Pfaffenberger, and B. Yood,* Calkin Algebras and Algebras of Operators on Banach Spaces
10. *E. O. Roxin, P.-T. Liu, and R. L. Sternberg,* Differential Games and Control Theory
11. *M. Orzech and C. Small,* The Brauer Group of Commutative Rings
12. *S. Thomeier,* Topology and Its Applications
13. *J. M. Lopez and K. A. Ross,* Sidon Sets
14. *W. W. Comfort and S. Negrepontis,* Continuous Pseudometrics
15. *K. McKennon and J. M. Robertson,* Locally Convex Spaces
16. *M. Carmeli and S. Malin,* Representations of the Rotation and Lorentz Groups: An Introduction
17. *G. B. Seligman,* Rational Methods in Lie Algebras
18. *D. G. de Figueiredo,* Functional Analysis: Proceedings of the Brazilian Mathematical Society Symposium
19. *L. Cesari, R. Kannan, and J. D. Schuur,* Nonlinear Functional Analysis and Differential Equations: Proceedings of the Michigan State University Conference
20. *J. J. Schäffer,* Geometry of Spheres in Normed Spaces
21. *K. Yano and M. Kon,* Anti-Invariant Submanifolds
22. *W. V. Vasconcelos,* The Rings of Dimension Two
23. *R. E. Chandler,* Hausdorff Compactifications
24. *S. P. Franklin and B. V. S. Thomas,* Topology: Proceedings of the Memphis State University Conference
25. *S. K. Jain,* Ring Theory: Proceedings of the Ohio University Conference
26. *B. R. McDonald and R. A. Morris,* Ring Theory II: Proceedings of the Second Oklahoma Conference
27. *R. B. Mura and A. Rhemtulla,* Orderable Groups
28. *J. R. Graef,* Stability of Dynamical Systems: Theory and Applications
29. *H.-C. Wang,* Homogeneous Banach Algebras
30. *E. O. Roxin, P.-T. Liu, and R. L. Sternberg,* Differential Games and Control Theory II
31. *R. D. Porter,* Introduction to Fibre Bundles
32. *M. Altman,* Contractors and Contractor Directions Theory and Applications
33. *J. S. Golan,* Decomposition and Dimension in Module Categories
34. *G. Fairweather,* Finite Element Galerkin Methods for Differential Equations
35. *J. D. Sally,* Numbers of Generators of Ideals in Local Rings
36. *S. S. Miller,* Complex Analysis: Proceedings of the S.U.N.Y. Brockport Conference
37. *R. Gordon,* Representation Theory of Algebras: Proceedings of the Philadelphia Conference
38. *M. Goto and F. D. Grosshans,* Semisimple Lie Algebras
39. *A. I. Arruda, N. C. A. da Costa, and R. Chuaqui,* Mathematical Logic: Proceedings of the First Brazilian Conference

40. *F. Van Oystaeyen,* Ring Theory: Proceedings of the 1977 Antwerp Conference
41. *F. Van Oystaeyen and A. Verschoren,* Reflectors and Localization: Application to Sheaf Theory
42. *M. Satyanarayana,* Positively Ordered Semigroups
43. *D. L. Russell,* Mathematics of Finite-Dimensional Control Systems
44. *P.-T. Liu and E. Roxin,* Differential Games and Control Theory III: Proceedings of the Third Kingston Conference, Part A
45. *A. Geramita and J. Seberry,* Orthogonal Designs: Quadratic Forms and Hadamard Matrices
46. *J. Cigler, V. Losert, and P. Michor,* Banach Modules and Functors on Categories of Banach Spaces
47. *P.-T. Liu and J. G. Sutinen,* Control Theory in Mathematical Economics: Proceedings of the Third Kingston Conference, Part B
48. *C. Byrnes,* Partial Differential Equations and Geometry
49. *G. Klambauer,* Problems and Propositions in Analysis
50. *J. Knopfmacher,* Analytic Arithmetic of Algebraic Function Fields
51. *F. Van Oystaeyen,* Ring Theory: Proceedings of the 1978 Antwerp Conference
52. *B. Kedem,* Binary Time Series
53. *J. Barros-Neto and R. A. Artino,* Hypoelliptic Boundary-Value Problems
54. *R. L. Sternberg, A. J. Kalinowski, and J. S. Papadakis,* Nonlinear Partial Differential Equations in Engineering and Applied Science
55. *B. R. McDonald,* Ring Theory and Algebra III: Proceedings of the Third Oklahoma Conference
56. *J. S. Golan,* Structure Sheaves over a Noncommutative Ring
57. *T. V. Narayana, J. G. Williams, and R. M. Mathsen,* Combinatorics, Representation Theory and Statistical Methods in Groups: YOUNG DAY Proceedings
58. *T. A. Burton,* Modeling and Differential Equations in Biology
59. *K. H. Kim and F. W. Roush,* Introduction to Mathematical Consensus Theory
60. *J. Banas and K. Goebel,* Measures of Noncompactness in Banach Spaces
61. *O. A. Nielsen,* Direct Integral Theory
62. *J. E. Smith, G. O. Kenny, and R. N. Ball,* Ordered Groups: Proceedings of the Boise State Conference
63. *J. Cronin,* Mathematics of Cell Electrophysiology
64. *J. W. Brewer,* Power Series Over Commutative Rings
65. *P. K. Kamthan and M. Gupta,* Sequence Spaces and Series
66. *T. G. McLaughlin,* Regressive Sets and the Theory of Isols
67. *T. L. Herdman, S. M. Rankin, III, and H. W. Stech,* Integral and Functional Differential Equations
68. *R. Draper,* Commutative Algebra: Analytic Methods
69. *W. G. McKay and J. Patera,* Tables of Dimensions, Indices, and Branching Rules for Representations of Simple Lie Algebras
70. *R. L. Devaney and Z. H. Nitecki,* Classical Mechanics and Dynamical Systems
71. *J. Van Geel,* Places and Valuations in Noncommutative Ring Theory
72. *C. Faith,* Injective Modules and Injective Quotient Rings
73. *A. Fiacco,* Mathematical Programming with Data Perturbations I
74. *P. Schultz, C. Praeger, and R. Sullivan,* Algebraic Structures and Applications: Proceedings of the First Western Australian Conference on Algebra
75. *L. Bican, T. Kepka, and P. Nemec,* Rings, Modules, and Preradicals
76. *D. C. Kay and M. Breen,* Convexity and Related Combinatorial Geometry: Proceedings of the Second University of Oklahoma Conference
77. *P. Fletcher and W. F. Lindgren,* Quasi-Uniform Spaces
78. *C.-C. Yang,* Factorization Theory of Meromorphic Functions
79. *O. Taussky,* Ternary Quadratic Forms and Norms
80. *S. P. Singh and J. H. Burry,* Nonlinear Analysis and Applications
81. *K. B. Hannsgen, T. L. Herdman, H. W. Stech, and R. L. Wheeler,* Volterra and Functional Differential Equations

82. *N. L. Johnson, M. J. Kallaher, and C. T. Long,* Finite Geometries: Proceedings of a Conference in Honor of T. G. Ostrom
83. *G. I. Zapata,* Functional Analysis, Holomorphy, and Approximation Theory
84. *S. Greco and G. Valla,* Commutative Algebra: Proceedings of the Trento Conference
85. *A. V. Fiacco,* Mathematical Programming with Data Perturbations II
86. *J.-B. Hiriart-Urruty, W. Oettli, and J. Stoer,* Optimization: Theory and Algorithms
87. *A. Figa Talamanca and M. A. Picardello,* Harmonic Analysis on Free Groups
88. *M. Harada,* Factor Categories with Applications to Direct Decomposition of Modules

Other Volumes in Preparation

Factor Categories with Applications to Direct Decomposition of Modules

MANABU HARADA
Osaka City University
Osaka, Japan

MARCEL DEKKER, INC. New York and Basel

Library of Congress Cataloging in Publication Data

Harada, Manabu.
Factor categories with applications to direct decomposition of modules.

(Lecture notes in pure and applied mathematics ; 88)
Includes bibliographical references and index.
1. Modules (Algebra) 2. Categories (Mathematics)
I. Title. II. Series: Lecture notes in pure and applied mathematics ; v. 88.
QA247.H265 1983 512'.522 83-14334
ISBN 0-8247-1897-6

MARCEL DEKKER, INC.
270 Madison Avenue, New York, New York 10016

Current printing (last digit):
10 9 8 7 6 5 4 3 2 1

PRINTED IN THE UNITED STATES OF AMERICA

Preface

Category theory was introduced in the early 1940s by Eilenberg and MacLane [48,49]. Since Grothendieck [81] succeeded in applying category theory to the study of algebraic geometry in 1957, it has been utilized in many branches of algebra. Torsion theory and localization of noncommutative rings, due to Gabriel [73], are well-known examples in noncommutative ring theory. In this book another application is given. Our approach is entirely different from that mentioned above. The usual categorical method in ring theory has been to generalize to several branches, whereas the material in this book is confined to a particular branch, following the Krull-Remak-Schmidt-Azumaya theorem. I would like to emphasize that category theory is, in a sense, a generalization of ring theory, so it is quite natural that it is applicable to ring theory, as shown in torsion theory and localization. On the other hand, the application of category theory in this book is a direct one: namely, we cannot find any interpretation of the method in ring theory itself. To pursue this idea, we must define new concepts in category theory.

This book is a portion of my work related to direct sums of completely indecomposable modules [90,118] and related topics. Various preliminaries covered in the papers are given in the first five chapters. Thus the book is essentially an exposition of my work concerning direct decompositions of modules.

We begin in Chap. 1 with the definition of group. Whenever the results are elementary, they are stated without proof. Readers familiar with the material may skip this chapter. We state the fundamental concept of category theory in Chap. 2. There is not enough space to describe this theory completely, so the requisite concepts are picked up from Mitchell's book [159] for use in later chapters. We refer the reader to Ref. 159 for details. One of the important results in this chapter is Freyd's criterion for abelian categories [65], which is restated in terms of ring theory in Chap. 6 for later use. The contents of Chap. 3 will also be familiar to readers who have more than an undergraduate course in ring theory. A few properties of the Jacobson radical are given, and the structure of small submodules and of essential submodules, studied in Chap. 3, will be used again in Chaps. 6 and 7. We also explore the ring of column-finite (summable) matrices, which is utilized frequently throughout the book. In Chap. 4 the general properties of projectives and injectives are stated and the well-known property of lifting idempotent is studied, which is generalized in Chap. 9 as the lifting property of submodules modulo radical. The fundamental notion of commutative rings is given to assist the beginner.

From Chap. 5 on, we study various branches of ring theory. The Morita context, a very important and useful tool in contemporary ring theory, and some related results are given in this chapter, which should be enough for the reader to study this branch of noncommutative ring theory. In order to introduce Matlis' work [156] for injective modules over a noetherian ring, the concept of completion of modules over such a ring is given. Some classical results of Nakayama's [173] on the quasi-Frobenius ring (QF-ring) are stated, and they are generalized in Chap. 10. We also define QF-1, 2, and 3 rings as generalizations of QF-rings following Thrall [215], and we go on to study particular QF-2 rings in Chap. 10. From Chap. 6 on, my own work concerning direct sums of completely indecomposable modules and related topics is explored. First we study the structure of semiperfect modules, due to Bass [20] and Mares [154]. The ring-theoretical interpretation of abelian categories is given, and the factor category of an additive category with respect to an ideal is defined similarly to the factor ring. If, in particular, we take the category induced from completely indecomposable modules, this factor category with respect to a certain ideal becomes a completely reducible Grothendieck category (Theorem 6.3.8), which is the starting point of this book. Factor categories induced from projectives and injectives are studied as examples, and their applications are given in Chap. 7. The ring-theoretical proof of (6.3.8) is added in the last section of Chap. 6, where we need not a new concept in ring theory, but the aid of row- or column-finite matrices over a division ring.

We apply (6.3.8) to get the Krull-Remak-Schmidt-Azumaya theorem in Chap. 7. The uniqueness of decomposition is an immediate consequence of (6.3.8). It is noted in the proof of the theorem that it is very important to know whether the ideal used for the factor category is equal to the Jacobson radical. This problem is studied by making use of a new concept--semi-T-nilpotency--and almost every known result concerning the Krull-Remak-Schmidt-Azumaya theorem is involved with this notion. The factor category suggests a new submodule of a direct sum of completely indecomposable modules, and this submodule--the full submodule--is studied. After finding a relationship between small submodules of a projective module and the Jacobson radical of its endomorphism ring, we apply it to study perfect modules and rings. The factor category induced from injectives is utilized to study the Goldie dimension of modules. The exchange property for modules is considered in Chap. 8. We study this property primarily for direct sums of completely indecomposable modules. The main result in this chapter is very technical and the proof is somewhat complicated. Many theorems on the exchange property are collected in Chap. 8. The terminology defined in Chaps. 9 and 10 will not be familiar to the reader, as it represents my own preferences. The basic concept discussed in Chap. 9 is very fundamental in ring theory and appears throughout the literature. We study this systematically and utilize these results in Chap. 10. It is shown that this concept is a generalization of generalized uniserial rings. The concept dual to the above is also studied. The results of Chap. 10 are also somewhat technical. The notions of injectives and quasi-projectives are generalized, and rings in a new class, which involve these notions and are related to QF-rings, are studied.

I would like to thank Mr. G. Bird of the University of Sydney for correcting the English in the first half of the draft manuscript, and the referee for his useful suggestions on the initial draft. I would also like to thank Mr. Z. Chen at Northern Traffic University of Beijing for correcting typographical errors in the draft manuscript, and Mrs. Marguerite Ellis for careful typing of the book. Finally, I would like to express my thanks to Professors K. Kelly and G. E. Wall of the University of Sydney for their kind hospitality during my six-week stay in 1981.

Manabu Harada

Contents

1
Preliminaries

In this chapter we develop fundamental concepts in ring theory. We assume that the reader is familiar with the elementary properties of group theory and set theory. Although we recall them in this chapter, we skip some of the proofs.

In Secs. 1.1 through 1.4 some fundamental properties of groups--the isomorphism theorem, direct products and sums of groups, and composition series--are considered. We state Zorn's lemma without proof in Sec. 1.5. After defining a ring R in Sec. 1.6, all results in Secs. 1.2 through 1.4 are extended to R-modules in Sec. 1.7. Many results in Secs. 1.2 through 1.4 will be generalized to properties of abelian categories in Chap. 2.

1.1 DEFINITIONS CONCERNING GROUPS

Let A and B be sets and g a mapping of A to B. We call A the *domain* of g and B the *codomain* of g. We denote the cartesian product of A and B by $A \times B$. Assume that there is a mapping $f : A \times A \to A$ satisfying the following conditions [we denote f(a,a') by aa' | a, a' $\in$ A]:

There exists a special element e in A such that ae = ea for all a $\in$ A. (1.1.1)

$a(bc) = (ab)c$ for a, b, and c in A. (1.1.2)

For each a in A there exists a special element a' in A such that aa' = a'a = e. (1.1.3)

In this case we call A a *group*. It is well known that e and a' are unique and we call them the *identity* of A and the *inverse element* of a, respectively. We denote a' by a^{-1}. Furthermore, if A satisfies

$ab = ba$ for all a, b in A (1.1.4)

we call A an *abelian group* or a *module*. In this case we usually denote the operation by addition +, and we call the identity element the *zero element*, which we denote by 0. Further, we denote a^{-1} by $-a$.

Let A, B, and C be groups. If a mapping f from A to B $(f : A \to B)$ satisfies

$$f(aa') = f(a)f(a') \qquad \text{for } a, a' \in A \tag{1.1.5}$$

we call f a *homomorphism*. If $g : B \to C$ is a homomorphism, the composition $g \circ f$ of f and g is also a homomorphism. We put $\ker f = \{a \in A \mid f(a) = e_B\}$ and $\operatorname{im} f = \{f(a) \mid a \in A\}$ and call them the *kernel* of f and the *image* of f, respectively. We often denote them by $f^{-1}(e)$ and $f(A)$, respectively. If $\ker f = \{e_a\}$, f is called a *monomorphism* (f is monic) and if $\operatorname{im} f = B$, we call f an *epimorphism* (f is epic). If f is a monomorphism and an epimorphism, we call f an *isomorphism*. Next, assume that A = B. Then we call f an *endomorphism* of A. In particular, if f is an isomorphism, we call f an automorphism. The special mapping $a \to a$ $(a \in A)$ is called the *identity mapping* and we denote it by 1_A. If there exists an isomorphism of A to B, we indicate this by $A \approx B$.

1.2 SUBGROUPS AND FACTOR GROUPS

Let G be a group and H a nonempty subset of G. If H is closed with respect to the operations in G, namely, $ab^{-1} \in H$ for $a,b \in H$, we call H a *subgroup* of G. If $gHg^{-1} = \{ghg^{-1} \mid h \in H\} \subseteq H$ for all $g \in G$, H is called a *normal subgroup* of G. If a group G does not contain any proper normal subgroups (i.e., which are not equal to either the whole group or the identity) then we call G a *simple group*. Let L_1 and L_2 be subsets of G. L_1L_2 means the set of all elements $\ell_1\ell_2$, where ℓ_i is in L_i. If H is a subgroup of G, we have a partition of G, $G = \bigcup Hg_\alpha$; $Hg_\alpha \cap Hg_\beta = \emptyset$ or $Hg_\alpha = Hg_\beta$. Hg is called a *coset*. Furthermore, if H is normal, $Hg_\alpha = g_\alpha H$, and if we define a product on $\{Hg_\alpha\}$ as follows,

$$HgHg' = Hgg' \qquad g, g' \in G \tag{1.2.1}$$

then $\{Hg_\alpha\}$ is a group and we denote it by G/H. It is called the *factor group* of G with respect to the normal subgroup H. We often denote Hg by $\bar{g}$ and $\{\bar{g}_\alpha\}$ is called a *representative* set of G/H. It is clear that $H = \dot{H}e = \bar{e}_{G/H}$ and $Hg^{-1} = (Hg)^{-1}$ in G/H. The mapping $\varphi : G \to G/H$ given by setting $\varphi(g) = Hg$ is an epimorphism and we call it the *natural epimorphism*.

Let G and G' be groups and $f : G \to G'$ a homomorphism. For a subgroup H' of G', we put $f^{-1}(H') = \{h \in G \mid f(h) \in H'\}$. Then $f^{-1}(H')$ is a subgroup of G and is called the *inverse image* of H' by f.

In Chap. 2 the following theorems are generalized to abelian categories, so here we only outline their proofs.

1.2.1 THEOREM Let G and G' be groups and $f : G \to G'$ a homomorphism. Then ker f is a normal subgroup of G and im f is a subgroup of G'. Further, $G/(\ker f) \approx \operatorname{im} f$.

Proof: $f(g^{-1}(\ker f)g) = f(g)e_{G'}f(g)^{-1} = e_{G'}$. We can define a mapping $\psi : G/(\ker f) \to \operatorname{im} f$ by setting $\psi((\ker f)g) = f(g)$. Then ψ is an isomorphism.

1.2.2 COROLLARY Let G be a group. Assume that $G_1 \supseteq G_2$ are normal subgroups of G. Then G_1/G_2 is a normal subgroup of G/G_2 and $(G/G_2)/(G_1/G_2) \approx G/G_1$.

Proof: Take the homomorphism $\psi : G/G_2 \to G/G_1$; $\psi(G_2g) = G_1g$ and use (1.2.1).

1.2.3 THEOREM Let G be a group and N a normal subgroup of G. Then there is a one-to-one correspondence between the set of subgroups K of G/N and the set of subgroups H of G containing N such that $K = H/N$. Furthermore, K is normal if and only if H is normal.

Proof: Consider the natural homomorphism $\varphi : G \to G/N$.

1.2.4 THEOREM Let G be a group and let H_1 and H_2 be subgroups of G. If H_1 is normal, $H_1H_2 = H_2H_1$ is a subgroup of G and $(H_1H_2)/H_1 \approx H_2/(H_1 \cap H_2)$.

Proof: Since $hH_1h^{-1} = H_1$ for $h \in H_2$, $H_1H_2 = H_2H_1$. Consider the homomorphism $\psi : H_2 \to (H_1H_2)/H_1$ given by setting $\psi(h_2) = H_1h_2$.

Let X be a subset of G. $\{x_{i_1}^{\varepsilon_1}x_{i_2}^{\varepsilon_2}\cdots x_{i_n}^{\varepsilon_n} \mid x_i \in X \text{ and } \varepsilon_i = \pm 1\}$ is a subgroup of G which is the unique minimal subgroup of G containing X. We denote it by $\langle X\rangle$ and call it the *subgroup generated by* X. X is called a *set of generators* of $\langle X\rangle$. If X is a finite set, $\langle X\rangle$ is called *finitely generated* (f.g.). In particular, if $X = \{a\}$, $\langle a\rangle$ is called a *cyclic group*.

1.2.5 PROPOSITION Let G and G' be groups and $f : G \to G'$ a homomorphism. Let H be a subgroup of G. If $f(H) = f(G)$, $G = H(\ker f)$.

Proof: Let g be in G. Then there exists h in H such that $f(g) = f(h)$. Hence $h^{-1}g$ is in ker f.

1.3 DIRECT PRODUCTS

Let I be a set and $\{G_i\}_{i\in I}$ a set of groups. We can consider the cartesian product $\Pi_I\ G_i$ of the set $\{G_i\}_I$. We indicate an element in $\Pi_I\ G_i$ by $(g_i)_I$, where g_i in G_i. We define a new operation on $\Pi_I\ G_i$ as follows:

$$(g_i)_I(h_i)_I = (g_ih_i)_I \tag{1.3.1}$$

Then $\Pi_I\ G_i$ is a group with this operation. $\Pi_I\ G_i$ is called the *product* of $\{G_i\}_I$. We can define a monomorphism $i_j : G_j \to G = \Pi_I\ G_i$ by setting $i_j(g_j) = (e_1,\ldots,e_{j-1},g_j,e_{j+1},\ldots)$ and an epimorphism $p_j : G \to G_j$ such that $p_j((g_i)_I) = g_j$. We call i_j and p_j an *injection* and a *projection*, respectively.

1.3.1 LEMMA Let G and G' be groups. Let $f : G \to G'$, $g : G' \to G$ be homomorphisms. If $gf = 1_G$, f is a monomorphism and g is an epimorphism. If $gf = 1_G$ and $fg = 1_{G'}$, f and g are isomorphisms.

Proof: $\ker f = 1_G(\ker f) = gf(\ker f) = e_G$. $\operatorname{im} g \supseteq \operatorname{im} gf = \operatorname{im} 1_G = G$.

1.3.2 THEOREM Let $\{G_i\}_{i\in I}$ be a set of groups. Then there exists a group G satisfying: (1) there exists a homomorphism $p_i : G \to G_i$ for each i; and (2) for any group A and homomorphisms $g_i : A \to G_i$ for $i \in I$, there exists a unique homomorphism $g : A \to G$ such that $p_ig = g_i$ for each i. This G is isomorphic to $\Pi_I\ G_i$.

Proof: First we show that $\Pi_I\ G_i$ satisfies (1) and (2). Take the projection p_i' on $\Pi_I\ G_i$. We can define a homomorphism $g : A \to \Pi_I\ G_i$ given by setting $g(a) = (g_i(a))_I$. Then $p_i'g = g_i$ by the definition and g is unique since $p_i'g = g_i$. Next we show that if G satisfies (1) and (2), then $G \approx \Pi_I\ G_i$. From (2) we obtain homomorphisms $f : G \to \Pi\ G_i$, $g : \Pi\ G_i \to G$ such that $p_i'f = p_i$ and $p_ig = p_i'$ for each $i \in I$. Then $p_i'fg = p_ig = p_i' = p_i'1_{\Pi G_i}$. Hence $fg = 1_{\Pi G_i}$ by (2). Similarly, we have $gf = 1_G$. Therefore, f and g are isomorphisms by (1.3.1).

Consider the subgroup $\Pi\ G_i$ consisting of all elements $(g_i)_I$, $g_j = e_{G_j}$ for almost all j, where "almost all" means all but a finite number. We call it the *direct* sum (*external* or *formal direct sum*) of $\{G_i\}_I$ and indicate it by $\Sigma_I \oplus G_i$. If I is a finite set, $\Pi_I\ G_i = \Sigma_I \oplus G_i$.

Each i_j is a monomorphism for $G = \Pi\ G_i$. Hence we may assume that each G_j is a (normal) subgroup of G via i_j, and (1) $\Sigma_I \oplus G_i$ is generated by $\{G_i\}_I$, (2) each element g in $\Sigma_I \oplus G_i$ is expressed uniquely as

$g = g_{i_1}g_{i_2}\cdots g_{i_n}$, and (3) $g_ig_j = g_jg_i$, $g_i \in G_i$, $g_j \in G_j$ for $i \neq j$. Conversely, let H be a group and $\{H_i\}_I$ a set of normal subgroups of H. If each h in H is expressed uniquely as $h = h_{i_1}h_{i_2}\cdots h_{i_n}$; $h_{i_j} \in H_{i_j}$ and $h_ih_k = h_kh_i$ $(i \neq k)$ for $h_p \in H_p$, then H is the direct sum of $\{H_i\}_I$ [we call H an *internal direct sum* of $\{H_i\}_I$; see (1.3.3) below]. H_i is called a *direct summand* of H.

If a group G is a direct sum of nontrivial subgroups, G is called *directly decomposable*; otherwise, we say that G is *directly indecomposable*.

1.3.3 THEOREM Let $\{G_i\}_I$ be a set of normal subgroups of a group G. Then the following conditions are equivalent:

1. $G = \Sigma_I \oplus G_i$.
2. G is generated by $\{G_i\}_I$ and every element g in G is expressed uniquely as $g = g_{i_1}g_{i_2}\cdots g_{i_n}$, $g_i \in G_i$.
3. G is generated by $\{G_i\}_I$ and, letting H_i be the subgroup of G generated by $\{G_j\}_{I-\{i\}}$, $H_i \cap G_i = \{e\}$ for all i.

Proof: Assume (2). Let g be in $H_i \cap G_i$. Then $g = g_{i_1}g_{i_2}\cdots g_{i_n}$, $g_j \in G_j$, by (1.2.4). Hence g = e by the uniqueness. Therefore, $H_i \cap G_i = \{e\}$. Since $G_j \subseteq H_i$, $G_j \cap G_i = \{e\}$ if $i \neq j$. Then $g_ig_jg_i^{-1}g_j^{-1} \in G_i \cap G_j = \{e\}$, for the G_i are normal. Hence $g_ig_j = g_jg_i$ for $i \neq j$. Therefore, we have shown that (1) ⇔ (2) and (2) ⇒ (3).

Assume (3). Let g be in G. Then $g = g_{i_1}g_{i_2}\cdots g_{i_n}$, $g_i \in G_i$, by (1.2.4). We assume that we have another expression $g = h_{i_1}h_{i_2}\cdots h_{i_n}$, $h_i \in G_i$. We know that $g_ig_j = g_jg_i$ for $i \neq j$ from the paragraph above. Hence

$$g_{i_1}h_{i_1}^{-1} = (g_{i_2}^{-1}h_{i_2})(g_{i_3}^{-1}h_{i_3})\cdots(g_{i_n}^{-1}h_{i_n}) \in G_{i_1} \cap H_{i_1} = \{e\}$$

Therefore, $h_i = g_i$. Thus the expression $g = g_{i_1}g_{i_2}\cdots g_{i_n}$ is unique.

The following is dual to (1.3.2).

1.3.4 THEOREM Let $\{G_i\}_I$ be a set of groups. Then there exists a group G satisfying: (1) there exists a homomorphism $i_j : G_j \to G$ for each $j \in I$; and (2) for any group H and any set of homomorphisms $h_i : G_i \to H$, there exists a unique homomorphism $h : G \to H$ such that $hi_j = h_j$ for each $j \in I$. This G is isomorphic to $\Sigma_I \oplus G_i$.

Proof: We leave the proof to the reader.

Let $\{G_\alpha\}$ be a set of normal subgroups of a group G. If the subgroup generated by $\{G_\alpha\}_I$ is the direct sum of $\{G_\alpha\}_I$, we call the set $\{G_\alpha\}_I$ *independent*.

1.4 COMPOSITION SERIES

Let G be a group and $\{N_i\}_{i=1}^r$ a set of subgroups of G. If $\{N_i\}_{i=1}^r$ satisfies

$$G = N_0 \supseteq N_1 \supseteq N_2 \supseteq \cdots \supseteq N_r = \{e\} \tag{1.4.1}$$

and N_{i+1} is a normal subgroup of N_i for each i, we call (1.4.1) a *normal chain* or *normal series* of G and r is called the *length of* this chain. We call N_i/N_{i+1} a factor group of the series. If $N_i \neq N_{i+1}$ for each i and N_i/N_{i+1} is a simple group, we call (1.4.1) a *composition series* of G. If N_i/N_{i+1} is not simple, we can find a subgroup H_i such that $N_i \supsetneq H_i \supsetneq N_{i+1}$ and H_i is normal in N_i. Repeating this process, we obtain a longer normal series. We call this process a *refinement* for (1.4.1).

1.4.1 PROPOSITION Let G be a group and $H \supset H'$, K subgroups of G. Then $(H'K) \cap H = H'(K \cap H)$.

Proof: Let $g = h'k$ be in H; $h' \in H'$, $k \in K \cap H$. Then $k = h^{-1}g$ is in $K \cap H$. Hence g is in $H'(K \cap H)$.

1.4.2 PROPOSITION Let G be a group. Let $H \supset H_1$ and $K \supset K_1$ be subgroups. If H_1 and K_1 are normal in H and K, respectively, then $H_1(H \cap K)/H_1(H \cap K_1) \approx K_1(H \cap K)/K_1(H_1 \cap K)$.

Proof: We can easily see that $H_1(H \cap K_1)$ and $K_1(H_1 \cap K)$ are normal in $H_1(H \cap K)$ and $K_1(H \cap K)$, respectively. $H_1(H \cap K_1) \cap (H \cap K) = (H \cap K_1)(H_1 \cap K)$ by (1.4.1). Hence $(H \cap K)/[(H_1 \cap K)(H \cap K_1)] \approx H_1(H \cap K)/H_1(H \cap K_1)$ by (1.2.4). Similarly, we obtain $(H \cap K)/[(H_1 \cap K)(H \cap K_1)] \approx K_1(H \cap K)/K_1(H_1 \cap K)$.

Let $G = H_0 \supset H_1 \supset H_2 \supset \cdots \supset H_r = \{e\}$ and $G = K_0 \supset K_1 \supset K_2 \supset \cdots \supset K_r = \{e\}$ be two normal series. If $H_i/H_{i+1} \approx K_{\rho(i)}/K_{\rho(i)+1}$, where ρ is a permutation, we say that the two series are *equivalent*.

1.4.3 THEOREM (*Schreier*) Let $G = H_0 \supset H_1 \supset \cdots \supset H_r = \{e\}$ and $G = K_0 \supset K_1 \supset \cdots \supset K_s = \{e\}$ be two normal series of G. Then after refining them suitably, we can obtain equivalent normal series.

Proof: Set $A_{ij} = H_i(H_{i-1} \cap K_j)$ and $B_{ji} = K_j(K_{j-1} \cap H_i)$. Then $A_{ij-1}/A_{ij} \approx B_{ji-1}/B_{ji}$ by (1.4.2).

1.4.4 COROLLARY (*Jordan-Hölder*) Two composition series of G are equivalent.

If G has a composition series, every composition series has the same length by (1.4.4). We denote this length by $\|G\|$ and we say that G has *finite length*. ($|G|$ means the cardinal of the set G.)

1.5 ZORN'S LEMMA

Let S be a set. If a relation $\geq$ on S satisfies

$a \geq a$ (1.5.1)

$a \geq b$ and $b \geq a$ imply that $a = b$ (1.5.2)

$a \geq b$ and $b \geq c$ imply that $a \geq c$ (1.5.3)

we call S a *partially ordered* set. If any two elements a and b are comparable (i.e., $a \geq b$ or $b \geq a$), we call S a *totally ordered set*. An element b in S is called an *upper bound* of a subset K of S if $b \geq k$ for all $k \in K$. An element t in K is called a *maximal element* of K whenever $t' \geq t$ implies that $t' = t$ for $t' \in K$.

Zorn's Lemma Let S be a partially ordered set. If every totally ordered subset in S has a (least) upper bound in S, then S contains a maximal element.

We shall use this lemma without proof throughout the book. We usually use it as follows. Let G be a group and $\underline{F}$ a family of subgroups $\{G_\alpha\}$ of G with certain properties. We can define a partial order in $\underline{F}$ by the inclusion relation. Let $A_1 \subseteq A_2 \subseteq \cdots \subseteq A_n \subseteq \cdots$ be any totally ordered subset of $\underline{F}$. If we can show $\cup A_i$ to be in $\underline{F}$, then $\cup A_i$ is a least upper bound. In this case $\underline{F}$ contains a maximal element. In the following, when we can easily see that $\cup A_i$ belongs to $\underline{F}$, we shall use a phrase such as "we obtain a maximal element by Zorn's lemma."

1.6 DEFINITIONS CONCERNING RINGS

Let R be an abelian group whose operation is denoted by addition (+). We assume further that R has another operation denoted by multiplication (·), which satisfies the following conditions:

$(ab)c = a(bc)$ (1.6.1)

$$(a + b)c = ac + bc \qquad a(b + c) = ab + ac \tag{1.6.2}$$

for any elements a, b, and c in R. We call such an abelian group R with product a *ring*. Further, if R satisfies

$$ab = ba \tag{1.6.3}$$

R is called a *commutative ring*. If there exists an element 1 in R such that

$$a1 = 1a \qquad \text{for all } a \in R \tag{1.6.4}$$

we call 1 the *identity element* of R. If R contains such an element 1, it is clearly unique. We denote the zero element for an abelian group by 0. Then $0a = a0 = 0$ for a in R.

An element a in R is called *nil* if $a^n = 0$ for some integer n, and *idempotent* if $a^2 = a$. For two nonzero elements a, b, if $ab = 0$, a (resp. b) is called a left (resp. right) *zero divisor*. In a ring with identity, if $ab = 1$, b (resp. a) is called a right (resp. left) *inverse* of a (resp. b). Finally, if every nonzero element is neither a left nor a right zero divisor, R is called a *domain* or an *integral domain*. We assume that a domain R contains an identity element. If an element a has a left and a right inverse, we call it a *unit* (or *regular*) element and we denote the common inverse by a^{-1}. R is called a *division ring* provided that every nonzero element is a unit. A commutative division ring is called a *field*.

In this book every ring R contains an identity element, unless otherwise stated. If a nonempty subset S of R is closed with respect to the operations + and $\cdot$ in R, then S is called a *subring* of R. Set $C(R) = \{x \in R \mid xa = ax \text{ for all } a \in R\}$. Then C(R) is a subring of R and we call C(R) the *center* of R. R is commutative if and only if $C(R) = R$.

Examples 1. The set of all integers Z is a commutative ring with the usual addition and product, and the set of all rationals Q is a field. Z is a domain but not a field.

2. Let R be the set of all continuous real-valued functions on the interval [0,1]. Then R is a commutative ring but not a domain with respect to the standard operations.

3. The set of square matrices of degree n over a ring R is a noncommutative ring with the usual operations. We denote it by $(R)_n$ or R_n. e_{ij} means the matrix whose (i,j)th component is 1 and whose other components are zero. We call $\{e_{ij}\}$ the *matrix units*. It is clear that $e_{ij}e_{jk} = e_{ik}$.

$e_{ij}e_{j'k} = 0$ if $j \neq j'$; i.e., $e_{ij}e_{mk} = \delta_{jm}e_{ik}$, where $\delta_{ij} = 0$ if $i \neq j$ and $\delta_{ii} = 1$ (δ_{ij} is called the *Kronecker delta*), and $1 = \Sigma_i e_{ii}$.

4. Let K be a ring and x an indeterminate. We set

$$K[x] = \{\text{polynomials } f(x) = a_0 + a_1x + \cdots + a_nx^n \mid a_i \in K\}$$

Then K[x] is a ring with the usual operations on polynomials.

5. Let G be a group. We denote the K-vector space with basis G by K[G]. We define $kg = gk$ and $(\Sigma_i k_i g_i)(\Sigma_j s_j h_j) = \Sigma_{i,j}(k_i s_j)(g_i h_j)$, where the s and the k are in K and the g and the h are in G. Then K[G] is a ring. We call it the *group ring* of G over K.

1.7 R-MODULES

Let R be a ring and M an abelian group (a module). We assume that there is a mapping f of M × R to M [we denote f(m,r) by mr for $m \in M$ and $r \in R$] satisfying

$$(m_1 + m_2)r = m_1r + m_2r \qquad m(r_1 + r_2) = mr_1 + mr_2 \tag{1.7.1}$$

$$(mr)r' = m(rr') \tag{1.7.2}$$

where the m's are in M and the r's in R. Then we call M a right *R-module* and denote it by M_R. We always assume that

$$m1 = m \qquad \text{for all } m \in M \tag{1.7.3}$$

unless otherwise stated. We can similarly define a left R-module.

Let S be another ring. We assume further that M is a right (resp. left) S-module. If

$$(mr)s = (ms)r \quad [\text{resp. } s(mr) = (sm)r] \qquad \text{for } s \in S,\ r \in R \tag{1.7.4}$$

is satisfied, M is called a right R-S (R,S-*bi*) *module* and we denote it by $M_{R,S}$ (resp. ${}_SM_R$). If, for any $r \neq 0$ in R, there exists an m in M with $mr \neq 0$, M is called a *faithful module*.

We shall make statements mainly for right R-modules. However, if these statements are valid for left or two-sided modules, we refer to the same theorem without comment.

When a submodule N of M is closed with respect to the operations of R, N is called an *R-submodule* and is denoted by $M_R \supset N_R$. If we take R itself, then R is a right, left, and two-sided R-module. An R-submodule of R is called a *right*, *left*, and *two-sided ideal* of R, respectively.

For an element m in M we put $r_R(m)$ [= $\mathrm{ann}_R^r(m)$] = $\{x \in R \mid mx = 0\}$. Then $r_R(m)$ is a right ideal in R and we call it the right *annihilator ideal* of m. We similarly define the left annihilator ideal $\ell_R(m)$ [= $\mathrm{ann}_R^{\ell}(m)$] = $\{x \in R \mid xm = 0\}$ for a left R-module N and an m in N. For subsets X of M and T of R, we denote the set of all elements of the form $\Sigma_i m_i t_i$, $m \in X$ and $t_i \in T$, by XT (cf. Sec. 1.2).

A right ideal A of R is called *idempotent* or *nilpotent* if $A^2 = A$ or $A^n = 0$ for some n, respectively. If each element in A is nil, we call A a *nil ideal*. Clearly, every nilpotent ideal is nil, but the converse is not true.

Let A be a two-sided ideal of R. Since A is a subgroup of the abelian group R, we can define the factor group R/A. Let r + A, r' + A be two elements in R/A. We define

$$(r + A)(r' + A) = rr' + A \tag{1.7.5}$$

It is clear that this definition does not depend on the choice of the representatives r and r'. Using this product, we can easily show that R/A is a ring. We call such a ring the *factor ring* of R with respect to A. Next, let $M_R \supset N_R$ be R-modules. In a manner similar to the above, we put

$$(m + N)r = mr + N \qquad \text{for } m + N \in M/N \text{ and } r \in R \tag{1.7.6}$$

Then M/N is a right R-module via the operation above.

Let M_{1R}, M_{2R} be modules and $f : M_1 \to M_2$. If

$$f(m_1 + m_1') = f(m_1) + f(m_1') \tag{1.7.7}$$

$$f(m_1 r) = f(m_1)r \tag{1.7.8}$$

where the m's are in M_1 and r in R, are satisfied, f is called an *R-homomorphism*. Let R and S be two rings. If a mapping f from R to S satisfies

$$f(r_1 + r_2) = f(r_1) + f(r_2) \tag{1.7.9}$$

$$f(r_1 r_2) = f(r_1)f(r_2) \tag{1.7.10}$$

where the r's are in R, then we call f a *ring homomorphism*. We usually assume that

$$f(1_R) = 1_S$$

We note that most elementary properties of abelian groups are valid for right R-modules provided that we slightly modify the argument in Sec. 1.2.

We shall state some of them.

1.7.1 THEOREM Let R be a ring and let M and M_i be right R-modules (i = 1, 2).

1. If f is an R-homomorphism of M_1 to M_2, then im f (resp. ker f) is an R-submodule of M_2 (resp. M_1) and im f $\approx M_1/\ker f$ as R-modules.
2. $(M/M_1)/(M_2/M_1) \approx M/M_2$ as R-modules, where $M \supset M_2 \supset M_1$.
3. $(M_1 + M_2)/M_1 \approx M_2/(M_1 \cap M_2)$ as R-modules.
4. For $M \supset M_1$ and $M \supset M_2 \supset M_3$, $(M_1 + M_3) \cap M_2 = M_3 + (M_1 \cap M_2)$.
5. $(R/A_1)/(A_2/A_1) \approx R/A_2$ as rings, where $A_2 \supset A_1$ are two-sided ideals of R.
6. Let S be another ring and f a ring homomorphism from R to S. Then ker f is a two-sided ideal of R and im f is a subring of S. Further, im f $\approx R/\ker f$ as rings.

Let $\{M_i\}_K$ be a set of R-modules. We can define the product $\Pi_K M_i$ and the direct sum $\Sigma_K \oplus M_i$ as abelian groups (Sec. 1.3). If we define

$$(m_i)_K r = (m_i r)_K \qquad (\Sigma_j m_{i_j})r = \Sigma_j m_{i_j} r \tag{1.7.11}$$

for the $m_i \in M$, $r \in R$, they are R-modules. In the above, if $M_i = R$ for all i, we call $\Sigma_K \oplus M_i$ a *free R-module*. We denote it by $\Sigma_K \oplus u_i R$ or $R^{(K)}$, where $\{u_i\}_K$ is a free basis and $u_i R \approx R$. Let $\{R_i\}_K$ be a set of rings. For the product $\Pi_K R_i$ we define

$$(r_i)_K (r_i')_K = (r_i r_i')_K \tag{1.7.12}$$

where r_i is in R_i. Then ΠR_i is a ring, which is called the *product* of $\{R_i\}_K$. If K is a finite set, we often call ΠR_i the *direct sum* of the R_i's.

1.7.2 THEOREM Every R-module M_R is a homomorphic image of a free R-module $\Sigma_X \oplus u_x R$, where X is a generating set of M.

Proof: Take a free module $\Sigma_X \oplus u_x R$ and define a mapping f from $\Sigma_X \oplus u_x R$ to M by setting $f(\Sigma_{i=1}^n u_{x_i} r_i) = \Sigma_{i=1}^n x_i r_i$. It is clear that f is an R-homomorphism and an epimorphism.

1.7.3 PROPOSITION Let $I = \{1,2,\ldots,\gamma\}$ be a set of indices and $\{M_{n\alpha(n)}\}_{I,\alpha(n)\in J(n)}$ a set of R-modules. Then

$$\Sigma_I \oplus (\Sigma_{J(n)} \oplus M_{n\alpha(n)}) = \Sigma_{I\times J(n)} \oplus M_{n\alpha(n)}$$

and

$$\Pi_I(\Pi_{J(N)} M_{n\alpha(n)}) = \Pi_{I\times J(n)} M_{n\alpha(n)}$$

Proof: This is clear from the definitions.

1.7.4 THEOREM Let M_R and M_{iR} be R-modules (i = 1, 2) and let $M = M_1 \oplus M_2$. If N is an R-submodule of M containing M_1, $N = M_1 \oplus (N \cap M_2)$.

Proof: $N = N \cap M = N \cap (M_1 \oplus M_2) = M_1 + (N \cap M_2)$ by (1.4.1). It is clear that $M_1 \cap (N \cap M_2) = 0$.

1.7.5 THEOREM [See (1.3.1)] Let M_R and N_R be R-modules and $f : M \to N$ an R-homomorphism. If there exists a $g : N \to M$ such that $fg = 1_N$, then $M = \ker f \oplus \operatorname{im} g$ and f (resp. g) is an epimorphism (resp. a monomorphism).

Proof: It is clear that $M = \operatorname{im} g + \ker f$ by (1.2.5), since $N = f(\operatorname{im} g)$. Let $x \in \operatorname{im} g \cap \ker f$. Then $x = g(y)$ for some $y \in N$. Hence $0 = f(x) = fg(y) = y$, so $x = 0$. The last statement is clear by (1.3.1).

In this case we say that f *splits*. Finally, let R be a ring and M_i be R-modules (i = 1, 2). We denote the set of all R-homomorphisms from M_1 to M_2 by $\operatorname{Hom}_R(M_1,M_2)$. For two elements f, g in $\operatorname{Hom}_R(M_1,M_2)$, define

$$(f \pm g)(m) = f(m) \pm g(m) \qquad m \in M_1 \tag{1.7.13}$$

Then $f \pm g \in \operatorname{Hom}_R(M_1,M_2)$. We put $0(m) = 0$ for every m and call it the *zero homomorphism*. With these definitions $\operatorname{Hom}_R(M_1,M_2)$ is an abelian group. For elements $f : M_1 \to M_2$ and $g : M_2 \to M_3$ we define

$$gf(m) = g(f(m)) \qquad m \in M_1 \tag{1.7.14}$$

Then $gf \in \operatorname{Hom}_R(M_1,M_3)$. It is clear that

$$g(f_1 + f_2) = gf_1 + gf_2 \qquad (g_1 + g_2)f = g_1f + g_2f \tag{1.7.15}$$

where the f's are in $\operatorname{Hom}_R(M_1,M_2)$ and the g's in $\operatorname{Hom}_R(M_2,M_3)$. In particular, if $M_1 = M_2 = M_3 = M$, $\operatorname{Hom}_R(M,M)$ is a ring with the operations (1.7.14) and (1.7.15), and 1_M is the identity element. We call it the *endomorphism ring* of M and denote it by $\operatorname{End}_R(M)$. If we put $S = \operatorname{End}_R(M)$, we may regard M as a left S- and a right R-module. An S-R submodule of M is called a *character* (*fully invariant*) R-submodule of M.

In this book, if M is a right (resp. left) R-module, the elements in $\operatorname{End}_R(M)$ operate from the left (resp. right) side.

2
Categories

In 1940, categories were first studied as a generalization of groups. By the latter half of 1950, they were being utilized to study algebraic geometry. Since then, category theory has been recognized as a useful tool in algebra.

Recently, category theory has been studied not only for applications to algebra but also as a discipline in itself. We shall apply category theory only to study the Krull-Remak-Schmidt-Azumaya theorem and the Goldie dimension of modules. Actually, category theory is of great use in the field of algebra. However, until now, category theory has been used essentially for terminology only, except for a few branches of noncommutative ring theory.

In Chaps. 6 and 7 we show that category theory is a very useful and powerful tool for the study of direct sums of indecomposable modules. However, our use is minimal and in this chapter we introduce the basic notions used in later chapters. For further study we refer the reader to the books by Mitchell [159] and Freyd [65].

In Sec. 2.1 we define some elementary concepts regarding categories, and we give categorical proofs of results in Secs. 1.2 through 1.4. We define the notion of an abelian category in Sec. 2.2 and give a criterion for a category to be abelian, following Freyd. We shall utilize this criterion in Chap. 6, so we develop a few concepts in this chapter to state it. In Sec. 2.3 we define functors. However, we use mainly the functors Hom and $\otimes$ except in Chaps. 5 and 6. We shall need Grothendieck categories as well as Freyd's criterion in Chap. 6. Hence we define Grothendieck categories in Sec. 2.4. In the final section, Sec. 2.5, we present some useful commutative diagrams in the category of modules.

2.1 DEFINITIONS CONCERNING CATEGORIES

Category $\underline{A}$ is a class, together with class $\underline{M}$, which is a disjoint union of the form

$$\underline{M} = \bigcup_{A,B\in\underline{A}} [A,B]_{\underline{A}}$$

satisfying the two conditions (2.1.1) and (2.1.2) below. Here we always assume that $[A,B]_{\underline{A}}$ is a set and we note that the A's are merely *objects* in the category and not necessarily sets. An element of $[A,B]_{\underline{A}}$ is called a *morphism*. If there is no confusion, we write $[A,B]$ instead of $[A,B]_{\underline{A}}$. Further, we denote a morphism in $[A,B]$ by $f : A \to B$.

For any three objects A, B, and C and any morphisms $f : A \to B$, $g : B \to C$, the composition gf is defined and gf is in $[A,C]$; namely, a composition map $[B,C] \times [A,B] \to [A,C]$ is defined and this composition satisfies the associative law, $h(gf) = (hg)f$, where $h : C \to D$. (2.1.1)

Each $[A,A]$ contains the identity element 1_A which satisfies $f1_A = f$ and $1_A g = g$ for any $f \in [A,B]$ and $g \in [C,A]$. (2.1.2)

For a morphism $f \in [A,B]$, A is called the *domain* of f and B is called the *codomain* of f. Let $\underline{A}$ be a category. Take a subclass $\underline{A}'$ of $\underline{A}$ and $\underline{M}' = \bigcup_{A,B\in\underline{A}'} [A,B]_{\underline{A}'}$, where $[A,B]_{\underline{A}'}$ is a subset of $[A,B]_{\underline{A}}$. If the composition of morphisms in $\underline{A}'$ is closed and $[A,A]_{\underline{A}'}$ contains 1_A for each A in $\underline{A}'$, then $\underline{A}'$ is called a *subcategory* of $\underline{A}$.

We shall define a new category $\underline{A}^o$ from $\underline{A}$. The objects of $\underline{A}^o$ are the same as those in $\underline{A}$ and we put $[A,B]_{\underline{A}^o} \equiv [B,A]_{\underline{A}}$. The composition $f \circ g$ in $\underline{A}^o$ is defined as the composition gf in $\underline{A}$, namely $f \circ g = gf$. Then $\underline{A}^o$ is a category, which we call the *dual category* of $\underline{A}$. If we have a statement P concerning $\underline{A}$ and we reverse the arrows, we obtain a statement concerning $\underline{A}^o$. We say that it is the dual of P and we denote the dual by the prefix "co."

It is clear that many mathematical theories study special categories. However, in this book we are interested only in the category of abelian groups $\underline{Ab}$. The objects of $\underline{Ab}$ consist of all abelian groups and $[A,B]_{\underline{Ab}} \equiv \mathrm{Hom}(A,B)$. It is clear that $\underline{Ab}$ is a category. Let R be a ring. If we take a subclass $\underline{M}_R$ of $\underline{Ab}$, whose objects are R-modules and whose morphisms are R-homomorphisms, $\underline{M}_R$ is a subcategory of $\underline{Ab}$.

We shall define some terminology for a general category $\underline{A}$. A morphism $f : A \to B$ is called a *coretraction* if there exists $g : B \to A$ such that

$gf = 1_A$. In this case A is called a *retract* of B (direct summand of B in $\underline{M}_R$). Dually, $f' : A \to B$ is called a *retraction* if there exists $g' : B \to A$ such that $f'g' = 1_B$. If f is a coretraction and a retraction, f is called an *isomorphism*. A morphism $\alpha : A \to B$ such that $\alpha f = \alpha g$ for any C and any $f, g : C \to A$ implies that $f = g$ is called a *monomorphism*. The dual of a monomorphism is called an *epimorphism*. In general, a monic and epic morphism is not an isomorphism in $\underline{A}$. If every such f is an isomorphism, $\underline{A}$ is called *balanced*.

If $\alpha : A \to B$ is a coretraction, $\alpha f = \alpha g$ implies that $\beta\alpha f = \beta\alpha g$ for β with $\beta\alpha = 1_A$. Hence $f = g$ and α is a monomorphism. Furthermore, if α is an epimorphism, α is an isomorphism, since $\alpha = \alpha 1_A = 1_B\alpha = \alpha(\beta\alpha) = (\alpha\beta)\alpha$, so $\alpha\beta = 1_B$. Dually, every retraction is an epimorphism.

If $\alpha : A' \to A$ is a monomorphism, A' is called a *subobject* of A via α and we denote it by $A' \subseteq A$. For $f : A \to B$, we denote $f\alpha$ by $f|A'$ (restriction of f to A'). (In $\underline{M}_R$, A and A' are sets and if A' is a submodule of A, A' is a subset of A. Hence we may sometimes neglect the inclusion mapping of A' into A. However, if we say that A' is a subobject of A in a category $\underline{A}$, we must specify the monomorphism α.)

We give an example. Let $\underline{Ab}'$ be the subcategory of $\underline{Ab}$ whose objects consist of all torsion-free groups (see Sec. 4.4). Let Z be the ring of integers and Q the field of rationals. Then Z and Q are in $\underline{Ab}'$. The inclusion mapping $i : Z \to Q$ is a monomorphism in $\underline{Ab}$ and thus in $\underline{Ab}'$. Let M be in $\underline{Ab}'$. Then every $f : Q \to M$ is determined by $f|Z$, since M is torsion free. Hence i is epic in $\underline{Ab}'$. It is clear that i is not epic in $\underline{Ab}$ and i is not an isomorphism in $\underline{Ab}'$. Hence $\underline{Ab}'$ is not balanced.

Let A, B, C, ... be objects and $\alpha, \beta, \gamma, \ldots$ be morphisms. For diagrams

assume $\alpha = \gamma\beta$ and $\delta\alpha = \gamma\beta, \ldots$; then we say that the diagrams are *commutative*. Let α and β be two morphisms in [A,B]. $u : K \to A$ is called an *equalizer* of α and β if u satisfies: (1) $\alpha u = \beta u$; and (2) for any $u' : K' \to A$ such that $\alpha u' = \beta u'$, there exists a unique morphism $\gamma : K' \to K$ such that $u\gamma = u'$.

In this case we shall show that u is monic and that any two equalizers of α and β are isomorphic to one another. Assume that $u\delta = u\varepsilon$. Then $\alpha u\delta = \beta u\delta$ and $\alpha u\delta = \beta u\varepsilon$. Hence $\delta = \varepsilon = \gamma$ from the uniqueness of γ. Therefore, u is monic. Let $u'' : K'' \to A$ be another equalizer. Then there exist $\delta'' : K'' \to K$ and $\delta : K \to K''$ such that $u'' = u\delta''$ and $u = u''\delta$. Then $u = u\delta\delta''$ and $u'' = u\delta''\delta$. Since u and u'' are monomorphisms, $\delta''\delta = 1_K$ and $\delta\delta'' = 1_{K''}$. We shall denote an equalizer by equ(α,β). Dually, we can define a *coequalizer* [coequ(α,β)] for α and β.

In the following we define certain objects and morphisms, and we shall be able to prove special properties from the definitions as above. Hence we state these properties without proof if their proofs are similar to the arguments above.

We take morphisms $\alpha_1 : A_1 \to A$ and $\alpha_2 : A_2 \to A$. A commutative diagram

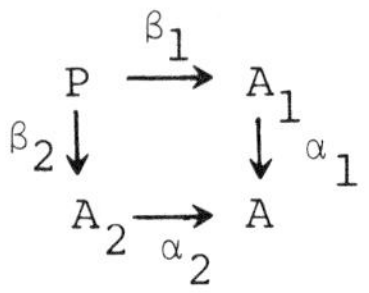

is called a *pullback* for α_1 and α_2 if for any commutative diagram

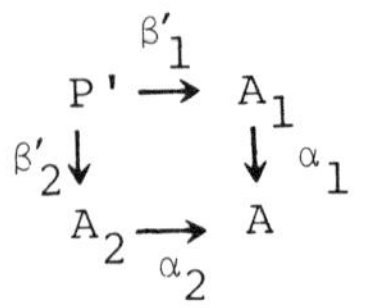

there exists a unique morphism $\gamma : P' \to P$ such that $\beta_i' = \beta_i\gamma$ for i = 1, 2. This P is uniquely determined up to isomorphism. [In $\underline{M}_R$ put $(A_1 \oplus A_2 \supseteq)\ P = \{a_1 + a_2 \in A_1 \oplus A_2 \mid \alpha_1(a_1) = \alpha_2(a_2)\}$ and $\beta_i(a_1 + a_2) = a_i$ for i = 1, 2. Then P gives a pullback for α_1 and α_2.]

2.1.1 LEMMA If α_1 is monic in the pullback above, so is β_2.

Proof: If $\beta_2 f = \beta_2 g$, $\alpha_2\beta_2 g = \alpha_1\beta_1 g$, and $\alpha_2\beta_2 g = \alpha_2\beta_2 f = \alpha_1\beta_1 f$. Hence $\alpha_1\beta_1 f = \alpha_1\beta_1 g$. Since α_1 is monic, $\beta_1 f = \beta_1 g$ (and $\beta_2 f = \beta_2 g$ from the first assumption). Hence we have f = g from the uniqueness.

Similarly, we have the following from the definition.

2.1.2 LEMMA In the commutative diagram

$$\begin{array}{ccccc} P & \longrightarrow & Q & \longrightarrow & B' \\ \downarrow & \text{I} & \downarrow & \text{II} & \downarrow \alpha \\ A & \longrightarrow & I & \underset{\beta}{\longrightarrow} & B \end{array}$$

assume that α is monic and that squares I and II are pullbacks. Then the outer rectangle is a *pullback.*

We call the dual of a pullback a *pushout.* [In $\underline{M}_R$, for $\alpha_i : A \to A_i$, $i = 1, 2$, put $(A_1 \oplus A_2 \supseteq)F = \{\alpha_1(a) - \alpha_2(a) \mid a \in A\}$ and $Q = (A_1 \oplus A_2)/F$. Then Q and the natural homomorphisms $\beta_i : A_i \to Q$ give a pushout.]

Let $\{u_i : A_i \to A\}_{i \in I}$ be a set of subobjects of A. If there exists a subobject $u : A' \to A$ satisfying the following conditions, we call $u : A' \to A$ the *intersection* of $\{u_i : A_i \to A\}$ and denote it by $\cap_I A_i$.

There exists a (unique) morphism $v_i : A' \to A_i$ such that $u = u_i v_i$. (2.1.3)

For any $f : B \to A$ and $f_i : B \to A_i$ such that $f = u_i f_i$ for all i, there exists a unique morphism $f : B \to A'$ such that $f = uf'$. (2.1.4)

$\cap_I A_i$ is uniquely determined up to isomorphism. If I is a finite set, we denote $\cap_I A_i$ by $A_1 \cap A_2 \cap \cdots \cap A_n$.

2.1.3 LEMMA Let $\alpha_i : A_i \to A$ be subobjects of A for $i = 1, 2$. Then a diagram

$$\begin{array}{ccc} P & \overset{\beta_1}{\longrightarrow} & A_1 \\ \beta_2 \downarrow & & \downarrow \alpha_1 \\ A_2 & \underset{\alpha_2}{\longrightarrow} & A \end{array}$$

is a pullback for α_1 and α_2 if and only if $P = A_1 \cap A_2$.

Proof: This is clear from the definition.

Consider a diagram

$$\begin{array}{ccc} A' & & B' \\ \alpha \downarrow & & \downarrow \beta \\ A & \underset{f}{\longrightarrow} & B \end{array}$$

with α and β monic. If there exists $f' : A' \to B'$ such that $\beta f' = f\alpha$, then we say that A' is *carried into* B' by f. Let $\{A_i\}_{i \in I}$ be a set of subobjects of A. A subobject A' of A which contains all A_i is called the *union* of A_i and is denoted by $\bigcup_I A_i$ if, for any B and any morphism $f : A \to B$ such that each A_i is carried into some subobject B' in B by f, A' is also carried into B' by f. This union is uniquely determined up to isomorphism.

Let $f : A \to B$ be a morphism. The *image* of f is the subobject $u : I \to B$ of B satisfying the following conditions: There exists $f' : A \to I$ such that $f = uf'$ and for any $B' (\subset_{u'} B)$ with $f'' : A \to B'$ and $f = u'f''$, $B \supseteq I$. We denote the image of f by im f. The dual of an image is a *coimage*.

2.1.4 LEMMA Let $f : A \to B$ be a morphism in a category with equalizers and let $f' : A \to I$ be as above, where I is the image of f. Then f' is epic.

Proof: Let $\alpha f' = \beta f'$. Then $I \subset \mathrm{equ}(\alpha,\beta)$ by the definition of image. Since $\mathrm{equ}(\alpha,\beta) \subseteq I$, $I = \mathrm{equ}(\alpha,\beta)$. Hence $\alpha = \beta$.

Consider the diagram

$$\begin{array}{ccc} P & \longrightarrow & B' \\ \downarrow & & \downarrow i \\ A & \xrightarrow[f]{} & B \end{array}$$

with i monic. If this is a pullback, we call P the *inverse image* of B' by f and denote it by $f^{-1}(B')$. In this case $f^{-1}(B')$ is a subobject of A by (2.1.1). We have the standard properties of inverse images from the definition.

If there is a special object 0 in $\underline{A}$ such that [0,A] and [A,0] contain only one element for every $A \in \underline{A}$, then we call 0 a *zero object*. If 0' is another zero object, 0' is isomorphic to 0. If $f : A \to B$ factors through 0, then we call f a *zero morphism*. As a special case of the above, if the diagram

$$\begin{array}{ccc} P & \longrightarrow & 0 \\ \downarrow & & \downarrow \\ A & \xrightarrow[f]{} & B \end{array}$$

is a pullback then we say that $i : P \to A$ is the *kernel* of $f : A \to B$ and denote it by ker f or $f^{-1}(0)$; namely, P = ker f if and only if $i : P \to A$ is a monomorphism with $fi = 0$ and, for any C and $\alpha : C \to A$ such that $f\alpha = 0$, there exists a (unique) morphism $\beta : C \to P$ such that $i\beta = \alpha$.

2.1.5 LEMMA Consider the diagram

$$\begin{array}{ccccc} A' & \xrightarrow{\alpha} & A & & \\ \downarrow & & \downarrow \gamma & & \\ B' & \xrightarrow[i]{} & B & \xrightarrow[\varepsilon]{} & B'' \end{array}$$

with $B' = \ker \varepsilon$. The left part is extendible to a pullback if and only if $\alpha : A' \to A$ is the kernel of $\varepsilon\gamma : A \to B''$.

Proof: We assume that the diagram

$$\begin{array}{ccccc} A' & \xrightarrow{\alpha} & A & & \\ \delta\downarrow & & \downarrow \gamma & & \\ B' & \xrightarrow[i]{} & B & \xrightarrow[\varepsilon]{} & B'' \end{array}$$

is a pullback. Then $\varepsilon\gamma\alpha = \varepsilon i\delta = 0$. Let $\beta : C \to A$ with $\varepsilon\gamma\beta = 0$. Then there exists $\theta : C \to B'$ such that $i\theta = \gamma\beta$. Hence we obtain $\varepsilon : C \to A'$ such that $\alpha\varepsilon = \beta$, so $\alpha : A' \to A = \ker(\varepsilon\gamma : A \to B'')$. We can show the converse similarly.

A *cokernel* is the dual notion to kernel and we denote it "coker." From the definition we have the following.

2.1.6 LEMMA Consider the diagram

$$\begin{array}{ccccc} K & \xrightarrow{u} & A & \xrightarrow{\alpha} & B \\ & & {\scriptstyle p}\searrow & & \\ & & & C & \end{array}$$

where $u : K \to A = \ker \alpha$ and $p : A \to C = \operatorname{coker} u$. Then $u : K \to A = \ker p$.

If every monomorphism $\alpha : A \to B$ in $\underline{A}$ is the kernel of some morphism $\beta : B \to C$, $\underline{A}$ is called *normal*. The dual is called *conormal*.

In the above we have defined several special morphisms and objects. We now show that they exist under certain assumptions.

2.1.7 COROLLARY Assume that $\underline{A}$ is normal. Let $u : K \to A$ be monic. If $v : A \to D$ is the cokernel of u, $u : K \to A$ is the kernel of v.

2.1.8 LEMMA Consider the diagram

$$\begin{array}{ccccc} A & \xrightarrow{\alpha} & B & \xrightarrow{p} & C \\ & {\scriptstyle q}\searrow & \uparrow v & & \\ & & I & & \end{array}$$

where $p : B \to C = \operatorname{coker} \alpha$ and $v : I \to B = \ker p$. So there exists a unique $q : A \to I$ with $vq = \alpha$. If $\underline{A}$ is normal and has cokernels, then $v : I \to B = \operatorname{im} \alpha$. Furthermore, if $\underline{A}$ has equalizers, then $q : A \to I = \operatorname{coim} \alpha$.

Proof: We shall now show that $v : I \to B = \operatorname{im} \alpha$. Let $B' \subseteq B$ via i and $f : A \to B'$ with $\alpha = if$.

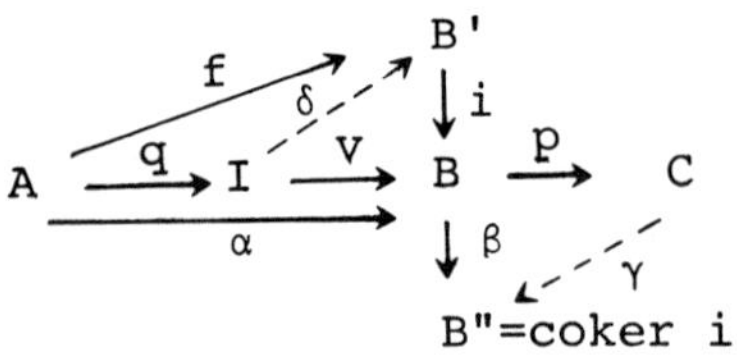

Then $i : B' \to B = \ker \beta$ by (2.1.7). Since $\beta\alpha = \beta if = 0$ and $p : B \to C = \operatorname{coker} \alpha$, there exists $\gamma : C \to B''$ with $I \to B \to B'' = I \to B \to C \to B'' = 0$. Hence we have $\delta : I \to B'$ such that $i\delta = v$. Therefore, $v : I \to B = \operatorname{im} \alpha$. If $\underline{A}$ has equalizers, then q is epic by (2.1.4). Using this fact, we can prove the remainder of (2.1.8).

A normal and conormal category $\underline{A}$ with kernels and cokernels is called an *exact category* provided that for any $v : A \to B$, there exists a factorization

$$v : A \xrightarrow{p} I \xrightarrow{q} B$$

where p is epic and q is monic. Put $\alpha : K \to A = \ker p = \ker v$ and $\beta : B \to C = \operatorname{coker} q = \operatorname{coker} v$. Then we have

$$K \xrightarrow{\alpha} \underbrace{A \xrightarrow{p} I \xrightarrow{q} B}_{\xrightarrow{\quad v \quad}} \xrightarrow{\beta} C$$

and $q : I \to B = \operatorname{im} v$, $p : A \to I = \operatorname{coim} v$ by (2.1.7) and (2.1.8).

A short sequence $A \xrightarrow{\alpha} B \xrightarrow{\beta} C$ is called *exact* if $\operatorname{im} \alpha = \ker \beta$. A long sequence $\cdots \to A_{i-1} \to A_i \to A_{i+1} \to \cdots$ is called *exact* if the short sequence $A_{i-1} \to A_i \to A_{i+1}$ is exact for each i. We note from (2.1.5) that if $\underline{A}$ is normal and has kernels, then $\underline{A}$ has pullbacks P of the form

$$\begin{array}{ccc} P & \longrightarrow & A_2 \\ \downarrow & & \downarrow \alpha_2 \\ A_1 & \xrightarrow[\alpha_1]{} & A \end{array}$$

with either α_1 or α_2 monic, and hence $\underline{A}$ has finite intersections and inverse images of morphisms.

From now on, we assume that $\underline{A}$ has a zero object.

2.1.9 LEMMA Let $\underline{A}$ be an exact category. Then we have:

1. $0 \to A \xrightarrow{\alpha} B$ is exact if and only if α is a monomorphism.
2. $A \xrightarrow{\alpha} B \to 0$ is exact if and only if α is an epimorphism.
3. $0 \to A \xrightarrow{\alpha} B \to 0$ is exact if and only if α is an isomorphism.

Proof: These are clear [note that if $\underline{A}$ is normal, then $\underline{A}$ is balanced (2.1.7)].

2.1.10 LEMMA (*9-lemma*) Let A be an exact category. Assume that every sequence is exact in the following commutative diagram:

$$\begin{array}{ccccccccc}
 & & 0 & & 0 & & 0 & & \\
 & & \downarrow & & \downarrow & & \downarrow & & \\
 & & A' & & A & & A'' & & \\
 & & \downarrow & & \downarrow & & \downarrow & & \\
0 & \to & B' & \to & B & \to & B'' & \to & 0 \\
 & & \downarrow & & \downarrow & & \downarrow & & \\
0 & \to & C' & \to & C & \to & C'' & \to & 0 \\
 & & \downarrow & & \downarrow & & \downarrow & & \\
 & & 0 & & 0 & & 0 & &
\end{array}$$

Then we have an exact sequence $0 \to A' \xrightarrow{u} A \xrightarrow{v} A'' \to 0$ which commutes in the diagram above.

Proof: It is clear that we obtain a sequence $A' \xrightarrow{u} A \xrightarrow{v} A''$ which commutes in the diagram. We shall show that it is exact. That u is monic is easily verified. First we show that a sequence $0 \to A' \to A \to B'' \to C'' \to 0$ is exact. Assume that $X \to A \to B'' = 0$. Then there exists $X \to B'$ such that $X \to B' \to B = X \to A \to B$.

$$X \to B' \to C' \to C = X \to B' \to B \to C = X \to A \to B \to C = 0$$

Hence $X \to B' \to C' = 0$ since $C' \to C$ is monic. Thus there exists $X \to A'$ such that $X \to A' \to B' = X \to B'$.

$$X \to A' \to A \to B = X \to A' \to B' \to B = X \to B' \to B = X \to A \to B$$

Since $A \to B$ is a monomorphism, $X \to A' \to A = X \to A$. Hence $u_1' : A' \to A = \ker(A \to B'')$, so $0 \to A' \to A \to B''$ is exact. From the dual we obtain an exact sequence $A \to B'' \to C'' \to 0$. Since $A'' = \ker(B'' \to C'')$, $0 \to A' \to A \to A'' \to 0$ is exact.

Let $A \supset B$ be two objects with $i : B \to A$. We denote coker i by A/B.

2.1.11 LEMMA Let $A \supset A_1 \supset A_2 \supset B$ be objects in an exact category $\underline{A}$. Then we have an exact sequence

$$0 \to A_2/B \to A_1/B \to A_1/A_2 \to 0$$

Proof: Consider the diagram

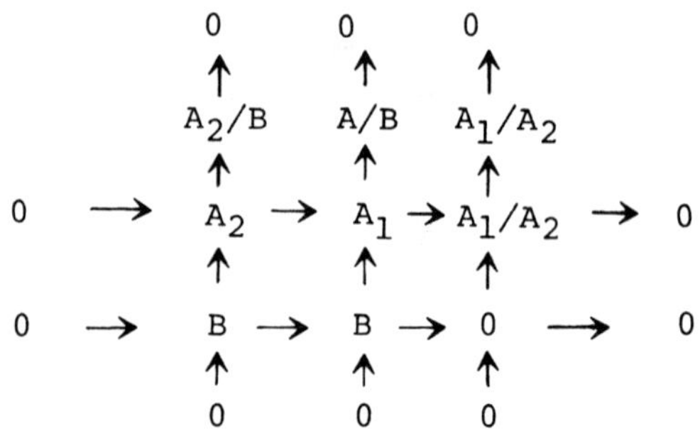

Then we have (2.1.11) from the dual of (2.1.10).

2.1.12 LEMMA Let

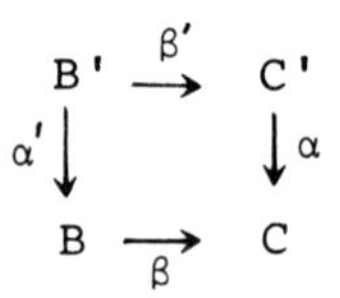

be a pullback in an exact category $\underline{A}$, where α is a monomorphism and β is an epimorphism. Then the diagram above can be extended to the exact diagram

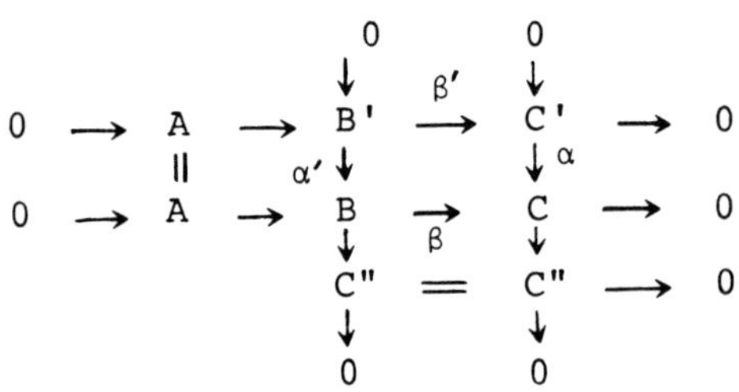

Proof: Put $A = \ker \beta$ and $C'' = \operatorname{coker} \alpha$. Then $\alpha' : B' \to B = \ker(B \to C \to C'')$ by (2.1.5). Hence the two columns are exact. Exactness of the top row follows from (2.1.10).

2.1.13 COROLLARY Let $f : A \to B$ be a morphism in an exact category $\underline{A}$ and $B' \subset B$. Put $I = \operatorname{im} f$. Then

$$f^{-1}(B') \to I \cap B' \to 0$$

and

$$0 \to f^{-1}(B') \to A \to I/I \cap B' \to 0$$

are exact.

Proof: By the definition of the pullback [cf. (2.1.2)]

$$\begin{array}{ccc} f^{-1}(B') & \longrightarrow & I \cap B' \\ \downarrow & & \downarrow \\ A & \longrightarrow & I \end{array}$$

is a pullback. Hence the result follows from (2.1.12).

2.1.14 LEMMA Let A_1 and A_2 be subobjects of A in an exact category $\underline{A}$. Then

$$0 \to A_2/A_1 \cap A_2 \to A/A_1 \to A/A_1 \cup A_2 \to 0$$

is exact.

Proof: We have the diagram

$$\begin{array}{ccccccccc} & & & & & & 0 & & \\ & & & & & & \downarrow & & \\ 0 & \longrightarrow & A_1 \cap A_2 & \longrightarrow & A_2 & \longrightarrow & A_2/A_1 \cap A_2 & \longrightarrow & 0 \\ & & \downarrow & \text{I} & \downarrow & \text{II} & \downarrow \alpha & & \\ 0 & \longrightarrow & A_1 & \longrightarrow & A & \longrightarrow & A/A_1 & & \\ & & & & \downarrow & \text{III} & \downarrow \beta & & \\ & & & & A/A_2 & \longrightarrow & A/A_1 \cup A_2 & & \end{array}$$

where I is the pullback by (2.1.3), III is a pushout obtained from the dual of (2.1.3), and since $A_1 \cap A_2 \to A_2 \to A \to A/A_1 = 0$, we obtain α, which gives the commutative diagram II. We shall show that the right column is exact. Since III is the pushout and $\operatorname{coker}(A_2 \to A) = A \to A/A_2$, $\beta : A/A_1 \to A/A_1 \cup A_2 = \operatorname{coker}(A_2 \to A \to A/A_1)$ by the dual of (2.1.5). Since $A_2 \to A_2/A_1 \cap A_2$ is epic, $\operatorname{coker} \alpha = \beta : A/A_1 \to A/A_1 \cup A_2$. Hence β is epic. On the other hand, since I is a pullback and $A_1 \to A$ is monic, $A_1 \cap A_2 \to A_2 = \ker(A_2 \to A \to A/A_1) = \ker(A_2 \to A_2/A_1 \cap A_2 \to A/A_1)$ by (2.1.5). Since $A_2 \to A_2/A_1 \cap A_2$ is epic, α is a monomorphism.

2.1.15 COROLLARY Let A_1 and A_2 be subobjects of A in an exact category. Then $A_1 \cup A_2/A_1 \approx A_2/A_1 \cap A_2$.

Proof: If we set $A = A_1 \cup A_2$ in (2.1.14), then $\alpha : A_2/A_1 \cap A_2 \to A_1 \cup A_2/A_1$ is an isomorphism.

Again we consider an arbitrary category $\underline{A}$ with zero. Let $\{A_i\}_{i \in I}$ be a set of objects in $\underline{A}$. If there exists an object A' in $\underline{A}$ satisfying the following conditions, then we call A' the *product* of $\{A_i\}_{i \in I}$.

There exists $p_i : A' \to A_i$ for each $i \in I$. (2.1.5)

For any set of morphisms $\{\alpha_i : B \to A_i\}$ there exists a unique morphism $\alpha : B \to A'$ such that $p_i\alpha = \alpha_i$ for each $i \in I$; namely,

$$\Pi_I([B,A_i]) \leftrightarrow [B,A'] \qquad \text{for any } B$$

gives us a one-to-one correspondence. (2.1.6)

It is clear that A' is uniquely determined up to isomorphism and we denote it by $\times_I A_i$ or $\Pi_I A_i$. If we take $B = A_i$ for all i, and $\alpha_i = 1_{A_i}$, $\alpha_j = 0$ for $j \neq i$, we have a monomorphism $u_i : A_i \to A'$ such that $p_iu_i = 1_{A_i}$ and $p_ju_i = 0$ if $i \neq j$. We call p_i a *projection* and u_i an *injection*. Then u_i is monic and p_i is epic.

2.1.16 LEMMA Let $\underline{A}$ be a category with zero and let A_i be objects in $\underline{A}$, $i = 1, 2$. If u_i, p_i are as above for $A_1 \times A_2$, then $\ker p_2 = u_1 : A_1 \to A_1 \times A_2$.

Proof: This is clear from the definition.

The dual of a product is called a *coproduct* and we denote it by $\oplus_I A_i$ or $\Sigma \oplus A_i$. We note that

$$[\oplus_I A_i,B] \leftrightarrow \Pi_I[A_i,B] \qquad \text{for any } B$$

gives us a one-to-one mapping. Let $f_i : A_i \to B_i$ $(i = 1, 2)$. Then we can define a unique morphism $f : A_1 \oplus A_2 \to B_1 \oplus B_2$ via $\{f_i\}$. We denote f by $f_1 \oplus f_2$. Let $\{A_i\}_I$ and $\{B_j\}_J$ be two sets of objects and let $\{f_{ji} = A_i \to B_j\}_{I \times J}$ be a set of morphisms. Then there exists a unique morphism $F_i : A_i \to \Pi_J B_j$ for each $i \in I$. Hence there exists a unique morphism $F : \oplus_I A_i \to \Pi_J B_j$. We can denote F by the (infinite) matrix (f_{ij}).

As we have shown for $\underline{M}_R$, we can easily prove the following.

2.1.17 LEMMA Let $\alpha_i : A_i \to A$ be morphisms in $\underline{A}$. Consider a diagram

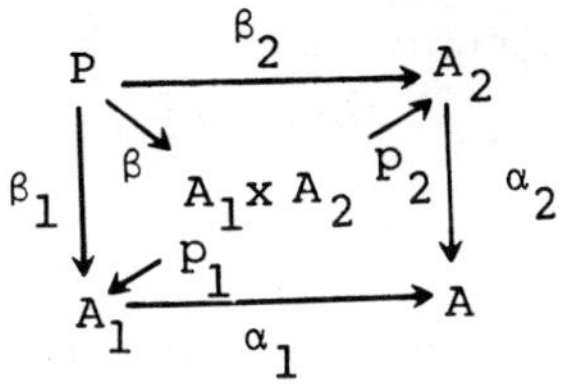

where the p_i's are projections and $\beta_i = p_i\beta$. Then β_1 and β_2 give a pullback for α_1 and α_2 if and only if β is an equalizer of $\alpha_1 p_1$ and $\alpha_2 p_2$.

2.1.18 LEMMA Let $\alpha, \beta : A \to B$ be morphisms in $\underline{A}$. Then for the diagram

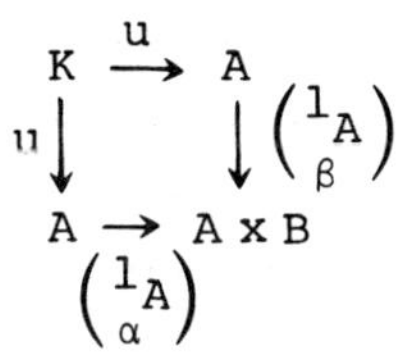

u is an equalizer of $\begin{pmatrix} 1_A \\ \alpha \end{pmatrix}$ and $\begin{pmatrix} 1_A \\ \beta \end{pmatrix}$ if and only if the diagram is a pullback.

We note that for A_1 and A_2 subobjects in A, $A_1 \cap A_2$ is the pullback for $u_i : A_i \to A$ and that the product of two objects A and B is the pullback for $A \to 0$ and $B \to 0$. Hence, from (2.1.17) and (2.1.18), we obtain the following.

2.1.19 PROPOSITION Let $\underline{A}$ be a category with zero. Then the following conditions are equivalent:

1. $\underline{A}$ has finite intersections and finite products.
2. $\underline{A}$ has pullbacks.
3. $\underline{A}$ has equalizers and finite products.

2.2 ABELIAN CATEGORIES

We shall deal with categories similar to $\underline{M}_R$ in this section. A category $\underline{A}$ is called an *additive category* if

Every [A,B] is an abelian group. (2.2.1)

$\alpha(\beta + \gamma) = \alpha\beta + \alpha\gamma \qquad (\beta + \gamma)\delta = \beta\delta + \gamma\delta$ (2.2.2)

if they are defined. Then

$$0\alpha = \alpha 0 = 0 \tag{2.2.3}$$

2.2.1 LEMMA Let $\underline{A}$ be an additive category and let $\{A_i\}_{i=1}^n$ be a finite set of objects in $\underline{A}$. Let $A = \oplus_i A_i$ and u_i, p_i be injections and projections, respectively. Then $1_A = \Sigma_i u_i p_i$ and $p_i u_j = \delta_{ij} 1_{A_i}$. Conversely, if an object A has morphisms satisfying these properties, $A = \oplus_i A_i$.

Proof: Since $p_i u_j = \delta_{ij} 1_{A_j}$ by the definition of p_i, $(\Sigma_i u_j p_j) u_i = u_i$ for all i and $1_A u_i = u_i$. Hence $1_A = \Sigma_i u_i p_i$ from the uniqueness. Conversely, assume that $1_A = \Sigma_i u_i p_i$. Let $\alpha_i : A_i \to B$ and put $\beta_i = \alpha_i p_i$. Then $\Sigma_i \beta_i : A \to B$ and $(\Sigma_j \beta_j) u_i = \alpha_i$. The uniqueness is obtained from $1_A = \Sigma_i u_i p_i$.

If we interchange projections and injections, we have the same situation on $\times_i A_i$. Hence $\oplus_i A_i \approx \times_i A_i$ in an additive category. Therefore, we can express morphisms in $[\oplus A_i, \oplus B_j]$ by matrices whose entries consist of morphisms in $[A_i, B_j]$. Furthermore, the composition of morphisms corresponds to the product of matrices.

We call a short exact sequence $0 \to A \xrightarrow{\alpha} B \xrightarrow{\beta} C \to 0$ split if there exists $\gamma : C \to B$ such that $\beta\gamma = 1_C$.

2.2.2 LEMMA Let $\underline{A}$ be an exact and additive category. If a short exact sequence $0 \to A \xrightarrow{\alpha} B \xrightarrow{\beta} C \to 0$ splits, then $B = A \oplus C$ via a projection β and an injection α.

Proof: Put $\theta = \gamma\beta$. Then θ and $1_B - \theta$ are idempotent. We can easily show that $\alpha : A \to B = \ker \theta$. Since $\theta(1_B - \theta) = 0$, there exists a $\rho : B \to A$ such that $1 - \theta = \alpha\rho$, and we have $\rho\alpha = 1_A$ $[\alpha\rho\alpha = (1_B - \theta)\alpha = \alpha]$.

2.2.3 LEMMA Let $\underline{A}$ be as in (2.2.2) and let A_1 and A_2 be subobjects of A. Then $A = A_1 \oplus A_2$ if and only if $A_1 \cap A_2 = 0$ and $A_1 \cup A_2 = A$.

Proof: We know from the diagram of the proof of (2.1.14) that $A = A_1 \cup A_2$ and $A_1 \cap A_2 = 0$, since $A_1 \approx A/A_2$ and $A_2 \approx A/A_1$. The converse is also clear from (2.1.14) and (2.2.2).

We call a category $\underline{A}$ *abelian* if $\underline{A}$ is an exact and additive category with finite products.

2.2.4 THEOREM Let $\underline{A}$ be an additive category. Then the following conditions are equivalent:

1. $\underline{A}$ is abelian.
2. $\underline{A}$ is normal and conormal and has kernels, cokernels, and finite products.

Proof: We show only that (2) $\Rightarrow$ (1). We know that $\underline{A}$ has finite intersections by (2.1.5). Hence $\underline{A}$ satisfies the conditions in (2.1.19), and $\underline{A}$ is exact by (2.1.4) and (2.1.8).

2.2.5 PROPOSITION Let $\underline{A}$ be abelian. If $A \subset B$ and C are subobjects of D in $\underline{A}$, then $A \cup (B \cap C) = B \cap (C \cup A)$.

Proof: $B \supset (A_1 =) (C \cap B) \cup A \subset (C \cup A) \cap B$ $(= A_2)$. Thus $A_1 \cap C \supset (C \cap A) \cup (C \cap B) = C \cap B$, so $A_1 \cap C = B \cap C$. Similarly, we have $A_2 \cap C = B \cap C$. $A_2 \cup C \subset (B \cup C) \cap (C \cup A) = A \cup C$. Hence $A_2 \cup C = A \cup C$. Similarly, $A_1 \cup C = A \cup C$. Therefore, $(A \cup C)/B \cap C = A_1/(B \cap C) \oplus C/(B \cap C) = A_2/(B \cap C) \oplus C/(B \cap C)$ and $A_1/(B \cap C) \supset A_2/(B \cap C)$. Hence $A_1 = A_2$.

We shall study further properties of abelian categories in Sec. 6.2.

2.3 FUNCTORS

Let $\underline{A}$ and $\underline{B}$ be two categories. A correspondence T from $\underline{A}$ to $\underline{B}$ satisfying the following conditions is called a *covariant* (*contravariant*) *functor* from $\underline{A}$ to $\underline{B}$:

T gives a correspondence: $A \in \underline{A} \to T(A) \in \underline{B}$ and a mapping: $[A,A'] \to [TA,TA']$ such that $\alpha : A \to A' \Rightarrow T(\alpha) : TA \to TA'$. (2.3.1)

$$T(\alpha\beta) = T\alpha T\beta, \quad [T(\alpha\beta) = T\beta T\alpha], \quad \text{and} \quad T(1_A) = 1_{TA} \tag{2.3.2}$$

If $\underline{A}$ and $\underline{B}$ are additive and

$$T(\alpha \pm \beta) = T(\alpha) \pm T(\beta) \tag{2.3.3}$$

then T is called *additive*. Let T and S be two functors from $\underline{A}$ to $\underline{B}$. If there exists ρ_A for each A in $\underline{A}$,

$$\rho_A : TA \to SA \tag{2.3.4}$$

and for each $\alpha : A \to A'$

$$\begin{array}{lcl} \rho_A : TA & \longrightarrow & SA \\ \quad\ \downarrow T\alpha & & \downarrow S\alpha \\ \rho_{A'} : TA' & \longrightarrow & SA' \end{array}$$

is commutative, we say that ρ is a *natural transformation* from T to S. If ρ_A is an isomorphism for each A, we say that T and S are *naturally equivalent*.

Let $\underline{A}$ be an additive category. For each object A in $\underline{A}$ we consider $H^A : B \to [A,B] \in \underline{\underline{Ab}}$. Then H^A gives a covariant and additive functor from $\underline{A}$ to $\underline{Ab}$. Similarly, $H_A : B \to [B,A]$ gives a contravariant additive functor.

Let P be a property in $\underline{A}$. If a functor T preserves this property P, we say that T is *P-preserving*. For example, we say that T is a *monofunctor* if $T\alpha$ is monic whenever α is so. We call an exactness preserving functor an *exact functor*. We shall define some important types of functors.

T is *faithful* if for $\alpha,\beta : A \to A'$ with $\alpha \neq \beta$, $T\alpha \neq T\beta$.

T is an *embedding* if T is faithful and $TA \neq TB$ whenever $A \neq B$.

T is *full* if $[A,B] \to [TA,TB]$ is an isomorphism.

T is *representative* if for each object B in $\underline{B}$ there exists an object A in $\underline{A}$ such that $TA \approx B$.

T is an *equivalence* if T is full and faithful and representative.

T is an *isomorphism* if T is an equivalence and T gives a one-to-one correspondence between the objects in $\underline{A}$ and $\underline{B}$.

Let $\underline{A}'$ be a subcategory of $\underline{A}$. Then we have a *natural embedding functor* $I : \underline{A}' \to \underline{A}$.

Let $0 \to A' \to A \to A'' \to 0$ be an exact sequence in an exact category. If $0 \to TA' \to TA \to TA'' \to 0$ is exact, T is called *exact*. If $0 \to TA' \to TA \to TA''$ ($TA' \to TA \to TA'' \to 0$) is exact, T is called *left* (*right*) *exact*.

2.3.1 PROPOSITION Let $A' \to A \to A'' \to 0$ be an exact sequence in an exact category $\underline{A}$ and T a covariant functor from $\underline{A}$ to an exact category $\underline{B}$. If T is right exact, $TA' \to TA \to TA'' \to 0$ is exact.

Proof: We can divide $A' \xrightarrow{f} A \to A'' \to 0$ into two exact sequences

$$0 \to \ker f = K \to A' \to \operatorname{im} f = L \to 0$$

and

$$0 \to L \to A \to A'' \to 0$$

Then we have two exact sequences

$$TK \to TA' \to TL \to 0$$

and

$$TL \to TA \to TA'' \to 0$$

Hence $TA' \to TA \to TA'' \to 0$ is exact.

2.4 GROTHENDIECK CATEGORIES

The category of abelian groups $\underline{Ab}$ and the category of right R-modules $\underline{M}_R$ over a ring R are very familiar to us. We generalize those categories in this section. The results of this section will be used primarily in Sec. 6.2. However, in Sec. 6.6 we show, by making use of an elementary method, that the category appearing in Sec. 6.2 is equivalent to $\underline{M}_\Delta$ where Δ is a division ring. In this section we give only those definitions that are used later. We refer the reader to [159] for the details.

A *diagram scheme* Σ is a triple (I,M,d), where I is a set, M is a set of arrows, and $d : M \to I \times I$ is a mapping. Let $d(m) = (i,j)$. Then i is called the *origin* and j the *extremity*. An element in I is called a *vertex*. Let $\underline{A}$ be a category and Σ a diagram scheme. We define a *diagram* D over Σ as follows. An object D_i in $\underline{A}$ is assigned for each $i \in I$ and a morphism $D(m) \in [D_i, D_j]$ is assigned for each m with $d(m) = (i,j)$. We denote the class of all diagrams over Σ by $[\Sigma, \underline{A}]$. We make $[\Sigma, \underline{A}]$ a category. We take all diagrams over Σ as the objects in $[\Sigma, \underline{A}]$. A morphism $\{f_i\}_{i \in I}$: $\{D_i, D(m)\} \to \{D_i', D(m)'\}$ is given by $f_i \in [D_i, D_i']$, where

$$\begin{array}{ccc} D_i & \xrightarrow{D(m)} & D_j \\ {\scriptstyle f_i}\downarrow & & \downarrow{\scriptstyle f_j} \\ D_i' & \xrightarrow{D(m)'} & D_j' \end{array}$$

is commutative.

If any composite $D(m_t)D(m_{t-1})\cdots D(m_1)$ depends only on the origin and the extremity of the composition $d(m_t)\cdots d(m_1)$, i.e., the origin and extremity of m_1 and m_t, respectively, and $d(m_0) = 1_{D_i}$ for $d(m_0) = (i,i)$, then we call D a *commutative diagram*. From now on we deal only with commutative diagrams. We may generalize the concept of products and coproducts. Let $\{f_i : X \to D_i\}_{i \in I}$ be a set of morphisms. If

$$\begin{array}{ccc} & \overset{f_i}{\nearrow} & D_i \\ X & & \downarrow D(m) \\ & \underset{f_j}{\searrow} & D_j \end{array}$$

is commutative for all D(m), we say that $\{f_i\}_I$ is *compatible*. If an object Y with morphisms $g_i : Y \to D_i$ in $\underline{A}$ satisfies the following two conditions, Y is called a *limit* of D:

$\{g_i : Y \to D_i\}$ is compatible. (2.4.1)

If $\{f_i : X \to D_i\}$ is compatible, there exists a unique morphism $f : X \to Y$ such that $g_i f = f_i$ for all i. (2.4.2)

If $D(m) = 0$ for all m $(i \neq j)$ ($\underline{A}$ has a zero object), then $\Pi_I D_i$ is a limit of D. If every diagram in $[\Sigma, \underline{A}]$ has a limit, we call $\underline{A}$ *Σ-complete*. If $\underline{A}$ is Σ-complete for any diagram scheme Σ, $\underline{A}$ is called *complete*. We can define dually *Σ-cocomplete* and *cocomplete*.

Let I be a partially ordered set. If for any two elements i, j in I, there exists k in I with $k \geq i$ and $k \geq j$, then I is called a *directed set*. An element D in $[I,\underline{A}]$ such that

$D(m)$ is unique for $d(m) = (i,j)$ with $i \leq j$ and $D(m) = \emptyset$ otherwise (2.4.3)

is called a *direct system*. A cocomplete of a direct system is called the *direct limit* and is denoted by $\varinjlim$.

2.4.1 PROPOSITION Let $\underline{A}$ be a category with images and inverse images. Let $\{D_i \to L\}$ be a colimit over a diagram D in $\underline{A}$. If $\{f_i : D_i \to A\}$ is a compatible system over D and $f : L \to A$ is the induced morphism, then $\bigcup \operatorname{im} f_i$ is defined and $\operatorname{im} f = \bigcup \operatorname{im} f_i$.

Proof: Since we have a commutative diagram

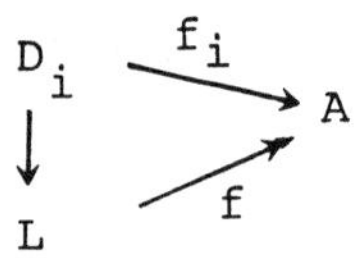

$\operatorname{im} f_i \subset \operatorname{im} f$ for all i. Let $g : A \to B$ and $B' \subset B$. We assume that all $\operatorname{im} f_i$ are carried into B' by g. Then we have a compatible system $\{D_i \to \operatorname{im} f_i \to B'\}$. Hence there exists the unique morphism $h : L \to B'$ such that

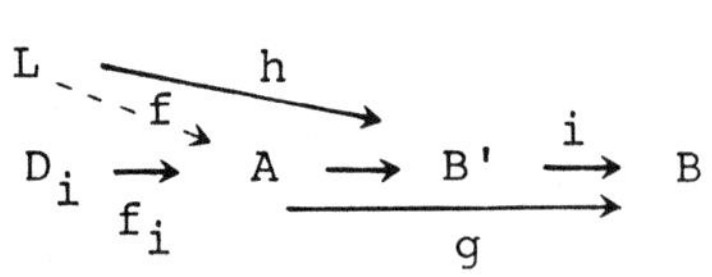

is commutative. Therefore, $ih = gf$, so $\operatorname{im} f \subset g^{-1}(B')$, which implies that $\operatorname{im} f$ is carried into B' by g. Thus $\operatorname{im} f = \bigcup \operatorname{im} f_i$.

Let $\Sigma_I \oplus A_\alpha$ be a coproduct in $\underline{A}$. Let $\{I_i\}$ be the set of all finite subsets of I. Then $\{I_i\}$ is a partially ordered set via the inclusion relation. Then we have, from the definition, the following.

2.4.2 LEMMA $\Sigma \oplus_I A_\alpha = \varinjlim \Sigma_{I_i} \oplus A_\gamma$.

2.4.3 COROLLARY Let $\underline{A}$ be as in (2.4.1). Assume that $\underline{A}$ has coproducts. Then, for $f : \Sigma_I \oplus A_\alpha \to B$, we have im $f = \cup$ im f_i, where $f_i : \Sigma_{I_i} \oplus A_\gamma \to \Sigma_I \oplus A_\alpha \xrightarrow{f} B$.

Proof: This is clear from (2.4.1) and (2.4.2).

A set $\{U_i\}_{i\in I}$ of objects in a category $\underline{A}$ is called a *family of generators* of $\underline{A}$ if for any A, B and $\alpha \neq \beta : A \to B$, there exists $i \in I$ and $u : U_i \to A$ such that $\alpha u \neq \beta u$. If {U} is a family of generators for a single object U, we call U a *generator* of $\underline{A}$.

2.4.4 THEOREM Let $\underline{A}$ be an abelian category with coproducts. Then an object U is a generator of $\underline{A}$ if and only if for each A, there exist $\oplus_I U = U^{(I)}$ and an epimorphism $\gamma : U^{(I)} \to A$, where $I = [U,A]$ and γ is given by $\alpha : U \to A$; $\alpha \in [U,A]$.

Proof: Let U be a generator. Then im $\gamma = \cup$ im α from (2.4.1). If im $\gamma \neq A$, there exists $\beta : U \to A$ such that $U \to A \to A/\text{im } \gamma \neq 0$. But im $\beta \subset$ im γ, which is a contradiction. The converse is clear.

If category $\underline{A}$ satisfies

$\underline{A}$ is abelian and cocomplete. (2.4.4)

$\underline{A}$ has a generator. (2.4.5)

Letting $A \supset B$ be objects in $\underline{A}$ and $\{A_i \subset A\}_{i\in I}$ be a direct system of subobjects of A, then

$$(\cup_i A_i) \cap B = \cup_i (A_i \cap B)$$

we call $\underline{A}$ a *Grothendieck category*. (2.4.6)

A cocomplete abelian category $\underline{A}$ is called a C_3-*category* if $\underline{A}$ satisfies (2.4.6).

2.4.5 PROPOSITION Let $\underline{A}$ be a C_3-category. Then the direct limit of every direct system of subobjects $\{A_i\}$ of an object A is $\cup A_i$.

Proof: Let $\varinjlim A_i = L$. Then we obtain a unique morphism $u : L \to A$ and im $u = \cup A_i$ by (2.4.1). We shall show that u is a monomorphism. Let K = ker u and $A_i' = \text{im } f_i(: A_i \to L)$. We know by (2.4.1) that $L = \cup A_i'$. Hence $K = K \cap L = \cup (K \cap A_i')$ by (2.4.6). If $K \neq 0$, then $K \cap A_i' \neq 0$ for some i, which contradicts the fact that $uf_i = i_{A_i} (: A_i \to A)$.

2.5 THE CATEGORY $\underline{M}_R$

Let R be a ring. We denote the category of all right R-modules by $\underline{M}_R$. Then $\underline{M}_R$ is a Grothendieck category. We give several lemmas which we shall utilize in the following chapters. Capital roman letters denote objects in $\underline{M}_R$.

2.5.1 LEMMA Consider a commutative diagram

$$\begin{array}{ccccc} A & \xrightarrow{u} & B & \xrightarrow{v} & C \\ a\downarrow & & b\downarrow & & c\downarrow \\ A' & \xrightarrow{u'} & B' & \xrightarrow{v'} & C' \end{array}$$

with exact rows. If c is monic, im b ∩ im u' = im u'a = im bu. If a is epic, ker b + im u = ker v'b = ker cv.

Proof: Assume that c is monic. It is clear that im u'a ⊂ im b ∩ im u'. Let x be in im b ∩ im u'. Then x = b(y) = u'(z); y ∈ B, z ∈ A'. Since cv(y) = v'b(y) = v'(x) = v'u'(z) = 0, y = u(w) for some w ∈ A, for c is monic. Hence x = bu(w) = u'a(w) ∈ im u'a. The second assertion is dual to the first, and the proof is left to the reader.

2.5.2 LEMMA We may extend the diagram in (2.5.1) as follows:

$$\begin{array}{ccccc} \ker a & \xrightarrow{u_1} & \ker b & \xrightarrow{v_1} & \ker c \\ \downarrow & & \downarrow & & \downarrow \\ A & \xrightarrow{u} & B & \xrightarrow{v} & C \\ a\downarrow & & b\downarrow & & c\downarrow \\ A' & \xrightarrow{u'} & B' & \xrightarrow{v'} & C' \\ \downarrow & & \downarrow & & \downarrow \\ \operatorname{coker} a & \xrightarrow{u_2} & \operatorname{coker} b & \xrightarrow{v_2} & \operatorname{coker} c \end{array}$$

where the u_i and v_i are the natural morphisms [cf. (2.1.10)].

1. If u' is monic, the first row is exact.
2. If v is epic, the fourth row is exact.
3. If (1) and (2) are satisfied, there exists a morphism d : ker c → coker a such that

$$\ker a \xrightarrow{u_1} \ker b \xrightarrow{v_1} \ker c \xrightarrow{d} \operatorname{coker} a \xrightarrow{u_2} \operatorname{coker} c$$

is exact.

Proof: 1. It is clear that $v_1u_1 = 0$. Let x be in ker b and $v_1(x) = 0$. Then, since $v(x) = 0$, we obtain y in A such that $x = u(y)$. $u'a(y) = bu(y) = b(x) = 0$, and hence $a(y) = 0$, for u' is monic. So $y \in \ker a$. Since $u_1 = u \mid \ker a$, $x = u_1(y)$.

2. This is dual to (1).

3. Assume (1) and (2). We shall define d as follows. For any element x in ker c, there exists y in B with $v(y) = x$. Since $v'b(y) = cv(y) = c(x) = 0$, there exists a unique element w in A' with $u'(w) = b(y)$. We set $d(x) = v_{A'}(w)$, where $v_{A'} : A' \to \operatorname{coker} a$ is the natural epimorphism. We shall show that d(x) is well defined. Assume that $v(y') = x$ for $y' \in B$. Since $v(y - y') = 0$, $y - y' \in \operatorname{im} u$. Hence there exists a z in A with $b(y) - b(y') = u'a(z)$. We take the unique element w' in A' for y' as above. Then $u'(w) - u'(w') = u'a(z)$. Hence $w - w' = a(z)$, for u' is monic. Therefore, $v_{A'}(w) - v_{A'}(w') = v_{A'}(a(z)) = 0$. We now show that

$$\ker b \to \ker c \to \operatorname{coker} a \to \operatorname{coker} b$$

is exact. Let x be in ker c with $d(x) = 0$. Then there exist $w \in A'$, $y \in B$, and $z \in A$ such that $x = v(y)$, $b(y) = u'(w)$, and $w = a(z)$. Since $b(y - u(z)) = 0$, $y - u(z) \in \ker b$. Hence $x = v(y - u(z)) = v_1(y - u(z)) \in \operatorname{im} v_1$. It is clear from the definition of d that $\operatorname{im} v_1 \subset \ker d$. We can establish the remaining assertions similarly.

2.5.3 COROLLARY (*5-lemma*) Consider the commutative diagram

$$\begin{array}{ccccccccc}
A_2 & \xrightarrow{f_2} & A_1 & \xrightarrow{f_1} & A_0 & \xrightarrow{f_0} & A_{-1} & \xrightarrow{f_{-1}} & A_{-2} \\
{\scriptstyle h_2}\downarrow & & {\scriptstyle h_1}\downarrow & & {\scriptstyle h_0}\downarrow & & {\scriptstyle h_{-1}}\downarrow & & {\scriptstyle h_{-2}}\downarrow \\
B_2 & \xrightarrow[g_2]{} & B_1 & \xrightarrow[g_1]{} & B_0 & \xrightarrow[g_0]{} & B_{-1} & \xrightarrow[g_{-2}]{} & B_{-2}
\end{array}$$

with exact rows. We have:

1. If $\operatorname{coker} h_2 = \ker h_1 = \ker h_{-1} = 0$, then $\ker h_0 = 0$.
2. If $\operatorname{coker} h_1 = \operatorname{coker} h_{-1} = \ker h_{-2} = 0$, then $\operatorname{coker} h_0 = 0$.
3. If $\operatorname{coker} h_2 = \ker h_{-2} = 0$ and h_1, h_{-1} are isomorphisms, then h_0 is an isomorphism.

Proof: 1. We can prove this directly or, after replacing A_0 and B_0 by $\operatorname{coker} f_2$ and $\operatorname{coker} g_2$, we can apply (2.5.2). We leave all proofs to the reader as exercises.

We note that the lemma above is valid in any small abelian category by Ref. 159 (Theorem 1.5, p. 97, and Theorem 2.6, p. 101).

3
General Modules and Rings

In this chapter we study some general properties of rings and modules that are used in later chapters. This chapter begins our study of ring theory proper.

In Sec. 3.1 we describe some important properties of idempotent elements. Throughout this book idempotent elements will prove to be very useful tools in ring theory. In Sec. 3.2 we study properties of mutually comaximal ideals. The Jacobson radical of a ring is a very important ideal. While it is defined in Sec. 3.3 and a few elementary properties are given, each chapter will reveal further properties. For the generalization of Nakayama's lemma, we define small submodules and the dual notion--essential submodules--in Sec. 3.4. They are also very important notions in the study of projective and injective modules. In this context, we shall see in Chaps. 5 and 6 that they are closely related to the Jacobson radicals of endomorphism rings. In Sec. 3.5 we consider limits and colimits in the category of modules. Most of their properties are valid for arbitrary Grothendieck categories when we modify the methods of Sec. 3.5 and use the results of Chap. 2. In ring theory the functors Hom and $\otimes$ are also very important and we explore their fundamental properties in Sec. 3.6. We make extensive use of these properties throughout the book. In the final section, Sec. 3.7, we examine the relationships between endomorphism rings of free modules and matrix rings.

3.1 IDEMPOTENTS

We have already defined idempotent elements in a ring R (Sec. 1.6). We shall study some of their properties. Let e and f be idempotents in R. If $ef = fe = 0$, we say that e and f are *orthogonal* and denote this by $e \perp f$. It is clear that $e \perp (1 - e)$.

3.1.1 THEOREM (*Peirce Decomposition*) Let R be a ring and e an idempotent in R. Then $R = eR \oplus (1 - e)R = Re \oplus R(1 - e)$ as R-modules, and $R = eRe \oplus (1 - e)Re \oplus eR(1 - e) \oplus (1 - e)R(1 - e)$.

Proof: For any x in R, $x = ex + (1 - e)x = xe + x(1 - e)$. Let y be in $eR \cap (1 - e)R$. Then $y = er = (1 - e)r'$ for some r, r' in R. Hence $y = ey = e(1 - e)r' = 0$. Therefore, $R = eR \oplus (1 - e)R$. Similarly, we have $R = Re \oplus R(1 - e)$. The last assertion now follows from the first.

3.1.2 PROPOSITION Let e be an idempotent. If $eR = L_1 \oplus L_2 \oplus \cdots \oplus L_n$ for right ideals L_i, then there exists a set of mutually orthogonal idempotents $\{e_i\}$ (i.e., $e_ie_j = \delta_{ij}e_i$) such that $e = \Sigma_{i=1}^n e_i$ and $L_i = e_iR$. Conversely, if there exists such a set of idempotents $\{e_i\}$, $\Sigma_{i=1}^n \oplus e_iR = eR$.

Proof: We assume that $eR = \Sigma_{i=1}^n \oplus L_i$. Then $e = \Sigma_{i=1}^n e_i$; $e_i \in L_i$. Since e_i is in $L_i \subset eR$, $e_i = ee_i = \Sigma_{j=1}^n e_je_i$. Hence $e_je_i = \delta_{ji}e_i$, by (1.3.3), since $e_je_i \in L_j$. It is clear that $L_i \supset e_iR$ and $eR = (\Sigma_i e_i)R \subset \Sigma_i e_iR \subset \Sigma_i \oplus L_i = eR$. Therefore, $L_i = e_iR$. Conversely, if $e_i \perp e_j$ for $i \neq j$, $e_i \perp (e_1 + \cdots + e_{i-1} + e_{i+1} + \cdots + e_n)$. Hence $e_iR \cap (e_1 + \cdots + e_{i-1} + e_{i+1} + \cdots + e_n)R = 0$, so $eR = \Sigma_i \oplus e_iR$ by induction on n.

Let e and f be two idempotents. If $eR \supset fR$ and $Re \supset Rf$, then $ef = fe = f$, and conversely. We denote this relation by $e \geq f$. This is equivalent to $f \perp (e - f)$ $[e = f + (e - f)$ and $(e - f)^2 = e - f]$.

3.1.3 PROPOSITION Let e and f be idempotents. Then $eR = fR$ if and only if $e - f \in e\ell_R(e)$. $e = f$ if and only if $eR = fR$ and $(1 - e)R = (1 - f)R$.

Proof: The fact that $eR = fR$ is equivalent to the fact that $fe = e$ and $ef = f$. This is also equivalent to the fact that $e - f = e(e - f) \in eR \cap \ell_R(e) = e\ell_R(e)$. If $eR = fR$ and $(1 - e)R = (1 - f)R$, then $1 - e = (1 - f)(1 - e) = 1 - f - e + fe = 1 - f$. Hence $e = f$.

3.1.4 COROLLARY Let e and f be idempotents such that $fe = 0$. Then $e_1 = e - ef$ is an idempotent and $f \perp e_1$, $eR = e_1R$.

Proof: $e_1^2 = e_1$ is clear. $fe_1 = fe - fef = 0$. $e_1f = ef - ef = 0$. Hence $f \perp e_1$. Since $e - e_1 = ef \in eR \cap \ell_R(e)$, $eR = e_1R$.

3.1.5 THEOREM Let e and f be idempotents. If eR contains fR, there exist idempotents e' and f' such that $eR = e'R$, $fR = f'R$, $e \geq f'$, and $e' \geq f$.

Proof: We obtain $R = fR \oplus (1 - f)R$ by (3.1.1). Since $eR \supset fR$, $eR = fR \oplus eR \cap (1 - f)R$ by (1.7.4). Hence $e = f' + f_1$, where $f_1^2 = f_1$,

$f'^2 = f'$, and $f_1 \perp f'$, and further $f_1R = eR \cap (1 - f)R$, $fR = f'R$ by (3.1.2). Therefore, $e \geq f'$ by (3.1.2). Since $f_1 \in eR \cap (1 - f)R$, $ff_1 = 0$. Put $f_1' = f_1 - f_1f$. Then $f_1'R = f_1R$ and $f_1' \perp f$ by (3.1.4). Put $e' = f + f_1'$. Then $e' \geq f$ and $e'R = fR \oplus f_1'R = fR \oplus f_1R = fR \oplus eR \cap (1 - f)R = eR$.

We have shown that idempotents in R give us direct decompositions of R. We shall generalize this to modules. Let M_R be an R-module and $S = \mathrm{End}_R(M)$. Then M is a left S- and a right R-module. First we take a direct decomposition $M = M_1 \oplus M_2 \oplus \cdots \oplus M_n$. Let p_j (resp. i_j) be the projections (resp. the injection) of the direct decomposition. Then $\Sigma_j i_j p_j = 1_M$. We put $e_j = i_j p_j$. Then $e_j^2 = e_j$ and $e_i \perp e_j$ if $i \neq j$. Hence $S = e_1S \oplus e_2S \oplus \cdots \oplus e_nS$ by (3.1.2). It is clear that $e_iS = \mathrm{Hom}_R(M,M_i)$ $[\subset \mathrm{Hom}_R(M,M)]$, since im $e_i = e_i(M) = e_i(M_i)$ and $e_i(M_j) = 0$ for $i \neq j$. We note that $\ker e_i = M_1 \oplus \cdots \oplus M_{i-1} \oplus M_{i+1} \oplus \cdots \oplus M_n$. Conversely, let $1_M = e_1 + e_2 + \cdots + e_n$ in S and $e_ie_j = \delta_{ij}e_i$. Put $M_i = \mathrm{im}\, e_i$. Then $M = M_1 + M_2 + \cdots + M_n$. Since $e_i \perp (e_1 + \cdots + e_{i-1} + e_{i+1} + \cdots + e_n)$, $M_i \cap (M_1 + \cdots + M_{i-1} + M_{i+1} + \cdots + M_n) = 0$. Hence $M = \Sigma_i \oplus M_i$. If we take the projections p_i and the injections i_j for this decomposition, then $e_j = i_jp_j$. Thus we have the following.

3.1.6 THEOREM Let M_R be an R-module and $S = \mathrm{End}_R(M)$. Then there exists a one-to-one mapping between the direct decompositions of M and the set of mutually orthogonal idempotents $\{e_i\}$ with $1 = \Sigma_i e_i$ in S [cf. (3.1.8) below].

We shall investigate further properties of idempotents in a ring R.

3.1.7 THEOREM Let R be a ring and let e and f be idempotents in R. Then $\mathrm{Hom}_R(R_R,R_R) \approx R$, $\mathrm{Hom}_R(eR,fR) \approx fRe$ (these isomorphisms are given by the left-sided multiplications of elements in R). More generally, let M_R be an R-module. Then $\mathrm{Hom}_R(R,M) \approx M$.

Proof: Let m be in M. We put $f(r) = mr$ for $r \in R$. Then $f \in \mathrm{Hom}_R(R_R,M)$. Conversely, if $f \in \mathrm{Hom}_R(R,M)$, then $f(1) = m \in M$ and $f(r) = f(1r) = mr$. Hence we have the final isomorphism. Similarly, if $g \in \mathrm{Hom}_R(eR,fR)$, then $g(e) = g(ee) = g(e)e \in fRe$.

3.1.8 COROLLARY Let M_R be an R-module and $S = \mathrm{End}_R(M)$. If $M = \Sigma_i \oplus M_i$, then $\mathrm{End}_R(M_i) = e_iSe_i$, where the e_i are the projections.

Proof: This is clear.

3.1.9 LEMMA Let R be a ring and e, f idempotents. Then $eR \approx fR$ if and only if there exist an $a \in fRe$, $b \in eRf$, such that $ab = f$ and $ba = e$.

Proof: This is clear from (3.1.7).

3.1.10 COROLLARY Let M_R be an R-module and $S = \text{End}_R(M)$. For idempotents e and f in S, $eS \approx fS$ if and only if $\text{im } e \approx \text{im } f$ as R-modules.

Proof: Let $M = eM \oplus (1 - e)M = fM \oplus (1 - f)M$. If $eS \approx fS$, there exist a and b as in (3.1.9). Then $a : eM \to fM$ gives the desired isomorphism. We note that $eS = \text{Hom}_R(M, eM)$. Hence we obtain (3.1.10).

Next we consider a ring R which is a direct sum of rings R_i as rings, namely, $R = R_1 \oplus R_2 \oplus \cdots \oplus R_n$. The R_i are two-sided ideals of R by (1.7.12). Hence $1 = \Sigma e_i$; $e_i e_j = \delta_{ij} e_i$ and $R_i = e_i R = Re_i$ by (3.1.2). Since $R_i = e_i R e_i$ and $R_i R_j = R_j R_i = 0$ for $i \neq j$, e_i is the identity element in R_i and is in the center of R (we call them *central idempotents*. Conversely, if $\{e_i\}$ is a set of mutually orthogonal central idempotents in R with $1 = \Sigma_i e_i$, then $R = \Sigma_i \oplus e_i R$ and $R_i = e_i R = e_i R e_i$ is a two-sided ideal of R. Hence R is a direct sum of rings R_i as rings. Thus we have the following.

3.1.11 THEOREM Let R be a ring. Then there is a one-to-one mapping between the direct decompositions of R as rings and the sets of mutually orthogonal central idempotents $\{e_i\}$ with $1 = \Sigma e_i$.

Let $R = \Sigma_{i=1}^n \oplus R_i$ be as above and M_R an R-module. We set $M_i = Me_i$. Since e_i is central, M_i is an R-module and $M = \Sigma_i \oplus M_i$. Let $N_R \subset M_R$. Then $N_i = Ne_i \subset M_i$ and $N = \Sigma_i \oplus N_i$. Hence the structure of M is completely determined by the M_i. Conversely, let M_{iR_i} be an R_i-module. Put $M = \Sigma_{i=1}^n \oplus M_i$. Then M is a right R-module and $M_i = Me_i$. Hence we can state the following:

3.1.12 THEOREM Let R and R_i be rings. Assume that $R = \Sigma_{i=1}^n \oplus R_i$ as rings. Then the structure of R-module is uniquely determined by the structure of the R_i-modules for all i.

Let ${}_SM_R$ be a module, where S is another ring. If $M = M_1 \oplus M_2$ as R-modules (resp. S-R modules) implies that $M_1 = 0$ or $M_2 = 0$, then we say that M is a directly indecomposable R- (resp. S-R) module (Sec. 1.3).

3.1.13 THEOREM Let R and R_i be rings. Assume that $R = \Sigma_{i=1}^n \oplus R_i$ as rings and the R_i are indecomposable as R-R modules. Then such a decomposition is unique.

Proof: Let $\{S_i\}_{i=1}^m$ be a set of rings like $\{R_i\}_{i=1}^n$; $R = \Sigma_{i=1}^m \oplus S_i$. Then $S_i = \Sigma_j \oplus (R_j \cap S_i)$ by the fact noted before (3.1.12). Hence $S_i = R_j \cap S_i$ for some j. Furthermore, S_i is a direct summand of R_j and an R-R module by (1.7.4). Hence $S_i = R_j$. Using the same argument for the R_i, we have (3.1.13).

3.2 RELATIVELY PRIME IDEALS (COMAXIMAL IDEALS)

Let R be a ring. We say that two ideals A and B of R are *relatively prime* (*comaximal*) if A + B = R. We denote this by (A,B) = 1. In general, let $\{A_i\}_{i=1}^n$ be a set of right ideals in R and put $B_i = A_1 \cap A_2 \cap \cdots \cap A_{i-1} \cap A_{i+1} \cap \cdots \cap A_n$. If $(A_i,B_i) = 1$ for all i, we say that $\{A_i\}$ are *mutually relatively prime*.

3.2.1 PROPOSITION Let $\{A_i\}_{i=1}^n$ and B be right ideals. If $(A_i,B) = 1$ for all i, $(A_1A_2\cdots A_n,B) = 1$.

Proof: Since $(A_i,B) = 1$, there exist an a_i and a b_i in A_i and B, respectively, such that $1 = a_i + b_i$. Hence $a_1a_2\cdots a_n = (1 - b_1)(1 - b_2)\cdots(1 - b_n) = 1 - b$; $b \in B$.

3.2.2 COROLLARY Let $\{A_i\}_{i=1}^n$ and $\{B_j\}_{j=1}^m$ be sets of right ideals. If $(A_j,B_i) = 1$ for all i and j, $(A_1A_2\cdots A_n,\ B_1B_2\cdots B_m) = 1$.

Proof: $(A_1A_2\cdots A_n,\ B_j) = 1$ for all j. Hence $(A_1A_2\cdots A_n,\ B_1B_2\cdots B_m) = 1$ by (3.2.1).

3.2.3 COROLLARY. Let $\{C_i\}_{i=1}^n$ be a set of two-sided ideals. Then $\{C_i\}$ are mutually relatively prime if and only if $(C_i,C_j) = 1$ for $i \neq j$.

Proof: We assume that $\{C_i\}$ are mutually relatively prime. Put $D_i = C_1 \cap \cdots \cap C_{i-1} \cap C_{i+1} \cap \cdots \cap C_n$. Then $C_i \supset D_j$ $(i \neq j)$. Hence $R = (C_i,D_i) \subset (C_i,C_j) \subset R$ for $i \neq j$. Conversely, if $(C_i,C_j) = 1$, $(C_i,C_1C_2\cdot\overset{i}{\vee}\cdot C_n) = 1$ by (3.2.1), where $\overset{i}{\vee}$ means omit the ith member. Since $C_1C_2\cdot\overset{i}{\vee}\cdot C_n \subset D_i$, $(C_i,D_i) = 1$.

3.2.4 COROLLARY Let $\{A_i\}_{i=1}^n$ be a set of mutually relatively prime right ideals and $B_i = A_1 \cap A_2 \cap \cdots \cap A_{i-1} \cap A_{i+1} \cap \cdots \cap A_n$. Then $A_1 \cap A_2 \cap \cdots \cap A_m = B_{m+1} + \cdots + B_n$. In particular, $R = B_1 + B_2 + \cdots + B_n$.

Proof: When m = n - 1, the result is obvious. We shall show (3.2.4) by induction on n. Assume that $A_1 \cap \cdots \cap A_k = B_{k+1} + \cdots + B_n$. In (1.7.1),4 put M = R, $M_1 = A_k$, $M_2 = A_1 \cap \cdots \cap A_{k-1}$ and $M_3 = B_k$. Then

$B_k + (B_{k+1} + \cdots + B_n) = B_k + (A_1 \cap \cdots \cap A_{k-1}) \cap A_k = (A_1 \cap \cdots \cap A_{k-1}) \cap (B_k + A_k) = A_1 \cap \cdots \cap A_{k-1}$. If $k = n$, $R = B_1 + B_2 + \cdots + B_n$.

3.2.5 THEOREM (*Chinese Remainder Theorem*) Let $\{A_i\}_{i=1}^n$ be a set of mutually relatively prime right ideals and $A = \cap_i A_i$. Then $R/A \approx \Sigma_i \oplus R/A_i$ as right R-modules (and as rings, if the A_i are two-sided ideals).

Proof: We put $f([r + A]) = \Sigma_i([r + A_i])$ for $r \in R$. Then f is a monomorphic element in $\mathrm{Hom}_R(R/A, \Sigma_i \oplus R/A_i)$, for $A = \cap_i A_i$. Since $R = A_i + B_i$, $1 = a_i + b_i$; $a_i \in A_i$, $b_i \in B_i$. Hence $f([a_i r + A]) = [r + A_i]$, since $b_i \in A_j$ $(i \neq j)$. Therefore, f is an isomorphism.

3.2.6 COROLLARY. Let $\{A_i\}$ be as in (3.2.5) and let $\{a_i\}_{i=1}^n$ be a set of elements in A. Then the equations

$$x \equiv a_i \pmod{A_i}$$

have a solution in R.

Proof: Take $f^{-1}(\Sigma_i [a_i + A_i])$.

3.3 JACOBSON RADICAL

When we study the structure of modules, we need special, but important, submodules. As we show later, some important properties of a ring R are inherited by R-modules. In this section we study one of these features, called the Jacobson radical.

Let R be a ring. We shall define a new operation $*$ as follows:

$$x * y = x + y - xy \qquad x,y \in R \tag{3.3.1}$$

We call it the *quasi-product*. It is clear that the quasi-product satisfies the associative law and 0 is the identity element with respect to $*$. From the definition,

$$1 - x * y = (1 - x)(1 - y) \tag{3.3.2}$$

Hence x has a right inverse with respect to $*$ if and only if $(1 - x)$ has a right inverse in R, and x is idempotent with respect to $*$ if and only if x is idempotent. If x has a right inverse with respect to $*$, we call x *right quasi-regular*. We define similarly a left quasi-regular element. We call x *quasi-regular* when x is both left and right quasi-regular.

3.3.1 PROPOSITION Let R be a ring and x an element in R. If $x^2 = x$ and x is right quasi-regular, $x = 0$. If x is nil, x is quasi-regular.

Proof: Assume that $x^2 = x$. Thus $x * x = x$. If $x * y = 0$, then $x = x * 0 = x * (x * y) = x * y = 0$. Assume that $x^n = 0$. Then $(1 - x)(1 + x + x^2 + \cdots + x^{n-1}) = (1 + x + x^2 + \cdots + x^{n-1})(1 - x) = 1$. Hence x is quasi-regular.

If every element in a one-sided ideal A of R is quasi-regular, we call A a *quasi-regular one-sided ideal.*

3.3.2 PROPOSITION Let A be a right ideal in R. If every element in A is right quasi-regular, A is quasi-regular.

Proof: Let a be in A and $a * b = 0$ for $b \in R$. Then $b = a - ab \in A$. Hence b is right quasi-regular. Therefore, b and a are quasi-regular.

3.3.3 LEMMA Let A be a quasi-regular right ideal and b a right quasi-regular element. Then a + b is right quasi-regular for any a in A.

Proof: $b * b' = 0$ for some b' and $a - ab' \in A$. Hence $(a - ab') * c = 0$ for some c. Then $(a + b) * b' * c = (a - ab') * c = 0$.

3.3.4 LEMMA If $ax * y = 0$ for a, x, and y in R, then $xa * (xya - xa) = 0$.

Proof: The result follows by direct computation.

3.3.5 THEOREM Every ring R contains a unique maximal quasi-regular right ideal N (which contains every quasi-regular right ideal). N is also a unique maximal quasi-regular left ideal and so N is a two-sided ideal.

Proof: Let N be the union of all quasi-regular right ideals. Then N is a quasi-regular right ideal by (3.3.2)-(3.3.4). We shall show that N is a left ideal. Let n be in N and x in R. We show that xnR is quasi-regular. Since $nR \subset N$, nrx is quasi-regular for any $r \in R$. Hence xnr is right quasi-regular by (3.3.4), so $xn \in xnR \subset N$. Accordingly, N is a left ideal. Similarly, we obtain a unique maximal quasi-regular left ideal N' which is two-sided. Hence N = N'.

We call N the *Jacobson radical* of R. We denote N by J(R) or simply J. If J(R) = 0, R is called *semisimple.*

3.3.6 PROPOSITION R/J(R) is semisimple.

Proof: Let A/J(R) be a quasi-regular right ideal in R/J(R). Then for $a \in A$ there exists a b such that $a * b \in J(R)$. Hence A is quasi-regular, so $A \subset J(R)$.

3.3.7 PROPOSITION Let $M_R \supset N_R$ be R-modules. If M is f.g. (finitely generated), M contains a maximal submodule containing N.

Proof: Let $\{m_1, m_2, \ldots, m_n\}$ be a set of generators of M. We put $\underline{S} = \{M'_R \mid M'$ is a submodule $\subsetneq M$ and $M' \supset N\}$. Then $\underline{S}$ is partially ordered by inclusion. Let $M_1 \subset M_2 \subset \cdots \subset M_n \subset \cdots$ be any linear subset in $\underline{S}$. Put $M_0 = \cup M_i$. If $M_0 = M$, there exists a t such that $M_t \ni m_i$ for all i, which contradicts the construction of $\underline{S}$. Therefore, $\underline{S}$ contains a maximal element M_0 by Zorn's lemma, which is the desired submodule.

3.3.8 THEOREM Let R be a ring. Then $J(R) = \cap M_\alpha$, where M_α runs through all maximal right (or left) ideals in R.

Proof: We know by (3.3.7) that R contains a maximal right ideal M_α. If $J(R) \not\subset M_\alpha$, $R = J(R) + M_\alpha$. Hence $1 = x + m$; $x \in J(R)$, $m \in M_\alpha$. Since $m = 1 - x$ is regular, this is a contradiction. Therefore, $J(R) \subset M_\alpha$. Let $x \in \cap M_\alpha$. Then the ideals $(1 - xr)R$ are not contained in any M_α for $r \in R$. Hence $(1 - xr)R = R$ by (3.3.7). Therefore, xR is quasi-regular, so $x \in xR \subseteq J(R)$. Hence $\cap M_\alpha \subset J(R)$ by (3.3.5).

We have defined the Jacobson radical of R. We shall generalize it to any module M_R, following (3.3.8).

If M contains no maximal submodules, then we put $J(M) = M$. Otherwise, $J(M) = \cap M_\alpha$, where M_α runs through all maximal R-submodules of M. M/M_α being a simple R-module, we have $J(M) = \cap_f \ker f$, where $f \in \mathrm{Hom}_R(M,S)$ and S runs through all simple R-modules. We call J(M) the *Jacobson radical* of M.

3.3.9 PROPOSITION Let M_R and N_R be R-modules and $f : M \to N$. Then $f(J(M)) \subset J(N)$.

Proof: Let g be an element of $\mathrm{Hom}_R(N,S)$. Then $gf : M \to S$. Hence $g(f(J(M))) = gf(J(M)) = 0$.

3.3.10 COROLLARY Let M_R and $\{M_{iR}\}_I$ be R-modules and $M = \Sigma_I \oplus M_i$. Then $J(M) = \Sigma_I \oplus J(M_i)$.

Proof: Consider the injection $i_j : M_j \to M$. Then $J(M_i) \subseteq J(M)$ by (3.3.9). Next, take the projection $p_j : M \to M_j$. Then $p_j(J(M)) \subset J(M_j)$. Hence $J(M) = \Sigma_I \oplus J(M_i)$.

3.3.11 COROLLARY $J(M) \supset MJ(R)$ for M_R.

Proof: We have an epimorphism $f : \Sigma_{m \in M} \oplus u_m R \to M$ by (1.7.2). Hence $J(M) \supset f(J(\Sigma_m \oplus u_m R)) = f(\Sigma_m \oplus u_m J(R)) = M(J(R))$.

3.3.12 THEOREM (*Nakayama's Lemma*) Let $M_R \supset N_R$ be R-modules. Assume that M/N is f.g. and $M = N + J(M)$. Then $N = M$.

Proof: Since M/N is f.g., M contains a maximal submodule $M' \supset N$ by (1.2.3) and (3.3.7), provided that $M \neq N$. Then $M' \supset J(M) + N = M$, which is a contradiction.

3.3.13 COROLLARY Let M_R be a f.g. R-module. If $MJ(R) = M$, then $M = 0$.

Proof: This is clear by (3.3.11) and (3.3.12).

We shall frequently use the following.

3.3.14 PROPOSITION Let R be a ring and x an element. Then x is a unit if and only if x is a unit in $R/J(R)$.

Proof: We assume that $\bar{x}$ is a unit in $\bar{R} = R/J$. Then there exists y in R such that $\bar{x}\bar{y} = \bar{y}\bar{x} = 1$. Since $1 - xy \in J$, $xy = 1 - (1 - xy)$ is a unit in R. Hence x is a left inverse. Similarly, x is a right inverse. The converse is clear.

3.4 SMALL SUBMODULES AND ESSENTIAL SUBMODULES

We shall generalize Nakayama's lemma to arbitrary R-modules. Let $M_R \supset T_R$ be R-modules. If $M = T + N$ for some R-submodule N of M implies that $N = M$, then we say that T is a *small submodule* in M or T is *superfluous* in M.

3.4.1 PROPOSITION 1. Let $M_R \supset N_R \supset T_R$ be R-modules. If T is small in N, then T is small in M.

2. Let $M_R \supset T_R$ and K_R be R-modules and $f : M \to K$. If T is small in M, $f(T)$ is small in K.
3. Let $M_R \supset T_{iR}$ $(i = 1, 2, \ldots, n)$ be R-modules. If T_i is small in M for all i, ΣT_i is small in M.

Proof: 1. If $M = M' + T$ for some R-submodule M' of M, $N = T + N \cap M'$ by (1.4.1). Hence $T \subset N = N \cap M' \subset M'$. Therefore, $M = M'$.

2. Let $f(T) + K' = K$ for some R-submodule K' of K. Then $M = f^{-1}(K') + T$. Hence $M = f^{-1}(K')$, so $f(T) \subset f(M) \subset K'$.

3. Let $M = \Sigma_{i=1}^{n} T_i + M'$. Then $M = \Sigma_{i=2}^{n} T_i + M' = \Sigma_{i=3}^{n} T_i + M' = \cdots = M'$.

3.4.2 COROLLARY Let $\{M_{iR}\}_I$ and $\{N_{iR}(\subset M_i)\}_I$ be sets of R-modules and $M = \Sigma_I \oplus M_i$, $N = \Sigma_I \oplus N_i$. If N is small in M, then N_i is small in M_i for all i. Conversely, if I is a finite set and N_i is small in M_i for all i, then N is small in M.

Proof: Take the projection $p_i : M \to M_i$ and use (3.4.1),2. Conversely, $N_i \subset N \subset M$, so $\Sigma_I N_i$ is small in M by (3.4.1).

3.4.3 THEOREM Let M_R be any R-module. Then J(M) is the union of all small R-submodules in M. If M is f.g., J(M) is the unique maximal small R-submodule in M.

Proof: First we assume that $M \neq J(M)$. Let T be a small R-submodule and N a maximal submodule. If $N \not\supset T$, $M = N + T$. Hence $N \supset T$. Therefore, $J(M) \supset T$. Let x be in J(M) and $xR + K = M$ for some $K_R \subsetneq M$. Put $\underline{S} = \{K'_R \mid M \supsetneq K' \text{ and } K' \not\ni x\}$. Then we can show by Zorn's lemma that $\underline{S}$ contains a maximal element N. It is clear that N is a maximal submodule of M. Hence $x \in N$, which is a contradiction. Therefore, xR is small in M. The argument above shows that xR is small in M for any $x \in M$ in case $M = J(M)$. If M is f.g., J(M) is small in M by (3.3.12).

We note that the union of infinitely many small submodules in an R-module M is not necessarily small in M. We shall study this problem later.

If every proper submodule of M is small in M, M is called a *hollow module*. If M is f.g. hollow, J(M) is the unique maximal submodule in M by (3.4.3). Hence M is a cyclic R-module and directly indecomposable.

Let M_R and N_R be R-modules and $f : M \to N$ epic. If fg being epic for any $g \in \mathrm{Hom}_R(T_R, M)$ implies that g is epic, then we say that f is a *minimal epimorphism*.

3.4.4 PROPOSITION Let $f : M \to N$ be epic. Then f is minimal if and only if ker f is small in M.

Proof: We assume that f is minimal and $M = L + \ker f$ for some $L_R \subset M$. Then fi is epic for the inclusion $i : L \to M$. Hence i is epic, so $L = M$. Conversely, we assume that ker f is small in M and fg is epic. Then $M = \ker f + \mathrm{im}\, g$ by (1.2.5). Hence $M = \mathrm{im}\, g$.

Next we study the concept dual to that of a small submodule. Let M be an R-module and T an R-submodule. If $T \cap N \neq 0$ for any nonzero submodule N of M, we say that T is *essential* in M or *large* in M.

The following is dual to (3.4.1).

3.4.5 PROPOSITION 1. Let $M_R \supset N_R \supset T_R$ be R-modules. If T is essential in M, then it is essential in N. If N is essential in M and T is essential in N, then T is essential in M.

2. Let M_R and $M'_R \supset K'_R$ be R-modules and $f : M \to M'$ a homomorphism. If K' is essential in M', $f^{-1}(K')$ is essential in M.
3. Let $\{T_i\}_{i=1}^n$ be a set of R-submodules of M_R. If T_i is essential in M for all i, $\bigcap_i T_i$ is essential in M.

Proof: The proof is dual to that in (3.4.1), so we leave it to the reader.

3.4.6 THEOREM Let $\{M_i\}_I$ be a set of R-modules and $M = \Sigma_I \oplus M_i \supset N$. Then N is essential in M if and only if $N_i = M_i \cap N$ is essential in M for each i.

Proof: Let $i_j : M_j \to M$ be the injection. If N is essential in M, $N_j = i_j^{-1}(N)$ is essential in M_i for each i. Let $m = \Sigma\, m_i \neq 0$ be in M. We may assume that $m = m_1 + m_2 + \cdots + m_n$, $m_i \neq 0$ in M_i. Then there exists r_1 in R such that $0 \neq m_1 r_1 \in N_1$; $mr_1 = m_1 r_1 + \cdots + m_n r_1$. We assume that $m_2 r_1 \neq 0$. There exists an r_2 in R such that $0 \neq m_2 r_2 \in N_2$. Noting that $m_1 r_1 r_2 \in N_1$ and repeating the argument, we finally obtain an element r in R such that $0 \neq mr \in N$. Hence N is essential in M.

3.4.7 PROPOSITION Let N_R and T_R be submodules of M_R. If $N \cap T = 0$, there exists a submodule $N_1 \supset N$ such that $N_1 \cap T = 0$ and $N_1 \oplus T$ is essential in M.

Proof: Let $\underline{S}$ be the set of R-submodules N' such that $N' \supset N$ and $N' \cap T = 0$. We can find a maximal element N_1 in $\underline{S}$ by Zorn's lemma. If $N_1 \oplus T$ is not essential, we can find a nonzero submodule K such that $(N_1 \oplus T) \cap K = 0$. Hence $(N_1 + K) \cap T = 0$ and $N_1 + K \supsetneq N_1$, which is a contradiction.

We call the above N_1 a *complement* of N and denote it by N^c. The union of all simple submodules of M is called the *socle* of M, and denoted by S(M). If there are no simple submodules, we set S(M) = 0.

3.4.8 LEMMA Let x be a nonzero element in M and L maximal among R-submodules in M not containing x. Then (xR + L)/L is simple.

Proof: We have a natural epimorphism φ of R onto (xR + L)/L given by setting $\varphi(r) = xr + L$, $r \in R$. Let K' be a right ideal such that $K = \varphi^{-1}(0) \subsetneq K' \subset R$. Since $xK' + L \supsetneq L$, $x \in xK' + L$ by the assumption. Let $x = xk' + y$, $k' \in K'$, $y \in L$. Then $1 - k' \in K$, so K' contains 1. Therefore, (xR + L)/L is simple.

3.4.9 THEOREM $S(M) = \cap N$, where N runs through all essential submodules in M.

Proof: Let S be a simple submodule and N an essential submodule in M. Then $S \cap N \neq 0$ implies that $S \subset N$. Hence $S(M) \subset \cap N$. Let x be not contained in S(M). Then we can find, by Zorn's lemma, a submodule L maximal among submodules L' such that $x \notin L'$ and $L' \supset S(M)$. Assume that $L \cap T = 0$ for a submodule T. If $T \neq 0$, $T + L \supsetneq L$, so L + T contains xR.

But $T \approx (T + L)/L \supset (xR + L)/L$ is simple by (3.4.8). Hence T contains a simple submodule, which contradicts the fact that $T \cap S(M) = 0$. Therefore, $T = 0$ and L is essential in M. The proof works even if $S(M) = 0$.

If every nonzero submodule of M is essential in M, we say that M is *uniform*. Every uniform module is directly indecomposable. If R is a commutative domain, R is itself a uniform R-module.

Let M_R be an R-module and $f : M \to N$ monic. If, whenever gf is monic for $g \in \mathrm{Hom}_R(N, T_R)$, then g is monic, we say that f is a *maximal monomorphism*.

3.4.10 PROPOSITION Let $f : M \to N$ be monic. Then f is maximal if and only if im f is essential in N.

Proof: This is dual to that of (3.4.4).

3.5 LIMITS

We studied limits and colimits in complete and cocomplete categories $\underline{A}$, respectively (Sec. 2.4). We shall recall some special limits in $\underline{M}_R$ and give complete proofs of some elementary properties. We skipped these proofs in Sec. 2.4. However, almost all proofs in this section work in general category $\underline{A}$ without much change.

Let I be a partially ordered set with relation $\geq$ (Sec. 1.5). If, for any two elements α, β in I, there exists γ in I such that $\gamma \geq \alpha$, $\gamma \geq \beta$, we say that I is a directed set. Now we assume that there are a set of R-modules $\{M_\alpha\}_I$ indexed by I and a set of R-homomorphisms $\pi_\alpha^\beta : M_\alpha \to M_\beta$ for $\alpha \leq \beta$. If $\{M_\alpha, \pi_\alpha^\beta\}_I$ satisfies

$$\pi_\beta^\gamma \pi_\alpha^\beta = \pi_\alpha^\gamma \qquad \text{for } \alpha \leq \beta \leq \gamma \tag{3.5.1}$$

$$\pi_\alpha^\alpha = 1_{M_\alpha} \qquad \text{for all } \alpha \tag{3.5.2}$$

we call $\{M_\alpha, \pi_\alpha^\beta\}_I$ a *direct system*. Dually, if we assume that $\pi_\alpha^\beta : M_\beta \to M_\alpha$ for $\alpha \leq \beta$ and

$$\pi_\alpha^\beta \pi_\beta^\gamma = \pi_\alpha^\gamma \qquad \text{for } \alpha \leq \beta \leq \gamma \tag{3.5.3}$$

and

$$\pi_\alpha^\alpha = 1_{M_\alpha} \qquad \text{for all } \alpha \tag{3.5.4}$$

are satisfied, $\{M_\alpha, \pi_\alpha^\beta\}_I$ is called a *projective* (*inverse*) *system*.

Let I and I' be directed sets. Let $\{M_\alpha, \pi_\alpha^\beta\}_I$, $\{M'_{\alpha'}, \pi'^{\beta'}_{\alpha'}\}_{I'}$ be direct systems over I and I', respectively. A morphism from $\{M_\alpha, \pi_\alpha^\beta\}_I$ to $\{M'_{\alpha'}, \pi'^{\beta'}_{\alpha'}\}_{I'}$ is given by the following. There exists $\varphi : I \to I'$ satisfying

$$\text{If } \alpha \leq \beta \text{ in I, } \varphi(\alpha) \leq \varphi(\beta) \text{ in I'.} \tag{3.5.5}$$

and there exists a set of R-homomorphisms $\phi^\alpha : M_\alpha \to M'_{\varphi(\alpha)}$ which give commutative diagrams

$$\begin{array}{ccc} M_\alpha & \xrightarrow{\pi_\alpha^\beta} & M_\beta \\ {\scriptstyle\phi^\alpha}\downarrow & & \downarrow{\scriptstyle\phi^\beta} \\ M'_{\varphi(\alpha)} & \xrightarrow[\pi_{\varphi(\alpha)}^{\varphi(\beta)}]{} & M'_{\varphi(\beta)} \end{array} \tag{3.5.6}$$

Dually, we can define a morphism of projective systems. Let X be an R-module and $\{f^\alpha : M_\alpha \to X\}_I$ or $\{f_\alpha : X \to M_\alpha\}_I$ a set of R-homomorphisms. If, for $\alpha \leq \beta$,

$$\begin{array}{ccc} M_\alpha & & \\ {\scriptstyle\pi_\alpha^\beta}\downarrow & \overset{f^\alpha}{\searrow} & X \\ M_\beta & \underset{f^\beta}{\nearrow} & \end{array} \qquad \begin{array}{ccc} & & M_\alpha \\ X & \overset{f_\alpha}{\nearrow} & \uparrow{\scriptstyle\pi_\alpha^\beta} \\ & \underset{f_\beta}{\searrow} & M_\beta \end{array} \tag{3.5.7}$$

is commutative, we say that $\{f^\alpha\}$ (resp. $\{f_\alpha\}$) is *compatible*.

3.5.1 THEOREM Every projective system in $\underline{M}_R$ has a limit.

Proof: Let $\{M_\alpha, \pi_\alpha^\beta\}_I$ be a projective system. Put $P = \pi_I M_\alpha$. We take the subset $M_\infty = \{(m_\alpha)_I \in P \mid \pi_\alpha^\beta(m_\beta) = m_\alpha \text{ for } \alpha \leq \beta\}$ of P. Then M_∞ is an R-module. Put $p_\alpha = p'_\alpha \mid M_\infty : M_\infty \to M_\alpha$, where p'_α is the projection of P. Then $\{p_\alpha\}$ is compatible by the construction of M_∞. Let $\{f_\alpha : X \to M_\alpha\}_I$ be a compatible set for X_R. Define $g : X \to P$ by setting $g(x) = (f_\alpha(x))_I$ [cf. (1.3.3)]. Then im $g \subset M_\infty$, $p_\alpha g = f_\alpha$ for all $\alpha \in I$, and g is unique with this property. Hence M_∞ is the limit.

We call M_∞ the *projective* (*inverse*) *limit* and denote it by $\varprojlim M_\alpha$.

Let $\{M_\alpha, \pi_\alpha^\beta\}_I$ and $\{M'_\alpha, \pi'^{\beta'}_{\alpha'}\}_I$ be two projective systems over directed sets I and I', respectively. If $\{\phi_\alpha : M_{\varphi(\alpha')} \to M'_{\alpha'}\}$ is a set of morphisms (note that $\varphi : I' \to I$), we can define a unique mapping $\phi_\infty : (m_\alpha)_I \in M_\infty \to (\phi_{\varphi(\alpha')}(m_{\varphi(\alpha')}))_{I'} \in M'_\infty$, which gives an R-homomorphism.

Next we study colimit. Let $\{M_\alpha, \pi_\alpha^\beta\}_I$ be a direct system.

3.5.2 THEOREM Every direct system in $\underline{M}_R$ has a limit.

Proof: Set $Q = \Sigma_I \oplus M_\alpha$. Let N be the subgroup of Q generated by $\{m_\alpha - \pi_\alpha^\beta m_\alpha \mid m_\alpha \in M_\alpha\}$ and set $M^\infty = Q/N$. Let $i^\alpha = \rho i'_\alpha$, where $i'_\alpha : M_\alpha \to Q$ is the injection and $\rho : Q \to Q/N$ the natural epimorphism. Then $\{i^\alpha\}_I$ is compatible. Let $\{f^\alpha : M_\alpha \to Y\}_I$ be compatible. Then we obtain a unique homomorphism $f' : Q \to Y$ such that $f'i'_\alpha = f^\alpha$ by (1.3.4). Since $\{f^\alpha\}_I$ is compatible, f' induces $f^\infty : M^\infty = Q/N \to Y$ and $fi^\alpha = f\rho i'_\alpha = f^\alpha$. Therefore, M^∞ is the colimit of the system and an R-module.

We shall call M^∞ the *direct limit* and denote it by $\varinjlim M_\alpha$ in $\underline{M}_R$. Let $\mathbb{A} = \{M_\alpha, \pi_\alpha^\beta\}_I$ and $\mathbb{B} = \{M'_{\alpha'}, \pi'^{\beta'}_{\alpha'}\}_{I'}$ be direct systems. Then a morphism $\{\phi^\alpha\} : \mathbb{A} \to \mathbb{B}$ gives an R-homomorphism $\phi^\infty : \varinjlim M_\alpha \to \varinjlim M'_{\alpha'}$.

3.5.3 LEMMA Let $M^\infty = \varinjlim M_\alpha$. For any element m in M^∞ there exists an m_α such that $i^\alpha(m_\alpha) = m$.

Proof: Let $m = \Sigma_{i=1}^n m_{\alpha_i}$ in $Q/N = M^\omega$. Since I is a directed set, there exists a γ in I with $\gamma \geq \alpha_i$ for all α_i. Then $m = \overline{\Sigma m_{\alpha_i}} = \Sigma_i \pi_{\alpha_i}^\gamma m_{\alpha_i}$, where $\pi_{\alpha_i}^\gamma m_{\alpha_i}$ are in M_γ.

3.5.4 LEMMA If $i^\alpha(m_\alpha) = 0$ for $m_\alpha \in M_\alpha$, there exists a γ in I with $\pi_\alpha^\gamma(m_\alpha) = 0$.

Proof: $i^\alpha(m_\alpha) = 0$ means that m_α is in N; i.e., $m_\alpha = \Sigma_{i=1}^n(m_{\beta_i} - \pi_{\beta_i}^{\gamma_i} m_{\beta_i})$. Let $\delta \geq \max\{\alpha, \gamma_i\}$ in I. Since $m_\beta - \pi_\beta^\gamma m_\beta = m_\beta - \pi_\beta^\delta m_\beta + \pi_\beta^\gamma(-m_\beta) - \pi_\gamma^\delta(\pi_\beta^\gamma(-m_\beta))$, $m_\alpha = \Sigma_{\alpha_i}(m'_{\alpha_i} - \pi_{\alpha_i}^\delta m'_{\alpha_i}) \in Q$; $m'_{\alpha_i} \in M_{\alpha_i}$. Hence $\alpha_i = \alpha$, $m_\alpha = m'_{\alpha_i}$ for some i and $m'_{\alpha_j} = 0$ for $j \neq i$. Therefore, $\pi_\alpha^\delta m_\alpha = 0$.

3.5.5 LEMMA $i^\alpha(m_\alpha) = i^\beta(m_\beta)$ if and only if there exists γ ($\geq \alpha, \beta$) such that $\pi_\alpha^\gamma(m_\alpha) = \pi_\beta^\gamma(m_\beta)$.

Proof: $i^\alpha(m_\alpha) = \pi_\alpha^\gamma(m_\alpha) = \pi_\beta^\gamma(m_\beta) = i^\beta(m_\beta)$. Conversely, if $i^\alpha(m_\alpha) = i^\beta(m_\beta)$, there exists a γ' ($\geq \alpha, \beta$) with $i^{\gamma'}(\pi_\alpha^{\gamma'} m_\alpha - \pi_\alpha^{\gamma'} m_\beta) = 0$. Hence there exists a $\gamma \geq \gamma'$ such that $\pi_{\gamma'}^\gamma(\pi_\beta^{\gamma'} m_\beta - \pi_\alpha^{\gamma'} m_\alpha) = 0$ by (3.5.4).

3.5.6 THEOREM Let I be a directed set and $\{M_\alpha, \pi_\alpha^\beta\}_I$, $\{M'_\alpha, \pi'^\beta_\alpha\}$, $\{M''_\alpha, \pi''^\beta_\alpha\}_I$ direct systems over I. Let $\{\varphi^\alpha\} : \{M'_\alpha, \pi'\}_I \to \{M_\alpha, \pi\}_I$, $\{\psi^\alpha\} : \{M_\alpha, \pi\}_I \to \{M''_\alpha, \pi''\}_I$ be morphisms. If $M' \xrightarrow{\varphi^\alpha} M_\alpha \xrightarrow{\psi^\alpha} M''_\alpha$ is exact for each $\alpha \in I$, then $M'^\infty \to M^\infty \to M''^\infty$ is exact.

Proof: $\psi^\infty\phi^\infty = 0$ is clear. Let $\psi^\infty(m) = 0$. Then there exists some m_α in M_α with $i^\alpha(m_\alpha) = m$ by (3.5.3) and $i''^\alpha(\psi^\alpha(m_\alpha)) = 0$. Hence we obtain a δ ($\geq \alpha$) such that $\pi_\alpha^\delta(\psi^\alpha(m_\alpha)) = \psi^\delta(\pi_\alpha^\delta(m_\alpha)) = 0$ by (3.5.4). From the assumption, we have an m'_δ in M'_δ with $\varphi^\delta(m'_\delta) = \pi_\alpha^\delta(m_\alpha)$. Hence $\phi^\infty(i^\delta(m'_\delta)) = i^\delta\phi^\delta(m'_\delta) = i^\delta(\pi_\alpha^\delta(m_\alpha)) = i^\alpha(m_\alpha) = m$.

3.5.7 THEOREM Let M_R be an R-module. Then M is the direct limit of the direct system of all f.g. submodules in M [cf. (2.4.5)].

Proof: Let $\underline{S}$ be the set of all f.g. submodules of M. $\underline{S}$ is a directed set via the inclusion relation. If $L_1 \subset L_2$ are in $\underline{S}$, we put $\pi_{L_2}^{L_1} = i : L_1 \to L_2$ the inclusion. Then $\{L_\alpha, \pi\}_{\underline{S}}$ is a direct system over $\underline{S}$. Put $Q = \Sigma_{\underline{S}} \oplus L_\alpha$ and $f : Q \to M$, where $f|L_\alpha$ is the identity mapping. Then f is epic and induces a homomorphism from $\varinjlim L_\alpha = Q/N$ onto M. Let m be in M. We take any L, containing m, in $\underline{S}$ and define a mapping $g : M \to Q/N$ by setting $g(m) = n + N$. It is clear that g is well defined, an R-homomorphism, and that $fg = 1_M$. Hence f is an isomorphism.

We give an example illustrating the dual of (3.5.6) for projective limits. Let I be the directed set of positive integers. Put $M_n = Z$ (the integers), $\pi_n^m a = 3^{m-n}a$ (a in M_n) for $m \geq n$, $M'_n = Z/2Z$ and $\pi'^m_n = 1_{Z/2Z}$. It is clear that $\varprojlim M'_n = Z/2Z$. On the other hand, let $x = (x_i)$ be in $\varprojlim M_n$. Then 3^{m-n} divides x_n for $m > n$. Hence $x = 0$, so ψ_∞ is not epic, where $\psi_n : Z \to Z/2Z$ is the natural epimorphism.

3.6 THE FUNCTORS Hom AND ⊗

Let R and S be rings and let M_R and N_R be R-modules. We defined an abelian group $\mathrm{Hom}_R(M,N)$ in Sec. 1.7. Let $f : N \to N'_R$ and $g : M'_R \to M$. Then $f\cdot\mathrm{Hom}_R(M,N) = \{fh \mid h \in \mathrm{Hom}_R(M,N)\}$ is a subgroup of $\mathrm{Hom}_R(M,N')$. Hence $\mathrm{Hom}_R(M,-) : \underline{M}_R \to \underline{Ab}$ is an additive covariant functor by (1.7.15) with $\mathrm{Hom}(M,f) : \mathrm{Hom}_R(M,N) \to \mathrm{Hom}_R(M,N')$ (see Sec. 2.3). Similarly, we can define an additive contravariant functor $\mathrm{Hom}_R(-,N) : \underline{M}_R \to \underline{Ab}$ with $\mathrm{Hom}(g,N) : \mathrm{Hom}_R(M,N) \to \mathrm{Hom}_R(M',N)$. Furthermore, if N is treated as a variable in $\mathrm{Hom}(g,N)$, g gives a natural transformation from $\mathrm{Hom}_R(M,-)$ to $\mathrm{Hom}_R(M',-)$. Similarly, $\mathrm{Hom}_R(M,f)$ gives a natural transformation from $\mathrm{Hom}_R(-,N)$ to $\mathrm{Hom}_R(-,N')$ (Sec. 2.3). We assume that M is a left S- and a right R-bimodule. Then each element s in S gives an element in $\mathrm{End}_R(M)$ by left multiplication. Hence $\mathrm{Hom}_R(s,N)$ gives an endomorphism of $\mathrm{Hom}_R(M,N)$. Using this fact, we can regard $\mathrm{Hom}_R(M,N)$ as a right S-module, since $\mathrm{Hom}_R(-,N)$

is contravariant. Similarly, we can regard $\mathrm{Hom}_S(M, {}_ST)$ as a left R-module. We may summarize the above as follows:

Let ${}_SM_R$ and $N_{R\text{-}S}$ be modules. For an f in $\mathrm{Hom}_R(M, T_R)$ and an s in S, we define

$$(fs)(m) = f(sm) \qquad \text{for } m \in M \tag{3.6.1}$$

Then $fs \in \mathrm{Hom}_R(M,T)$ and $\mathrm{Hom}_R(M,T)$ is a right S-module via (3.6.1). For an $f' \in \mathrm{Hom}_R(T,N)$ we define

$$(f's)(t) = f'(t)s \qquad \text{for } t \in T \tag{3.6.2}$$

Then $\mathrm{Hom}_R(T,N)$ is a right S-module. If $h : T \to T'$ is an R-homomorphism, we can easily check that $\mathrm{Hom}_R(M,h)$ and $\mathrm{Hom}_R(h,N)$ are S-homomorphisms with respect to the above operations of S. We can similarly define new modules for other situations. We showed $\mathrm{Hom}_R(R,M) \approx M$ in (3.1.7). Since R is a two-sided R-module, this isomorphism is an R-isomorphism in the above sense.

Next we consider $\mathrm{Hom}_R(-,-)$ as a functor.

3.6.1 THEOREM Let M_R be an R-module. Then $\mathrm{Hom}_R(M,-)$ and $\mathrm{Hom}_R(-,M)$ are left exact. If ${}_SM_R$, $\mathrm{Hom}_R(M,-)$ is a covariant functor from $\underline{M}_R$ to $\underline{M}_S$ and $\mathrm{Hom}_R(-,M)$ is a contravariant functor from $\underline{M}_R$ to ${}_S\underline{M}$

Proof: Let $0 \to N' \xrightarrow{f} N \xrightarrow{g} N''$ be an exact sequence of R-modules. Then $\mathrm{Hom}(M,g)\mathrm{Hom}(M,f) = \mathrm{Hom}(M,gf) = 0$. If $\mathrm{Hom}(M,f)h = 0$ for $h \in \mathrm{Hom}_R(M,N')$, $fh = 0$. Since f is monic, $h = 0$. Hence $\mathrm{Hom}(M,f)$ is monic. Assume that $\mathrm{hom}(M,g)k = 0$ for $k \in \mathrm{Hom}_R(M,N)$. Then $gk = 0$. Hence $\mathrm{im}\, k \subset \ker g = \mathrm{im}\, f \approx N'$. Set $k' = (f \mid \mathrm{im}\, f)^{-1}k \in \mathrm{Hom}_R(M,N')$. Then $\mathrm{Hom}(M,f)k' = k$. Therefore, $\mathrm{Hom}(M,-)$ is left exact. We can show similarly that $\mathrm{Hom}(-,M)$ is left exact.

3.6.2 PROPOSITION Let $\{M_\alpha\}_I$ and N be R-modules. Then $\mathrm{Hom}_R(\Sigma_I \oplus M_\alpha, N) \approx \Pi_I \mathrm{Hom}_R(M_\alpha, N)$.

Proof: This is clear from (1.3.2) or Sec. 2.1.

Next we shall describe another functor. Let M_R, ${}_RN$ be modules. A mapping f of $M \times N$ to G (an abelian group) satisfying

$$\begin{aligned} f(m_1 + m_2, n) &= f(m_1, n) + f(m_2, n) \\ f(m, n_1 + n_2) &= f(m, n_1) + f(m, n_2) \\ f(mr, n) &= f(m, rn) \end{aligned} \tag{3.6.3}$$

where $m_i \in M$, $n_i \in N$, and $r \in R$, is called *balanced*.

3.6.3 THEOREM Let M_R and $_RN$ be R-modules. Then there exists a module G satisfying the following conditions:

1. There is a balanced mapping $f : M \times N \to G$.
2. G is generated by $\{f(m,n) \mid m \in M, n \in N\}$.
3. For any module H and balanced mapping $h : M \times N \to H$, there exists a unique $k : G \to H$ such that $h = kf$.

Two such groups are isomorphic to one another.

Proof: Let $F = \Sigma_{M\times N} \otimes (m,n)Z$ be a free module with basis $M \times N$. Let K be the subgroup generated by $\{(m_1 + m_2, n) - (m_1, n) - (m_2, n), (m, n_1 + n_2) - (m, n_2), (mr, n) - (m, rn) \mid m_i \in M, n_i \in N, \text{ and } r \in R\}$. Put $G = F/K$. The mapping $f(m,n) = (m,n) + K$ is clearly a balanced mapping and G is generated by $f(m,n)$; $m \in M$, $n \in N$. For h in (3), we define $k' : F \to H$ by setting $k'(m,n) = h(m,n)$. Since h is balanced, k' induces a homomorphism $k : G \to H$. It is clear that $kf = h$. Hence G is the desired group. We can show the uniqueness up to isomorphism in a manner similar to (1.3.2).

We denote G by $M \otimes_R N$ and call it the *tensor product* of M and N (over R). We shall denote $(m,n) + K$ in G by $m \otimes n$. Hence $M \otimes_R N$ is generated by $\{m \otimes n \mid m \in M, n \in N\}$ and $(m_1 + m_2) \otimes n = m_1 \otimes n + m_2 \otimes n$, $m \otimes (n_1 + n_2) = m \otimes n_1 + m \otimes n_2$, and $mr \otimes n = m \otimes rn$. We shall show that $M \otimes_R -$ and $- \otimes_R N$ are covariant functors. Let $f : M \to M'_R$, $g : N \to N'_R$ be R-homomorphisms. Then $(m,n) \to (f(m),g(n)) \to \varphi(f(m),g(n))$ is balanced for a balanced mapping $\varphi : M' \times N' \to M' \otimes_R N'$. Hence we have a homomorphism $f \otimes g : M \otimes_R N \to M' \otimes_R N'$ such that $(f \otimes g)(m \otimes n) = f(m) \otimes g(n)$ by (3.6.3). It is clear that defining $(- \otimes N)(f) = f \otimes 1_N$ and $(M \otimes -)(g) = 1_M \otimes g$ give additive covariant functors from $\underline{M}_R$ and $_R\underline{M}$ to $\underline{Ab}$, respectively. In a manner similar to Hom, we can define an S-module structure on $M \otimes_R N$ for $_SM_R$ or $_RN_S$, namely,

$$s(m \otimes n) = sm \otimes n \qquad [(m \otimes n)s = m \otimes ns] \qquad \text{for } m \in M, n \in N, s \in S \qquad (3.6.4)$$

It is clear from the properties of additive functors or the construction of $M \otimes_R N = G/K$ that (3.6.4) is well defined and $M \otimes_R N$ can be regarded as an S-module.

3.6.4 PROPOSITION Let M_R, $_RN_S$, and $_SP$ be modules. Then

$$(M \otimes_R N) \otimes_S P \approx M \otimes_R (N \otimes_S P)$$

Proof: The mapping induced by $(m \otimes n) \otimes p \to m \otimes (n \otimes p)$ is well defined and an isomorphism.

3.6.5 THEOREM For M_R and ${}_RN$, $M \otimes_R -$ and $- \otimes_R N$ are right exact covariant additive functors. If ${}_SM_R$, $M \otimes -$ is a functor from ${}_R\underline{M}$ to ${}_S\underline{M}$.

Proof: We shall show that $- \otimes_R N$ is right exact. Let $M'_R \xrightarrow{f} M_R \xrightarrow{g} M'' \to 0$ be an exact sequence of R-modules. It is clear that $(g \otimes 1_N)(f \otimes 1_N) = 0$ and $g \otimes 1_N$ is epic. We define a mapping $\phi : M'' \otimes N \to (M \otimes N)/\text{im}\,(f \otimes 1_N)$ by setting $\phi(m'' \otimes n) = \overline{m \otimes n}$, where $m \in M$ and $g(m) = m''$. We have to show that ϕ is well defined. Let m_1 be in M with $g(m_1) = m''$. Then we obtain an m' in M' such that $f(m') = m - m_1$. Hence $m \otimes n - m_1 \otimes n = f(m') \otimes n \in \text{im}\,(f \otimes 1_N)$. We can show easily that ϕ is a homomorphism. Since $(g \otimes 1_N)(f \otimes 1_N) = 0$, $g \otimes 1_N$ gives a homomorphism of $M \otimes N/\text{im}\,(f \otimes 1_N)$ to $M'' \otimes N$. Since $(g \otimes 1_N)\phi = 1_{M''\otimes N}$ and $\phi(g \otimes 1_N) = 1_{M\otimes N}/\text{im}\,(f \otimes 1_N)$, ϕ is an isomorphism by (1.3.1). Hence $\text{im}\, f \otimes 1_N = \ker g \otimes 1_N$.

3.6.6 PROPOSITION $M \otimes_R R \approx M$.

Proof: $m \otimes r = mr \otimes 1$. Hence $m \leftrightarrow m \otimes 1$ gives an isomorphism.

3.6.7 COROLLARY Let A be a left ideal of R. Then $M \otimes_R (R/A) \approx M/MA$.

Proof: From the exact sequence $0 \to A \xrightarrow{i} R \to R/A \to 0$, we have the exact sequence $M \otimes A \to M \otimes R \to M \otimes R/A \to 0$. Hence $M \otimes_R R \approx M$ and $\text{im}\,(1_M \otimes i) \approx MA$.

3.6.8 THEOREM Let $\{M_{\alpha R}, \pi_\alpha^\beta\}_I$ be a direct system of R-modules over a directed set I. Then $\{M_\alpha \otimes_R N, \pi_\alpha^\beta \otimes 1_N\}_I$ is a direct system and $(\varinjlim M_\alpha) \otimes_R N \approx \varinjlim (M_\alpha \otimes_R N)$ for ${}_RN$.

Proof: Since $- \otimes_R N$ is covariant, $\{M_\alpha \otimes N, \pi_\alpha^\beta \otimes 1_N\}_I$ is a direct system. Further, $\{i^\alpha \otimes 1_N : M_\alpha \otimes N \to (\varinjlim M_\alpha) \otimes_R N\}$ is compatible. Hence we have a homomorphism $\phi : \varinjlim (M_\alpha \otimes_R N) \to (\varinjlim M_\alpha) \otimes_R N$. Let x be in $M^\infty \otimes N$ with $x = \Sigma_{i=1}^t m_{\alpha_i} \otimes n_i$. Then there exists a γ in I such that $m_{\alpha_i} = \pi_{\alpha_i}^\gamma m_{\alpha_i}$ in M^∞. Hence $\phi(\Sigma\, \pi_{\alpha_i}^\gamma m_{\alpha_i} \otimes n_i) = x$ and ϕ is epic. We shall define $\psi : M^\infty \otimes_R N \to \varinjlim (M_\alpha \otimes_R N)$. Since $x = \Sigma_i\, \pi_{\alpha_i}^\gamma m_{\alpha_i} \otimes n_i$, put $\psi(x) = i^\gamma(\Sigma_i\, \pi_{\alpha_i}^\gamma m_{\alpha_i} \otimes n_i)$. We can easily prove that ψ is well defined and is a homomorphism. Since $\psi\phi = 1$, ϕ is an isomorphism.

3.6.9 COROLLARY Let $\{M_{\alpha R}\}_I$ and ${}_RN$ be R-modules. Then $(\Sigma_I \oplus M_\alpha) \otimes_R N \approx \Sigma_I \oplus (M_\alpha \otimes_R N)$.

Proof: If I is finite, then the mapping $(\Sigma_{i=1}^{t} m_1) \otimes n \to \Sigma_{i=1}^{t} m_i \otimes n$ gives an isomorphism (cf. Sec. 3.7). If I is infinite, let $\underline{S}$ be the set of all finite subsets of I. Then $\underline{S}$ is a directed set. As in the proof of (3.5.7), $\{\Sigma_{I_i} \oplus M_\alpha, \pi\}_{\underline{S}}$ is a direct system and $\Sigma_I \oplus M_\alpha = \varinjlim \Sigma_{I_i} \oplus M_\alpha$. Hence $(\Sigma_I \oplus M_\alpha) \otimes_R N = (\varinjlim \Sigma_{I_i} \oplus M_\alpha) \otimes_R N = \varinjlim \Sigma_{I_i} \oplus (M_\alpha \otimes_R N) = \Sigma_I \oplus (M_\alpha \otimes N)$ [see (2.4.2)].

Let K be a commutative ring and R a ring. If there is a K-homomorphism φ from K to the center C(R) of R, we say that R is a K-*algebra*. Since $R \supset K/\ker \varphi$, we may assume that K is a subring of R contained in C(R) and $1_K = 1_R$. Let S be another K-algebra. Then $R \otimes_K S$ is a module. If we define the product in $R \otimes_R S$ as follows:

$$(r \otimes s)(r' \otimes s') = rr' \otimes ss' \qquad r \in R,\ s \in S \tag{3.6.5}$$

then (3.6.5) is well defined and $R \otimes_K S$ is a K-algebra ($K = 1_R \otimes K = K \otimes 1_S$). When R is a K-algebra, we assume that every R-module M satisfies $km = mk$ for $k \in K$ and $m \in M$. Let R and S be K-algebras and $M_{R,S}$ a right module over R and S. Then if we define

$$m(r \otimes s) = mrs \qquad m \in M,\ r \in R, \text{ and } s \in S \tag{3.6.6}$$

M is a right $R \otimes_K S$-module from (1.7.4) and (3.6.5). We can define a new K-algebra R^o from R. R^o has the same structure as R as an abelian group. We define a new product $a^o \cdot b^o = ba$ for $a^o = a$ and $b^o = b$ in R^o. Then R^o is a K-algebra. We call R^o the *opposite algebra* of R. Furthermore, $R \otimes_K R^o$ is called the *enveloping algebra* of R and we denote it by R^e. If ${}_RM$ is a left R-module, we define $mr^o = rm$. Then M is a right R^o-module from this definition. Therefore, M is a two-sided R-module if and only if M is a left (or right) R^o-module.

Finally, we shall give two relations between Hom and $\otimes$.

3.6.10 THEOREM Let A, B, and C be K-algebras. Then, for $M_{A,B}$, ${}_BN_C$, and ${}_BP_C$, we have

$$\mathrm{Hom}_{A\otimes B}(M, \mathrm{Hom}_C(N,P)) \approx \mathrm{Hom}_{B\otimes C}(M \otimes_A N, P)$$

where $\otimes$ means $\otimes_K$.

Proof: We define $\varphi : X = \mathrm{Hom}_{A\otimes B}(M, \mathrm{Hom}_C(N,P)) \to \mathrm{Hom}_{B\otimes C}(M \otimes_A N, P) = Y$. Let $m \in M$, $n \in N$, and $f \in X$. Put $\varphi(f)(m \otimes n) = f(m)(n)$ [note that $f(m) \in \mathrm{Hom}_C(N,P)$]. We can show easily that φ is well defined and a homomorphism.

Next, we define $\psi : Y \to X$. Let g be in $\mathrm{Hom}_{B\otimes C}(M \otimes_A N, P)$. Set $(\psi(g)(m))(n) = g(m \otimes n)$. Then $\psi\phi = 1_X$ and $\phi\psi = 1_Y$. (We leave precise proofs to the reader; cf. the following proof.)

3.6.11 PROPOSITION For ${}_AM$, ${}_AN_B$, and P_B, we have a homomorphism

$$\theta : \mathrm{Hom}_B(N,P) \otimes_A M \to \mathrm{Hom}_B(\mathrm{Hom}_A(M,N),P)$$

Proof: Put $\theta(f \otimes m)(g) = f(g(m))$; $f \in \mathrm{Hom}_B(N,P)$, $g \in \mathrm{Hom}_A(M,N)$, and $m \in M$. Then $\theta(f \otimes am)(g) = f(g(am)) = f(ag(m)) = (fa)(g(m)) = \theta(fa \otimes m)(g)$, where a is in A. $(f \otimes m)(gb) = f((gb)(m)) = f(g(m)b) = (f(g(m)))b = \theta(f \otimes m)(g)b$, where b is in B. Hence θ is well defined.

3.7 MATRIX RINGS

Let R be a ring. Recall the notion of a direct sum of modules (Sec. 1.3). Let $\{M_{iR}\}_{i=1}^n$ be a set of R-modules and $M = \Sigma_{i=1}^n \oplus M_i$. Let $i_j : M_j \to M$ and $p_j : M \to M_j$ be the injection and the projection, respectively. Then

$$p_j i_j = 1_{M_j} \qquad p_k i_j = 0 \quad \text{if } k \neq j \qquad \sum_{j=1}^n i_j p_j = 1_M \tag{3.7.1}$$

Conversely, if we have sets of R-homomorphisms $\{i_j' : M_j \to M\}$ and $\{p_j' : M \to M_j\}$ satisfying (3.7.1), then each i_j' is monic, $M = \Sigma_{j=1}^n \oplus i_j'(M_j)$, and i_j' and p_j' are the injections and the projections of this direct sum by (2.2.1). Since $- \otimes_R N$ is an additive functor for ${}_RN$, $p_j \otimes_R N$ and $i_j \otimes_R N$ satisfy (3.7.1). Hence $(\Sigma_{i=1}^n \oplus M_i) \otimes_R N \approx \Sigma_{i=1}^n \oplus (M_i \otimes_R N)$.

Put $S = \mathrm{End}_R(M)$ and let s be in S. Then $s = 1_M s 1_M = \Sigma_{j,k} i_j p_j s i_k p_k$ and $f_{ik} = p_j s i_k : M_k \to M_j$. Conversely, for $g_{jk} \in \mathrm{Hom}_R(M_k, M_j)$ we put $g = \Sigma_{j,k} i_k g_{kj} p_j$. Then g is in S and $p_r g i_t = g_{rt}$. Hence

$$s \leftrightarrow \begin{pmatrix} f_{11} & f_{12} & f_{13} & \cdots & f_{1n} \\ f_{21} & f_{22} & f_{23} & \cdots & f_{2n} \\ \vdots & & & & \\ f_{n1} & f_{n2} & f_{n3} & \cdots & f_{nn} \end{pmatrix}$$

is a one-to-one mapping, and $\mathrm{End}_R(M) \approx \{(f_{ij})\}$ as rings, where we define the ring operation of (f_{ij}) as for matrices. If $M_i = M_0$ for all i, then $\mathrm{End}_R(M) \approx (\mathrm{End}_R(M_0))_n$ (Sec. 1.6), where S_n denotes $n \times n$ matrices over S. In particular, we have the following.

3.7.1 PROPOSITION Let R be a ring and $F = \Sigma_{i-1}^{n} \oplus u_i R$ a free R-module. Then $R_n \approx \mathrm{End}_R(F)$.

3.7.2 PROPOSITION Let $N_R = N_{1R} \oplus N_{2R}$. If $\mathrm{Hom}_R(N_i, N_j) = 0$ for $i \neq j$, $\mathrm{End}_R(N) \approx \mathrm{End}_R(N_1) \oplus \mathrm{End}_R(N_2)$.

Put $T = \mathrm{End}_Z(M)$, where Z is the ring of integers. Then for $r \in R$, the mapping: $m \to mr$ is in T. Hence there is a homomorphism $R \to T$. We denote the image of R in T by R also. Then $V_T(R) = \{t \in T \mid rt = tr$ for all $r \in R\} = \mathrm{End}_R(M)$. Hence we sometimes call $\mathrm{End}_R(M)$ the centralizer of M. We can consider $V_T(\mathrm{End}_R(M)) \supset R$, and we call it the *double centralizer* of M.

3.7.3 THEOREM Let M_R be an R-module and $M^{(n)} = \Sigma_{i=1}^{n} \oplus M_i$, $M_i = M$. Then the double centralizer of M is isomorphic to the double centralizer W of $M^{(n)}$.

Proof: Put $S = \mathrm{End}_R(M)$. Then $\mathrm{End}_R(M^{(n)}) = S_n$. Let t be in $V_T(S)$ and put $(\Sigma_{i=1}^{n} m_i)t^{(n)} = \Sigma_{i=1}^{n} m_i t$ for $\Sigma m_i \in M^{(n)}$, $m_i \in M_i$. Hence $t^{(n)} \in W$. Then

$$(s_{ij})\left[\begin{pmatrix} m_1 \\ m_2 \\ \vdots \\ m_n \end{pmatrix} t^{(n)}\right] = \left[(s_{ij})\begin{pmatrix} m_1 \\ m_2 \\ \vdots \\ m_n \end{pmatrix}\right] t^{(n)} \qquad \text{for } (s_{ij}) \in S$$

Conversely, let t' be in W. Since e_{11}, the projection of M onto M_1, is in S_n, $M_1 t' = e_{11}Mt' = e_{11}(Mt') \subset M_1$ and $(rM_1)t' = (re_{11}M)t' = r(e_{11}Mt') = r(M_1 t')$ for $r \in S$. Hence $t'|M_1$ is in $V_T(S)$. Further, $M_j t' = (e_{j1}M_1)t' = e_{j1}[(e_{11}M)t']$ for any j. Hence $t' = (t|M_1)^{(n)}$, where $s^{(n)} = s \oplus s \oplus \cdots \oplus s : M^{(n)} \to M^{(n)}$ for $s \in \mathrm{End}_R(M)$. Therefore, $V_S(T) \approx W$ via the correspondence above.

Next, we consider the above argument for infinite sums. Let $\{M_\alpha\}_I$ and $\{N_\beta\}_J$ be sets of R-modules. Then $\mathrm{Hom}_R(\Sigma_I \oplus M_\alpha, N_\beta) = \Pi_I \mathrm{Hom}_R(M_\alpha, N_\beta)$ by (3.6.2). On the other hand, $\Sigma_J \oplus N_\beta \subset \Pi_J N_\beta$, so $\mathrm{Hom}_R(M_\alpha, \Sigma_J \oplus N_\beta) \subset \mathrm{Hom}_R(M_\alpha, \Pi_J N_\beta) = \Pi_I \mathrm{Hom}_R(M_\alpha, N_\beta)$ by the dual of (3.6.2). We shall characterize $\mathrm{Hom}_R(M_\alpha, \Sigma_J \oplus N_\beta)$ in $\Pi_J \mathrm{Hom}_R(M_\alpha, N_\beta)$. Let $f = (f_\beta : M_\alpha \to N_\beta)_J$ be an element of $\mathrm{Hom}_R(M_\alpha, \Sigma_J \oplus N_\beta)$ and m in M. Since $f(m) \in \Sigma_J \oplus N_\beta$, $f(m) =$

$f_{\beta_1}(m) + \cdots + f_{\beta_n}(m)$ and $f_\beta(m) = 0$ if $\beta \notin \{\beta_i\}_{i=1}^n$. Conversely, if $(f_\beta)_J \in \Pi_J \operatorname{Hom}_R(M_\alpha, N_\beta)$ satisfies the condition above for any $m \in M$, $\Sigma_J f_\beta : M_\alpha \to \Sigma_J \oplus N_\beta$, $(\Sigma_J f_\beta)(m) = \Sigma f_\beta(m)$, has a meaning and $\Sigma_J f_\beta \in \operatorname{Hom}_R(M_\alpha, \Sigma_J \oplus N_\beta)$. Noting this fact, we call a set of homomorphisms $\{f_\beta : M_\alpha \to N_\beta\}_J$ *summable* if

For $m \in M_\alpha$, $f_\beta(m) = 0$ for almost all β in J (3.7.2)

(depending on m) and $(\Sigma_J f_\beta)(m) = \Sigma f_\beta(m)$ has meaning.

Combining these, we have the following.

3.7.4 THEOREM Let $\{M_\alpha\}_I$ and $\{N_\beta\}_J$ be sets of R-modules. Then $\operatorname{Hom}_R(\Sigma_I \oplus M_\alpha, \Sigma_J \oplus N_\beta)$ is isomorphic to the set of matrices $(f_{\beta\alpha} : M_\alpha \to N_\beta)_{I\times J}$ for which each column $\{f_{\beta\alpha}\}_{\beta\in J}$ is summable. If $\{M_\alpha\}_I = \{N_\beta\}_J$, $\operatorname{End}_R(\Sigma_I \oplus M_\alpha)$ is isomorphic to the above set of matrices as a ring.

We call such a matrix a *column-summable matrix*. If M_α is f.g., (3.7.2) is to $f_\beta = 0$ for almost all β. In this case each column of $(f_{\beta\alpha})$ has only finitely many nontrivial entries and we call it a *column-finite matrix*. The ring of column-finite matrices over R with index I is denoted by R_I. Similarly, we can define a row-summable and row-finite matrix, respectively.

The following are immediate consequences of (3.7.1) and (3.7.3).

3.7.5 THEOREM Let R be a ring and let $F = \Sigma_I \oplus u_\alpha R$ be a free R-module with basis $\{u_\alpha\}_I$. Then $\operatorname{End}_R(F) \approx R_I$.

3.7.6 THEOREM Let M_R be an R-module and $M^{(I)} = \Sigma_I \oplus M_\alpha$, $M_\alpha = M$. Then the double centralizer of M is isomorphic to that of $M^{(I)}$.

Using the terminology above, we have the following.

3.7.7 THEOREM Let $\{M_\alpha\}_I$ and N be R-modules. We put $M = \Sigma_I \oplus M_\alpha$. Let i_α and p_α be the injection and the projection, respectively. Then $\{e_\alpha\}_I$ is summable and $1_M = \Sigma_I e_\alpha$, where $e_\alpha = i_\alpha p_\alpha$. Conversely, if $i'_\alpha : M_\alpha \to N$, $p'_\alpha : N \to M_\alpha$ satisfy $p'_\alpha i'_\alpha = 1_{M_\alpha}$, $p'_\alpha i'_\beta = 0$ for $\alpha \neq \beta$ with $\{i'_\alpha p'_\alpha\}_\alpha$ summable, and $1_N = \Sigma\, i'_\alpha p'_\alpha$, then $N = \Sigma_I \oplus M_\alpha$. If ${}_RT$ is an R-module, $\{i_\alpha \otimes_R T, p_\alpha \otimes T\}$ satisfies the conditions above. Hence $(\Sigma_I \oplus M_\alpha) \otimes_R T \approx \Sigma_I \oplus (M_\alpha \otimes_R T)$ [cf. (3.6.9)].

4
Special Modules and Rings

In Chap. 3 we studied some general properties of modules and rings. In order to study the deeper structure of modules, we have to restrict ourselves to particular types of modules and rings.

This chapter is preparatory to later chapters and introduces general features of ring theory. We study semisimple modules and chain conditions in Secs. 4.1 and 4.2, respectively. Projective modules (flat modules) and injective modules dominate contemporary ring theory. We study the former in Sec. 4.3 and the latter in Sec. 4.4. They both play an important role in homological algebra. We sketch a little of this in those sections, and we shall be able to recognize their importance in Chap. 5.

The most convenient rings to work with are the artinian rings. However, as we indicate in Sec. 4.5, we know few of their properties at present, except in the case of semisimple artinian rings. We shall try to reduce properties of semisimple rings to those of artinian rings. The importance of idempotent elements and the Jacobson radical are evident from this point of view.

We apply this method to lifting idempotents in Sec. 4.6. These results will be utilized in Sec. 8.5. We study the fundamental properties of lifting idempotents in a more general form, namely, the lifting properties of submodules modulo the Jacobson radical, in Chap. 9.

In Sec. 4.7, commutative rings, of which noetherian commutative rings form an important subclass, are studied. We consider certain aspects of noetherian rings which we shall make use of later, namely, primary decomposition of ideals and Dedekind domains. We remark that almost all statements in this section are for commutative rings, except for a few theorems. We study the completion of commutative noetherian rings in Sec. 5.5.

4.1 SEMISIMPLE MODULES

If M is a vector space over a field K, then M has a basis $\{x_\alpha\}_I$, $M = \Sigma_I \oplus x_\alpha K$, and $x_\alpha K$ is a simple K-module. We shall generalize this property to any module.

Let R be a ring and M_R an R-module. If M has no proper submodules, M is called *simple, minimal*, or *irreducible*.

4.1.1 THEOREM Let M_1 and M_2 be simple modules. Then for any element $f \in \mathrm{Hom}_R(M_1,M_2)$, f is either an isomorphism or zero. Therefore, $\mathrm{End}_R(M)$ is a division ring.

Proof: Since im f and ker f are submodules of M_2 and M_1, respectively, (4.1.1) is obvious.

If M is a direct sum of simple submodules, we call M a *semisimple module*.

4.1.2 THEOREM The following conditions are equivalent:

1. M is semisimple.
2. M is a sum of simple submodules; i.e., M = S(M).
3. Every submodule of M is a direct summand of M (and every factor module of M is semisimple).

Proof: (1) ⇒ (3). Let N be a submodule of a semisimple module M; $M = \Sigma_I \oplus M_\alpha$, where the M_α are simple. If $N \neq M$, there exists M_α such that $M_\alpha \not\subset N$. Hence $N \cap M_\alpha = 0$, since M_α is simple. Let $\underline{S}$ be the set of R-submodules $\{\Sigma_J \oplus M_\beta \mid J \subset I, (\Sigma_J \oplus M_\beta) \cap N = 0\}$. Then $\underline{S} \neq \emptyset$ if $N \neq M$. We define a partial order in $\underline{S}$ by the inclusion relation on the subsets J. Then $\underline{S}$ contains a maximal element $\Sigma_{J_0} \oplus M_\beta$ by Zorn's lemma. $\{N,M_\beta\}_{J_0}$ is independent (Sec. 1.3), and if $N \oplus \Sigma_{J_0} \oplus M_\beta \neq M$, we have M_δ with $(N \oplus \Sigma_{J_0} \oplus M_\beta) \cap M_\delta = 0$ by the earlier remark. Hence $(\Sigma_{J_0} \oplus M_\beta \oplus M_\delta) \cap N = 0$, which is a contradiction. Therefore, $M = N \oplus \Sigma_{J_0} \oplus M_\beta$ and $M/N \approx \Sigma_{J_0} \oplus M_\beta$.

(3) ⇒ (2). Let $N \neq 0$ be a submodule and $n \neq 0$ in N. Then nR contains a maximal submodule A by (3.3.7). A is a direct summand of M and hence $nR = A \oplus B$ by (1.7.4). Therefore, N contains a simple submodule B. If $M \neq S(M)$, $M = S(M) \oplus M'$ by (3) and $M' \neq 0$. Then $M' \cap S(M) \neq 0$ from the above, which is a contradiction.

(2) ⇒ (1). We assume that $M = S(M) = \cup_J N_\beta$; N_β simple. We put $\underline{T} = \{\cup_K N_\beta \mid \{N\}_K$ is independent$\}$. Then $\underline{T}$ contains a maximal element

$N = \Sigma_{K_0} \oplus N_\beta$ by Zorn's lemma [cf. the proof of (1) ⇒ (3)]. If $N \neq M$, there exists N_γ with $N_\gamma \not\subset N$, and hence $\{N_\beta, N_\gamma\}_{K_0}$ is independent, which contradicts the maximality of N.

If $M = \Sigma_I \oplus M_\alpha$ is semisimple (M_α is simple), we can find a representative set of isomorphism classes $\{M_{\alpha_i}\}_{I'}$ of $\{M_\alpha\}_I$. Put $M(i) = \Sigma \oplus M_\rho (\subseteq M)$, where M_ρ runs through all M_γ isomorphic to M_{α_i}. Then $M = \Sigma_{I'} \oplus M(i)$. M(i) is called a *homogeneous component* of M. If $I' = \{1\}$, we call M *homogeneous*.

4.1.3 THEOREM Let M_R be semisimple. Then the homogeneous component M(i) is the sum of all simple submodules of M isomorphic to M_{α_i}. Hence M(i) does not depend on the direct decomposition. Let $M = \Sigma_{I'} \oplus M(i)$. Then $\mathrm{End}_R(M) = \Pi_{J'} \mathrm{End}_R(M(i))$ and hence M(i) is a character R-submodule of M.

Proof: Let $M = \Sigma_{I'} \oplus M(i)$ and $M(i) = \Sigma_{I_i} \oplus M_\gamma$; $M_\gamma \approx M_{\alpha_i}$. Let T be a simple submodule of M and $p_\alpha : M \to M_\alpha$ the projection. If $p_\alpha(T) \neq 0$, $T \approx M_\alpha$ by (4.1.1). Hence $T \approx M_{\alpha_i}$ if and only if $p_\delta(T) \neq 0$ for some δ in I_i and $p_\gamma(T) = 0$ for $\gamma \in I - I_i$. Hence $M(i) = \cup T$, where T runs through all simple submodules isomorphic to M_{α_i}. It is clear that $\mathrm{Hom}_R(M(i), M(j)) = 0$ if $i \neq j$ by (3.7.4) and (4.1.1). Hence $\mathrm{Hom}_R(\Sigma_{I'} \oplus M(i), \Sigma_{I'} \oplus M(i)) = \Pi_{I'} \mathrm{Hom}_R(M(i), \Sigma_{I'} \oplus M(j)) = \Pi_{I'} \mathrm{Hom}_R(M(i), M(i))$ by (3.6.2).

If M is semisimple, we may assume without loss of generality that M is homogeneous, by (4.1.3).

4.1.4 THEOREM Let $M = \Sigma_I \oplus M_\alpha = \Sigma_J \oplus N_\beta$ be semisimple and homogeneous. Then $|I| = |J|$, where the M_α and the N_β are simple and $|I|$ is the cardinal number of I.

Proof: We show that $|I| \geq |J|$. Each M_α and N_β is cyclic; i.e., $M_\alpha = x_\alpha R \approx N_\beta = y_\beta R$. If $|I| < \infty$, $|I| = |J|$ by (1.4.4). We assume that $|I|, |J| \geq \aleph_0$. Let $x_\alpha = \Sigma_{i=1}^{t(\alpha)} y_{\beta_i(\alpha)} r_i$. Put $F(\alpha) = \{\beta_i(\alpha)\}_{i=1}^{t(\alpha)}$ and $X = \cup_I F(\alpha)$. Let $y_\beta = \Sigma_{j=1}^{t'} x_{\alpha_j} r'_j = \Sigma_j \Sigma_k y_{\beta_k(\alpha_j)} r'_j r_k$. Hence β is contained in $\cup F(\alpha_i) \subset X$. Therefore, $|J| \leq |X| = |I|\aleph_0 = I$, since $|I| \geq \aleph_0$ and $F(\alpha)$ is finite.

4.2 CHAIN CONDITIONS

Let M_R be an R-module. A series of R-submodules

$$M_1 \supset M_2 \supset \cdots \supset M_n \supset \cdots \tag{4.2.1}$$

is called a *descending chain* and

$$N_1 \subset N_2 \subset \cdots \subset N_n \subset \cdots \tag{4.2.2}$$

is called an *ascending chain*. If for any (4.2.1) [resp. (4.2.2)], there exists an integer t such that $M_t = M_{t+s}$ (resp. $N_t = N_{t+s}$) for all $s > 0$, we say that M satisfies the *descending* (resp. *ascending*) *chain condition* or *minimal* (resp. *maximal*) *condition*. If M satisfies the minimal (resp. maximal) condition, we say that M is *artinian* (resp. *noetherian*).

4.2.1 PROPOSITION M is artinian and noetherian if and only if M has finite composition length.

Proof: This is clear from (1.4.4) and the definition.

4.2.2 THEOREM The following conditions are equivalent for an R-module M_R:

1. M is noetherian.
2. Every R-submodule is f.g.

Proof: (1) ⇒ (2). Let N be an R-submodule. Let $x_1 \neq 0$ be in N. If $x_1R \neq N$, we take x_2 in $N - x_1R$. Then $x_1R \subsetneq x_1R + x_2R$. If $x_1R + x_2R \neq N$, take x_3 in $N - (x_1R + x_2R)$. Then $x_1R \subsetneq x_1R + x_2R \subsetneq x_1R + x_2R + x_3R$. Repeating this argument, we obtain $N = \langle x_1, x_2, \ldots, x_n \rangle$ for some n by (1).

(2) ⇒ (1). Let $N_1 \subset N_2 \subset \cdots \subset N_n \subset \cdots$ be a chain. Then $\bigcup_i N_i = \langle y_1, y_2, \ldots, y_n \rangle$. Hence there exists a t such that $N_t \ni y_j$ for all j. Thus $N_t = N_{t+s} = \bigcup N_i$ for all $s > 0$.

4.2.3 THEOREM Let $M_R \supset N_R$ be R-modules. Then M is noetherian (resp. artinian) if and only if N and M/N are noetherian (resp. artinian).

Proof: We show the artinian case. If M is artinian, then N and M/N are, by the definition and (1.2.3). We assume that N and M/N are artinian. Let $M_1 \supset M_2 \supset \cdots \supset M_n \supset \cdots$ be a descending chain. Then $M_1 \cap N \supset M_2 \cap N \supset \cdots \supset M_n \cap N \supset \cdots$ is a descending chain in N and $(M_1 + N)/N \supset (M_2 + N)/N \supset \cdots \supset (M_n + N)/N \supset \cdots$ is a descending chain in M/N. Hence there exists a t such that $M_t \cap N = M_{t+s} \cap N$ and $(M_t + N)/N = (M_{t+s} + N)/N$ for all $s > 0$. Then $M_{t+s} + N = M_t + N$ and $M_t = M_t \cap (M_{t+s} + N) = M_{t+s} + (M_t \cap N) = M_{s+t} + (M_{s+t} \cap N) = M_{s+t}$ by (1.4.1).

4.2.4 COROLLARY Let $M = \Sigma_{i=1}^n \oplus N_i$. Then M is noetherian (resp. artinian) if and only if N_i is so for all i.

Proof: Use induction on n and (4.2.3).

If the ring R is itself noetherian (resp. artinian) as a right R-module, we say that R is *right noetherian* (resp. *artinian*).

4.2.5 COROLLARY Let R be right noetherian (resp. artinian). Then every f.g. R-module is noetherian (resp. artinian).

Proof: Let $M = \Sigma_{i=1}^{t} m_i R$. Then we have an epimorphism of a free R-module $\Sigma_{i=1}^{t} \oplus u_i R$ to M by (1.7.2). Hence (4.2.5) is clear by (4.2.3) and (4.2.4).

We note the following fact from the proof above. Let $\underline{L}(M)$ be the set of all submodules of M. We define a partial order in $\underline{L}(M)$ by inclusion. Then M is noetherian (resp. artinian) if and only if every subset of $\underline{L}(M)$ contains a maximal (resp. minimal) element.

In order to obtain the dual of (4.2.2), we define the concept dual to f.g. modules. Let M_R be an R-module. If for any set $\{N_\alpha\}_I$ of R-submodules with $\cap_I N_\alpha = 0$ there exists a finite subset J of I such that $\cap_J N = 0$, then we say that M is *finitely cogenerated* (we abbreviate this by f.cog.).

4.2.6 THEOREM Let M_R be an R-module. Then the following conditions are equivalent:

1. M is artinian.
2. M/N is f.cog. for every R-submodule N of M.

Proof: We note that if M is artinian, M is f.cog. For let $\cap_I N_\alpha = 0$. If $N_1 \neq 0$ and there exists N_2 such that $N_2 \not\supset N_1$, then $N_1 \cap N_2 \subsetneq N_1$. Repeating this argument, we have an infinite chain $N_1 \supsetneq N_1 \cap N_2 \supsetneq \cdots \supsetneq N_1 \cap N_2 \cap \cdots \cap N_n \supsetneq \cdots$ if $N_1 \cap N_2 \cap \cdots \cap N_n \neq 0$ for any n. Hence (1) ⇒ (2) is obvious by (4.2.3).

(2) ⇒ (1). Let $N_1 \supset N_2 \supset \cdots \supset N_n \supset \cdots$ be a chain of R-submodules. Set $N_0 = \cap_i N_i$ and $\bar{M} = M/N_0$. Then $\cap_i \bar{N}_i = 0$. Hence there exists an m such that $\cap_{i=1}^{m} \bar{N}_i = 0$, so $N_m = N_{m+s}$.

4.2.7 PROPOSITION Let M_R be an R-module and f an element in $\mathrm{End}_R(M)$. If M is noetherian (resp. artinian) and f is epic (resp. monic), then f is an isomorphism.

Proof: Since $(f^n)^{-1}(0) \supset (f^{n-1})^{-1}(0)$, there exists a t such that $(f^{t+1})^{-1}(0) = (f^t)^{-1}(0)$. Hence since f^n is epic for any n, $f^{-1}(0) = (f^t(f^t)^{-1})(f^{-1}(0)) = f^t(f^{t+1})^{-1}(0) = 0$.

Let Z be the ring of integers and Q the rationals. We put $Z_{p^\infty} = \{a/p^n$ in $Q/Z \mid a \in Z$ and p is a fixed prime$\}$. Then every proper subgroup of Z_{p^∞} is a cyclic group $\langle 1/p^n \rangle$. Hence Z_{p^∞} is artinian but not noetherian. Since every submodule of Z is cyclic, Z is noetherian. However, $\langle p \rangle \supsetneq \langle p^2 \rangle \supsetneq \cdots \supsetneq \langle p^n \rangle \supsetneq \cdots$ is an infinite descending chain, so Z is not artinian.

4.3 PROJECTIVE MODULES AND FLAT MODULES

We have studied the covariant functors $\mathrm{Hom}_R(M,-)$ and $M \otimes_R -$ in Sec. 3.6. Next we study some modules that give exact functors $\mathrm{Hom}_R(M,-)$ or $M \otimes_R -$. In general, $\mathrm{Hom}_R(M,-)$ is an additive covariant left exact functor by (3.6.1). If $\mathrm{Hom}_R(P,-)$ is exact, P_R is called *projective*. We can state this as follows:

For any diagram

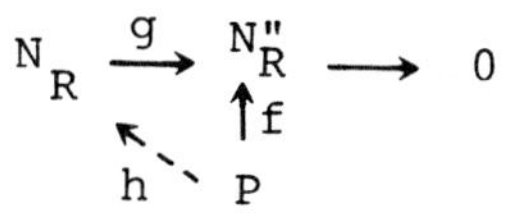

there exists an $h : P \to N$ such that $gh = f$. (4.3.1)

4.3.1 THEOREM Let $\{P_{\alpha R}\}_I$ be a set of R-modules and $P = \Sigma_I \oplus P_\alpha$. Then P is projective if and only if all P_α are projective.

Proof: Since $\mathrm{Hom}_R(\Sigma_I \oplus P_\alpha, -) = \Pi_I \mathrm{Hom}_R(P_\alpha,-)$ by (3.6.2), (4.3.1) is clear.

4.3.2 PROPOSITION Let P_R and M_R be R-modules. Assume that $g : M \to P$ is epic and P is projective. Then there exists $h : P \to M$ such that $gh = 1_P$, and $M = \mathrm{im}\, h \oplus \ker g$ with $\mathrm{im}\, h \approx P$.

Proof: We apply (4.3.1) to a case $f = 1_P$. Then $gh = 1_P$ for some h. Hence $M = \mathrm{im}\, h \oplus \ker g$ with $\mathrm{im}\, h \approx P$ by (1.7.5).

4.3.3 THEOREM An R-module P is projective if and only if P is a direct summand of a free R-module.

Proof: Since $\mathrm{Hom}_R(R,M) \approx M$, R_R is projective. Hence free R-modules and their direct summands are projective by (4.3.1). Assume that P is projective. Then we have an exact sequence $F \to P \to 0$ with F a free R-module by (1.7.2). Hence P is a direct summand of F by (4.3.2).

Let M_R and N_R be R-modules and S another ring. We put $\tau_N(M) = \cup\, \mathrm{im}\, f$, where f runs through all elements in $\mathrm{Hom}_R(M,N)$. If N is a left S-module, $\tau_N(M)$ is an S-R bimodule by (3.6.2). If in particular, we set $N = R$, $\tau(M) = \tau_R(M)$ is a two-sided ideal of R. We call it the *trace ideal* of M. Now, set $S = \mathrm{End}_R(M)$ and regard M as a left S- and a right R-bimodule. Since $\mathrm{Hom}_R(M_R,R_R)$ is a left R-module and a right S-module, we can define

$$\mu_M : M \otimes_R \mathrm{Hom}_R(M,R) \to S \qquad (4.3.2)$$

$$\rho_M : \mathrm{Hom}_R(M,R) \otimes_S M \to R \tag{4.3.3}$$

by setting $\mu_M(m \otimes f)(m') = mf(m')$ and $\rho_M(f \otimes m) = f(m)$ for $m, m' \in M$ and $f \in \mathrm{Hom}_R(M,R)$. It is clear that μ_M and ρ_M are well defined and μ_M (resp. ρ_M) is an S-S (resp. R-R) homomorphism. We shall denote $\mu_M(m \otimes f)$, $\rho_M(f \otimes m)$ by $[m,f]$ and (f,m), respectively. Then $[m,f]$ is an element of S and (f,m) is an element of R. We have two relations:

$$[m,f]m' = m(f,m') \qquad (f,m)f' = f[m,f'] \tag{4.3.4}$$

for m, m' in M and f, f' in $\mathrm{Hom}_R(M,R)$. The first equation is clear. The two elements in the second equation are in $\mathrm{Hom}_R(M,R)$ and $((f,m)f')(m') = (f,m)f'(m') = f(m)f'(m) = f(mf'(m')) = (f[m,f'])(m')$ for any $m' \in M$.

4.3.4 THEOREM Let P_R be an R-module. Then P is projective if and only if there exist sets $\{f_i \in \mathrm{Hom}_R(P,R)\}_I$ and $\{m_i \in M\}_I$ such that
1. $\{f_i\}_I$ is summable.
2. $m = \Sigma_I\, m_i(f_i,m)$ for any m in M.

Proof: Assume that there exist such sets $\{f_i\}_I$ and $\{m_i\}_I$. Put $F = \Sigma_I \oplus u_iR$, a free R-module. We define $g : M \to F$ by setting $g(m) = \Sigma_I\, u_if_i(m)$. Then g is well defined by (1) and an R-homomorphism. Further, if we define $h : F \to M$ by $h(u_i) = m_i$, then $hg = 1_M$. Hence $F = \mathrm{im}\, g \oplus \ker h$ and $\mathrm{im}\, g \approx M$ by (1.7.5). So M is projective. Conversely, assume that M is projective. M is the homomorphic image of a free R-module $F' = \Sigma_{I'} \oplus u_iR$ by (1.7.2) via h'. Thus there exists $g' : M \to F'$ such that $h'g' = 1_M$ by (4.3.2). Put $m_i = g'(u_i)$ and $f_i = p_ig'$, where $p_i : F' \to u_iR \approx R$ is the projection. Them $m_i \in M$, $f_i \in \mathrm{Hom}_R(M,R)$, $m = 1_M(m) = h'g'(m) = h'(\Sigma_{I'} u_if_i(m)) = \Sigma_{I'} m_if_i(m)$, and $\{f_i\}_{I'}$ is summable.

4.3.5 COROLLARY Let P_R be projective. Then $\tau_R(P)$ is idempotent and $P = P\tau_R(P)$.

Proof: Since $m = \Sigma\, m_i(f_i,m)$, $f(m) = \Sigma\, f(m_i)(f_i,m) \in \tau_R(P)^2$. Hence $\tau_R(P) = \tau_R(P)^2$.

4.3.6 THEOREM Let M_R be an R-module. Then the following conditions are equivalent:
1. M is f.g. projective.
2. μ_M is epic.
3. μ_M is an isomorphism.

Proof: (1) ⇒ (2). Since μ_M is an S-S homomorphism, it is sufficient to show that im μ_M contains 1_M. In the proof of (4.3.3), if M is f.g., we can take a f.g. free R-module F'. Hence we can find sets $\{m_i\}_{i=1}^n$ and $\{f_i\}_{i=1}^n$ such that $m = \Sigma_i\ m_i(f_i, m)$ for all m in M. Hence $\mu_M(\Sigma\ m_i \otimes f_i) = 1_M$.

(2) ⇒ (1). Since μ_M is an epimorphism, there exist finite sets $\{m_i\}_{i=1}^n$ and $\{f_i\}_{i=1}^n$ such that $1_M = \mu_M(\Sigma_i\ m_i \otimes f_i)$. Hence $m = 1_M(m) = \Sigma\ m_i(f_i, m)$. Accordingly, M is generated by the finite set $\{m_i\}$ and is an R-projective module by (4.3.4).

(2) ⇒ (3). Assume that $\mu_M(\Sigma_{i=1}^{n'}\ m'_i \otimes f'_i) = 0$; i.e., $\Sigma[m'_i, f'_i] = 0$. Now $m'_i = \Sigma\ m_i(f_i, m'_i)$ by (2) as above. Hence $\Sigma_i\ m'_i \otimes f'_i = \Sigma\ m_j \otimes (\Sigma_{i=1}^{n'}(f_j, m'_i)f'_i) = \Sigma_{j=1}^{t}\ m_j \otimes f_j(\Sigma_{i=1}^{n'}[m'_i, f'_i]) = 0$.

Let M_R be an R-module. Then we have an exact sequence

$$0 \to K_0 \to P_0 \to M \to 0$$

with P_0 projective by (1.7.2) and (4.3.3). Since K_0 is an R-module, we obtain an exact sequence

$$0 \to K_1 \to P_1 \to K_0 \to 0$$

with P_1 projective. Hence we have the exact sequence

$$0 \to K_1 \to P_1 \to P_0 \to M \to 0$$

with P_i projective, i = 1, 2. Continuing in this manner, we obtain the exact sequence

$$\cdots \to P_n \to P_{n-1} \to \cdots \to P_1 \to P_0 \to M \to 0 \tag{4.3.5}$$

with P_i projective. We call such an exact sequence (4.3.5) a *projective resolution* of M. Consider two exact sequences

$$\begin{array}{ccccccccc} 0 & \longrightarrow & K & \xrightarrow{i} & P & \xrightarrow{f} & M & \longrightarrow & 0 \\ & & & & & & \| \, 1_M & & \\ 0 & \longrightarrow & K' & \xrightarrow{i'} & P' & \xrightarrow{f'} & M & \longrightarrow & 0 \end{array}$$

with P, P' projective. Since $1_M f \in \mathrm{Hom}_R(P, M)$ and f' is epic, there exists $g : P \to P'$ such that $f'g = 1_M f$. Further, since $0 = fi = f'gi$, $g(K) \subset K'$. Set $g' = g|K$. Then $g' \in \mathrm{Hom}_R(K, K')$. Hence we have the commutative diagram

$$\begin{array}{ccccccccc} 0 & \longrightarrow & K & \xrightarrow{i} & P & \xrightarrow{f} & M & \longrightarrow & 0 \\ & & g' \downarrow & & g \downarrow & & \downarrow 1_M & & \\ 0 & \longrightarrow & K' & \xrightarrow{i'} & P' & \xrightarrow{f'} & M & \longrightarrow & 0 \end{array}$$

4.3.7 THEOREM Let M_R be an R-module and consider the commutative diagram above. Then $K \oplus P' \approx K' \oplus P$.

Proof: We shall show that $0 \to K \xrightarrow{\alpha} K' \oplus P \xrightarrow{\beta} P' \to 0$ is exact, where $\alpha(k) = g'(k) + i(k)$ and $\beta(k' + p) = i'(k') - g(p)$; $k \in K$, $k' \in K'$, and $p \in P$. Since $f'\beta(K' \oplus P) = f'g(P) = f(P) = M$ and $\beta(K' \oplus P) \supset K'$, β is epic. It is clear that $\beta\alpha = 0$ $[\beta\alpha(k) = \beta(g'(k) + i(k)) = g'(k) - g'(k) = 0]$. Let $k' + p \in \ker \beta$. Then $g(p) = i'(k')$. Hence $f(p) = f'g(p) = f'(i'(k')) = 0$, so $p = i(k)$ for some $k \in K$. On the other hand, $i'(k') = g(p) = gi(k) = i'g'(k)$. Hence, since i' is monic, $k' = g'(k)$. Therefore, $k' + p = \alpha(k) \in \operatorname{im} \alpha$. Accordingly, the sequence above is exact and, since P' is projective, $K' \oplus P \approx P' \oplus K$ by (4.3.2).

4.3.8 PROPOSITION Let A be a right ideal of R. Then $A = eR$ for an idempotent e if and only if R/A is R-projective.

Proof: If $A = eR$, $R = A \oplus (1 - e)R$ by (3.1.1). Hence $R/A \approx (1 - e)R$ is projective. We have an exact sequence $0 \to A \to R \to R/A \to 0$. If R/A is projective, $R = A \oplus B$; $B \approx R/A$ by (4.3.2). Hence $A = eR$ by (3.1.1).

We have shown above that for every module M, there exists an exact sequence $P_0 \xrightarrow{\mu} M \to 0$ with P_0 projective. If $\mu : P_0 \to M$ is a minimal epimorphism (Sec. 3.4), we call (P_0, μ) a *projective cover* of M. Let $P \to M \to 0$ be exact, where P is projective. Then there exists a homomorphism $h' : P \to P_0$ such that $\mu h = f$. Hence h is epic. Since P_0 is projective, $P = \ker h \oplus P'$, $P' \approx P_0$, and $f(\ker h) = 0$, $f|P' = \mu$ by (4.3.2).

4.3.9 THEOREM Let $P \xrightarrow{\mu} M \to 0$ be an exact sequence, with P projective, for an R-module M. Then (P, μ) is a projective cover of M if and only if $\ker \mu$ is a small submodule in P. In this case if $P' \to M \to 0$ is another exact sequence with P' projective, $P' \approx P \oplus P''$ and $f|P = \mu$.

Proof: This is clear from the above and (3.4.4).

4.3.10 COROLLARY All projective covers of M are isomorphic to one another; namely, if (P_i, μ_i) are projective covers of M for i = 1, 2, then there exists $h : P_1 \to P_2$ such that

$$\begin{array}{ccccc} P_1 & \xrightarrow{\mu_1} & M & \longrightarrow & 0 \\ h \downarrow & & \| & & \\ P_2 & \xrightarrow{\mu_2} & M & \longrightarrow & 0 \end{array}$$

is commutative and h is an isomorphism.

Proof: This is clear from (4.3.9).

Next we shall study modules giving exact functors $M \otimes_R -$. In general, $M \otimes_R -$ is a right exact functor by (3.6.5). If $M \otimes_R -$ is an exact functor, we call M a *flat module*.

4.3.11 PROPOSITION Let $\{M_\alpha\}_I$ be a set of R-modules. Then $M = \Sigma_I \oplus M_\alpha$ is flat if and only if each M_α is flat.

Proof: Since $(\Sigma_I \oplus M_\alpha) \otimes_R - \approx \Sigma_I \oplus (M_\alpha \otimes -)$ by (3.6.9), (4.3.11) is clear.

4.3.12 COROLLARY Every projective module is flat.

Proof: Since $R \otimes_R N = N$, R is flat as a left R-module. Hence we have (4.3.12) by (4.3.3) and (4.3.11).

When we consider projective modules P, we may first consider free modules F. If a statement is true for F and its proof is categorical, we can usually see that the statement is true for P by making use of an argument like the above. In this case we shall use a phrase "we may assume that P is free" [see the proof of (4.3.18) below].

4.3.13 LEMMA Let $0 \to {}_RN' \to {}_RN \to {}_RN'' \to 0$ be an exact sequence of R-modules. Take exact sequences $0 \to K' \to P' \to N' \to 0$ and $0 \to K'' \to P'' \to N'' \to 0$ with P', P'' projective. Then there is a commutative diagram

$$\begin{array}{ccccccccc}
 & & 0 & & 0 & & 0 & & \\
 & & \downarrow & & \downarrow & & \downarrow & & \\
0 & \longrightarrow & K' & \longrightarrow & K & \longrightarrow & K'' & \longrightarrow & 0 \\
 & & \downarrow & & \downarrow & & \downarrow & & \\
0 & \longrightarrow & P' & \xrightarrow{f'} & P' \oplus P'' & \xrightarrow{g'} & P'' & \longrightarrow & 0 \\
 & & \alpha' \downarrow & & \alpha \downarrow & & \alpha'' \downarrow & & \\
0 & \longrightarrow & N' & \xrightarrow{f} & N & \xrightarrow{g} & N'' & \longrightarrow & 0 \\
 & & \downarrow & & \downarrow & & \downarrow & & \\
 & & 0 & & 0 & & 0 & &
\end{array}$$

where each row and each column are exact.

Proof: From the diagram

$$\begin{array}{ccccc}
 & \overset{\beta}{\swarrow} & P'' & & \\
 & & \downarrow \alpha'' & & \\
N & \xrightarrow[g]{} & N'' & \longrightarrow & 0
\end{array}$$

we have $\beta : P'' \to N$ such that $g\beta = \alpha''$, since P'' is projective. We define $\alpha : P' \oplus P'' \to N$, $f' : P' \to P' \oplus P''$, and $g' : P' \oplus P'' \to P''$ by setting

$\alpha(p' + p'') = f\alpha'(p') + \beta(p'')$, $f'(p') = p' + 0$, and $g'(p' + p'') = p''$, where $p' \in P'$ and $p'' \in P''$. Then $\alpha f' = f\alpha'$ and $\alpha'' g' = g\alpha$. Since $\operatorname{im} \alpha \supset \operatorname{im} f = \ker g$ and $g(\operatorname{im} \alpha) = N''$, α is an epimorphism by (1.2.5). Put $K = \ker \alpha$. Then we have the above commutative diagram by (2.1.10).

4.3.14 THEOREM Let M_R be an R-module. Then the following conditions are equivalent:

1. M is flat.
2. Let $\{m_1, m_2, \ldots, m_n \mid m_i \in M\}$ and $\{\mu_1, \mu_2, \ldots, \mu_n \mid \mu_i \in R\}$ be sets with $\Sigma_{i=1}^n m_i\mu_i = 0$. Then there exist sets $\{x_1, x_2, \ldots, x_m \mid x_i \in M\}$ and $\{\lambda_{ij} \in R\}_{i=1}^{m}{}_{j=1}^{n}$ such that
$$m_i = \sum_{j=1}^{n} x_j\lambda_{ji} \quad \text{for each } i \qquad \sum_{i=1}^{n} \lambda_{ji}\mu_i = 0 \quad \text{for each } j$$
3. For any exact sequence $0 \to K \to F \xrightarrow{\varphi} M \to 0$ with F free and any element k in K, there exists a u in $\operatorname{Hom}_R(F,K)$ such that $u(k) = k$.
4. In (3) take a finite set $\{k_1, k_2, \ldots, k_t\}$ of elements in K. Then there exists a u in $\operatorname{Hom}_R(F,K)$ such that $u(k_i) = k_i$ for each i.

Proof: (1) ⇒ (2). Put $A = \Sigma R\mu_i$. Then we have an exact sequence $0 \to K \to \Sigma_{i=1}^n \oplus Ru_i \to A \to 0$, where $f(u_i) = \mu_i$. Since M is flat, $0 \to M \otimes_R K \to \Sigma \oplus M \otimes_R Ru_i \to M \otimes_R A \to 0$ and $0 \to M \otimes_R A \to M \otimes_R R = M$ are exact. Let $\Sigma m_i\mu_i = 0$. Then $\Sigma m_i \otimes u_i \in M \otimes_R K$ from the sequence above. Hence $\Sigma_{i=1}^n m_i \otimes u_i = \Sigma_{j=1}^m x_j \otimes k_j$, where the x_j are in M and the k_j in K. Let $k_j = \Sigma \lambda_{ji}u_i$, $\lambda_{ji} \in R$. Then $0 = f(k_j) = \Sigma \lambda_{ji}\mu_i$ and $m_i = \Sigma_{j=1}^{m} x_j\lambda_{ji}$.

(2) ⇒ (3). Put $m_\alpha = \varphi(u_\alpha)$, where $F = \Sigma \oplus u_\alpha R$. Let $k = \Sigma_{i=1}^n u_{\alpha_i} r_{\alpha_i}$. We may assume that $\{\alpha_1, \alpha_2, \ldots, \alpha_n\} = \{1, 2, \ldots, n\}$. Then $\Sigma_{i=1}^m m_i r_i = 0$. From (2) we obtain x_i in M and λ_{ij} in R such that $m_i = \Sigma_j x_j\lambda_{ji}$ and $\Sigma_j \lambda_{ij} r_j = 0$. Take any element $v_i \in F$ such that $\varphi(v_i) = x_i$ for i = 1, 2, ..., m. We define θ in $\operatorname{Hom}_R(F,K)$ by setting $\theta(u_i) = u_i - \Sigma v_j\lambda_{ji}$ and $\theta(u_\alpha) = 0$ for $\alpha \notin \{1, 2, \ldots, n\}$, since $\varphi(u_i - \Sigma v_j\lambda_{ji}) = m_i - \Sigma x_j\lambda_{ji} = 0$. Then $\theta(k) = \Sigma_i(u_i - \Sigma v_j\lambda_{ji})r_i = \Sigma u_i r_i = k$.

(3) ⇒ (4). We shall prove this by induction on t. We assume that there exists a θ_{t-1} in $\operatorname{Hom}_R(F,K)$ such that $\theta_{t-1}(k_i) = k_i$ for $i \le t - 1$. Since $\theta_{t-1}(k_t)$ is in K, $k_t - \theta_{t-1}(k_t)$ is in K. Hence there exists a $\theta' : F \to K$ with $\theta'(k_t - \theta_{t-1}(k_t)) = k_t - \theta_{t-1}(k_t)$. Hence $k_t = (\theta' + \theta_{t-1} - \theta'\theta_{t-1})(k_t)$. Put $\theta_t = \theta' + \theta_{t-1} - \theta'\theta_{t-1}$. Then $\theta_t(k_i) = \theta'(k_i) + \theta_{t-1}(k_i) - \theta'\theta_{t-1}(k_i) = \theta_{t-1}(k_i) = k_i$ for $i \le t - 1$.

(4) ⇒ (3). This is clear.

(3) ⇒ (2). Let $0 \to K' \to F' = \Sigma_{m\in M} \oplus u_m R \xrightarrow{\varphi} M \to 0$ be an exact sequence with F' a free module. We assume $\Sigma_{i=1}^{n} m_i\mu_i = 0$. Then $\Sigma_{i=1}^{n} u_{m_i}\mu_i = k$ is in K'. Thus there exists a $\theta : F' \to K'$ such that $\theta(k) = k$. Let $u_{m_i} - \theta(u_{m_i}) = \Sigma\, u_{x_j}\lambda_{ji}$; $\lambda_{ji} \in R$ for all i. Since $\theta(u_{m_i})$ is in K', $m_i = \varphi(u_{m_i}) = \Sigma_j\, x_j\lambda_{ji}$ and $k = \theta(k) = \Sigma\, \theta(u_{m_i})\mu_i = \Sigma(u_{m_i} - \Sigma\, u_{x_j}\lambda_{ji})\mu_i = k - \Sigma_{i,j}\, u_{x_j}\lambda_{ji}\mu_i$. Hence $\Sigma_i\, \lambda_{ji}\mu_i = 0$ for all j.

(2) ⇒ (1). First we take an exact sequence

$$0 \to K \xrightarrow{i} F = \Sigma \oplus Ru_\alpha \to N \to 0$$

for left R-modules N and show that $0 \to M \otimes_R K \to M \otimes_R F = \Sigma \oplus Mu_\alpha$ is exact. Let $k = \Sigma_{i=1}^{n} m_i \otimes k_i$ be in $M \otimes_R K$. Assume that $(M \otimes i)(k) = 0$. Put $k_i = \Sigma_{j=1}^{t} \mu_{ij}u_{\alpha_j}$ and assume that $\{\alpha_1, \alpha_2, \ldots, \alpha_t\} = \{1, 2, \ldots, t\}$. We may assume that the number t of elements of this set is minimal for all expressions $k = \Sigma\, m_i' \otimes k_i'$. We shall show that $t = 0$. Let $t \neq 0$. Since $Mu_i \approx M$ and $\Sigma_j(\Sigma_i\, m_i\mu_{ij}u_j) = 0$, $\Sigma\, m_i\mu_{ij} = 0$ for each j. We shall take the case $j = 1$. Then there exist elements $x_j \in M$ and $\lambda_{ij} \in R$ such that $m_i = \Sigma_{j=1}^{m} x_j\lambda_{ji}$ and $\Sigma_i\, \lambda_{ji}\mu_{i1} = 0$. Hence $(\Sigma_{i=1}^{n} m_i \otimes k_i) = \Sigma_{i=1}^{n}(\Sigma_j\, x_j\lambda_{ji} \otimes k_i) = \Sigma\, x_j \otimes (\Sigma_i\, \lambda_{ji}k_i)$. Put $k_j' = \Sigma_i\, \lambda_{ji}k_i \in K$. Then $k_j' = \Sigma_i\, \Sigma_{s=1}^{t}\, \lambda_{ji}\mu_{is}u_s - \Sigma_i\, \Sigma_{s=2}^{t}\, \lambda_{ji}\mu_{is}u_s$. So $k = \Sigma\, x_j \otimes k_j'$ gives a smaller set $\{2, \ldots, t\}$, which is a contradiction. Thus $k = 0$. Next we consider any exact sequence $0 \to {}_RN' \to {}_RN \to {}_RN'' \to 0$. Then we have the commutative exact diagram as in (4.3.13). Hence we have the commutative diagram

$$\begin{array}{ccccccccc}
 & & 0 & & 0 & & 0 & & \\
 & & \downarrow & & \downarrow & & \downarrow & & \\
 & & M \otimes K' & \to & M \otimes K & \to & M \otimes K'' & \to & 0 \\
 & & \downarrow & & \downarrow & & \downarrow & & \\
0 & \to & M \otimes F' & \to & M \otimes (F' \oplus F'') & \to & M \otimes F'' & \to & 0 \\
 & & \downarrow & & \downarrow & & \downarrow & & \\
0 & \to & M \otimes N' & \to & M \otimes N & \to & M \otimes N'' & \to & 0 \\
 & & \downarrow & & \downarrow & & \downarrow & & \\
 & & 0 & & 0 & & 0 & &
\end{array}$$

where F' and F'' are free modules. Now the middle row is exact by (3.6.9) and each column is exact from the first assertion. Hence the lowest row is exact by (2.5.2) and (3.6.5).

Let $0 \to K \to P \to M \to 0$ be exact for a f.g. R-module M. If P and K are f.g. and P is projective, M is called *finitely presented*; i.e., there exists a projective resolution $P_1 \to P_0 \to M \to 0$ of M with P_i finitely generated.

4.3.15 COROLLARY Every finitely presented flat module M is projective.

Proof: (4.3.14),3 shows that M is a direct summand of F, where F is a free module.

4.3.16 PROPOSITION Let $\{M_\alpha, \pi_\alpha^\beta\}$ be a direct system over a directed set I. If each M_α is flat, so is $\varinjlim M_\alpha$.

Proof: This is clear from (3.5.6) and (3.6.8).

4.3.17 COROLLARY If every f.g. submodule of an R-module M is flat, so is M.

Proof: We have (4.3.17) from (3.5.7) and (4.3.16).

We shall finish this section with some homological theorems.

4.3.18 THEOREM Let R and S be algebras over a commutative ring K. Let M_R and N_R be R-modules. If M is f.g. and S is K-flat, then

$$\mathrm{Hom}_R(M,N) \otimes_K S \approx \mathrm{Hom}_{R\otimes_K S}(M \otimes_K S,\ N \otimes_K S)$$

Proof: We shall denote $\mathrm{Hom}_R(-,-)$ by $[-,-]$ and $\otimes_K$ by $\otimes$, respectively. Let $P_1 \to P_0 \to M \to 0$ be a projective resolution of M with P_i f.g. Since $[-,-]$ is left exact by (3.6.1), we obtain the exact sequence

$$0 \to [M,N] \to [P_0,N] \to [P_1,N]$$

Define $\varphi_M : [M,N] \otimes S \to [M \otimes S,\ N \otimes S]_{R\otimes S}$ by setting $\varphi_M(f \otimes s)(m \otimes s') = f(m) \otimes ss'$, where $f \in [M,N]$, $m \in M$, and $s, s' \in S$. It is clear that φ_M is well defined and a homomorphism. First we shall show that φ_P is an isomorphism if P is f.g. projective. If $P = R$, φ_P is an isomorphism. Since P is a direct summand of a f.g. free R-module F, $\varphi_F = \varphi_P \oplus \varphi_{P'}$, where $F = P \oplus P'$. Hence φ_P is an isomorphism; cf.

$$\begin{array}{ccc}
[F,N]\otimes S & \xrightarrow{\ \varphi_F\ } & [F\otimes S,\ N\otimes S]_{P\otimes S} \\
\| & & \| \\
[P,N]\otimes S \oplus [P',N]\otimes S & \xrightarrow{\ \varphi_P \oplus \varphi_{P'}\ } & [P\otimes S,\ N\otimes S]_{R\otimes S} \oplus [P'\otimes S,\ N\otimes S]_{R\otimes S}
\end{array}$$

Noting that S is K-flat, we obtain a commutative diagram

$$\begin{array}{ccccccc}
0 \to & [M,N]\otimes S & \to & [P_0,N]\otimes S & \to & [P_1,N]\otimes S \\
 & \downarrow \varphi_M & & \wr\!\wr\ \varphi_{P_0} & & \wr\!\wr\ \varphi_{P_1} \\
0 \to & [M\otimes S,\ N\otimes S] & \to & [P_0\otimes S,\ N\otimes S] & \to & [P_1\otimes S,\ N\otimes S]
\end{array}$$

Hence φ_M is an isomorphism by (2.5.2).

Let A_R and B_R be R-modules. Let

$$\begin{array}{l}
\cdots \to P_n \xrightarrow{d_n} P_{n-1} \xrightarrow{d_{n-1}} \cdots \to P_1 \xrightarrow{d_1} P_0 \xrightarrow{\varepsilon} A \to 0 \\
\to P'_n \xrightarrow{d'_n} P'_{n-1} \xrightarrow{d'_{n-1}} \cdots \to P'_1 \xrightarrow{d'_1} P'_0 \xrightarrow{\varepsilon'} B \to 0
\end{array} \qquad (4.3.6)$$

be projective resolutions of A and B, respectively. Take $f \in \mathrm{Hom}_R(A,B)$. If there exists a family of homomorphisms $\{f_i : P_i \to P'_i\}$ such that

$$\begin{array}{ccc}
P_n & \xrightarrow{d_n} & P_{n-1} \\
f_n \downarrow & & \downarrow f_{n-1} \\
P'_n & \xrightarrow{d'_n} & P'_{n-1}
\end{array} \qquad\qquad
\begin{array}{ccc}
P_0 & \xrightarrow{\varepsilon} & A \\
f_0 \downarrow & & \downarrow f \\
P'_0 & \xrightarrow{\varepsilon'} & B
\end{array}$$

are commutative, we say that $\{f_i\}$ is a *set of morphisms over* f. Let $\{g_i : P_i \to P'_i\}$ be another set of morphisms over f. If there exists a family of homomorphisms $\{s_i : P_i \to P'_{i+1}\}$,

$$\begin{array}{ccccc}
P_{n+1} & \longrightarrow & P_n & \longrightarrow & P_{n-1} \\
 & \swarrow s_n & & \swarrow s_{n-1} & \\
P'_{n+1} & \longrightarrow & P'_n & \longrightarrow & P'_{n-1}
\end{array}$$

such that $f_0 - g_0 = d'_1 s_0$, $f_n - g_n = s_{n-1}d_n + d'_{n+1}s_n$, we say that $\{f_i\}$ and $\{g_i\}$ are *homotopic*.

4.3.19 LEMMA Let P_R be a projective module and let $A'' \xrightarrow{\varphi} A \xrightarrow{\psi} A'$ be an exact sequence of R-modules. If there exists $\tau : P \to A$ with $\psi\tau = 0$, then there exists $h : P \to A''$ with $\varphi h = \tau$.

Proof: Since im φ = ker ψ, we have a diagram

$$\begin{array}{ccccc}
 & & P & & \\
 & \overset{h}{\swarrow} & \downarrow \tau & & \\
A'' & \longrightarrow & \mathrm{im}\,\varphi & \longrightarrow & 0
\end{array}$$

Hence we have $h : P \to A''$ with $\varphi h = \tau$.

4.3.20 THEOREM Let A and A' be R-modules and $f : A \to A'$. Then we can find a set of morphisms $\{f_i : P_i \to P_i'\}$ over f for (4.3.6) and any two such sets are homotopic.

Proof: We shall show the existence of $\{f_i\}$ by induction. First we take a diagram

$$\begin{array}{ccccc} P_0 & \xrightarrow{\varepsilon} & A & & \\ & & \downarrow f & & \\ P'_0 & \xrightarrow{\varepsilon'} & A' & \longrightarrow & 0 \end{array}$$

Then there exists $f_0 : P_0 \to P_0'$ with $f\varepsilon = \varepsilon' f_0$ by (4.3.19). Now assume that we have obtained $f_i : P_i \to P_i'$ for $i \leq n - 1$. Consider the diagram

$$\begin{array}{ccccc} P_n & \xrightarrow{d_n} & P_{n-1} & \xrightarrow{d_{n-1}} & P_{n-2} \\ & & \downarrow f_{n-1} & & \downarrow f_{n-2} \\ P'_n & \xrightarrow{d'_n} & P'_{n-1} & \xrightarrow{d'_{n-1}} & P'_{n-2} \end{array}$$

Then $d'_{n-1}f_{n-1}d_n = f_{n-2}d_{n-1}d_n = 0$. Hence there exists $f_n : P_n \to P_n'$ with $d'_n f_n = f_{n-1}d_n$ by (4.3.19). Let $\{f_i' : P_i \to P_i'\}$ be another set of morphisms over f. We shall show that $\{f_i\}$ and $\{f_i'\}$ are homotopic. Take the diagram

$$\begin{array}{ccccc} P_1 & \longrightarrow & P_0 & \xrightarrow{\varepsilon} & A \\ & \swarrow & \downarrow f_0 - f'_0 & & \downarrow f - f = 0 \\ P'_1 & \xrightarrow{d'} & P'_0 & \xrightarrow{\varepsilon'} & A' \end{array}$$

Since $\varepsilon'(f_0 - f_0') = 0$, we have $s_0 : P_0 \to P_1'$ with $d_1's_0 = f_0 - f_0'$. Assume that we have obtained s_i for $i \leq n - 2$. Consider the diagram

$$\begin{array}{ccccc} P_n & \xrightarrow{d_n} & P_{n-1} & \xrightarrow{d_{n-1}} & P_{n-2} \\ f_n - f'_n \downarrow & & \downarrow f_{n-1} - f'_{n-1} & & \downarrow f_{n-2} - f'_{n-2} \\ P'_n & \xrightarrow[d'_n]{} & P'_{n-1} & \xrightarrow[d'_{n-1}]{} & P'_{n-2} \end{array}$$

Put $\tau = f_{n-1} - f'_{n-1} - s_{n-2}d_{n-1}$. Then $d'_{n-1}\tau = (f_{n-2} - f'_{n-2})d_{n-1} - d'_{n-1}s_{n-2}d_{n-1} = (f_{n-2} - f'_{n-2})d_{n-1} - (f_{n-2} - f'_{n-2} - s_{n-3}d_{n-2})d_{n-1} = 0.$

Hence we can obtain $s_{n-1} : P_{n-1} \to P'_n$ with $d'_n s_{n-1} = f_{n-1} - f'_{n-1} - s_{n-2}d_{n-1}$.

4.3.21 LEMMA Let $0 \leftarrow A \leftarrow L_0 \leftarrow L_1$ and $0 \leftarrow A' \leftarrow L'_0 \leftarrow L'_1$ be exact sequences. If we have a commutative diagram

$$\begin{array}{ccccccc} 0 & \leftarrow & A & \overset{\varepsilon}{\leftarrow} & L_0 & \leftarrow & L_1 \\ & & & & \downarrow f_0 & & \downarrow f_1 \\ 0 & \leftarrow & A' & \overset{\varepsilon'}{\leftarrow} & L'_0 & \leftarrow & L_1 \end{array}$$

then there is a unique homomorphism $f : A \to A'$ with $f\varepsilon = \varepsilon' f_0$. Furthermore, if $\{g_i\}$ give a similar commutative diagram and there exists $s_0 : L_0 \to L'_1$ with $f_0 - g_0 = d'_1 s_0$, i.e., $\{f_i\}$ and $\{g_i\}$ are homotopic, then $\{f_i\}$ and $\{g_i\}$ induce the same homomorphism.

Proof: This is clear.

4.4 INJECTIVE MODULES

In Sec. 4.3 we studied modules M that gave the exact covariant functors $\mathrm{Hom}_R(M,-)$. We now study modules that give the exact contravariant functors $\mathrm{Hom}_R(-,M)$.

Let E_R be an R-module. If $\mathrm{Hom}_R(-,E)$ is an exact functor, E is called *injective*; namely, for any exact sequence

$$\begin{array}{ccccc} 0 & \longrightarrow & M'_R & \overset{g}{\longrightarrow} & M_R \\ & & f \downarrow & \swarrow h & \\ & & E & & \end{array} \qquad (4.4.1)$$

there exists $h : M \to E$ such that $f = hg$.

4.4.1 THEOREM Let $\{E_{\alpha_R}\}_I$ be a set of R-modules and $E = \Pi_I E_\alpha$. Then E is injective if and only if each E_α is.

Proof: This is clear from the fact that $\mathrm{Hom}_R(-, \Pi_I E_\alpha) \approx \Pi_I \mathrm{Hom}_R(-,E_\alpha)$ by (1.3.2) (cf. Sec. 2.1).

4.4.2 THEOREM (Baer's Condition) Let E_R be an R-module. Then E is injective if and only if for any right ideal I of R and any homomorphism $f : I \to E$, there exists e in E such that $f(x) = ex$ for all $x \in I$.

Proof: We assume that the condition stated in (4.4.2) is satisfied and take a diagram (4.4.1). We may assume that $M' \subset M$ and g is the inclusion. Let $M \supset M_0 \supset M'$ and $f_0 : M_0 \to E$ with $f_0|M' = f$. Let F be the set of all such extensions (M_1, f_1) of M'. We define a partial order $\geq$ in $\underline{F}$ as follows: $(M_\alpha, f_\alpha) \geq (M_\beta, f_\beta) \leftrightarrow M_\alpha \supset M_\beta$ and $f_\alpha | M_\beta = f_\beta$. For any totally ordered subset $(M_1, f_1) \leq (M_2, f_2) \leq \cdots \leq (M_n, f_n) \leq \cdots$, put $M^* = \cup_i M_i$. We define $f^* : M^* \to E$ by $f^*(m) = f_n(m)$, where $m \in M^*$ (note that $m \in M_n$ for some n). It is clear that f^* does not depend on the choice of n and $(M^*, f^*) \in \underline{F}$. Hence $\underline{F}$ contains a maximal element (M^+, f^+). We shall show $M^+ = M$. If $M^+ \neq M$, take an $m' \in M - M^+$ and put $I = \{r \in R \mid m'r \in M^+\}$. Then I is a right ideal of R. We define $t : I \to E$ by setting $t(x) = f^+(m'x)$. Then there exists an e in E such that $t(x) = ex$. Consider $M^+ + m'R \subset M$ and define $p : M^+ + m'R \to E$ by $p(m + m'r) = f^+(m) + er$. We can easily see that p is well defined and an R-homomorphism, which contradicts the maximality of (M^+, f^+). Conversely, if E is injective, we take $M' = I$, $M = R$, and g the inclusion in (4.4.1). Then for $e = g(1)$ we have $f(x) = g(1x) = ex$.

We can define the injective envelope of M_R as the concept dual to the projective cover (Sec. 4.3). First we need some preparation.

Let T be a commutative domain and A_T a T-module. If $At = A$ for any $t \neq 0$ in T, A is called *divisible*. An element a in A is called *torsion* if $at' = 0$ for some $t' \neq 0$ in T. Let $t(A)$ be the set of all torsion elements in A. Then $t(A)$ is a T-submodule. If $A = t(A)$ [resp. $t(A) = 0$], we call A a *torsion* (resp. *torsion-free*) *module*. $A/t(A)$ is torsion free.

4.4.3 PROPOSITION Let T be a commutative domain and M_T a T-module. If M is injective, M is divisible. Conversely, if M is torsion free and divisible, M is injective.

Proof: Assume that M is injective. Let t be a nonzero element in T and m in M. Then, for $f : tT \to M$ $[f(t) \to m]$, there exists an m' in M such that $f(t) = m't$. Hence $Mt = M$. Conversely, let $f' : I \to M$ be a homomorphism of a right ideal I to M. For any x in I there exists an m_x in M such that $f(x) = m_x x$ (since M is torsion free, m_x is unique, where $x \neq 0$). Let y be another element in I. Then $m_x xy = f(x)y = f(xy) = f(yx) = m_y yx$. Hence if $xy \neq 0$, $m_x = m_y$ $(= m)$. Therefore, $f(z) = mz$ for all z in I and M is injective by (4.4.2).

4.4.4 LEMMA Let T be as in (4.4.3). Then every T-module M is contained in a divisible T-module.

Proof: Let K be the quotient field of T (see Sec. 4.7 below). We have exact sequences $0 \to A \to F \to M \to 0$, with F a free R-module, and $0 \to T \to K$. Then $0 \to F \otimes_T T = F \to F \otimes_T K$ is exact by (4.3.12). Hence

$M = F/A \subset (F \otimes K)/A$. $F \otimes K$ is divisible, so $(F \otimes K)/A$ is divisible.

4.4.5 THEOREM Let R and S be rings, and let $A_{R\text{-}S}$, B_R be modules. If A is S-flat and B is R-injective, then $\mathrm{Hom}_R(A,B)$ is S-injective.

Proof: Let $0 \to {}_SC' \to {}_SC \to {}_SC'' \to 0$ be exact. Then

$$\text{(a)}\quad 0 \to \mathrm{Hom}_S(C'',\mathrm{Hom}_R(A,B)) \to \mathrm{Hom}_S(C,\mathrm{Hom}_R(A,B)) \to \mathrm{Hom}_S(C',\mathrm{Hom}_R(A,B))$$

is exact by (3.6.1). Furthermore

$$\text{(b)}\quad 0 \to A \otimes_S C' \to A \otimes_S C \to A \otimes_S C'' \to 0$$

is exact by the assumption on A. On the other hand, $\mathrm{Hom}_S(C,\mathrm{Hom}_R(A,B)) \approx \mathrm{Hom}_R(A \otimes_S C, B)$ by (3.6.10). Using this fact, we have the commutative diagram

$$\begin{array}{ccccc}
0 \longrightarrow \mathrm{Hom}_S(C'',\mathrm{Hom}_R(A,B)) & \longrightarrow & \mathrm{Hom}_S(C,\mathrm{Hom}_R(A,B)) & \longrightarrow & \mathrm{Hom}_S(C',\mathrm{Hom}_R(A,B)) \\
\wr\wr & & \wr\wr & & \wr\wr \\
0 \longrightarrow \mathrm{Hom}_R(A \otimes_S C'', B) & \longrightarrow & \mathrm{Hom}_R(A \otimes_S C, B) & \longrightarrow & \mathrm{Hom}_R(A \otimes_S C', B)
\end{array}$$

Since B_R is injective, we know that the last mapping in the above is an epimorphism from (b). Hence $\mathrm{Hom}_R(A,B)$ is S-injective.

4.4.6 THEOREM Every R-module M is contained in an injective R-module.

Proof: Let Z be the ring of integers. Then M is contained in a divisible abelian group N by (4.4.4). We shall show that N is an injective Z-module. Let $I = \langle n \rangle$ be an ideal in Z and $f : I \to N$. Since N is divisible, we have an x in N such that $f(n) = xn$. Hence N is Z-injective. Put $L = \mathrm{Hom}_Z({}_RR_Z,N)$. Then L_R is R-injective by (4.4.5). For each n in M we define $f_n \in L$ by $f_n(r) = nr$. Since $f(1) = n$, $f_n \neq 0$ provided that $n \neq 0$. Therefore, M is contained in L as an R-module via $n \to f_n$ [see (3.6.1) and (3.6.2)].

Now we can define the injective envelope of an R-module M_R. Let E_R be an injective module containing M. If the inclusion mapping $i : M \to E$ is maximal (Sec. 3.4), i.e., M is an essential submodule of E, we say that E is an *injective envelope* (or *injective hull*) of M. We denote E by E(M). We shall show existence of E(M) in (4.4.8) below.

4.4.7 THEOREM Let E_R be an R-module. Then the following conditions are equivalent:

1. E is injective.
2. If $M_R \supset E$ and E is essential in M, then E = M.
3. If $N_R \supset E$, E is a direct summand of N.

Proof: (1) ⇒ (2). Consider a diagram

$$\begin{array}{ccccc} 0 & \longrightarrow & E & \xrightarrow{i} & M \\ & & \| \; 1_E & & \\ & & E & & \end{array}$$

Then there exists h : M → E such that 1_E = hi. Hence h is an isomorphism.

(2) ⇒ (3). Let $\underline{F}$ be the set of all submodules K' of N with K' ∩ E = 0. Then $\underline{F}$ contains a maximal element K by Zorn's lemma. Consider N/K. Then N/K contains a submodule E' isomorphic to E. Let L/K ≠ 0 be a submodule of N/K. Then L ∩ (E ⊕ K) = K ⊕ (L ∩ E) by (1.7.1) and L ∩ E ≠ 0 by the maximality of L. Hence L ∩ (E ⊕ K) $\supsetneq$ K. Therefore, E ⊕ K = N by (2).

(3) ⇒ (1). E is contained in an injective module E_0 by (4.4.6). Since E is a direct summand of E_0 by (3), E is injective by (4.4.1).

4.4.8 THEOREM Let M_R be an R-module. Then there exists an injective envelope of M and all injective envelopes of M are isomorphic to one another.

Proof: Let E be an injective module containing M by (4.4.6). By $\underline{F}$ we denote the set of all submodules K of E such that K contains M and M is essential in K. Then $\underline{F}$ contains a maximal element K_0 by (3.4.5) and Zorn's lemma. We shall show that K_0 satisfies the condition (2) of (4.4.7). Let L_R be an R-module containing K_0 such that K_0 is essential in L. Then we have the diagram

$$\begin{array}{ccccc} 0 & \longrightarrow & K_0 & \xrightarrow{i_L} & L \\ & & \downarrow i_E & & \\ & & E & & \end{array}$$

where the i are inclusions. Since E is injective, we obtain an h : L → E with $hi_L = i_E$. K_0 being essential in L, h is a monomorphism. Hence K_0 = L by the maximality of K_0, so K_0 is injective by (4.4.7). Therefore, K_0 is an injective envelope of M. Let E_i be an injective envelope of M (i = 1, 2). Then there exists a homomorphism f : $E_1 \to E_2$ with $f|M = 1_M$

(cf. the proof above). Hence f is a monomorphism by (3.4.10). Further, $E_2 = \text{im } f \oplus T$ by (4.4.7). Since im $f \supset M$ and M is essential in E_2, $T = 0$.

We have obtained the following result from the proof above.

4.4.9 PROPOSITION Let M_R be an R-module and let E be an injective module containing M. A submodule E_1 of E containing M is an injective envelope of M if and only if E_1 is a maximal member in the set $\underline{F}$ of all submodules $F \supset M$ such that M is essential in F.

4.4.10 COROLLARY Let M_R be an R-module and $M = M_1 \oplus M_2$. Then $E(M) = E(M_1) \oplus E(M_2)$.

Proof: We can obtain $E(M_i)$ in $E(M)$ ($i = 1, 2$). Since $M_1 \cap M_2 = 0$, $E(M_1) \cap E(M_2) = 0$. M is essential in $E(M_1) \oplus E(M_2)$ by (3.4.6). Hence $E(M_1) \oplus E(M_2) = E(M)$.

4.4.11 COROLLARY Let $M_R \supset N_R$ be R-modules. If N is essential in M, $E(M) = E(N)$.

Proof: Apply (4.4.9) to E(N).

4.5 ARTINIAN RINGS

We defined artinian rings in Sec. 1.2. We now study some structures associated with an artinian ring R that are fundamental and which gave rise to much of ring theory. We start with nilpotent ideals.

4.5.1 THEOREM Let A be a minimal right ideal in a ring R. Then either $A^2 = 0$ or $A = cR$, where c is idempotent.

Proof: We assume that $A^2 \neq 0$. Since A is minimal, there exists an x in A with $xA = A$. Hence we obtain an a in A such that $xa = x$, so $xa^2 = xa = x$. Put $I = \{y \in A \mid xy = 0\}$. Then $a^2 - a \in I$. If $a^2 - a \neq 0$, $I = A$, for A is minimal, and $0 = xa = a$, which is a contradiction. Hence $a^2 = a$ and $A = aR$.

4.5.2 THEOREM Let R be a right artinian ring and A a right ideal. If A is not nilpotent, A contains a nonzero idempotent.

Proof: Since R is right artinian, there exists a minimal element A among the set of nonnilpotent ideals. Then $A^2 = A$. Next we take an ideal B minimal among right ideals B' ($\subset A$) with $B'A \neq 0$. Then there exists x in B such that $xA \neq 0$ ($\subset B$). Further, $xAA = xA \subset B$, so $B = xA$. Hence we have an a in A with $xa = x$. Since $xa^p = xa^2 = xa \neq 0$, $a^p \neq 0$. Put $a^2 - a = n$; then $xn = 0$. Let $I = \{y \in A \mid xy = 0\}$. Then $n \in I$ and $a \notin I$, which implies that $I^m = 0$ for some m by the minimality of A. Hence $n^m = 0$. If we put $a_1 = a + n - 2an$, a_1 and n are in <a>, so $an = na$, $aa_1 = a_1a$. Since $a^p \neq 0$ for any p and $n - 2an$ is nil, a_1 is not nil. Further, $(n_1 =)$ $a_1^2 - a_1 = n^2(4n - 3)$ and $n_1^{t_1} = 0$ for some $t_1 < m$. Repeating this, we obtain finally an a_k ($\in$ A) such that $0 = n_k = a_k^2 - a_k$ and $a_k = 0$.

4.5.3 COROLLARY Let R be right artinian. Then J(R) is nilpotent, and every nil ideal is nilpotent.

Proof: Put $J = J(R)$. We have a chain $J \supset J^2 \supset \cdots \supset J^m \supset \cdots$. Hence $J^n = J^{2n}$ for some n. If $J^n \neq 0$, J contains a nonzero idempotent e by (4.5.2), which contradicts (3.3.1). Every nil ideal is contained in J(R) by (3.3.1) and (3.3.5).

Semisimple Rings

If R is a right artinian ring, J(R) is the unique maximal nilpotent ideal in R by (4.5.3). If $J(R) = 0$, we call R a *semisimple ring* (Sec. 3.3). We shall study the structure of an artinian semisimple ring R.

4.5.4 THEOREM Let R be a ring. Then the following conditions are equivalent:

1. R is a right artinian semisimple ring.
2. Every R-module is semisimple.
3. Every R-module is projective
4. Every R-module is injective.

Proof: (1) $\Rightarrow$ (2). Let A be a minimal right ideal of R. Then $A = eR$ and $e^2 = e$ by (4.5.1). Hence $R = eR \oplus (1 - e)R$. We take a minimal right ideal A_1 contained in $(1 - e)R$. Then A_1 is a direct summand of R, so $(1 - e)R = A_1 \oplus B$ by (1.7.4). Continuing in this way, we obtain $R = \Sigma \oplus A_i$; the A_i are simple. Every R-module M is a homomorphic image of a free R-module $\Sigma \oplus R$, which is semisimple by the above, and hence M is semisimple by (4.1.2).

(2) $\Rightarrow$ (3), (4). This is clear by (4.1.2).

(3) $\Rightarrow$ (1). We take an exact sequence $0 \to A \to R \to R/A \to 0$ for a right ideal A. Since R/A is projective, $R = A \oplus C$ by (4.3.2). Hence R is semisimple as an R-module by (4.1.2). Since R contains the identity, $R = \Sigma_{i=1}^n \oplus A_i$, where the A_i are simple. Hence R is right artinian. Further, J is a direct summand of R. Hence, $J = eR$; $e^2 = e$, so $e = 0$.

(4) $\Rightarrow$ (1). We can use the same argument as in (3) $\Rightarrow$ (1).

We know from the proof of (4.5.4) that R is a semisimple ring if and only if R is semisimple as a right R-module. Let $R = \Sigma \oplus A_i$, A_i a minimal right ideal. Then we can take the homogeneous component $R(i) = \Sigma_i \oplus A_{ij}$. If a ring S contains no proper two-sided ideals, S is called a *simple ring*.

4.5.5 THEOREM (*Wedderburn*) Let R be a right artinian semisimple ring and let R(i) be as above. Then R(i) is a two-sided ideal of R and R(i) is a simple ring. Further, $R = \Sigma_i \oplus R(i)$ as rings.

Proof: It is clear that $R = \Sigma_i \oplus R(i)$ as R-modules. Let A be a simple right ideal in R(i). Then rA is either a simple right ideal isomorphic to A or zero for $r \in R$. Hence $rA \subset R(i)$ by (4.1.3), so R(i) is a two-sided ideal. Therefore, $R(i)R(j) \subset R(i) \cap R(j) = 0$ provided that $i \neq j$. Let B be a simple right ideal in R(i). Then $A \approx B$, $A = eR$, and $B = fR$, where e and f are idempotents. Since $\mathrm{Hom}_R(eR, fR) = fRe$ by (3.1.7), $B = xA$ for some $x \in fRe$. Therefore, $R(i) = RAR$. Accordingly, R(i) is simple.

4.5.6 THEOREM Let R be a ring. Then R is simple and right artinian if and only if $R \approx K_n$, where K is a division ring. In this case all minimal right ideals are isomorphic to one another

Proof: If R is simple, $J = 0$. Hence $R = \Sigma_{i=1}^n \oplus e_iR$ is homogeneous by (4.5.5). Thus $R \approx \mathrm{End}_R(R) \approx K_n$ by (3.7.4), where $K = \mathrm{End}_R(e_1R)$. Since e_1R is simple, K is a division ring by (4.1.1). We assume that $R \approx K_n$. Then $R = \Sigma_{i=1}^n \oplus c_{ii}R$, where $[c_{ij}]$ is the set of matrix units. Let $x \neq 0$ be in $e_{11}R$. Then $x = e_{11}x$ and xe_{jj} $(= x_j) \neq 0$ for some j. Hence $x(x_j^{-1}c_{j1}) = c_{11} \in xR$. Therefore, $c_{11}R$ is simple and R is semisimple and artinian by (4.2.1). Further, e_{i1} gives us an isomorphism of $e_{11}R$ to $e_{ii}R$, so R is homogeneous. The remaining statements are clear by (4.1.3) and (4.5.5).

Since K_n is symmetric with respect to right side and left side, we have the following.

4.5.7 COROLLARY Let R be a semisimple ring. Then R is left artinian if and only if R is right artinian.

4.5.8 PROPOSITION Let K and K' be division rings. If $K_n \approx K'_{n'}$ as rings, $K \approx K'$ and $n = n'$.

Proof: Put $R = K_n$. Then $K \approx \mathrm{End}_R(e_{11}R)$, $e_{11}R$ is the unique simple R-module, and n the length of a composition series $\{\Sigma_{i=1}^t e_{ii}R\}_{t=0}^n$ of R. Hence, if $K_n \approx K'_{n'}$, $K \approx K'$ and $n = n'$.

Semiprimary Rings

If R' is a right artinian ring, J = J(R') is nilpotent. Further, R'/J is semisimple and artinian by (3.3.6) and (4.2.3). We shall generalize this situation. If J is nilpotent and R/J is (right) artinian for a ring R, where J = J(R), we call R a *semiprimary ring*. By (4.5.5) and (4.5.6) we know the structure of R/J completely. We reduce some important properties of R/J to those of R. Note that many results in this section are valid for more general rings, e.g., the perfect rings in Sec. 6.3 below.

First we consider an artinian ring.

4.5.9 THEOREM If R is right artinian, then R is right noetherian.

Proof: If R is right artinian, we have the chain $R = J^0 \supset J \supset J^2 \supset \cdots \supset J^{n-1} \supset J^n = 0$. Since J^i/J^{i+1} is an R/J-module, J^i/J^{i+1} is semisimple as an R/J-module and hence an R-module by (4.5.4). Further, J^i/J^{i+1} is artinian and every submodule of J^i/J^{i+1} is a direct summand. Hence $\|J^i/J^{i+1}\| < \infty$ (where $\|\ \|$ is length of composition series) for each i. Therefore, R is noetherian by (4.2.1).

Next we take a nil ideal A and consider some relationships between idempotents in R and R/A.

4.5.10 THEOREM Let R be a ring and A a nil ideal. For an element a in R such that $a^2 \equiv a \pmod{A}$ there exists an idempotent e in R such that $e \equiv a \pmod{A}$. We can find such an e in the ring generated by a.

Proof: We consider some polynomial identities with integer coefficients:

$$1 = [x + (1 - x)]^{2n} = \sum_{j=0}^{2n} \binom{2n}{j} x^{2n-j}(1 - x)^j$$

Put $f_n(x) = \Sigma_{j=0}^{n} \binom{2n}{j} x^{2n-j}(1 - x)^j$. Then

$$\text{(a)} \quad f_n(x) \equiv 0 \pmod{x^n} \qquad f_n(x) \equiv 1 \quad [\text{mod } (1 - x)^n]$$

Hence we have

$$\text{(b)} \quad f_n(x) \equiv f_{n-1}(x) \quad [\text{mod } (x - x^2)^{n-1}]$$
$$f_1(x) = x^2 + 2x(1 - x) \quad [\text{mod } (x^2 - x)]$$

Put $a_n = f_n(a)$. Then a_n is an element in the ring generated by a, from (a), and $a \equiv a_1 \equiv a_n$ [mod $(a^2 - a)$], from (b). Since $a^2 - a \in A$,

$(a^2 - a)^m = 0$ for some integer m. Hence $a_m^2 - a_m = 0$ from (a). Thus a_m is the desired element.

4.5.11 COROLLARY Let R and A be as in (4.5.10). Assume that $\{a_i\}_{i=1}^n$ is a set of mutually orthogonal idempotents modulo A. Then there exists a set of mutually orthogonal idempotents $\{e_i\}_{i=1}^n$ in the ring generated by $a_1, a_2, \ldots, a_n$ such that $a_i \equiv e_i \pmod{A}$.

Proof: We shall show (4.5.11) by induction on n. Theorem (4.5.10) is the case n = 1. Assume that there exists a set of mutually orthogonal idempotents $\{e_i\}$ in the ring generated by $a_1, a_2, \ldots, a_{n-1}$ such that $e_i \equiv a_i \pmod{A}$ for $i \le n - 1$. We put $a_n' = a_n - (e_1 + \cdots + e_{n-1})a_n - a_n(e_1 + \cdots + e_{n-1}) + (e_1 + \cdots + e_{n-1})a_n(e_1 + \cdots + e_{n-1})$. Then $a_n' \equiv a_n \pmod{A}$ and $e_i a_n' = a_n' e_i = 0$ for $i \le n - 1$. Further, a_n' is an element of the ring generated by $a_n, e_1, \ldots, e_{n-1}$ and hence by $a_1, a_2, \ldots, a_n$. There exists an idempotent e_n in the ring generated by a_n' (and hence by $a_1, \ldots, a_n$) such that $e_n \equiv a_n' \equiv a_n \pmod{A}$ by (4.5.10). Since $e_i a_n' = a_n' e_i = 0$ for $i \le n - 1$, $e_n e_j = e_j e_n = 0$ for $j \le n - 1$.

4.5.12 COROLLARY Let R and A be as in (4.5.10) and let e be an idempotent in R. Assume that $e \equiv a_1 + a_2 + \cdots + a_n \pmod{A}$ and $a_i a_j \equiv \delta_{ij} a_i \pmod{A}$, where δ_{ij} is the Kronecker delta. Then there exists a set of mutually orthogonal idempotents $\{e_i\}_{i=1}^n$ such that $e = \Sigma_{i=1}^n e_i$ and $e_i \equiv a_i \pmod{A}$ for all i.

Proof: Put $a_i' = ea_i e$. Then $a_i'^2 \equiv a_i' a_i' \equiv a_i \pmod{A}$. There exists a set of mutually orthogonal idempotents $\{e_i\}_{i=1}^n$ in the ring generated by $a_1', a_2', \ldots, a_n'$ such that $e_i \equiv a_i' \pmod{A}$. Since $ea_i' = a_i'e = a_i'$ for all i, $ee_i = e_i e = e_i$. Put $e' = e_1 + e_2 + \cdots + e_n$. Then $e'e = ee' = e'$ and $e - e' \in A$. $(e - e')^2 = e - e'$ implies that $e = e'$.

If an idempotent e is minimal with respect to the order $\ge$ in Sec. 3.1, then we call e *primitive*.

4.5.13 COROLLARY Let A be a nil ideal. Then an idempotent e in R is primitive if and only if e + A is primitive in R/A.

Proof: This is clear from (4.5.12) and (3.1.2).

Finally, we consider a semiprimary ring R. Then R/J ($= \bar{R}$) is semisimple.

4.5.14 PROPOSITION Let R be semiprimary. Then J(M) is the unique maximal small submodule in any R-module M_R (Sec. 3.4) and J(M) = MJ.

Proof: Let $M = N + MJ$ for $N_R \subset M$. Then $M = N + (N + MJ)J = N + MJ^2 = \cdots = N + MJ^n = N$ for $J^n = 0$. Hence MJ is small in M. Further, since M/MJ is an R/J-module, M/MJ is a semisimple R-module. Accordingly, $MJ = J(M)$ and $J(M)$ is the unique maximal small submodule in M by (3.4.3).

4.5.15 PROPOSITION Let e_i be an idempotent in a semiprimary ring R ($i = 1, 2$). Then $e_1R \approx e_2R$ if and only if $e_1R/e_1J \approx e_2R/e_2J$.

Proof: The natural epimorphism $e_iR \to e_iR/e_iJ$ is a projective cover of e_iR/e_iJ by (3.4.3). Hence we have (4.5.15) by (4.3.10).

We can express $\bar{R}$ as $\bar{R} = \Sigma_{i=1}^{n}\Sigma_{j=1}^{\rho(i)} \oplus \overline{e_{ij}R}$ for a semiprimary ring R, where $\overline{e_{ij}R}$ ($\approx \overline{e_{i1}R}$) is simple and $\overline{e_{i1}R} \not\approx \overline{e_{j1}R}$ if $i \neq j$. From (4.5.12) we can find a set of mutually orthogonal idempotents $\{e_{ij}\}$ in R such that $1 = \Sigma_{i=1}^{n}\Sigma_{j=1}^{\rho(i)} e_{ij}$. Thus we have from (4.5.15),

$$R = \sum_{i=1}^{n} \sum_{j=1}^{\rho(i)} \oplus e_{ij}R \tag{4.5.1}$$

where $\{e_{ij}\}$ is a set of mutually orthogonal primitive idempotents with $1_R = \Sigma\Sigma e_{ij}$ and $e_{ij}R \approx e_{i1}R$, $e_{i1}R \not\approx e_{j1}R$ if $i \neq j$. Since $\overline{e_{i1}Re_{j1}} = 0$, $e_{i1}Re_{j1} = e_{i1}Je_{j1}$ for $i \neq j$. The expression (4.5.1) will be used in the later chapters.

Let M_R be an R-module. We can define the chain of submodules by transfinite induction as follows:

$S_0(M) = 0$, $S_1(M) = S(M)$ (the socle of M), ..., $S_\alpha(M)/S_{\alpha-1}(M) = S(M/S_{\alpha-1}(M))$ if α is not a limit ordinal and $S_\alpha(M) = \cup_{\beta<\alpha} S_\beta(M)$ if α is a limit ordinal.

We call the series $S_0 \subset S_1 \subset \cdots \subset S_\alpha \subset \cdots$ the *lower Loewy series* of M.

4.5.16 THEOREM Let R be a semiprimary ring. Then $M = S_\gamma(M)$ for every M_R, where γ is a suitable ordinal depending on M. Hence $S(M)$ is essential in M.

Proof: We show $S(M) \neq 0$ for every $M_R \neq 0$. Let $J^n = 0$ and $J^{n-1} \neq 0$ for some n. Then $MJ^t = 0$ and $MJ^{t-1} \neq 0$ for some $t \leq n$. Since MJ^{t-1} is an R/J module, $MJ^{t-1} = S(MJ^{t-1}) \subset S(M)$ by (4.5.4). If $S(M)$ is not essential in M, there exists a nonzero submodule N of M such that $S(M) \cap N = 0$. However, $0 \neq S(N) \subset S(M)$, which is a contradiction.

We shall give some examples of semiprimary rings.

4.5.17 LEMMA Let M_R be an R-module and $\|M\| = r < \infty$. Let s_i be in $\mathrm{End}_R(M)$ for $i = 1, 2, \ldots, r$. Assume that there exists a submodule $N \neq 0$ in M such that $s_i N \subset N$ for all i and $s_r s_{r-1} \cdots s_1 N \neq 0$. Then there exists $N^* \neq 0$ in N such that $N^* = \Sigma_i\, s_i N^*$.

Proof: Put $N_0 = N$ and we define inductively $N_i = \Sigma_{j=1}^r s_j N_{i-1}$. Then $N_r \supset s_r s_{r-1} \cdots s_1 N \neq 0$ and $s_k N_i \subset N_i$ (use induction on i). Hence $N_0 \supset N_1 \supset \cdots \supset N_r \supset s_r \cdots s_1 N \neq 0$. Since $\|M\| = r$, $N_k = N_{k+1}$ for some $k \leq r - 1$. Then $N_{k+1} = \Sigma_{i=1}^r s_i N_k = N_k$.

4.5.18 LEMMA Let M_R be a directly indecomposable R-module and $\|M\| < \infty$. Then every element in $\mathrm{End}_R(M) = S$ is either a unit or nilpotent. Furthermore, the set of nonunit elements in S is the Jacobson radical J(S), S/J(S) is a division ring, and J(S) is nilpotent.

Proof: Let s be in S. Then by considering the chain $M \supset sM \supset s^2M \supset \cdots \supset s^nM \supset \cdots$, we obtain a t such that $s^tM = s^{2t}M$ by assumption. Hence $M = s^tM + \ker s^t$ by (1.2.5). Further, since $s^t(s^tM) = s^tM$, $s^t|s^tM$ is either an isomorphism or zero by (4.2.7). Hence $M = s^tM \oplus \ker s^t$. Accordingly, s is either an isomorphism or $s^t = 0$. We denote the set of nonisomorphisms in S by J'. Then J' is closed with respect to the product. We shall show that J' is a two-sided ideal of S. Let s_i be in J' for i = 1, 2. If $s_1 + s_2 \notin J'$, $s = s_1 + s_2$ is a unit. Hence $1_M = s^{-1}s_1 + s^{-1}s_2$ and $s^{-1}s_i \in J'$. We put $s_i' = s^{-1}s_i$. Since $s_1'1 = s_1'^2 + s_1's_2' = 1s_1' = s_1'^2 + s_2's_1'$, $s_1's_2' = s_2's_1'$. Therefore, $1 = 1^k = (s_1' + s_2')^k = 0$ for a large k by the first observation, which is a contradiction. Thus we have known that J' is a two-sided ideal. It is clear that $J' \subset J = J(S)$ and S/J' is a division ring. Hence J' = J. Finally, we shall show that $J^r = 0$ for $r = \|M\|$. Assume that $J^r \neq 0$. Then there exist s_i in S with $s_1 s_2 \cdots s_r \neq 0$. Thus we obtain M_0 in M such that $M_0 = \Sigma_i\, s_i M_0$ by (4.5.17). Let $\underline{F}$ be the set of all nonzero submodules N of M such that $N = \Sigma\, t_i'N$ for some finite set $\{t_i'\}$ in J. Consider the minimal element N' in $\underline{F}$. We can find

$$\text{(c)} \quad N_0 = \sum_{i=1}^{k} t_i N_0$$

among the N' such that k is minimal. Then $N_0 = \Sigma_{i,j}\, t_i t_j N_0 = \cdots = \Sigma\, t_{i_1} t_{i_2} \cdots t_{i_p} N_0$. Since $N_0 \neq 0$, there exist t_{i_j} such that $t_{i_1} t_{i_2} \cdots t_{i_p} \neq 0$ for any large p. Hence we obtain some $t = t_{i_a}$ such that t appears r times in $t_{i_1} t_{i_2} \cdots t_{i_p}$ for a large p. Let $t_{i_1} t_{i_2} \cdots t_{i_p} = \cdots (t \cdots)(t \cdots) \cdots (t \cdots)$ and put $q_1 = (t \cdots)$, $q_2 = (t \cdots) \cdots$ with the occurrences of t determined from the left side. Then $q_1q_2\cdots q_r \neq 0$ and $q_iN_0 \subset N_0$. Hence there exists $N^* \neq 0$ $(\subset N_0)$ such that $N^* = \Sigma\, q_iN^*$. If $k \neq 1$ in (c), $N_0 \supsetneq tN_0 \supset q_iN^*$ for all i, so $N^* \subset tN_0 \subsetneq N_0$, which contradicts the minimality of N_0. Hence $t_1N_0 = N_0$. However, t_1 is nil, so $N_0 = t_1^nN_0 = 0$, which is a contradiction.

4.5.19 THEOREM Let M be an R-module with $\|M\| < \infty$. Then $\mathrm{End}_R(M)$ is semiprimary, where $\|M\|$ is the length of composition series of M.

Proof: M is a direct sum of directly indecomposable R-modules M_{ij}. We assume that $M_{ij} \approx M_{i1}$ and $M_{i1} \not\approx M_{j1}$ for $i \neq j$. Furthermore, we may assume that $M_{i1} = M_{ij}$ $(= M_i)$ for all j. Let $f_{ij} \in \mathrm{Hom}_R(M_j,M_i)$. Then $f_{ij}f_{ji}$ is not an isomorphism if $i \neq j$. For if $f_{ij}f_{ji} \in \mathrm{End}_R(M_i)$ is an isomorphism, f_{ij} and f_{ji} are isomorphisms by (1.7.5), but $M_i \not\approx M_j$. Now $S = \mathrm{End}_R(M)$ is isomorphic to the ring of matrices $(f_{ij,kp})$ by (3.7.4), where $f_{ij,kp} \in \mathrm{Hom}_R(M_{kp},M_{ij})$. Put $J' = \{(f_{ij,kp}) \mid f_{ij,kp}$ is not an isomorphism$\}$. Then J' is a two-sided ideal by the above and (4.5.18), and $S/J' \approx \Sigma \oplus \mathrm{End}_R(M_i)_{n_i}/J(\mathrm{End}_R(M_i)_{n_i}$ is semisimple by (4.5.18). Hence $J' \supset J(S)$. On the other hand, $[J(\mathrm{End}_R(M_i)]^{m_i} = 0$ for some m_i. Hence every entry $g_{ij,kp}$ of an element of J'^a has the form $xh_{sk,st}y$ with $h_{sk,st} \in [J(\mathrm{End}_R(M_s))]^{m_s} = 0$ for a large a. Hence J' is nilpotent and $J' = J$.

Let $\{M_{iR}\}_I$ be a set of R-modules and $\{f_i : M_i \to M_{i+1}\}$ a set of homomorphisms. Put $\theta(n,m) = f_nf_{n-1}\cdots f_m \in \mathrm{Hom}_R(M_m,M_{n+1})$ for $m \leq n$.

4.5.20 LEMMA Let $\{M_i\}_I$ be a family of indecomposable R-modules with $\|M_i\| = n < \infty$ for all i. Let f_i be a homomorphism of M_i to M_{i+1} which is not an isomorphism (M_i may be equal to M_j). Then $\|\theta(2^m,1)(M_1)\| \leq \max(0, n - m - 1)$ for all m.

Proof: Since $\|M_i\| = n$ for all i, each f_i is neither a monomorphism nor an epimorphism. We shall prove the result by induction on m.

1. If $m = 0$, $\|f_1(M_1)\| \leq n - 1 = n - 0 - 1$.
2. Assume that $\|\theta(2^m,1)(M_1)\| \leq n - m - 1$ and $n - m - 1 \neq 0$.
3. It is clear that $\|\theta(2^{m+1},1)(M_1)\| \leq n - m - 1$.

If $\|\theta(2^{m+1},1)(M_1)\| = n - m - 1$, then $\|\theta(2^m,1)(M_1)\| = n - m - 1$. Hence $\theta(2^{m+1},2^m + 1) \mid \theta(2^m,1)(M_1)$ is a monomorphism. Furthermore, we have $\|\theta(2^{m+1},2^m + 1)(M_{2^m+1})\| \leq n - m - 1$ from the induction hypothesis (2). Since $\|\theta(2^{m+1}, 2^m + 1)(M_{2^m+1})\| \geq \|\theta(2^{m+1},1)(M_1)\| = n - m - 1$,

$\theta(2^{m+1}, 2^m + 1)(M_{2^m+1}) = \theta(2^{m+1},1)(M_1)$. Hence $M_{2^m+1} = \theta(2^m,1)(M_1) \oplus \ker \theta(2^{m+1}, 2^m + 1)$ and $\theta(2^m,1)(M_1) \neq 0$, $\ker \theta(2^{m+1}, 2^m + 1) \neq 0$, which is a contradiction. Therefore, $\|\theta(2^{m+1},1)(M_1)\| \leq n - m - 2$.

4.5.21 THEOREM Let $\{M_i\}_I$ be a family of indecomposable R-modules with $\|M_i\| \leq n < \infty$ for all i and let $\{f_i : M_i \to M_{i+1}\}$ be a set of homomorphisms which are not isomorphisms. Then $\theta(n_0,1) = 0$ for some n_0.

Proof: It is clear from the assumption that at least one f_{i+j} among f_{i+k}, $k = 0, 1, \ldots, n$, is not a monomorphism. Assume $f_{i_1}, f_{i_2}, \ldots$ are not monomorphisms. Then $g_1 = \theta(i_2 - 1, i_1)$, $g_2 = \theta(i_3 - 1, i_2)$, ... are not monomorphisms. We take a family $\{M_{i_j}\}$. Since $\|M_{i_j}\| \leq n$, there exist some $r \leq n$ and an infinite subset $\{M_{i_k}\}$ such that $\|M_{i_k}\| = r$ for all k. We put $h_1 = g_{k_2-1}g_{k_2-2}\cdots g_{k_1}$, $h_2 = g_{k_2-1}\cdots g_{k_2}$, ... and apply (4.5.20) for the set $\{h_i\}$. Then we have a large fixed n_0 such that $\theta(n_0,1) = 0$. (It is clear that we can find such n_0 independently of the choice of M_i and f_i.)

4.6 LIFTING IDEMPOTENTS

If a two-sided ideal A of a ring R is nil, every idempotent in R/A is lifted to that in R by (4.5.10). We shall generalize the above property.

Let A be a right ideal. We consider the following condition:

If $x^2 \equiv x \pmod{A}$ as a module, there exists e in R such that $e = e^2$ and $e \equiv x \pmod{A}$. (4.6.1)

In this case we say that idempotents can be *lifted modulo* A. If (4.6.1) is valid for every right ideal, then we call R (right) *suitable*, following Nicholson [179]. In Sec. 8.5 we shall show that this definition is left- and right-sided symmetric, so we omit the word "right."

4.6.1 LEMMA R is suitable if and only if (4.6.1) is satisfied for every principal right ideal of the form $(x^2 - x)R$.

Proof: Let A be a right ideal and $x^2 \equiv x \pmod{A}$. Then $(x^2 - x)R \subset A$. Hence we have (4.6.1).

4.6.2 THEOREM The following conditions are equivalent (x is any element in R):

1. R is suitable.
2. There exists an e in xR such that $e = e^2$ and $e \equiv x\ [\mathrm{mod}\ (x - x^2)R]$.

3. There exists an e in xR such that $e^2 = e$ and $(1 - e) \equiv (1 - x)c$ [mod J(R)] for some c in R.
4. There exists an e in xR such that $e = e^2$ and $R = eR + (1 - x)R$.
5. There exists $e = e^2$ in xR such that $1 - e$ is in $(1 - x)R$.

Proof: Since $e \equiv x$ [mod $(x^2 - x)R$] implies that $e \in xR$, (1) $\Leftrightarrow$ (2) is clear from (4.6.1).

(2) $\Rightarrow$ (3). Since $e - x = (x^2 - x)r$ for some $r \in R$, $1 - e = (1 - x)(1 - xr)$.

(3) $\Rightarrow$ (4). Since $1 - (1 - e) - (-c + xc) = e + (1 - x)c$ is a unit, $R = eR + (1 - x)R$.

(4) $\Rightarrow$ (5). Let $1 = ea + (1 - x)b$ and put $f = e + ea(1 - e)$. Then $f^2 = f \in eR \subset xR$ and $1 - f = (1 - ea)(1 - e) = (1 - x)b(1 - e) \in (1 - x)R$.

(5) $\Rightarrow$ (2). $1 - e = (1 - x)r$ and $e = xa$. Hence $e - x = (1 - x)e - x(1 - e) \in (x^2 - x)R$.

For any x in R, consider $1 = x + 1 - x$. Then (5) above means that there exist idempotents e and f such that $1 = e + f$, $e \in xR$, and $f \in (1 - x)R$.

4.6.3 COROLLARY Every homomorphic image of a suitable ring is suitable.

Proof: This is clear.

4.6.4 THEOREM A ring R is suitable if and only if R/J(R) is suitable and idempotents can be lifted module J(R).

Proof: We need to prove the "if" part. Let x be in R and $\bar{x}$ in $R/J = \bar{R}$. Then there exists an $\bar{a}$ in $\overline{xR}$ ($a \in xR$) such that $\bar{a} = \bar{a}^2$ and $\bar{1} - \bar{a} = (\bar{1} - \bar{x})\bar{c}$. Take $f^2 = f$ with $\bar{f} = \bar{a}$ from the assumption and put $u = 1 - f + a$. Then $uf = af$. Since $\bar{u} = \bar{1}$, u is a unit, so $e = ufu^{-1} = afu^{-1} \in xR$ and $e^2 = e$. Further, $\bar{1} - \bar{e} = \bar{1} - \bar{f} = \bar{1} - \bar{a} = (\bar{1} - \bar{x})\bar{c}$. Hence $1 - e \equiv (1 - x)c$ [mod J(R)].

4.6.5 PROPOSITION If a ring R is suitable, so is eRe for every idempotent e.

Proof: Let x be in eRe. Then take $f = f^2$ in xR and $1 - f \in (1 - x)R$. Put $f' = fe$. Then $f' = ef \in xeRe$ and $f'^2 = f'$. Further, $e - f' = e - fe = e(1 - f) \in (e - x)eRe$.

Next we shall consider orthogonal idempotents.

4.6.6 LEMMA Let R be a suitable ring. Assume that $1 = \Sigma_{i=1}^{n} x_i$, where the x_i are in R. Then there exists a set of mutually orthogonal idempotents e_i in x_iR such that $1 = \Sigma\, e_i$.

Proof: We shall show the result by induction on n. Assume that $1 = x_1 + x_2 + \cdots + x_n + x_{n+1}$. Then there exists $f^2 = f$ in $(x_1 + \cdots + x_n)R$ and $1 - f \in [1 - (x_1 + \cdots + x_n)]R = x_{n+1}R$. Let $f = (x_1 + \cdots + x_n)r$. fRf is suitable by (4.6.5). We may assume that $x_i r = fx_i rf$. Then, by the induction hypothesis, there exists a set of mutually orthogonal idempotents f_i in $x_i rfRf \subset x_iR$ such that $f = \Sigma_{i=1}^n f_i$. Put $f_{n+1} = 1 - f$ $(\in x_{n+1}R)$. Then $1 = \Sigma_{i=1}^{n+1} f_i$ and $f_i f_j = \delta_{ij} f_i$.

4.6.7 THEOREM. Let R be a suitable ring and let A be a right ideal. If $\{x_i\}_{i=1}^n$ is a set of mutually orthogonal idempotents modulo A, then there exists a set of mutually orthogonal idempotents e_i in x_iR with $e_i \equiv x_i$ (mod A).

Proof: Put $x = \Sigma\, x_i$. Then there exists a set of mutually orthogonal idempotents e_i in x_iR $(i \le n)$ and $1 - e \in (1 - x)R$ by (4.6.6), where $e = e_1 + \cdots + e_n$. From the assumption we have $x_i x \equiv x_i$ (mod A). Since $e_i \in x_iR$, $x_j e = x_j(\Sigma\, x_i r_i) \equiv x_j r_j \equiv e_j$ (mod A), where $e_i = x_i r_i$. Let $1 - e = (1 - x)r$. Then $x_i - e_i \equiv x_i(1 - e) \equiv x_i(1 - x)r \equiv 0$ (mod A).

4.7 COMMUTATIVE RINGS

We develop some fundamental properties of commutative ring theory in this section. We assume that the reader is familiar with the elementary concepts of commutative ring theory. Only sketches of proofs will be given.

Let R be a commutative ring. An ideal P is called *prime* if R/P is a domain. Let Q be an ideal. If every zero divisor in R/Q is nilpotent, Q is called a ***primary ideal.*** If Q is primary, we put $P = \{a \in R \mid a + Q$ is nil in $R/Q\}$. Then P is prime and we say that P *belongs to* Q or P is *associated with* Q. We denote this by $P = r(Q)$. Q is called a P-primary ideal. If P is f.g., $P^n \subset Q$ for some n.

Let S be a *multiplicatively closed set* in R; i.e., (1) $0 \notin S$, (2) if $x, y \in S$, $xy \in S$. We take the cartesian product $R \times S$ and define an equivalence relation in $R \times S$ as follows:

$$(r_1,s_1) \equiv (r_2,s_2) \leftrightarrow \text{There exists an } s \text{ in } S \text{ such that } (r_2s_1 - r_1s_2)s = 0, \text{ where the } r_i \text{ are in R and the } s \text{ in S.} \tag{4.7.1}$$

We define $(r_1,s_1) + (r_2,s_2) = (r_1s_2 + r_2s_1, s_1s_2)$ and $(r_1,s_1)(r_2,s_2) = (r_1r_2,s_1s_2)$. Then these operations are well defined and the set of equivalence classes $\{(r,s)\}$ has a ring structure. We denote it by R_S. There is a natural mapping $\rho : R \to R_S$ given by $\rho(r) = (rs,s)$. $N = \ker \rho =$

$\{r \in R \mid \text{there exists an } s \text{ in } S \text{ with } rs = 0\}$ is an ideal of R. Hence R_S contains an isomorphic image of R/N. We sometimes denote R/N by R if there is no confusion. It is clear that every element $\rho(s)$ in $\rho(S)$ is a unit in R_S. Hence we shall express (r,s) in R_S as rs^{-1}. We call R_S the *quotient ring* of R with respect to S.

4.7.1 THEOREM Let R be a commutative ring and S a multiplicatively closed set in R. Let T be a ring and $f : R \to T$ a ring homomorphism. If f(s) is a unit in T for all $s \in S$, then there exists a unique homomorphism $h : R_S \to T$ such that $f = h\rho$.

Proof: This is clear from the definition [cf. the proof of (3.6.3)].

Let $\rho : R \to R_S$ be the natural mapping. Let A be an ideal in R. Then $\rho(A)R_S$ is an ideal in R_S. We denote it by A^e or A_S. Conversely, if B is an ideal in R_S, then $\rho^{-1}(B)$ is an ideal in R. We denote it by B^c or $B \cap R$.

4.7.2 THEOREM Let R and S be as in (4.7.1) and let Q be a P-primary ideal. Then we have the following:

1. If $S \cap P \neq \emptyset$, $Q^e = R_S$.
2. If $S \cap P = \emptyset$, P^e is prime, Q^e is P^e-primary and $Q^{ec} = Q$.
3. For an ideal A in R_S, $A^{ce} = A$.

Proof: 1. If $s \in S \cap P$, $\rho(s)^n \in Q^e$ for some n. Since $\rho(s)$ is a unit, $Q^e = R_S$.

2. Let $ab \in P^e$ for a, $b \in R_S$ and $a = r_1s_1^{-1}$, $b = r_2s_2^{-1}$. Then $r_1r_2s \in P$ for some $s \in S$. Since $s \notin P$, r_1 or r_2 is in P. Hence P^c is prime. Let x be in p^{ec}. Then $\rho(x) = as^{-1}$, $a \in P$. Since $xs's \in P$ for some $s' \in S$, $x \in P$. Therefore, $P^{ec} = P$. We can similarly prove the remaining assertions.

4.7.3 COROLLARY If R is noetherian, then so is R_S for any multiplicatively closed set S.

Proof: This is clear from (4.7.2),3.

4.7.4 THEOREM Let R and S be as in (4.7.1). Then there is a one-to-one mapping between the set of prime ideals P' in R_S and the set of prime ideals P in R with $P \cap S = \emptyset$ given by $P'^c = P$ and $P^e = P'$.

Proof: This is clear from (4.7.2).

Let P be a prime ideal in R. Then $R - P = S$ is a multiplicatively closed set. We denote R_S by R_P in this case. If R is a domain, R_0 is called the *quotient field* of R. If a ring R contains a unique maximal ideal, R is called a *local ring*.

4.7.5 COROLLARY Let R be a ring and P a prime ideal. Then R_P is a local ring and P_P is the unique maximal ideal in R_P.

Next, we take an R-module. For a multiplicatively closed set S in R, consider $M \times S$. In a manner similar to (4.7.1) we define a relation

$(m_1,s_1) \equiv (m_2,s_2) \leftrightarrow$ There exists an s in S such that $(m_1s_2 - m_2s_1)s = 0$, where the m are in M and the s in S, and $(m_1,s_1) + (m_2,s_2) = (m_1s_2 + m_2s_1, s_1s_2)$. (4.7.2)

Then the set of equivalence classes $\{(m,s)\}$ has an R_S-module structure. We denote it by M_S and express each element in M_S as ms^{-1}. It is clear that $ms^{-1} = 0$ in M_S if and only if there exists s' in S such that $ms' = 0$. We can define a natural mapping $\rho : M \to M_S$ by setting $\rho(m) = (m,s)s^{-1}$, and we have $\ker \rho = \{m \in M \mid$ there exists an s in S with $ms = 0\}$.

4.7.6 LEMMA Let S be a multiplicatively closed set in a ring R and let M be an R-module. Then $M_S \approx M \otimes_R R_S$.

Proof: Define homomorphisms $f : M \otimes R_S \to M_S$, $g : M_S \to M \otimes R_S$ by setting $f(m \otimes rs^{-1}) = mrs^{-1}$ and $g(ms^{-1}) = m \otimes s^{-1}$. We can easily show that f and g are well defined and $fg = 1_{M_S}$, $gf = 1_{M\otimes R_S}$.

The following is obtained from (4.7.6) and the remark above.

4.7.7 THEOREM Let S be a multiplicatively closed set in R. Then $- \otimes_R R_S$ is an exact covariant functor.

4.7.8 PROPOSITION Let S be as in (4.7.7) and let M be an R-module. Then, for any R-submodules M_i $(i = 1, 2)$, $(M_1 \cap M_2)_S = M_{1S} \cap M_{2S}$.

Proof: We may assume that all submodules in the equation are in M_S by (4.7.7). Then $(M_1 \cap M_2)_S \subset M_{1S} \cap M_{2S}$. Let $x = m_1s_1^{-1} = m_2s_2^{-1}$ in $M_{1S} \cap M_{2S}$. Then there exists an s in S with $(m_1s_2 - m_2s_1)s = 0$. Hence $m_1s_2s \in M_1 \cap M_2$ and $x = m_1s_1^{-1} = (m_1ss_2)(ss_2s_1)^{-1}$ is in $(M_1 \cap M_2)_S$.

4.7.9 THEOREM Let $M_R \supset N_R$ be R-modules. If $M_P = N_P$ for all maximal ideals, P, N = M.

Proof: From the exact sequence $0 \to N \to M \to M/N \to 0$ we have the exact sequence $0 \to N_P \to M_P \to (M/N)_P \to 0$ by (4.7.7). If $M/N \neq 0$, $r_R(x) \neq R$ for $x \neq 0$ in M/N. Let P be a maximal ideal containing $r_R(x)$. Then $0 \neq xR_P \subset (M/N)_P$. Hence M = N.

Primary Decomposition

Let M_R be an R-module. If for a submodule N of M, $N = N_1 \cap N_2$ implies that $N = N_1$ or $N = N_2$, then N is called *irreducible*, where the N_1 are submodules of M. This fact is equivalent to M/N being uniform. Assume that R is a commutative ring.

4.7.10 LEMMA Let R be a commutative noetherian ring. Then every ideal A is a finite intersection of irreducible ideals.

Proof: If there exists an ideal A without the above property, we can find a maximal ideal B among such ideals. Then B is not irreducible. Hence there exist C, D such that $B = C \cap D$, $C \supsetneq B$, and $D \supsetneq B$. Then C and D are finite intersections of irreducible ideals by the maximality of B, which is a contradiction.

4.7.11 REMARK The proof in (4.7.10) is valid for any ring T and a noetherian T-module.

4.7.12 LEMMA If R is noetherian, every irreducible ideal is primary.

Proof: Let A be irreducible. We may show that the zero ideal of R/A is primary. Hence assume that A = 0. Let ab = 0 for a, b $\in$ R and a $\neq$ 0. We have a chain $r_R(b) \subset r_R(b^2) \subset \cdots \subset r_R(b^m) \subset \cdots$, and so $r_R(b^n) = r_R(b^{n+1})$ for some n. We shall show that $(a) \cap (b^n) = 0$. Let $x = ya = zb^n$ be in $(a) \cap (b^n)$. Then $xb = yab = 0$. Hence, since $0 = xb = zb^{n+1} = 0$, $zb^n = 0$. Therefore, x = 0. Since 0 is irreducible and $(a) \neq 0$, $b^n = 0$.

4.7.13 LEMMA Let R be a commutative noetherian ring, and $Q_1, Q_2, \ldots, Q_n$ be P-primary. Then $\cap_{i=1}^n Q_i$ is P-primary.

Proof: We may assume that n = 2 and $Q = Q_1 \cap Q_2$. Let ab be in Q with $a \notin Q$. We may assume without loss of generality that $a \notin Q_1$. Then $b \in r(Q_1) = P$. Hence $b^m \in Q_1 \cap Q_2 = Q$ ($P^n \subset Q_1 \cap \cdots \cap Q_n = Q$ for some n).

Let A be an ideal in R. If $A = Q_1 \cap Q_2 \cap \cdots \cap Q_n$ with Q_i P_i-primary, we call this a *primary decomposition*. Further, if $Q_i \not\supset Q_1 \cap Q_2 \cap \overset{i}{\vee} \cap Q_n$ for all i, we call this intersection *irredundant*. If R is noetherian, every ideal A has a primary decomposition by (4.7.10) and (4.7.12). Further, we can take an irredundant primary decomposition of A satisfying $r(Q_i) \neq r(Q_j)$ for $i \neq j$ by (4.7.13).

4.7.14 THEOREM (Uniqueness) Let R be a commutative noetherian ring and A an ideal. Then A has an irredundant primary decomposition $A = Q_1 \cap Q_2 \cap \cdots \cap Q_n$ such that $r(Q_i) \neq r(Q_j)$ for $i \neq j$. The set of prime ideals

$\{r(Q_i)\}$ does not depend on the decomposition.

Proof: We know that A has an irredundant primary decomposition as stated. Let $A = Q_1' \cap Q_2' \cap \cdots \cap Q_{n'}'$ be another decomposition. We note that if $x \notin Q_1$, then $x \notin Q_{1P_1}$, where $P_1 = r(Q_1)$. We assume that $P = r(Q_1) \notin \{r(Q_i')\}_{i=1}^{n'}$. Then $A_P = Q_{1P} \cap Q_{2P} \cap \cdots \cap Q_{nP} = Q_{1P}' \cap Q_{2P}' \cap \cdots \cap Q_{n'P}'$ by (4.7.8). Since either $Q_{iP} = R_P$ or Q_{iP} is $r(Q_i)_P$-primary by (4.7.2), we may assume that R is local and $r(Q_1)$ is the maximal ideal, from the remark above. Since $A = Q_1 \cap Q_2 \cap \cdots \cap Q_n$ is irredundant, we obtain a y in $Q_2 \cap \cdots \cap Q_n - Q_1$ with $yQ_1 \subset A \subset Q_i'$. Since $r(Q_1)$ is the unique maximal and the unique minimal prime ideal containing Q_1, $r(Q_i') \not\supset Q_1$. Hence $Y \in \cap Q_i' = A$, which is a contradiction.

In the following few theorems, we consider rings that are not necessarily commutative. We shall study indecomposable injective modules.

4.7.15 LEMMA Let R be any ring and M_R an R-module. Then the following conditions are equivalent:

1. M is uniform
2. E(M) is indecomposable.

Proof: (1) ⇒ (2). If $E = E(M) = E_1 \oplus E_2$, $E_i \cap M$ is essential in E_i by (3.4.6), so $E_i \cap M \neq 0$. Since $(E_1 \cap M) \cap (E_2 \cap M) = 0$, E_1 or E_2 is equal to zero.

(2) ⇒ (1). If M is not uniform, we obtain two nonzero submodules N_i with $N_1 \cap N_2 = 0$. Then $E \supset E(N_1) \oplus E(N_2)$ by (4.4.10) and $E = E(N_1) \oplus E(N_2) \oplus E'$, which is a contradiction.

4.7.16 COROLLARY Let R be any ring. For a right ideal K of R, K is irreducible if and only if E(R/K) is indecomposable.

Proof: K is irreducible if and only if R/K is uniform, by the definition.

4.7.17 THEOREM Let R be any ring and let $M_R \supset N_R$ be R-modules. If $N = N_1 \cap \cdots \cap N_n$ is an irredundant intersection of irreducible modules N_i, then $E(M/N) \approx \Sigma_{i=1}^{n} \oplus E(M/N_i)$ and $E(M/N_i)$ is indecomposable.

Proof: Let $\varphi : M/N \to \Sigma_{i=1}^{n} \oplus M/N_i$ be the natural monomorphism given by setting $\varphi(m + N) = \Sigma_{i=1}^{n}(m + N_i)$. Since $N = N_1 \cap N_2 \cap \cdots \cap N_n$ is irredundant, there exists $m_i \in (N_1 \cap N_2 \cap \overset{\vee}{\cdots} \cap N_n) - N_i$ for each i. Then $0 \neq \varphi(m_i) \in \varphi(M/N) \cap M/N_i$. Hence since M/N_i is uniform, $\varphi(M/N)$ is essential in $\Sigma_{i=1}^{n} \oplus M/N_i$. Therefore, $E(M) \approx \Sigma_{i=1}^{n} \oplus E(M/M_i)$ by (4.4.10) and (4.4.11), and $E(M/N_i)$ is indecomposable.

4.7.18 THEOREM Let R be a ring. Then the following conditions are equivalent:

1. R is right noetherian.
2. Any direct sum of injective modules is injective.
3. Every injective module is a direct sum of indecomposable R-submodules.

Proof: (1) ⇒ (2). Let I be a right ideal. Let $\{E_\alpha\}_K$ be a set of injective modules and $E = \Sigma_K \oplus E_\alpha$. Take an f in $\mathrm{Hom}_R(I,E)$. Since I is f.g. by (4.2.2), $f(I) \subset \Sigma_{i=1}^n \oplus E_{\alpha_i}$ for some n. Then $\Sigma \oplus E_{\alpha_i}$ is injective by (4.4.1). Hence there exists an e in $\Sigma \oplus E_{\alpha_i} \subset E$ such that $f(x) = ex$ for all $x \in I$ by (4.4.2). Accordingly, E is injective, again by (4.4.2).

(1) ⇒ (3). Let E be an injective module and let $e \neq 0$ be in E. Then $eR \approx R/r_R(e)$. Since $r_R(e)$ is an irredundant intersection of irreducible right ideals by (4.7.11), $E \supset E(eR)$ contains an indecomposable injective submodule by (4.7.17). Let $\underline{F}$ be the set of all submodules which are direct sums of indecomposable injective submodules. Then $\underline{F}$ contains a maximal element N by Zorn's lemma. Since N is injective by the above argument for (1) ⇒ (2), $E = N \oplus E_0$. If $E_0 \neq 0$, E_0 contains an indecomposable injective module by the first remark, which contradicts the maximality of N.

(2) ⇒ (1). Let $I_1 \subset I_2 \subset \cdots \subset I_n \subset \cdots$ be a chain of right ideals. Put $I = \cup_i I_i$ and $E_i = E(R/I_i)$. We define a mapping $g : I \to E' = \Sigma_i \oplus E_i$ as follows. Each element x in I is contained in I_n for almost all n. Hence $x + I_m = 0$ for almost all m. Set $g(x) = \Sigma\, x + I_j$. Then g is well defined. Since E' is injective, there exists e in $\Sigma_{k=1}^m \oplus E_{i_k}$ such that $g(x) = ex$ for all $x \in I$ by (4.4.2). Hence $\mathrm{im}\, g \subset \Sigma_{k=1}^m \oplus E_{i_k}$ and so $I_{i_m+1} = I_{i_m+p}$ for all $p > 0$. Therefore, R is right noetherian.

(3) ⇒ (1). See (6.4.2).

Similarly, we have the following.

4.7.19 THEOREM A ring R is right artinian if and only if every injective module is a direct sum of $E(S_i)$'s, where the S_i are simple R-modules.

Proof: We assume that R is right artinian. Then we know by (4.5.9) that R is right noetherian. Hence every injective module E is a direct sum of indecomposable modules E_α by (4.7.18). Since R is artinian, $S_\alpha = S(E_\alpha)$ is nonzero simple by (4.7.15). Further, $E_\alpha = E(S_\alpha)$. Conversely, let $A_1 \supset A_2 \supset \cdots \supset A_n \supset \cdots$ be a chain of right ideals. Put $A = \cap_i A_i$.

If $A_i/A \neq 0$, $S(A_i/A) \neq 0$, since $S(E(A_i/A)) \neq 0$. Let $E(R/A) = \Sigma_I \oplus E(S_\alpha)$. Since $R/A \subset \Sigma_{i=1}^n \oplus E(S_{\alpha_i})$, I is finite. Hence $S(R/A) \approx \Sigma_{i=1}^n \oplus S_{\alpha_i}$. However, $\cap_i (A_i/A) = 0$, so $A_i/A = 0$ for some i from the above. Therefore, $A_i = A_{i+1} = \cdots$.

Again we return to the special case of commutative rings.

4.7.20 THEOREM Let R be a commutative noetherian ring. Then we have the following:

1. Every indecomposable injective module is of the form $E(R/P)$.
2. If $E(R/P) \approx E(R/P')$, then $P = P'$, where P and P' are prime.
3. Let $E = E(R/P)$ and $A_n = \{e \in E \mid eP^n = 0\} = r_E(P^n)$; then $E = \cup A_n$.

Proof: 1. Let E be an indecomposable injective module. Then $E = E(R/Q)$ as in the proof of (4.7.18), where Q is P-primary for some prime P. Since $P^n \subset Q$, there exists an $\bar{x}$ in R/Q with $r_R(\bar{x}) \supset P$. On the other hand, $r_R(\bar{x}) \subset P$ if $\bar{x} \neq 0$. Hence $r(\bar{x}) = P$. $E \supset \bar{x}R \approx R/P$ and $E = E(R/P)$.

2. If $g : E(R/P) \approx E(R/P')$, $g(R/P) \cap R/P' \neq 0$. Hence there exists an x in R such that $P = r_R(x) = r_R(g(x)) = P'$.

3. Let e be in $E = E(R/P)$. Then $r_R(e)$ is irreducible by (4.7.16), so P'-primary by (4.7.12), where P' is prime. Since $E = E(R/P')$ from (1), $P' = P$. Hence $r_R(e) \supset P^n$ for some n and $e \in A_n$

4.7.21 THEOREM Let R be a commutative noetherian ring and P prime. Then $E = E(R/P)$ is an R_P-injective R_P-module and $\mathrm{Hom}_R(E,E) = \mathrm{Hom}_{R_P}(E,E)$.

Proof: Put $S = R - P$ and let $s \in S$. Then the mapping $f_s : R/P \to R/P$ given by multiplication by s is a monomorphism. Then f_s can be extended to an element F_s in $T = \mathrm{End}_R(E)$. Since R/P is essential in E, F_s is unique and monic. Hence since $F_s(E)$ is injective, $F_s(E) = E$. Thus we can define the multiplication of s^{-1} on E via F_s^{-1}. Now we set $e(r,s) = (er)s^{-1}$ for $(r,s) \in R_P$. We can easily see that this product is well defined and E is an R_P-module. Next we show that E is R_P-injective. Let I be an ideal in R_P and $f : I \to E$. Then for $f\rho : I^c \to E$, we obtain an e in E with $f\rho(r) = er$ $(r \in R)$ by (4.4.2). Hence $f(r') = er'$ for $r' \in I$, since $I = I^{ce}$. It is clear that $\mathrm{Hom}_R(E,E) \supset \mathrm{Hom}_{R_S}(E,E)$. Since $f(es^{-1})s = f(e)$ for $f \in \mathrm{Hom}_R(E,E)$, $e \in E$, and $s \in S$, f is in $\mathrm{Hom}_{R_S}(E,E)$.

4.7.22 COROLLARY Let R and E be as in (4.7.21). Then A_1 is isomorphic to the quotient field K of R/P.

Proof: $A_1 \supset R/P$. We know from the proof above that each element in K induces an automorphism of A_1 (note that $A_1P = 0$). Hence $A_1 \supset K$. Let $a \neq 0$ be in A_1. Since A_1 is uniform, there exists an r in R with $0 \neq ar \in R/P$. Hence $r \notin P$, so $a \in K$.

Since $E = E(R/P)$ is an indecomposable R_P-injective module by (4.7.21), we may assume in the following that R is a local noetherian ring with maximal ideal P and E is an indecomposable R-injective module. From the chain $R = P^0 \supset P \supset P^2 \supset \cdots \supset P^n \supset \cdots$ we obtain the chain $0 \subset A_1 \subset A_2 \subset \cdots \subset A_n = r_E(P^n) \subset \cdots$. Now take the exact sequence

$$0 \to P^{n-1}/P^n \to R/P^n \to R/P^{n-1} \to 0$$

Since E is injective, we obtain the exact sequence

$$0 \to \mathrm{Hom}_R(R/P^{n-1},E) \to \mathrm{Hom}_R(R/P^n,E) \to \mathrm{Hom}_R(P^{n-1}/P^n,E) \to 0$$

On the other hand, for any (right) ideal A and a (right) R-module M, we have the exact sequence

$$\mathrm{Hom}_R(A,M) \leftarrow \mathrm{Hom}_R(R,M) = M \leftarrow \mathrm{Hom}_R(R/A,M) \leftarrow 0$$

Hence we have $r_M(A) \approx \mathrm{Hom}_R(R/A,M)$ in general. Thus there is an exact commutative diagram

$$\begin{array}{ccccccc} 0 \longrightarrow & \mathrm{Hom}_R(R/P^{n-1},E) & \longrightarrow & \mathrm{Hom}_R(R/P^n,E) & \longrightarrow & \mathrm{Hom}_R(P^{n-1}/P^n,E) \\ & \wr\wr & & \wr\wr & & \downarrow \\ 0 \longrightarrow & A_{n-1} & \longrightarrow & A_n & \longrightarrow & A_n/A_{n-1} \end{array}$$

Hence $\mathrm{Hom}_R(P^{n-1}/P^n,E) \approx A_n/A_{n-1}$ by (2.5.2). On the other hand, since $(P^{n-1}/P^n)P = 0$, $\mathrm{Hom}_R(P^{n-1}/P^n,E) = \mathrm{Hom}_R(P^{n-1}/P^n,A_1) = \mathrm{Hom}_R(P^{n-1}/P^n, R/P)$ by (4.7.22). Hence, noting that R/P is a field, we obtain the following.

4.7.23 THEOREM Let R be a commutative local ring with maximal ideal P and $E = E(R/P)$. Then $A_n/A_{n-1} \approx P^{n-1}/P^n$ as R/P-modules, and $r_R(A_n) = P^n$.

Proof: Take r not in P^n. Then $r + P^n/P^n \neq 0$ ($\in R/P^n$). Thus there exists an f in $\mathrm{Hom}_R(R/P^n,E)$ such that $f(r + P^n/P^n) \neq 0$ for $E = E(R/P)$ and $rR + P^n/P^n$ contains a simple submodule isomorphic to R/P (note that R/P^n is artinian). Hence $A_n \approx \mathrm{Hom}_R(R/P^n,E)$ implies that $r_R(A_n) = P^n$.

We note that $\mathrm{Hom}_R(A_n/A_{n-1},E) = \mathrm{Hom}_R(A_n/A_{n-1},A_1)$ and $\mathrm{Hom}_R(A_n/A_{n-1},R/P) \approx P^{n-1}/P^n$ by (4.7.23). Indeed, we have the last isomorphism as follows.

Let $r \in P^{n-1}$. Then $rA_n \subset A_1$ and $rA_{n-1} = 0$. Since $r_R(A_n) = P^n$, $r \to (x + A_{n-1} \to rx)$ gives us a monomorphism from P^{n-1}/P^n to $\mathrm{Hom}_R(A_n/A_{n-1}, R/P)$. Hence it is an isomorphism by (4.7.23).

Finally, we shall add here a useful lemma for noetherian rings.

4.7.24 LEMMA Let R be a commutative (noetherian) ring and A an ideal of R. If a f.g. R-module M satisfies M = MA, then $R = A + r_R(M)$.

Proof: Let $M = \Sigma_{i=1}^n m_i R$. M = MA implies that $m_i = \Sigma\, m_j a_{ji}$; $a_{ji} \in A$ for i = 1, 2, ..., n. We consider the matrix (a_{ij}). Then

$$\begin{pmatrix} m_1 \\ m_2 \\ \vdots \\ m_n \end{pmatrix} [I - (a_{ij})] = 0$$

where I is the identity matrix. Therefore, det $(I - (a_{ij})) \subset r_R(M)$ by a well-known theorem for matrices. It is clear that det $(I - (a_{ij})) = 1 + a'$; $a' \in A$.

Dedekind Domains

We shall study a special but important class of rings called Dedekind domains. Let R be a commutative ring. Consider a chain of prime ideals $R \supsetneq P_0 \supsetneq P_1 \supsetneq \cdots \supsetneq P_n$ with length n. The largest length among all such chains is called the *Krull dimension* of R and is denoted by Krull dim R. We denote the set of all prime ideals in R by spec(R).

4.7.25 PROPOSITION Any commutative ring contains a minimal prime ideal.

Proof: Let $P_1 \supset P_2 \supset \cdots \supset P_n \supset \cdots \supset P_\alpha \supset \cdots$ be a chain of prime ideals. Put $\cap P_\alpha = P$. If $ab \in P$ and $a \notin P$, then there is an α with $a \notin P_\beta$ for all $\beta \geq \alpha$. Thus $b \in P_\beta$ for all $\beta \geq \alpha$. Therefore, P is prime. If we define a partial order in spec(R) by $P_1 \geq P_2 \leftrightarrow P_1 \subset P_2$, we can use Zorn's lemma to give the result.

4.7.26 THEOREM Let R be a commutative artinian ring. Then Krull dim R = 0 and R is a direct sum of artinian local rings R_i with nilpotent maximal ideal.

Proof: Since R is noetherian by (4.5.9), $0 = Q_1 \cap Q_2 \cap \cdots \cap Q_n$ is an irredundant intersection of P_i-primary ideals Q_i by (4.7.14). Since R/P_i is artinian simple, P_i is maximal. Further, $P_i^{n_i} \subset Q_i$ for some n_i, so $\{Q_i\}$ are mutually relatively prime. Hence $R \approx \Sigma \oplus R/Q_i$ by (3.2.5). R/Q_i contains a unique maximal ideal P_i/Q_i and $(P_i/Q_i)^{n_i} = 0$.

We now assume that R is an integral domain and K is the quotient field of R. For any ideal I in R we put $I^{-1} = \{x \in K \mid xI \subset R\}$. Then I^{-1} is an R-module in K and $I^{-1}I = II^{-1}$ is an ideal in R. Hence I^{-1} is f.g. if R is noetherian. If $II^{-1} = R$, I is called *invertible.*

4.7.27 THEOREM Let R be a noetherian integral domain. Then the following conditions are equivalent:

1. Every ideal in R is projective.
2. Every ideal in R is invertible.
3. Every ideal in R is expressible uniquely as a product of prime ideals and Krull dim $R \leq 1$.

Proof: First we note that $I^{-1} = \mathrm{Hom}_R(I,R)$ for an ideal I in R. It is clear that $I^{-1} \subset \mathrm{Hom}_R(I,R)$ via the multiplication of elements in I^{-1}. Let f be in $\mathrm{Hom}_R(I,R)$. We define $f(xy^{-1}) = f(x)y^{-1} \in K$ for any $z = xy^{-1} \in I$; $0 \neq y \in R$. Then we can see that $f \in \mathrm{Hom}_R(IK,K) = \mathrm{Hom}_K(K,K) = K$. Hence f is given by an element in I^{-1}.

(1) ⇒ (2). Let I be an ideal in R. Since I is f.g., $\mu(I^{-1} \otimes I)$ contains $1_I = 1_R$ by (4.3.6). Hence $I^{-1}I$ contains 1_R.

(2) ⇒ (1). This is similar to the above by (4.3.6) and the initial observation.

(2) ⇒ (3). We assume that $R \neq K$. Let I be a nonmaximal ideal in R and let P_1 be a maximal ideal containing I. Then $R = P_1^{-1}P_1 \supset P_1^{-1}I \supsetneq I$. If $P_1^{-1}I$ is not maximal, we have a maximal ideal P_2 containing $P_1^{-1}I$. Hence $R \supset P_2^{-1}P_1^{-1}I \supset P_1^{-1}I$. Since R is noetherian, we finally obtain a maximal ideal $P_n = P_{n-1}^{-1}\cdots P_1^{-1}I$. Hence $I = P_1P_2\cdots P_n$. Therefore, every prime ideal ideal $\neq 0$ is maximal, so Krull dim R = 1. The uniqueness is clear since P_i is prime.

(3) ⇒ (2). Let M be a maximal ideal. We shall show that $(I^{-1})_M = (I_M)^{-1}$ for any ideal I. It is clear that $I_M^{-1} \subset (I_M)^{-1}$. Let x be in $(I_M)^{-1}$. Then $xI_M \subset R_M$. Let $I = \langle y_1, y_2, \ldots, y_t \rangle$. Then there exists an s in R - M with $xy_is \in R$ for all i. Hence $xs \in I^{-1}$ and $x \in (I^{-1})_M$. We put $A = MM^{-1}$. Then $A_N = M_N(M^{-1})_N = M_N(M_N^{-1})$ for every maximal ideal N. If $N \neq M$, $A_N = R_NR_N^{-1} = R_N$. We shall show $A_M = R_M$. We may assume that R is a local ring with maximal ideal M and satisfies (3) by (4.7.2). Since $M^2 \neq M$ by (3), we obtain $m \in M - M^2$. Then $\langle m \rangle = M^t$ by (3) and hence $M = \langle m \rangle$. Therefore, $M^{-1} = \langle m^{-1} \rangle$ and $A = MM^{-1} = R$. Accordingly, in the general situation, $MM^{-1} = R$ by (4.7.9). Therefore, every ideal is invertible by (3).

If R satisfies one of the conditions above, we call R a *Dedekind domain*.

4.7.28 COROLLARY Let R be a commutative noetherian domain. Then the following conditions are equivalent:

1. R is a Dedekind domain.
2. R_P is a Dedekind domain.
3. R_P is a principal ideal domain (every ideal is cyclic).

Here P runs through all prime ideals.

Proof: (1) ⇒ (2), (3). We obtain this in the above.

(3) ⇒ (2). Since R_P (= R) is principal, every prime is maximal and is equal to P. Let I be an ideal. Then $I \supset P^n$ for some n by (4.7.14), provided that $I \neq 0$. Hence $I = P^m$ for some m and is invertible.

(2) ⇒ (1). The proof of (4.7.27) showed that every ideal is invertible assuming (2).

Let R be a Dedekind domain and M a torsion R-module. Put M(P) = $\{m \in M \mid r_R(m) = P^n$ for some n$\}$, where P is a nonzero prime. We call M(P) the P-*primary component* of M.

4.7.29 THEOREM Let R be a Dedekind domain and let M be a torsion R-module. Then $M = \Sigma \oplus M(P_i)$, where P_i runs through all nonzero primes.

Proof: Let P_1, P_2 be distinct primes. Then $(P_1,P_2) = 1$. Hence $(P_1^{n_1},P_2^{n_2}) = 1$ for any $n_i > 0$ by (3.2.2). Let m be in $M(P_1) \cap M(P_2)$ and $mP_1^{n_1} = mP_2^{n_2} = 0$. We take x_i in $P_i^{n_1}$ with $1 = x_1 + x_2$. Then m = ml = $mx_1 + mx_2 = 0$. Using the same argument, we know that $\{M(P_i)\}$ is independent. Let m be in M and $r_R(m) = \cap_{i-1}^{n} Q_i$, where the Q_i are P_i-primary. Put $B_i = Q_1 \cap Q_2 \cap \cdot\overset{i}{\vee}\cdot \cap Q_n$. Then $R = \Sigma B_i$ by (3.2.4), since $Q_i \supset P_i^{n_i}$. Let $1 = \Sigma x_i$; $x_i \in B_i$. Then $m = \Sigma mx_i$ and $mx_iP_i^{n_i} \subset mr_R(m) = 0$; i.e., $mx_i \in M(P_i)$.

4.7.30 THEOREM Let R be a Dedekind domain and M_R an R-module. If M is divisible, M is injective.

Proof: Let P be a prime ideal. Then M_P is R_P-divisible and R_P is a principal ideal domain by (4.7.28). Hence M_P is R_P-injective by the proof of (4.4.6). Let I be an ideal of R (i : I → R). Consider $\mathrm{Hom}_R(i,M)$: $\mathrm{Hom}_R(R,M) \to \mathrm{Hom}_R(I,M)$. Since M_P is R_P-injective, $\mathrm{Hom}(i,M)_P$ is an isomorphism by (4.4.2), (4.3.18), and (4.7.7). Hence Hom(i,M) is epic by (4.7.9). Therefore, M is injective by (4.4.2).

4.7.31 COROLLARY Let R be as above and M an R-module. Then M contains a unique maximal injective module M_0, say $M = M_0 \oplus M_1$, and M_1 contains no nonzero divisible submodules.

Proof: The union of divisible submodules which are linearly ordered by inclusion is divisible. Hence M contains a maximal divisible submodule M_0. Then M_0 is injective by (4.7.30), so $M = M_0 \oplus M_1$. The remaining assertions are clear from the decomposition above.

We say that M_1 is a *reduced submodule* of M.

We shall study the dual of (4.7.30).

4.7.32 LEMMA Let R be a Dedekind domain. Then every ideal in R is projective.

Proof: Let A be an ideal in R. Then A is finitely presented. So $\mathrm{Hom}_R(A,-) \otimes_R R_P = \mathrm{Hom}_{R_P}(A_P, - \otimes_R R_P)$ for every prime P by (4.3.18) and (4.7.7). Since $R_P \approx A_P$ by (4.7.28), $\mathrm{Hom}_R(A,-)$ is an exact functor by (4.7.9) and (3.6.1).

4.7.33 THEOREM Let R be a Dedekind domain. Then every submodule of a free module is isomorphic to a direct sum of ideals in R and so is projective.

Proof: Let $F = \Sigma_I \oplus R_\alpha$, where $R_\alpha \approx R$ is a free R-module. We assume that I is a well-ordered set. We denote $\Sigma_{\beta<\alpha} \oplus R_\beta$ (resp. $\Sigma_{\beta\le\alpha} \oplus R_\beta$) by F_α (resp. $\bar{F}_\alpha$). Let A be an R-submodule of F. We have the projection p_α : $A \cap \bar{F}_\alpha \to R_\alpha$. Since im p_α is projective by (4.7.32), p_α splits, namely, $A \cap \bar{F}_\alpha = B_\alpha \oplus (A \cap F_\alpha)$ and B_α is isomorphic to an ideal in R. We shall show that $A = \Sigma B_\alpha = \Sigma \oplus B_\alpha$. Let $\Sigma x_{\alpha_i} = 0$ for $x_{\alpha_i} \in B_{\alpha_i}$ ($\alpha_1 < \alpha_2 < \cdots < \alpha_n$). Then $x_{\alpha_n} = -\Sigma_{i=1}^{n-1} x_{\alpha_i} \in F_{\alpha_n} \cap B_{\alpha_n} = 0$. Hence we have inductively $x_{\alpha_i} = 0$ for all i. Therefore, $\{B_\alpha\}$ is independent. Finally, we assume that $A \supsetneq \Sigma \oplus B_\alpha$. Let a' be in $A - \Sigma \oplus B_\alpha$. Then there exists β' such that $a' \in A \cap \bar{F}_{\beta'}$. We can take the minimal β among the β' when a' runs through all elements in $A - \Sigma \oplus B_\alpha$. Assume that $a \in \bar{F}_\beta \cap A$ and $a \notin \Sigma \oplus B_\alpha$. Let $a = a_1 + a_2$, where a_1 is in $A \cap F_\beta$ and a_2 in B_β. Then $a_1 \notin \Sigma \oplus B_\alpha$ and $a_1 \in A \cap F_{\beta''}$ $(\beta'' < \beta)$, which is a contradiction.

4.7.34 LEMMA Let R be a domain and M an R-module. Let K be the quotient field of R and i : R → K the inclusion. Then the kernel of $1_M \otimes i$: $M \otimes_R R \to M \otimes_R K$ is equal to t(M).

Proof: K is the direct limit of $\{a^{-1}R\}_a$ by (2.4.4). Now, consider the mapping

$$i_a - 1 : M \to M \otimes_R a^{-1}R;\ (m \to m \otimes 1)$$

and

$$M \otimes_R a^{-1}R \to M;\ (m \otimes a^{-1}r \to mr)$$

We can show easily, using these homomorphisms, that $\ker (i_a - 1) = r_M(a)$. Hence $\ker i \otimes 1_M = \cup \ker (i_a - 1) = t(M)$ by (3.5.6).

4.7.35 THEOREM Let R be a Dedekind domain. Then every f.g. torsion-free R-module is projective.

Proof: Let P be a f.g. torsion-free module. Then $P \subset P \otimes_R Q$, where Q is the quotient field of R by (4.7.34). Since P is f.g., $P \otimes_R Q$ is of finite dimension over Q. Let $P \otimes_R Q = \Sigma_{i=1}^{t} \oplus u_iQ$; $u_i \in P \otimes_R Q$. If $P = \langle p_1, p_2, \ldots, p_s \rangle$, $p_i = \Sigma_{j=1}^{t} u_j q_{ji}$ $(i = 1, \ldots, s)$. Then there exists an r in R such that $rq_{ji} \in R$ for all j. Then $Pr \subseteq \Sigma \oplus u_iR$, so $Pr \approx P$ is projective by (4.7.33).

4.7.36 COROLLARY Let R be as above and M a f.g. R-module. Then $M = t(M) \oplus M_0$, where M_0 is a torsion-free R-submodule.

Proof: Since M/t(M) is f.g. torsion free, M/t(M) is projective by (4.7.35). Hence the natural epimorphism: $M \to M/t(M)$ splits.

Quotient Rings and Large Extensions

We shall introduce a new concept adapted from noncommutative ring theory.

Let R, T be commutative rings and let $f : R \to T$ be a ring homomorphism. Let M be a T-module. Then M is regarded as an R-module as follows:

$$mr = mf(r), \quad \text{where m is in M and r in R} \tag{4.7.3}$$

In particular, T is an R-module. If f(R) is a small submodule in T as an R-module, we say that T is a *large extension* of R and f is a *small homomorphism.* For example, let Z be the ring of integers and Q the quotient field (rationals). We take the inclusion mapping for f. Let $Q = Z + N$ for a submodule N. Put $N \cap Z = (n) \neq 0$. Then $n^{-1} = x + y$; $x \in Z$, $y \in N$. Hence $1 = nx + ny \in (n) + N = N$, so $N = Q$. Therefore, Q is a large extension of Z. Any ring containing a large extension is itself a large extension. We shall define the smallest large extension of R. Let $\tilde{R}$ be a ring and $f : R \to \tilde{R} \neq 0$ a ring homomorphism. If $\tilde{R}$ and f satisfy the following:

1. $\tilde{R}$ is a large extension of R via f.
2. If $R \xrightarrow{h} T$ is a large extension of R, then there exists a unique homomorphism $g : \tilde{R} \to T$ such that $h = gf$. (4.7.4)

then we say that $R \xrightarrow{f} \hat{R}$ is a *small hull* of R.

In the above, if $\tilde{R}$ and T are always local rings, we say that $\tilde{R}$ is a *local small hull* of R. It is clear that if R has small hulls, they are isomorphic to one another. In general, R does not have a small hull. We know from below that only special rings have small hulls.

Let R and R' be rings and $f : R \to R'$ a ring homomorphism. If P' is prime in R', $R/f^{-1}(P')$ is isomorphic to a subring of R'/P'. Hence $f^{-1}(P')$ is prime in R. From this correspondence we have a mapping $f_* : \operatorname{spec}(R') \to \operatorname{spec}(R)$.

4.7.37 THEOREM Let R and R' be commutative rings and $f : R \to R'$ a ring homomorphism. Then the following conditions are equivalent:

1. f is small.
2. $f_*(M')$ is not maximal in R for any maximal ideal M' in R'.

Proof: (1) ⇒ (2). Let M' be a maximal ideal in R' and $M = f_*(M')$. If M is maximal, R/M is a subfield of R'/M'. f is not epic, so there exists an R-submodule L in R' such that $R' \supsetneq L \supset M'$, $R' = L + R$, which is a contradiction [where we denote f(R) by R].

(2) ⇒ (1). Let M be a maximal ideal in R. If $R'M \neq R'$, we can take a maximal ideal $M' \supset R'M$ in R'. Then $M = f_*(M')$. Hence $R'M = R'$ by (2). Now assume that $R' = R + T$ for an R-module T in R'. Then $R'_M = R'_M M_M = (R_M + T_M)M_M = R_M M_M + T_M$. Since R'_M/T_M is a cyclic R_M-module, $R'_M = T_M$ for any maximal ideal M in R by (3.3.12). Hence $R' = T$ by (4.7.9).

4.7.38 COROLLARY Let P be prime. Then $\rho_P : R \to R_P$ is a small homomorphism if and only if P is not maximal.

Proof: This is clear, since $P^{ec} = P$.

Let S be a multiplicatively closed set in R. If $\rho_S : R \to R_S$ is a large extension, we call S *large*.

4.7.39 LEMMA Let S be a multiplicatively closed set in R and let P be maximal among ideals A with $A \cap S = \emptyset$. Then P is prime.

Proof: Let $ab \in P$. Then $(aR + P)(bR + P) \subset P$. If $a \notin P$, $b \notin P$, $(aR + P) \cap S \neq \emptyset$ and $(bR + P) \cap S \neq \emptyset$, which is a contradiction.

4.7.40 THEOREM Let R be a commutative ring and let S, S' be multiplicatively closed sets in R. Then there is a one-to-one correspondence φ between the set of maximal ideals M in R and the set of S such that S' is not large for any $S' \subset S$ and S'' is large for any $S'' \supsetneq S$, given by $\varphi(M) = R - S$.

Proof: Let $S = R - M$. Then R_S is not a large extension of R and hence $R_{S'}$ is not large for $S' \subset S$ by (3.4.1), since we have mappings

$R \to R_{S'} \to R_S$. On the other hand, if $S \subsetneq S''$, $M'^c \subsetneq M$ for any maximal ideal M' in $R_{S''}$. Hence $R_{S''}$ is a large extension. Conversely, we assume that S has the above property. Then there exists a prime ideal P in R with $P \cap S = \emptyset$ as in (4.7.39). Since we have mappings $R \to R_S \to R_P$, P is maximal by (4.7.37). Set $S' = R - P$. Then $S' \supset S$. Hence since $R_{S'} = R_P$ is not large, $S' = S$.

Finally, we consider the existence of small hulls.

4.7.41 THEOREM Let R be a commutative ring. Then R has a local small hull if and only if there exists a unique maximal P among nonmaximal prime ideals in R. In this case $\rho_P : R \to R_P$ is a local small hull and also a small hull.

Proof: Assume that there is a unique nonmaximal prime ideal P as in (4.7.41). Then $\rho_P : R \to R_P$ is small by (4.7.37). Let R' be a local extension ring with maximal ideal M' and $g : R \to R'$ a small homomorphism. Since $p = R \cap M'$ is not maximal by (4.7.37), $p \subset P$. Therefore, there exists a unique homomorphism $h : R_P \to R'$ such that $h\rho_P = g$ by (4.7.1), since $R' = R'_{M'}$. Hence R_P is a local small hull of R. Conversely, assume that $f : \tilde{R} \to R$ is a local small hull. Let p be a nonmaximal prime ideal in R. Then $\rho_p : R \to R_p$ is small. Hence there exists a unique homomorphism $h : \tilde{R} \to R_p$ such that $\rho_p = hf$. Put $R_0 = h(\tilde{R})$ and $M_0 = h(\tilde{M})$, where $\tilde{M}$ is the maximal ideal of $\tilde{R}$. Since f is small, $P = f^{-1}(M)$ is not maximal by (4.7.37). Furthermore, $p = \rho_p^{-1}(pR_p) = (hf)^{-1}(pR_p) = (hf)^{-1}(pR_p \cap R_0) \subset (hf)^{-1}(M_0) = f^{-1}(\tilde{M}) = P$. We shall show that R_P is also a small hull of R. Let $g : R \to R'$ be a small homomorphism and $s \in R - P$. If $g(s)R' \neq R'$, we can take a maximal ideal M' in R' containing $g(s)R'$. Then $s \in R \cap M'$. Since g is small, $R \cap M' \subset P$, which is a contradiction. Hence $g(s)R' = R'$; i.e., g(s) is a unit in R'. Therefore, we obtain a unique homomorphism $h : R_P \to R'$ such that $h\rho_P = g$.

The following theorem shows that the existence of a small hull does not guarantee the existence of a local small hull. We need some results for the theorem which are beyond this section. Hence we shall only state the theorem (cf. [102]).

4.7.42 THEOREM Let R be a commutative noetherian ring. Then R has a small hull if and only if Krull dim R = 1. In this case R_S is a small hull of R, where $S = R - \bigcup_i p_i$ and $\{p_i\}_i$ is the set of minimal, but not maximal, prime ideals.

5

Morita Contexts

When we study algebras of finite dimension over a field, or finite abelian groups, we use the concepts of equivalence and duality. However, in the past they could not be treated systematically. Since 1958, when K. Morita and G. Azumaya succeeded in finding methods to deal with the problem, we have been able to work with them explicitly. The Morita-Azumaya method provides one of the starting points in the present-day study of algebra. Because it is such a useful and powerful tool, we freely make use of the ideas. Therefore, we shall summarize the Morita-Azumaya techniques. Beginning in this chapter we embark upon a study of some special but very important branches of ring theory. We study Morita contexts in Sec. 5.1 as a generalization of Sec. 4.3, which provides a foundation for Morita equivalence. In order to obtain Morita equivalence, we study generators in the category of modules in Sec. 5.2. In Sec. 5.3 we deal with equivalence, which is presently one of the important tools in the study of rings. As the dual concept, we study duality. First we explore quasi-injective modules in Sec. 5.4. We investigate the structure of their endomorphism rings in Sec. 5.5, which contains useful results for obtaining duality for artinian rings. Here we introduce the concept of completion of commutative noetherian rings which we shall use in Chap. 9. In Sec. 5.6 we study duality with the aid of results in Secs. 5.5 and 5.6. Finally, we deal with quasi-Frobenius rings and some generalizations of them in Secs. 5.7 and 5.8, respectively.

5.1 MORITA CONTEXTS: GENERAL

Let R be a ring with identity and M a right R-module. Put $S = \mathrm{End}_R(M)$. Then we may regard M as a left S- and right R-bimodule. Thus $\mathrm{Hom}_R(M,R)$ is a left R- and right S-bimodule.

We defined mappings in (4.3.2),

$$\mu_M : M \otimes_R \mathrm{Homr}_R(M,R) \to S; \quad ([m,f])$$
$$\tau_M : \mathrm{Hom}_R(M,R) \otimes_S M \to R; \quad ((f,m))$$

and obtained (4.3.4),

$$[m,f]m' = m(f,m') \qquad (f,m)f' = f[m,f']$$

The above situation may be generalized. Let R and S be rings. Let M be a right R- and left S-bimodule and let N be a right S- and left R-bimodule. If there exist a two-sided S-homomorphism φ and a two-sided R-homomorphism ψ as follows:

$$\varphi : M \otimes_R N \to S; \quad \varphi(m,n) = [m,n] \tag{5.1.1}$$
$$\psi : N \otimes_S M \to R; \quad \psi(n,m) = (n,m) \tag{5.1.2}$$

satisfying

$$[m,n]m' = m(n,m') \quad \text{and} \quad (n,m)n' = n[m,n'] \tag{5.1.3}$$

we say that (R,S,M,N,φ,ψ) is a *Morita context.* $(R,S = \mathrm{End}_R(M), M, \mathrm{Hom}_R(M,R), \mu_M, \tau_M)$ is one example of a Morita context. We shall study Morita contexts, where M is a f.g. projective R-module and a generator in $\underline{M}_R$, and $N = \mathrm{Hom}_R(M,R)$. In the above, im τ_M is a two-sided ideal in R. We call it the *trace ideal* of M and denote it by $\tau_M(M)$.

5.2 GENERATORS

Let R be a ring with identity. By $\underline{M}_R$ we shall denote the category of all right R-modules. For a right R-module P put $S = \mathrm{End}_R(P)$. Then P may be regarded as a left S- and right R-bimodule. We are going to seek some relationships between $\underline{M}_R$ and $\underline{M}_S$. Since we are studying the category $\underline{M}_R$, we shall denote $\mathrm{Hom}_R(M,N)$ by $[M,N]$ for M and N in $\underline{M}_R$.

The functor $[M,-]$ is a covariant left exact functor from $\underline{M}_R$ to $\underline{\mathrm{Ab}}$ (cf. Sec. 3.6) and we denote it by h^M. Similarly, we shall denote the contravariant functor $[-,M]$ by h_M. Furthermore, h^M is an exact functor if and only if M is projective, and h_N is exact if and only if N is injective (Secs. 4.3 and 4.4).

We defined a generator in a category $\underline{A}$ in Sec. 2.4. The following gives a characterization for generators in $\underline{M}_R$.

5.2.1 THEOREM Let U be an R-module. Then the following conditions are equivalent:

1. U is a generator in $\underline{M}_R$.
2. Every R-module is a homomorphic image of a direct sum $U^{(I)}$ of I copies of U.
3. R is isomorphic to a direct summand of a (finite) direct sum of copies of U.

Proof: (1) $\Leftrightarrow$ (2) is clear from (2.4.4).

(2) $\Rightarrow$ (3). Since R is R-projective, this is clear.

(3) $\Rightarrow$ (1). Assume that $F = U^{(I)} = R \oplus Q$. Let M be a nonzero R-module. Then $h^R(M) \neq 0$. Hence $h^U(M) \neq 0$ by the assumption, so U is a generator.

Let $S = [U,U]$ for U in $\underline{M}_R$. Then h^U is a covariant functor from $\underline{M}_R$ to $\underline{M}_S$ by Sec. 3.6.

5.2.2 THEOREM Let U, S be as before. Then the following conditions are equivalent:

1. U is a generator in $\underline{M}_R$.
2. h^U is a faithful and full functor from $\underline{M}_R$ to $\underline{M}_S$.

Proof: (2) $\Rightarrow$ (1) is clear from the definition.

(1) $\Rightarrow$ (2). h^U is faithful by the definition. Let f be in $[h^U(M), h^U(N)]_S$. Consider the epimorphism $\alpha : \Sigma_{\lambda \in [U,M]} U_\lambda \to M$ as in the proof of (2.4.4). Since $\lambda \in h^U(M)$, we have $f(\lambda) \in h^U(N) = [U,N]$. There is a unique R-homomorphism $\beta : \Sigma \oplus U_\lambda \to N$ obtained from $\{f(\lambda)\}$. If we can find $\gamma \in [M,N]_R$ giving a commutative diagram

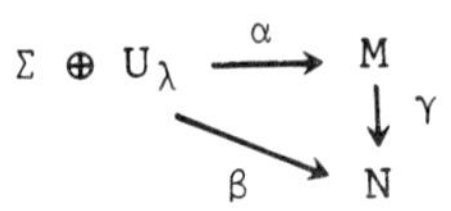

then $\gamma\lambda = f(\lambda)$ for all λ. Hence $h^U(\gamma) = f$. Put $K = \ker \alpha$. In order to show the existence of γ, it is sufficient to show that $\beta | K = 0$. Since $K = \cup_{I'} \ker (\alpha \mid \Sigma_{I'} \oplus U_{\lambda'})$, where I' runs through all finite subsets of $[U,M]$, we need only show that $\beta | K' = 0$, where $K' = \ker (\alpha \mid \Sigma_{I'} \oplus U_{\lambda'})$ and $K' \xrightarrow{i} \Sigma_{I'} \oplus U_{\lambda'}$. Let g be an element in $[U,K']$. Then $ig \in [U, \Sigma_{I'} \oplus U_{\lambda'}] = \Sigma_{I'} \oplus [U,U] = \Sigma_{I'} \oplus S$. Let $ig = \Sigma_{I'} s_i$; the s_i are in S. Since $0 = \alpha ig = \Sigma_{I'} \alpha s_i = \Sigma_{I'} \lambda_i s_i$, $0 = \Sigma_{I'} f(\lambda_i s_i) = \Sigma_{I'} f(\lambda_i) s_i$. Hence $\beta ig = \Sigma_{I'} \beta s_i = \Sigma_{I'} f(\lambda_i) s_i = 0$. Therefore, $\beta(K') = 0$, since U is a generator.

We now give a relation between a generator and its trace ideal.

5.2.3 THEOREM Let U be a right R-module. Then U is a generator in $\underline{M}_R$ if and only if $\tau(U) = R$.

Proof: If U is a generator, R is a direct summand of $F = \Sigma \oplus U$ by (5.2.1). Let $p : F \to R$ be the projection. Then $1_R = p(1_R) = p(\Sigma x_i) = \Sigma p(x_i)$, where the x_i are in U. Since $p|U_i$ is in [U,R], 1_R is in $\tau(U)$. Conversely, if $\tau(U) = R$, $1_R = \Sigma_{I'} f_i(y_i)$, where the f_i are in [U,R], and the y_i in U. Put $F' = \Sigma_{I'} \oplus U$. Then we obtain a unique homomorphism $f : F' \to R$ from $\{f_i\}$. Since $f(\Sigma y_i) = \Sigma f_i(y_i) = 1_R$, f is an epimorphism. Hence U is a generator by (5.2.1).

5.2.4 COROLLARY Let P_R be an R-module and $S = [P,P]_R$. Then ${}_SP_R$ is f.g. S-projective if and only if P_R is a generator. ${}_SP$ is a generator if and only if P_R is f.g. R-projective.

Proof: Assume that P_R is a generator. Then $1_R = \Sigma (q_i, p_i)$, where the q_i are in [P,R] and the p_i in P. Hence $p = \Sigma p(q_i, p_i) = \Sigma [p, q_i] p_i$ and $[p, q_i] \in [P,S]$. Therefore, ${}_SP$ is f.g. S-projective by (4.3.4). Similarly, we can prove the remaining assertions.

5.2.5 COROLLARY Let R be a commutative ring and P a f.g. projective module. Then P is faithful if and only if P is a generator.

Proof: Since P is projective, $P = P\tau(P)$ by (4.3.5). Since R is commutative and P is f.g., $R = r_R(P) + \tau(P)$ by (4.7.24). Therefore, $r_R(P) = 0$ if and only if $\tau(P) = R$ [note that $0 = \tau(Pr_R(P)) = \tau(P)r_R(P)$].

5.3 EQUIVALENCES

Let R be a ring as before. Let M_R be an R-module and $S = \mathrm{End}_R(M)$. We shall study the equivalence of $\underline{M}_R$ and $\underline{M}_S$.

5.3.1 THEOREM Let P_R be an R-module and $Q = [P,R]$, $S = [P,P]_R$, and $T = [P,P]_S$. Then

1. The diagram

$$\begin{array}{ccc} Q \otimes_S P & \xrightarrow{\theta \otimes 1} & [P,S] \otimes_S P \\ \tau_{P_R}\downarrow & & \downarrow \mu_{{}_SP} \\ R & \xrightarrow{i} & T \end{array}$$

is commutative, where $\theta(q)(p) = [p,q]$ and i is the natural mapping.

2. $\tau(P) r_R(P) = 0$.

3. If $R = r_R(P) + \tau(P)$, P is a f.g. projective S-module and $R = r_R(P) \oplus \tau(P)$.
4. If P is a generator, $r_R(P) = 0$ and $T = R$.

Furthermore, θ is a monomorphism, all maps in (1) are isomorphisms, and Q is f.g. S-projective.

Proof: 1. $(\mu_P(\theta \otimes 1))(q \otimes p)(p') = (\mu_P(\theta(q) \otimes p))(p') = (\theta(q)(p'))p = [p',q]p$. On the other hand, $(i\tau_P(q \otimes p))(p') = p'(q,p)$. Hence the diagram is commutative by (5.1.3).

2. Let x be in $r_R(P)$ and (q,p) in $\tau(P)$. Then $(q,p)x = q(px) = 0$.

3. Assume that $R = r_R(P) + \tau(P)$. Let $1_R = y + z$, where y is in $r_R(P)$ and z in $\tau(P)$. Then $z = \Sigma\ (q_i,p_i)$, $q_i \in Q$ and $p_i \in P$, so $i\tau(\Sigma\ q_i \otimes p_i) = 1_P$. Since $Q \otimes_S P$ is a right T-module, $\alpha : T \to Q \otimes P$ given by $\alpha(t) = (\Sigma\ q_i \otimes p_i)t$ is a T-homomorphism. First we note that $i\tau\alpha = \mu(\theta \otimes 1)\alpha = 1_T$, since $(i\tau\alpha(t))p = i\tau(\Sigma\ q_i \otimes p_it)(p) = \Sigma\ p(q_i,p_it) = \Sigma\ [p,q_i](p_it) = (\Sigma\ [p,q_i]p_i)t = i\tau(\Sigma\ q_i \otimes p_i)(p)t = pt$. Therefore, μ_P is an epimorphism, P is a f.g. projective S-module, so μ_P is an isomorphism by (4.3.6). Since $r_R(P)$ and $\tau(P)$ are two-sided ideals in R, $R = r_R(P) \oplus \tau(P)$ from (1).

4. If P is a generator, then $r_R(P) = 0$ from (2) and (5.2.3). Since i is an epimorphism from the proof of (3), i is an isomorphism. On the other hand, if $\theta(q) = 0$, $[P,q] = 0$. Hence $Pq(p') = 0$ for all $p' \in P$. Since $r_R(P) = 0$, $q(p) = 0$. Therefore, θ is a monomorphism. Since ${}_SP$ is S-projective, $\theta \otimes 1$ is a monomorphism. Therefore, all maps in (1) are isomorphisms. Now, $T = R$, so $Q = [P,R] = [P,[P,P]_S]_R$. Since P is f.g. S-projective, Q is a direct summand of a free S-module $[P,[\Sigma_I \oplus S,P]_S]_R = \Sigma_I \oplus [P,P]_R = \Sigma_I \oplus S$, where I is finite.

5.3.2 LEMMA Let A and B be rings. For modules ${}_AM$, ${}_AN_B$, and P_B we defined in (3.6.11)

$$\theta_M : \mathrm{Hom}_B(N,P) \otimes_A M \to \mathrm{Hom}_B(\mathrm{Hom}_A(M,N),P))$$

where $\theta(f \otimes m)(g) = f(g(m))$. If M is a f.g. projective A-module, θ is an isomorphism.

Proof: If M is a f.g. free A-module, θ is an isomorphism. Let $F = M \oplus M'$ be a f.g. free A-module. Then $\theta_F = \theta_M \oplus \theta_{M'}$. Since θ_F is an isomorphism, θ_M and $\theta_{M'}$ are [cf. the proof of (4.3.18)].

5.3.3 LEMMA Let P_R, S, and ${}_SQ$ be as in (5.3.1). If P is f.g. R-projective, $Q \otimes_S P \approx R$, $P \otimes_R Q \approx S$, and $[Q,S]_S \approx P$.

Proof: The first assertions are clear from (5.3.1) and (4.3.4). $[Q,S]_S = [Q,[P,P]_R]_S \approx [P \otimes_R Q, P]_S \approx [S,P]_S \approx P$ by (3.6.10). We can easily check that the above isomorphisms are S-R isomorphisms (we note that the above isomorphism is given by $p \to \{q \to [p,q]\}$).

Now we can state the Morita equivalence theorem.

5.3.4 THEOREM Let P_R be a f.g. projective R-module and a generator in $\underline{M}_R$. Put $Q = \mathrm{Hom}_R(P,R)$ and $S = \mathrm{End}_R(P)$. Then $\underline{M}_R$ and $\underline{M}_S$ are equivalent via functors $h^P : \underline{M}_R \to \underline{M}_S$ and $h^Q : \underline{M}_S \to \underline{M}_R$.

Proof: We know from (5.3.3) that $h^Q h^P$ is naturally isomorphic to the identity functor $1_{\underline{M}_R}$ on $\underline{M}_R$. ${}_RQ_S$ has the same property as ${}_SP_R$ by (5.2.4). Hence $h^P h^Q$ is naturally equivalent to $1_{\underline{M}_S}$. We easily see from these remarks that $\underline{M}_R$ and $\underline{M}_S$ are equivalent to each other.

Since ${}_RQ_S$ is an S-R-bimodule, $h'^Q : {}_R\underline{M} \to {}_S\underline{M}$ is a functor. $h'^Q(M) = {}_R[Q,M] = {}_R[[P,R],M] \approx P \otimes_R [R,M] \approx P \otimes_R M$ by (3.6.10). Therefore, h'^Q is naturally isomorphic to $P \otimes_R -$. Similarly, $h'^P : {}_S\underline{M} \to {}_R\underline{M}$ is naturally isomorphic to $Q \otimes_S$.

We shall give the converse of (5.3.4).

5.3.5 THEOREM Let R and R' be rings. If $\underline{M}_R$ and $\underline{M}_{R'}$ are equivalent, there exist ${}_{R'}P_R$ and ${}_RQ_{R'}$ which give the equivalent functors in (5.3.4).

Proof: Since $\underline{M}_R$ and $\underline{M}_{R'}$ are equivalent, there exist functors $T : \underline{M}_R \to \underline{M}_{R'}$ and $S : \underline{M}_{R'} \to \underline{M}_R$ such that ST and TS are naturally isomorphic to $1_{\underline{M}_R}$ and $1_{\underline{M}_{R'}}$, respectively. We may assume that $ST = 1_{\underline{M}_R}$ and $TS = 1_{\underline{M}_{R'}}$. Put $Q = T(R)$. Then Q is a right R'-module. Since $R = [R,R]_R \xrightarrow{\theta} [TR,TR] = [Q,Q]_{R'}$ is a one-to-one mapping, we may assume that Q is a left R-module by $rq = \theta(r)q$ for $r \in R$ and $q \in Q$. Hence $R = [Q,Q]_{R'}$. Since R_R is projective and a generator, $Q_{R'}$ is R'-projective and a generator. We note that an R-module N is f.g. if and only if for any subset of submodules N_α of N with $N = \cup_I N_\alpha$, there exists a finite subset I' of I such that $N = \cup_{I'} N_\alpha$. This statement is categorical, so $Q_{R'}$ is f.g. Similarly, if we put $P = S(R')$, then ${}_{R'}P_R$ is f.g. R-projective and a generator in $\underline{M}_R$. $[P,P]_R = R'$ and $[P,R] = [S(R'),R] \approx [S(R'), ST(R)] \xrightarrow{\theta} [R',T(R)] = Q$. We can check easily that this is an R' - R isomorphism by the definition. Finally, we shall show that T and h^P are naturally isomorphic. We know that $h^P(R) = [P,R] = Q = T(R)$. Let A be in $\underline{M}_R$ and $0 \leftarrow A \leftarrow F_0 \leftarrow F_1$ a projective resolution of A with $F_i = \Sigma \oplus u_{i\alpha}R$, a free R-module. Since P is f.g., $[P, \Sigma_\alpha \oplus u_{i\alpha}R] = \Sigma_\alpha \oplus [P,R]$. Hence we have two exact sequences

$$\begin{array}{ccccccc}
0 & \leftarrow & h^P(A) & \leftarrow & h^P(F_0) & \leftarrow & h^P(F_1) \\
 & & & & \downarrow \varphi_0 & & \downarrow \varphi_1 \\
0 & \leftarrow & T(A) & \leftarrow & T(F_0) & \leftarrow & T(F_1)
\end{array}$$

Since the φ_i are isomorphisms, we obtain an isomorphism $\gamma_A : h^P(A) \to T(A)$ which commutes in the above diagram by (2.5.2). Further, γ_A is uniquely determined and is a natural isomorphism from h^P to T by (4.3.20) and (4.3.21).

Let R be a right artinian ring. Then we have the standard decomposition in Sec. 4.5,

$$R = \sum_{i=1}^{n} \sum_{j=1}^{\rho(i)} \oplus\, e_{ij}R$$

where $\{e_{ij}\}$ is a set of mutually orthogonal primitive idempotents such that $e_{ij}R \approx e_{i1}R$ and $e_{i1}R \not\approx e_{j1}R$ if $i \neq j$. We can easily see that e_0R is a f.g. projective generator in $\underline{M}_R$ by (5.2.1), where $e_0 = e_{11} + e_{21} + \cdots + e_{n1}$. Furthermore, $\mathrm{End}_R(e_0R) = e_0Re_0$ by (3.1.8). We call e_0Re_0 a *basic ring* of R. Note that $e_0Re_0/J(e_0re_0)$ is a direct sum of division rings.

5.3.6 COROLLARY Let R be a right artinian ring and R_0 a basic ring of R. Then $\underline{M}_R$ is equivalent to $\underline{M}_{R_0}$.

If R is the ring $(R_0)_n$ of matrices over a ring R_0, then R and R_0 are in the situation above.

5.3.7 COROLLARY Let $T = (R)_n$. Then $\underline{M}_T$ is equivalent to $\underline{M}_R$.

Let U be an R-module and $S = \mathrm{End}_R(U)$. For an R-submodule V of U, $S(V) = \mathrm{Hom}_R(U,V)$ is a right ideal in S. Conversely, for a right ideal A of S, $U(A) = \bigcup_{a \in A} \mathrm{im}\, a$ is an R-submodule of U. It is clear that $S(U(A)) \supset A$ and $U(S(V)) \subset V$. Hence $S(V) \subset S(U(S(V))) \subset S(V)$, so $S(V) = S(U(S(V)))$.

5.3.8 PROPOSITION Let P be a f.g. projective R-module and $S = \mathrm{End}_R(P)$. Then $A = S(P(A))$ for a right ideal A of S. If P is a projective generator in $\underline{M}_R$, $P(S(Q)) = Q$ for an R-submodule Q of P.

Proof: Put $F = \Sigma_{a \in A} \oplus P_a$; $P_a \approx P$. Take the exact sequence

$$F \xrightarrow{f} P(A) \to 0$$

such that $fi_a = a$, where $i_a : P_a \to F$ is the injection ($p_a : F \to P_a$ is the projection). Let g be an element in $S(P(A)) = \mathrm{Hom}_R(P,P(A))$. There exists $h : P \to F$ such that $fh = g$. Since P is f.g., $\mathrm{im}\ h \subset \Sigma_{i=1}^{n} \oplus P_{a_i}$, and hence $h = \Sigma\, p_{a_i} h$. We may assume that $h_i = p_{a_i} h$ is in S. Then $g = fh = \Sigma\, a_i h_i \in$ A. Hence $A = S(P(A))$. Next, we assume that P is a projective generator. Let Q be an R-submodule of P. From the exact sequence

$$0 \to P(S(Q)) \to Q \to Q/P(S(Q)) \to 0$$

we have the exact sequence

$$0 \to \mathrm{Hom}_R(P,P(S(Q))) \to S(Q) \to \mathrm{Hom}_R(P,Q/P(S(Q))) \to 0$$

It is clear from the initial observation that $\mathrm{Hom}_R(P,P(S(Q))) = S(P(S(Q))) = S(Q)$. Hence $Q/P(S(Q)) = 0$ for P is a generator.

5.4 QUASI-INJECTIVE MODULES

Let R be a ring with identity. If, for an element a in R, there exists an x in R such that $axa = a$, then $(ax)^2 = ax$ $(= e)$. Hence $aR = eR$ is a direct summand of R, so aR is projective (flat). Conversely, if $aR = eR$ for an idempotent e, $e = ax$ and $ea = a$ for some x in R. Hence $a = axa$. If every element in R satisfies the condition above, R is called a *regular ring* (in the sense of von Neumann). It is clear that regular rings are semisimple (3.3.1).

5.4.1 THEOREM Let R be a ring. Then the following conditions are equivalent:

1. R is a regular ring.
2. Every f.g. right (resp. left) ideal is generated by an idempotent element.
3. Every right (resp. left) R-module is flat.

Proof: The definition of regular ring is left- and right-sided symmetric and hence we consider only right R-modules.

(1) ⇒ (2). Let $I = \langle x_1, x_2, \ldots, x_n \rangle$ be a right ideal. We show (1) ⇒ (2) by induction on n. Assume that $\langle x_1, x_2, \ldots, x_{n-1} \rangle = eR$ and $e^2 = e$. Then $R = eR \oplus (1 - e)R$ by (3.1.1). Let $x_n = a + b$, where a is in eR and b in $(1 - e)R$. We know that $bR = fR$; $f^2 = f$ from (1). Since b is in $(1 - e)R$ and f in bR, $ef = 0$. Then $e_1 = e - ef$ is idempotent, $eR = e_1R$ and $e_1 \perp f$ by (3.1.4). Hence $I = e_1R + x_nR = e_1R + bR = e_1R \oplus fR = (e_1 + f)R$.

(2) ⇒ (3). First we show that every f.g. submodule M of a free R-module $F = \Sigma_I \oplus u_\alpha R$ is a direct summand of F. Since M is f.g., we may assume that I is finite, say $F = \Sigma_{i=1}^{n} \oplus u_i R$. Put $M_{n-1} = M \cap (\Sigma_{i=1}^{n-1} \oplus u_i R)$ and $A = \{r_n \in R \mid$ there exists an m in M such that $m = u_n r_n + \Sigma_{i=1}^{n-1} u_i r_i\}$. Since M is f.g., so is A. Hence A = eR, where $e^2 = e$ by (2). Take m_1 in M with $m_1 = u_n e + \Sigma_{i=1}^{n-1} u_i r_i'$. We may assume that $m_1 e = m_1$, so $r_i' e = r_i'$. Put $v_n = u_n + \Sigma_{i=1}^{n-1} u_i r_i'$. Then $F = v_n R \oplus \Sigma_{i=1}^{n-1} \oplus u_i R$. Let m be in M and $m = \Sigma_{i=1}^{n} u_i x_i$, where the x_i are in R. Since x_n is in A, $m - m_1 y \in M_{n-1}$, where $x_n = ey$. Hence $M = M_{n-1} + m_1 R$. Let $m' \in M_{n-1} \cap m_1 R$. Then $m' = m_1 ex = u_n ex + (\Sigma_{i=1}^{n-1} u_i r_i')x \in M_{n-1}$. Hence ex = 0, so m' = 0. Accordingly, $M = M_{n-1} \oplus m_1 R$. $m_1 R \subset v_n eR \subset v_n R$ and $M_{n-1} \subset \Sigma_{i=1}^{n-1} \oplus u_i R$. Hence M is a direct summand of F. Therefore, M is flat by (4.3.14),4.

(3) ⇒ (1). Take an exact sequence $0 \to xR \to R \to R/xR \to 0$ for x in R. Since R/xR is flat, xR is a direct summand of R by (4.3.14). Hence xR = eR by (3.1.2), where $e^2 = e$.

Next, we study quasi-injective modules. First we generalize the concept of an injective module. Let M_R be an R-module. Let N be any R-submodule of M and f in $\mathrm{Hom}_R(N,M)$. If f can be extended to $\mathrm{End}_R(M)$ for any N, M is called *quasi-injective*. It is clear that every injective module is quasi-injective.

5.4.2 THEOREM Let M_R be an R-module. Then M is quasi-injective if and only if M is a character R-submodule of E(M).

Proof: Put $E = E(M) \supset M$ and $S = \mathrm{End}_R(E)$. We assume that M is an S-R bimodule. Let $f : N \to M$ be a homomorphism for N a submodule of M. Then f can be extended to an element s in S, since E is injective. Hence $s|M \in \mathrm{End}_R(M)$ is an extension of f. Conversely, assume that M is quasi-injective. Let s be in S. Put $N = s^{-1}(M)$. Then $s|N$ is extendible to an s_0 in $\mathrm{End}_R(M)$ and, further, to an s' in S. Assume that $(s - s')(M) \neq 0$. Then $M \cap (s - s')(M) \neq 0$. Let $m = (s - s')m'$ be in $M \cap (s - s')(M)$, where m and m' are in M. Then $sm' = m + s'm' \in M$. Hence m' is in N. Therefore, $m = sm' - s_0 m' = 0$, which is a contradiction. Thus $s(M) = s'(M) \subset M$.

5.4.3 COROLLARY Let M be quasi-injective. Assume that $E(M) = E_1 \oplus E_2$. Then $M = (E_1 \cap M) \oplus (E_2 \cap M)$.

Proof: Let $e_i : E \to E_i$ be the projection. Then $e_i(M) \subset M$ by (5.4.2). Hence $M = e_1(M) \oplus e_2(M)$ and $e_i(M) = M \cap E_1$.

5.4.4 THEOREM Let M be quasi-injective and $S = \mathrm{End}_R(M)$. Then we have the following.

1. $J' = \{f \in S \mid \ker f \text{ is essential in } M\} = J(S)$.
2. $S/J(S)$ is a regular ring.
3. Let $\{\bar{u}_i\}_{i=1}^n$ be a set of mutually orthogonal idempotents in $S/J(S)$. Then there exists a set $\{e_i\}_{i=1}^n$ of mutually orthogonal idempotents in S such that $\bar{e}_i = \bar{u}_i$.

Proof: First we show that J' is right ideal in S. Let s_1, s_2 be in J'. Since $\ker(s_1 + s_2) \supset \ker s_1 \cap \ker s_2$, $s_1 + s_2$ is in J' by (3.4.5). Furthermore, $\ker(ss_1) \supset \ker s_1$ for s in S. Hence J' is a left ideal. Next we show that $J' \subset J(S)$. Let s be in J' and put $\ker s = M_1$. Since $(1 - s) \mid M_1 = 1_{M_1}$, $1 - s$ is a monomorphism. Consider $f = (1 - s)^{-1} \mid \mathrm{im}\,(1 - s)$. Then f is extendible to an element f' in S, so $E = \ker f' \oplus \mathrm{im}\,(1 - s)$ by (1.7.5). Further, $\mathrm{im}\,(1 - s) \supset M_1$ and hence $\ker f = 0$. Therefore, $1 - s$ is a unit in S for any s in J'. Since J' is a right ideal, $J' \subset J$ by (3.3.5). Now, let s be in S and put $K = \ker s$. Taking a complement K^c of K, $K \oplus K^c$ is essential in M by (3.4.7). Since $s|K^c$ is a monomorphism, we can extend $s^{-1}|s(K^c)$ to $s' \in S$. Put $t = ss's - s$. Then $\ker t \supset K^c \oplus K$ and hence t is in J'. Finally, we take an n in J. Then there exists an s' in S such that $ns'n - n \in J'$ from the above. Hence since $s'n - 1$ is a unit in S, n is in J'. Thus we have established (1) and (2).

(3) We may assume that $u_i \in S$ in (3). Since $\bar{1} - \Sigma_{i=1}^n \bar{u}_i = \bar{u}_{n+1}$ is idempotent and $u_i \perp u_{n+1}$ for $i \le n$, we may assume that $\bar{1} = \Sigma_{i=1}^n \bar{u}_i$. Put $C = \ker(1 - \Sigma_i u_i) \cap \ker(u_i^2 - u_i) \cap \cap_{i \neq j}(\ker u_i u_j)$. Since each homomorphism in the above is in J', C is essential in M. Put $U = \Sigma_{i-1}^n u_i(C)$. Then $U \supset C$, since $C \subset \ker(1 - \Sigma_{i=1}^n u_i)$. Further, since $u_i^2(x) = u_i(x)$ and $u_i u_j(x) = 0$ for x in C, $\{u_i(C)\}_{i=1}^n$ is independent [cf. (3.1.2)]. Hence $E(M) = E(C) = \Sigma_{i=1}^n \oplus E(u_i(C))$. Put $M_i = M \cap E(u_i(C)) \supset u_i(C)$. Then $\Sigma_{i=1}^n \oplus u_i(C) \subset \Sigma_{i=1}^n \oplus M_i = M$ by (5.4.3). Let $e_i : M \to M_i$ be the projection. Then $(e_i - u_i)(C) \in (e_i - u_i)(\Sigma_j u_j(C)) = (e_i - u_i)(u_i(C)) = 0$, so $e_i \equiv u_i \pmod{J'}$ and $1 = \Sigma_{i=1}^n e_i$.

5.4.5 COROLLARY Let K be a division ring. Then the endomorphism ring of a free K-module (i.e., the ring of column-finite matrices K_I) is a regular ring.

Proof: Since K is a division ring, every K-module is semisimple. Hence an essential submodule is always equal to the whole module.

Therefore, $J(\mathrm{End}_K(F)) = 0$ and $\mathrm{End}_R(F)$ is a regular ring, where F is a free K-module.

5.4.6 PROPOSITION Let M_R be quasi-injective. If M contains an isomorphic copy of R_R, then M is injective.

Proof: This is clear by (4.4.2).

5.4.7 PROPOSITION Let M_{iR} be R-modules for i = 1, 2 and $M = M_1 \oplus M_2$. Then M is quasi-injective if and only if M_i is quasi-injective and $f_{ji}(M_i) \subset M_j$ for any f_{ji} in $\mathrm{Hom}_R(E(M_i), E(M_j))$.

Proof: If M is quasi-injective, M_i is clearly quasi-injective from the definition. Put $E_i = E(M_i)$ and $E = E(M)$. Then $\mathrm{End}_R(E) = \{(f_{ij}) \mid f_{ij} \in \mathrm{Hom}_R(E_j, E_i)\}$ by (3.7.4) and (4.4.10). Hence we have (5.4.7) by (5.4.2).

5.4.8 COROLLARY Let M_R be quasi-injective and $M^{(n)}$ the direct sum of n copies of M. Then $M^{(n)}$ is quasi-injective.

5.4.9 PROPOSITION Let M_R be quasi-injective. If M is indecompodable, then $\mathrm{End}_R(M)$ is a local ring.

Proof: Let $E = E(M)$ and $S = \mathrm{End}_R(E)$. Then E is indecomposable by (5.4.3). Let s be in S. If s is a monomorphism, s is a unit. If s is not a monomorphism, ker s is essential in E. Hence J(S) consists of all nonisomorphisms in S, so S is a local ring. Since $\mathrm{End}_R(M)$ is a homomorphic image of S by (5.4.2), it is also local.

Finally, we shall define quasi-projective module as the dual concept of quasi-injective module. Let M_R be an R-module. M is called to be *quasi-projective* if for every diagram

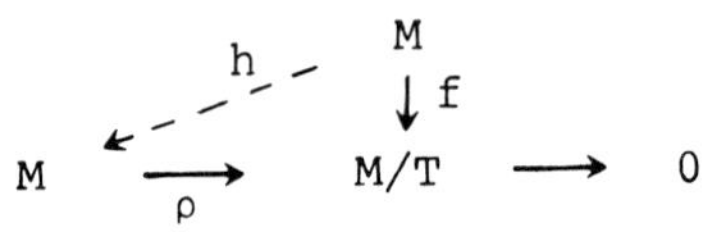

there exists $h : M \to M$ with $\rho h = f$, where T is a submodule of M and ρ is the natural epimorphism. The following theorem is the dual to (5.4.1). Hence the proof will be left to the reader.

5.4.10 PROPOSITION Let P be a projective module and K an S-R bimodule of P, where $S = \mathrm{End}_R(P)$. Then P/K is quasi-projective.

5.4.11 THEOREM Assume M_R has a projective cover P : $(P \xrightarrow{\varphi} M)$ (Sec. 4.3). If M is quasi-projective, ker φ is a character submodule of P.

Proof: Let $S = \mathrm{End}_R(P)$ and s be in S. We shall show that $s(K) \subset K$, where $K = \ker \varphi$. Put $T = K + s(K)$. Then s induces $s' : P/K \to P/T$. Hence we have a diagram

$$\begin{array}{ccccccc} & & & & & & M \\ & & & \overset{t}{\dashrightarrow} & & & \downarrow \pi \\ s'': M & \xrightarrow{(\varphi|P/K)^{-1}} & P/K & \xrightarrow{s'} & P/T & \xrightarrow{\varphi'} & M/\varphi(T) \\ & & & & & & \downarrow \\ & & & & & & 0 \end{array}$$

where π is the natural epimorphism and φ' is induced from φ, and also from $\pi\varphi$. Since M is quasi-projective, we have $t : M \to M$ which commutes in the above. Further, P is projective, and we obtain s_1 in S such that

$$\begin{array}{ccccc} P & \xrightarrow{\varphi} & M & & \\ s_1\downarrow & & \downarrow t & & \\ P & \xrightarrow{\varphi} & M & \xrightarrow{\pi} & M/\varphi(T) \end{array}$$

is commutative [and hence $s_1(K) \subset K$]. Then $\pi\varphi(s_1 - s)(p) = s''\varphi(p) - \pi\varphi s(p) = 0$, where p is in P, from the next-to-last diagram. Hence $(s_1 - s)(P) \subset T$, since $\varphi^{-1}(\varphi(T)) = T$. Then $s_1(p) - s(p) = k_1 + s(k_2)$, where the k_i are in K. Put $X = \{p \in P \mid s(p) - s_1(p) \in K\}$. Now $s(p + k_2) = -k_1 + s_1(p)$, so $s(p + k_2) - s_1(p + k_2) = -k_1 - s_1(k_2) \in K$ for $s_1(K) \subset K$. Accordingly, $P = X + K$, and $P = X$ since K is small in P. Therefore, $s(k) - s_1(k) \in K$ implies that $s(K) \subset K$.

5.4.12 THEOREM Let M and P be as in (5.4.11). If $P = \Sigma \oplus P_i$, then $\ker \varphi = \Sigma \oplus K_i$; $K_i \subset P_i$ and $0 \to \Sigma \oplus K_i \to \Sigma \oplus P_i \to M \to 0$ is exact.

Proof: This is clear from (5.4.11) and the proof of (5.4.3).

5.5 ENDOMORPHISM RINGS OF QUASI-INJECTIVE MODULES

Let R be a ring and A a right R-module. Put $S = \mathrm{End}_R(A)$. Then A is an S-R bimodule. For a subset B of A we put $\ell_S(B)$ $[= \mathrm{ann}_S^{\ell}(B)] = \{f \in S \mid f(B) = 0\}$. Then $\ell_S(B)$ is a left ideal in S. Conversely, for a subset T of S, put $r_A(T)$ $[= \mathrm{ann}_A^r(T)] = \cap_t \ker t$, where t runs through all elements in T; this is an R-submodule of A.

We have the following relations:

1. If $A \supset B_1 \supset B_2$, then $\ell_S(B_1) \subset \ell_S(B_2)$. If $S \supset T_1 \supset T_2$, then $r_A(T_1) \subset r_A(T_2)$.
2. $r_A\ell_S(B) \supset B$ and $\ell_S r_A(T) \supset T$.
3. If B_1' and B_2' are R-submodules of A, $\ell_S(B_1' + B_2') = \ell_S(B_1') \cap \ell_S(B_2')$, $\ell_S r_A \ell_S(B) = \ell_S(B)$ and $r_A\ell_S r_A(T) = r_A(T)$.

Let $\underline{L}$ be the set of left ideals in S and $\underline{M}$ the set of right R-submodules of A. Consider mappings $T \in \underline{L} \to r_A(T) \in \underline{M}$ and $B \in \underline{M} \to \ell_S(B) \in \underline{L}$. If $\ell_S r_A(T) = T$ for all $T \in \underline{L}$ and $r_A\ell_S(B) = B$ for all $B \in \underline{M}$, we say that $\underline{L}$ and $\underline{M}$ are *anti-isomorphic*.

5.5.1 PROPOSITION Let A be a quasi-injective module and $S = \mathrm{End}_R(A)$. Assume that $\ell_S r_A(L) = L$ for a left ideal L in S. Then $\ell_S r_A(L + Sx) = L + Sx$ for any x in S.

Proof: Let y be any element in $\ell_S r_A(L + Sx) = \ell_S(r_A(L) \cap r_A(Sx))$. Then $r_A(y) \supset r_A\ell_S(r_A(L) \cap r_A(Sx)) \supset r_A(L) \cap r_A(Sx)$. Hence we have a commutative diagram

$$\begin{array}{ccccccccc} 0 & \longrightarrow & r_A(L) \cap r_A(Sx) & \longrightarrow & r_A(L) & \xrightarrow{x} & x(r_A(L)) & \longrightarrow & 0 \\ & & \downarrow & & \downarrow i & & \downarrow \theta & & \\ 0 & \longrightarrow & r_A(y) & \longrightarrow & A & \xrightarrow{y} & y(A) & \longrightarrow & 0 \end{array}$$

since $\ker x = r_A(Sx)$. Hence we obtain a $\theta : x(r_A(L)) \to y(A)$ such that $\theta x = yi$. Since A is quasi-injective, θ is extendible to an element s in S. Hence $(y - s)(r_A(L)) = 0$, so $y - sx \in \ell_S r_A(L) = L$. Therefore, y is in $L + Sx$. The converse is clear.

5.5.2 COROLLARY Let A be as in (5.5.1). If L is a f.g. left ideal in S, then $L = \ell_S r_A(L)$. Hence S is left noetherian if and only if A satisfies the descending chain condition on $\{r_A(X) \mid X \text{ is a left ideal}\}$.

Proof: We note that S is left noetherian if and only if S satisfies the ascending chain condition on the set of f.g. left ideals. Let $L_1 \subset L_2 \subset \cdots \subset L_n \subset \cdots$ be a chain of f.g. left ideals in S. Then we obtain the chain $r_A(L_1) \supset r_A(L_2) \supset \cdots \supset r_A(L_n) \supset \cdots$ of R-submodules of A and $\ell_S r_A(L_n) = L_n$ for all n by (5.5.1). Hence if $\{r_A(X)\}$ satisfy the descending chain condition, then S is left noetherian. Conversely, consider a descending chain $r_A(X_1) \supset r_A(X_2) \supset \cdots \supset r_A(X_n) \supset \cdots$, where the X_i are left ideals in S. Then $\ell_S r_A(X_m) = \ell_S r_A(X_{m+1}) = \cdots$ for some m, provided that S is left noetherian. Hence $r_A(X_m) = r_A(\ell_S r_A(X_m)) = r_A\ell_S r_A(X_{m+1}) = r_A(X_{m+1})$.

Next we consider relationships between right ideals K of R and S-submodules B of A. Put $r_R(B) = \{r \in R \mid Br = 0\}$ and $\ell_A(K) = \{a \in A \mid aK = 0\}$.

5.5.3 PROPOSITION Let A be a quasi-injective R-module and $S = \mathrm{End}_R(A)$. Assume that for an S-submodule B of A, $\ell_A r_R(B) = B$. Then, for any element x in A, $\ell_A r_R(B + Sx) = B + Sx$, where $Sx = \{f(x) \mid f \in S\}$.

Proof: We shall use the same technique as in the proof of (5.5.1). Let y be in $\ell_A r_R(B + Sx) = \ell_A r_R(B) \cap r_R(Sx)$. Then the mapping $\theta : xr_R(B) \to A$ given by setting $\theta(xa) = ya$ is well defined and is an R-homomorphism. Since A is quasi-injective, θ can be extended to an element s in S. Hence $(sx - y)r_R(B) = 0$ implies that $sx - y \in \ell_A r_R(B) = B$. Therefore, y is in $B + Sx$.

5.5.4 COROLLARY Let A be as in (5.5.3). For every f.g. S-submodule B of A, $B = \ell_A r_R(B)$.

5.5.5 PROPOSITION Let A be a quasi-injective R-module and $S = \mathrm{End}_R(A)$. A is noetherian as a left S-module if and only if A satisfies the ascending chain condition on left S-modules $\ell_A(P)$, where P runs through all right ideals in R.

Proof: The "if" part is clear by (5.5.4) and the remark in the proof of (5.5.2). Assume the latter condition. Let $r_R(B_1) \supset r_R(B_2) \supset \cdots \supset r_R(B_n) \supset \cdots$ be a chain of right ideals of R, where the B_i are S-submodules of A. Then $\ell_A r_R(B_1) \subset \ell_A r_R(B_2) \subset \cdots \subset \ell_A r_R(B_n) \subset \cdots$. From the assumption, $\ell_A r_R(B_m) = \ell_A r_R(B_{m+k})$ for some m and all k. Hence $r_R(B_m) = r_R \ell_A r_R(B_m) = r_R \ell_A r_R(B_{m+k}) = r_R(B_{m+k})$. Therefore, R satisfies the descending chain condition on $r_R(B)$. Let T be an S-submodule of A. Consider the set $\underline{T}$ of right ideals $r_R(T^*)$, where T^* is a f.g. S-submodule of T. Then there exists a minimal right ideal $r_R(T')$ by the argument above. Let t be any element of T. Then $r_R(T' + St) \subset r_R(T')$. Hence $r_R(T' + St) = r_R(T')$. Therefore, $T' = \ell_A r_R(T') = T' + St$ by (5.5.4), which means that T (= T') is f.g. Thus A is S-noetherian.

5.5.6 COROLLARY Let A and S be as in (5.5.5). If R is right artinian, A is S-noetherian.

Proof: We note that $\ell_A(P) = \ell_A r_R \ell_A(P)$. Hence we may assume that $P = r_R \ell_A(P)$ in the statement of (5.5.5). Therefore, if $\ell_A(P) \supset \ell_A(P')$, $P \subset P'$. Thus (5.5.6) is clear from (5.5.5).

5.5.7 COROLLARY Let R be right artinian and E an injective R-module. Then $EJ = \{x \in E \mid xr_R(J) = 0\} = \ell_E r_R(J)$, where $J = J(R)$.

Proof: $EJr_R(J) = E0 = 0$. Put $D = \text{ann}_R(E)$. Then R/D is right artinian and $J(R/D) = (J + D)/D$. E is also R/D-injective by (4.4.2). $EJ = E(J + D)$ and $\ell_E(r_R(J + D/D)) = \ell_E r_R(J)$. Hence we may assume $D = 0$. Since E is S-noetherian by (5.5.6), $EJ = \ell_E r_R(EJ)$ by (5.5.4). Let r be in $r_R(EJ)$. Then $EJr = 0$. Hence $Jr = 0$, since $D = 0$. Therefore, r is in $r_R(J)$, so $r_R(EJ) \subset r_R(J)$. Thus $EJ = \ell_E r_R(EJ) \supset \ell_E r_R(J)$.

5.5.8 THEOREM Let A be a quasi-injective R-module and $S = \text{End}_R(A)$. If A satisfies the ascending chain condition on $r_A(X)$, where X is a left ideal in S, then S is semiprimary.

Proof: Let $A = A_1 \oplus A_2$ as R-modules and e_i the projection of A onto A_i. Then $A_1 = r_A(S(1 - e_2))$. Since A satisfies the ascending chain condition with respect to $\{r_A(X)\}$, $A = \Sigma_{i=1}^n \oplus A_i$, where the A_i are indecomposable. First we assume that A is indecomposable. We use the same argument as in the proof of (4.5.1). Let $\underline{F}$ be the set of R-submodules B such that $A \neq B = r_A(P) = r_A(P^2)$ for a left ideal P in S. We assume that $\underline{F} \neq \emptyset$. From the assumption, there exists a maximal submodule $C = r_A(P) = r_A(P^2)$ in $\underline{F}$. Since $A \neq C$, $P^2 \neq 0$. Hence we have z in P with $Pz \neq 0$. Consider $r_A(z) = r_A(Sz)$. Again from the assumption, we obtain a maximal $r_A(x)$ such that $Px \neq 0$; $x \in P$. If $PPx = 0$, $\text{im}\, x \subset r_A(P^2) = r_A(P)$, so $Px = 0$. Therefore, there exists y in P such that $Pyx \neq 0$. Since yx is in P, $r_A(yx) \supset r_A(x)$ and $r_A(yx) = r_A(x)$ by the maximality of $r_A(x)$. Hence $Sx = Syx$ by (5.5.2). Accordingly, there exists an a in $Sy \subset P$ with $x = xa$. Hence $x = a^n x$. If $a^2 \neq a$, $(a^2 - a)x = 0$ implies that $0 \neq P' = \{z \in P \mid zx = 0\} \subsetneq P$. Furthermore, $r_A(P') \supset \text{im}\, x$ and $r_A(P) \not\supset \text{im}\, x$. Thus $[r_A(P) \subset]$ $r_A(P') \subsetneq r_A(P'^2)$ by the maximality of $r_A(P)$. However, from the assumption, $r_A(P'^t) = r_A(P'^{2t})$ for some t. Hence $r_A(P'^t) = A$ and $P'^t = 0$. Now we can show from the facts that (1) $x = a^n x$ and (2) $P'^t = 0$, and the method in the proof of (4.5.1), that P contains an idempotent. Since A is indecomposable, $P = S$, which is a contradiction. Hence $\underline{F} = \emptyset$. Now let P be a proper ideal of S. Then $r_A(P^n) = r_A(P^{2n})$ for some n, so $P^n = 0$ by the above assertion. S is a local ring by (5.4.9). Therefore, S is semiprimary. Next assume that $A = \Sigma_{i=1}^n \oplus A_i$, where the A_i are indecomposable and quasi-injective [and hence uniform by (5.4.3)]. Then

$$S = \begin{bmatrix} \mathrm{Hom}_R(A_1,A_1) & \mathrm{Hom}_R(A_2,A_1) & \cdots & \mathrm{Hom}_R(A_n,A_1) \\ \cdots & \cdots & \cdots & \cdots \\ \mathrm{Hom}_R(A_1,A_n) & \mathrm{Hom}_R(A_2,A_n) & \cdots & \mathrm{Hom}_R(A_n,A_n) \end{bmatrix}$$

In order to show that S is semiprimary, we may assume that $A_i \not\approx A_j$ for $i \neq j$ by (5.3.6). Then

$$J(S) = \begin{bmatrix} J(S_{A_1}) & \mathrm{Hom}_R(A_2,A_1) & \cdots & \cdots \\ \mathrm{Hom}_R(A_1,A_2) & J(S_{A_2}) & \cdots & \cdots \\ \cdots & \cdots & \cdots & \cdots \\ \mathrm{Hom}_R(A_1,A_n) & \cdots & \cdots & J(S_{A_n}) \end{bmatrix}$$

by (5.4.4), where $S_{A_i} = \mathrm{End}_R(A_i)$ and J(S) is nilpotent [see the proof of (4.5.19)].

5.5.9 COROLLARY Let R be a right artinian ring and A quasi-injective. If A is R-noetherian, $S = \mathrm{End}_R(A)$ is left artinian and A is S-artinian.

Proof: S is semiprimary by (5.5.8). Since R is right artinian, A is R-artinian. Hence S is left noetherian by (5.5.2). Therefore, S is left artinian and A is S-artinian by (5.5.6).

Let M be an R-module. If, for two submodules $N_1 \supset N_2$ of M with N_1/N_2 simple, $\mathrm{Hom}_R(N_1/N_2,M) \neq 0$, we say that M is *weakly distinguished.* If M is a cogenerator (see Sec. 5.6), then M is weakly distinguished.

5.5.10 LEMMA Let A be a quasi-injective R-module. Then the following conditions are equivalent:

1. A is weakly distinguished.
2. For any submodules $A_{1R} \supsetneq A_{2R}$, $\ell_S(A_1) \subsetneq \ell_S(A_2)$.
3. For any submodule B_R of A, $B = r_A\ell_S(B)$.

Proof: (1) ⇒ (2). Since any cyclic module contains a maximal submodule, A_1/A_2 contains modules $A_3/A_2 \supset A_4/A_2$ with A_3/A_4 simple. Let $f \neq 0$ be in $\mathrm{Hom}_R(A_3/A_4,A)$. Then $g : A_3 \to A_3/A_4 \xrightarrow{f} A$ is nonzero and can be extended to an h in $S = \mathrm{End}_R(A)$ with $0 = h(A_4) \supset h(A_2)$. Therefore, h is in $\ell_S(A_2) - \ell_S(A_1)$.

(2) ⇒ (3). If $B \subsetneq r_A\ell_S(B)$, then $\ell_S(B) \supsetneq \ell_S r_A \ell_S(B) = \ell_S(B)$, which is a contradiction.

(3) ⇒ (2). This is clear.

(2) ⇒ (1). Let s be in $\ell_S(A_2) - \ell_S(A_1)$. Then s induces a nonzero $\bar{s}$ in $\mathrm{Hom}_R(A_1/A_2, A)$.

5.5.11 LEMMA Let A be a quasi-injective R-module with $S = \mathrm{End}_R(A)$. Assume that A satisfies the ascending chain condition on $r_A(X)$, where $X \subset S$ and

$$r_A(P \cap P') = r_A(P) + r_A(P') \quad \text{for left ideals } P \text{ and } P' \text{ in } S \quad \underline{(5.5.1)}$$

Then S is left artinian.

Proof: S is semiprimary by (5.5.8). Hence S has the nonzero socle $T = \Sigma_i \oplus P_i$, where the P_i are simple. Put $L_i = \Sigma_{j \geq i} \oplus P_j$. Then $L_1 \supset L_2 \supset \cdots$ and $r_A(L_1) \subset r_A(L_2) \subset \cdots \subset r_A(L_n) \subset \cdots$. From the assumption we have an n such that $r_A(L_n) = r_A(L_{n+1})$. Hence $A = r_A(P_n \cap L_{n+1}) = r_A(P_n) + r_A(L_{n+1}) = r_A(P_n) + r_A(L_n) = r_A(P_n)$, since $P_n \subset L_n$. Therefore, $P_n = 0$. Thus $T = \Sigma_{i=1}^{n-1} \oplus P_i$. Put $A_1 = r_A(T)$. Since T is a two-sided ideal in S, A_1 is an S-R bimodule. Hence A_1 is quasi-injective by (5.4.2). Furthermore, $T = \ell_S(A_1)$ by (5.5.2). Put $S_1 = \mathrm{End}_R(A_1)$. Then S_1 is naturally isomorphic to S/T, since A is quasi-injective. We may regard $S_1 = S/T$. For a left ideal $\bar{P}$ in S_1, we can find a left ideal $P \supset T$ with $P/T = \bar{P}$ and $r_A(P) \subset r_A(T) = A_1$. Hence $r_{A_1}(\bar{P}) = r_A(P)$. Therefore, A_1 satisfies the same hypothesis as A. Hence S_1 contains the socle T_1/T with finite length, by the above argument. Repeating this process, we obtain a chain of ideals $T = T_0 \subset T_1 \subset T_2 \subset \cdots \subset T_n \subset \cdots \subset S$ such that T_i/T_{i-1} is the socle with finite length of S/T_{i-1}. Now $J^n = 0$ [$J = J(S)$] and $J^{n-1} \subset T_0$. Further, $JT_i \subset T_{i-1}$. Hence, by induction, $J^i \subset T_{n-i-1}$, so $S = T_{n-1}$ is left artinian.

5.5.12 THEOREM Let A be a weakly distinguished and quasi-injective R-module and $S = \mathrm{End}_R(A)$. Then the following two conditions are equivalent:

1. S is left noetherian.
2. A is R-artinian.

(1) or (2) implies that

3. A is S-injective.

Further, if A satisfies the ascending chain condition on $r_A(X)$, where $X \subset S$, then (3) implies (1) and (2), S is left artinian, and A is R-noetherian.

Proof: (1) ⇒ (2). This is clear by (5.5.10).

(1) ⇒ (3). Assume that S is left noetherian. Let P_1, P_2 be left ideals in S. Then $\ell_S(r_A(P_1) + r_A(P_2)) = \ell_S r_A(P_1) \cap \ell_S r_A(P_2) = P_1 \cap P_2$ by

(5.5.2). Hence (5.5.1) is satisfied by (5.5.10). We shall show by induction on the number of generators of a left ideal in S that A satisfies Baer's condition (4.4.2). Let $P = Sx \subset S$. Then $\ell_S(xA) = \ell_S(x)$. Hence $xA = r_A\ell_S(xA) = r_A\ell_S(x)$ by (5.5.10). Let f be in $\mathrm{Hom}_S(P,A)$. Then $\ell_S(x)f(x) = f(\ell_S(x)x) = 0$, so f(x) is in $r_A\ell_S(x) = xA$. Therefore, there exists an a in A such that $f(x) = xa$. Let $P = \Sigma_{i=1}^{n} Sx_i$ and $P_1 = \Sigma_{i=1}^{n-1} Sx_i$. From the exact sequence $0 \to P_1 \to P \to P/P_1 \overset{\varphi}{\approx} Sx_n/(Sx_n \cap P_1) \to 0$, we have the exact sequence $\mathrm{Hom}_S(P_1,A) \leftarrow \mathrm{Hom}_S(P,A) \leftarrow \mathrm{Hom}_S(P/P_1,A) \approx \mathrm{Hom}_S(Sx_n/(Sx_n \cap P_1), A) \leftarrow 0$. Let f be in $\mathrm{Hom}_S(P_1,A)$. Then there exists an a in A such that $f(x) = xa$ for $x \in P_1$ by the induction hypothesis. Define f_a in $\mathrm{Hom}_S(P,A)$ by setting $f_a(x) = xa$ for x in P. Then $g = f - f_a$ is in $\mathrm{Hom}_S(P/P_1,A)$. Since $\mathrm{Hom}_S(Sx_n,A) \leftarrow \mathrm{Hom}_S(Sx_n/(Sx_n \cap P_1), A)$ is a monomorphism, there exists an a' in A such that $g(\varphi^{-1}(\overline{sx_n})) = sx_na'$ by the initial observation, where $\overline{sx_n}$ denotes the equivalence class of sx_n in $Sx_n/(Sx_n \cap P_1)$. Hence a' is in $r_A(Sx_n \cap P_1) = r_A(Sx_n) + r_A(P_1)$ by (5.5.1). Let $a' = (a_1 + a_2)$, where a_1 is in $r_A(x_n)$ and a_2 in $r_A(P_1)$. Define f_{a_2} as above. Then, for any $x = x_1 + x_2$ in P ($x_1 \in P_1$, $x_2 \in Sx_n$), $g(x) = g(\varphi^{-1}(\bar{x}_2)) = x_2a' = x_2a_2 = f_{a_2}(x)$. Therefore, $f = f_{a+a_2}$.

(3) ⇒ (1). Let A be S-injective and P_1, P_2 left ideals in S. Then we have the exact sequence

$$0 \leftarrow \mathrm{Hom}_S(S/(P_1 \cap P_2), A) \leftarrow \mathrm{Hom}_S(S/P_1,A) \oplus \mathrm{Hom}_S(S/P_2,A)$$

from the exact sequence $0 \to S/(P_1 \cap P_2) \to S/P_1 \oplus S/P_2$. Hence A satisfies (5.5.1). Therefore, S is left artinian by (5.5.11). The remaining assertions are clear from (5.5.10).

We shall study the endomorphism rings of injective modules over a commutative noetherian ring. First we consider the Artin-Rees lemma for a commutative noetherian ring.

5.5.13 THEOREM Let R be a noetherian ring and M a f.g. R-module. Consider an ideal A of R and an R-submodule N of M. Then there exists an integer r such that $A^nM \cap N = A^{n-r}(A^rM \cap N)$ for all $n > r$.

Proof: Put $R^* = R \oplus M$ as modules and define

$$(r + m)(r' + m') = rr' + rm' + rm' \qquad \text{for } r, r' \text{ in R and } m, m' \text{ in M}$$

Then M is an ideal in R*. Hence we may assume without loss of generality that M is an ideal in R. It is clear that $A^{n-r}[A^r(M \cap N)] \subset A^nM \cap N$ for any $n > r$. Let $(a_1,a_2,\ldots,a_t)$ be a set of generators of A. Then each element in A^m is expressible as a homogeneous polynomial of degree m over (a_i). Hence

$$a = f^{(m)}(a_1, a_2, \dots, a_t) \qquad \text{for } a \in A^m M \cap N \subset A^m$$

where $f^{(m)}(x_1, x_2, \dots, x_t)$ is an mth-degree homogeneous polynomial in $R[x_1, x_2, \dots, x_t]$. Let S be the set of such polynomials, where a and m run through all possibilities, and P the ideal generated by S. Note that R^* is also noetherian by (4.2.3), since $R \approx R^*/M$ and M is noetherian. Thus P is f.g., say $P = \langle s_1(x), s_2(x), \dots, s_k(x)\rangle$, by (4.2.2). Let n_i be the degree of $s_i(x)$ and $r = \max(n_i)$. Then we have $f^{(n)}(x) = \Sigma\, h_i(x)s_i(x)$ for $n > r$, where $h_i(x) \in R[x]$. Since $f^{(n)}(x)$ is homogeneous, we may assume that

$$f^{(n)}(x) = \Sigma\, g_i^{(n-n_i)}(x)s_i(x)$$

where $g_i^{(n-n_i)}(x)$ is homogeneous. Hence $a = f^{(n)}(a_1, a_2, \dots, a_t) = \Sigma\, g_i^{(n-n_i)}(a_j)s_i(a_j)$ and $g_i^{(n-n_i)}(a_1, \dots, a_t)s_i(a_1, \dots, a_t)$ is in $A^{n-n_i}(A^{n_i}M \cap N) = A^{n-r}A^{r-n_i}(A^{n_i}M \cap N)$. On the other hand, $A^{r-n_i}(A^{n_i}M \cap N) \subset A^rM \cap N$. Hence a is in $A^{n-r}(A^rM \cap N)$.

5.5.14 COROLLARY Let M, A, and R be as in (5.5.13). If A is in J(R), then $\cap_n (A^nM) = 0$.

Proof: Put $N = \cap A^nM$. Then $N \subset A^nM \cap N = A^{n-r}(A^rM \cap N) \subset A^{n-r}N \subset N$. Hence $N = AN = J(R)N$, so $N = 0$ by (3.3.13).

Let R be a commutative ring. If R contains a unique maximal ideal, then we call R a local ring. More generally, if R contains a finite number of maximal ideals, we call R a *semilocal* ring. Let $\{P_1, P_2, \dots, P_k\}$ be a set of prime ideals. Then $S = R - \cup_i P_i$ is a multiplicatively closed set and R_S is a semilocal ring with P_{iS} as maximal ideals by (4.7.2).

From now on, we assume that R is a commutative noetherian ring. Let A be an ideal and M an R-module. We shall define a topology on M as follows. Let m be in M. We take $\{m + A^nM\}_n$ as a basic neighborhood system; namely, any subset V of M containing m is a neighborhood of m if and only if V contains $m + A^nM$ for some n. We call such a topology on M the *A-adic topology*. Let m and m' be two elements in M. For a neighborhood $m + m' + A^nM$ of $m + m'$, we have $(m + A^nM) + (m' + A^nM) = m + m' + A^nM$. Hence M is a topological module. If $M = R$, we know similarly that the product operation is continuous with respect to the topology. Therefore, R becomes a topological ring.

5.5.15 PROPOSITION Let R be a noetherian ring and A an ideal contained in J(R). If M is a f.g. R-module, then the A-adic topology is Hausdorff.

Proof: Let m and m' be two elements in M. Then $m - m' \notin A^nM$ for some n by (5.5.14). Hence $(m + A^nM) \cap (m' + A^nM) = \emptyset$.

5.5.16 LEMMA Let R, A, and M be as in (5.5.13). Then we have $N = \bigcap_n (N + A^nM)$ for any R-submodule N of M.

Proof: Since M/N is f.g., $0 = \bigcap_n A^n(M/N) = \bigcap (N + A^nM)/N$ by (5.5.14).

5.5.17 PROPOSITION Let M, R, and A be as in (5.5.13). Then every R-submodule is closed.

Proof: We have a partition of $M = \bigcup_x (x + N + A^nM)$. Hence $N + A^nM = M - \bigcup_{x \neq 0} (x + N + A^nM)$ is closed, so $N = \bigcap_n (N + A^nM)$ is also closed.

5.5.18 PROPOSITION Let R be a commutative noetherian ring and A an ideal. Let M be a f.g. R-module and $N_R \subset M_R$. Then the topology on N induced from that of M is also the A-adic topology on N.

Proof: $A^nM \cap N = A^{n-r}(A^rM \cap N) \subset A^{n-r}N$ for all $n > r$ for some r, by (5.5.13). Conversely, $A^nN \subset A^nM \cap N$.

Next we shall define the completion of the A-adic topology on M. We assume that R is a commutative noetherian ring, A is an ideal in J(R), and M_R is f.g.

Let $\{m_i\}$ be a sequence of elements in M. If, for any integer n, there exists an r such that

$$m_i - m_j \in A^nM$$

for all $i > r$ and $j > r$, then we say that $\{m_i\}$ is a *Cauchy sequence*. Let $\{m_i\}$ be a Cauchy sequence and m an element in M. If, for any n there exists r such that

$$m - m_i \in A^nM$$

for all $i > r$, then m is called a *limit* of $\{m_i\}$ and we denote it by $\lim m_i = m$ [m is uniquely determined by (5.5.14)]. If every Cauchy sequence in M has a limit in M, we say that M is *complete* (with respect to the A-adic topology). It is clear from the definition that if a Cauchy sequence $\{m_i\}$ has a limit, then every infinite subsequence of $\{m_i\}$ is a Cauchy sequence and has the same limit.

Now we shall define the completion of M. Let $\underline{M}$ be the set of all Cauchy sequences in M. We define a relation on $\underline{M}$ as follows.

$$\{m_i\} \equiv \{m_i'\}$$

if and only if for any n there exists an r such that

$$m_i - m_i' \in A^n M \qquad \text{for all } i > r$$

It is clear that the above relation is an equivalence relation. We denote the equivalence class of $\{m_i\}$ by $\{\overline{m_i}\}$. Let $\{m_i\}$ and $\{m_i'\}$ be two elements in $\underline{M}$. We define

$$\{m_i\} \pm \{m_i'\} = \{m_i \pm m_i'\}$$

Then $\{m_i \pm m_i'\}$ is in $\underline{M}$. In case M = R we define

$$\{r_i\}\{r_i'\} = \{r_i r_i'\}$$

Then $\{r_i r_i'\}$ is in $\underline{R}$. Finally, we define

$$\{r_i\}\{m_i\} = \{r_i m_i\}$$

We can easily check that the above operations extend to operations on the equivalence classes. We shall denote the set of all equivalence classes by $\hat{M}$ and $\hat{R}$, respectively. Then $\hat{R}$ is a commutative ring and $\hat{M}$ is an $\hat{R}$-module. We note, from the definition of $\hat{M}$, that $\bigcap_n A^n\hat{M} = 0$.

Let $\{m_i\}$ be in $\underline{M}$ and let $\{m_{i_k}\}$ be an infinite subsequence of $\{m_i\}$. Then $\{\overline{m_i}\} = \{\overline{m_{i_k}}\}$ by the definition. Hence we may assume that, for fixed n,

$$m_i - m_{i+1} \in A^{n+i}M \qquad \text{for all } i$$

5.5.19 LEMMA $\hat{R}$ contains a subring isomorphic to R. Hence we may assume that R is a subring of $\hat{R}$. $\hat{M}$ contains a submodule R-isomorphic to M.

Proof: We define a mapping $\varphi : R \to \hat{R}$ by setting $\varphi(r) = \{\overline{r_i}\}_{r_i = r}$. If $\varphi(r) = 0$, r is in $\bigcap_n A^n = 0$ by (5.5.14). It is clear that φ is a ring homomorphism. We may prove the last assertion similarly.

From (5.5.19) we may assume that $R \subset \hat{R}$ and $M \subset \hat{M}$.

5.5.20 LEMMA $A^n\hat{M} = \{\{\overline{m_i}\} \in \hat{M} \mid m_i \in A^nM\}$.

Proof: It is clear from the definition that $A^n\hat{M} \subset \{\{\overline{m_i}\} \in \hat{M} \mid m_i \in A^nM\}$. Let $A^n = \langle x_1, x_2, \ldots, x_t\rangle$. Let $\{\overline{m_i}\}$ be in $\hat{M}$ and m_i in A^nM. Then

$$m_s = m_1 + (m_2 - m_1) + \cdots + (m_s - m_{s-1})$$

and we may assume that $m_1 \in A^nM$ and $m_{t+1} - m_t \in A^{n+t}M = A^n(A^tM)$. Let $m_1 = \Sigma_{j=1}^t x_jm_{j1}$ and $m_p - m_{p-1} = \Sigma_{j=1}^t x_jm_{jp}$, where $m_{ip} \in A^{p-1}M$ ($A^0 = R$). Put $m(s,p) = \Sigma_{q=1}^s m_{pq}$ ($p = 1, \ldots, t$). We obtain $m(s',p)$ inductively for all s' ($p = 1, \ldots, t$). Then $\{m(s,p)\}_s$ is in $\underline{M}$ and $\{\overline{m_i}\} = \Sigma_{p=1}^t x_p\overline{\{m(s,p)\}}_s \in A^n\hat{M}$.

For any $\{m_i\}$ in $\hat{M}$, we put $m_1' = m_1, \ldots, m_t' = m_{t+1} - m_t$ ($\in A^tM$). Then $\{\overline{m_i}\} = \lim\,(m_1' + m_2' + \cdots + m_t') = m_1' + m_2' + \cdots + m_n' + \cdots$. Hence, if $M = \langle n_1, n_2, \ldots, n_s\rangle$, $A^pM = \Sigma_{i=1}^s A^pn_i$, so $\hat{M} = \Sigma_{i=1}^s \hat{R}n_i$, namely, $\hat{M}$ is a f.g. $\hat{R}$-module.

5.5.21 COROLLARY $A^n\hat{M} \cap M = A^nM$ and $\hat{M}/A^n\hat{M} \approx M/A^nM$.

Proof: Let $\{\overline{m_i}\}$ be in $A^n\hat{M} \cap M$. Then $\{\overline{m_i}\} = \{\bar{m}\}$ for some m in M and m_i in A^nM by (5.5.20). Hence m is in A^nM. Thus $A^n\hat{M} \cap M = A^nM$. Let f be the composition of $\varphi : M \to \hat{M}$ and $\nu : \hat{M} \to \hat{M}/A^n\hat{M}$, where ν is the natural epimorphism. Then $\ker f = A^nM$ from the first observation. Let $m = \{\overline{m_i}\}$ be in $\hat{M}$. Then we can express m as $m = m_1' + m_2' + \cdots + m_n' + \cdots$, where m_i' is in A^iM. Then $\Sigma_{k\geq n} m_k'$ is in $A^n\hat{M}$ by (5.5.20). Hence $f(\Sigma_{p=1}^{n-1} m_p') = m + A^n\hat{M}$.

5.5.22 THEOREM $\hat{M}$ is a completion with respect to the A-adic topology.

Proof: Let $\{\tilde{m}_i\}$ be a Cauchy sequence in $\hat{M}$ and let $\tilde{m}_i = \{m_{ij}\}_j$ be in $\underline{M}$. Then $\{m_{ii}\}_i$ is in $\underline{M}$ and $\{\overline{m_{ii}}\} = \lim \tilde{m}_i$ by the definition.

We shall show in (5.5.26) that $\widehat{J(R)} = J(\hat{R})$ and $\hat{R}$ is noetherian provided that R is semilocal. We call $\hat{M}$ the *completion* of M with respect to the A-adic topology. Finally, consider the ring of polynomials $R[x_1, x_2, \ldots, x_n]$ over R. As the completion of the $(x_1, x_2, \ldots, x_n)$-adic topology of $R[x_1, x_2, \ldots, x_n]$ we can consider the class of formal infinite series

$$\sum_0^\infty a_{i_1i_2\cdots i_n} x_1^{i_1}x_2^{i_2}\cdots x_n^{i_n} \qquad a_{i_1i_2\cdots i_n} \in R$$

We can define a ring structure on $\{\Sigma\, a_{i_1i_2\cdots i_n} x_1^{i_1}\cdots x_n^{i_n}\}$ in a manner analogous to that for polynomials. We denote this ring by $R[[x_1,x_2,\ldots,x_n]]$ and call it the ring of *formal power series*. It is clear that $R[[x_1,x_2]] \approx (R[[x_1]])[[x_2]]$ and we shall identify the two rings. We note that every element of the form $1 + \Sigma_{i\geq 1} a_i x^i$ in $R[[x]]$ is a unit in $R[[x]]$, since $(1 + \Sigma_{i\geq 1} a_i x^i)^{-1} = 1 - \Sigma_{i\geq 1} a_i x^i - (\Sigma_{i\geq 1} a_i x^i)^2 - \cdots$.

5.5.23 THEOREM Let R be a commutative noetherian ring. Then so is $R[[x_1,x_2,\ldots,x_n]]$.

Proof: It is sufficient to show that $R[[x]]$ is noetherian. For any element a in $R[[x]]$, we denote the homogeneous term of minimal degree by $\bar{a}$. Let A be an ideal in $R[[x]]$. Let A_0 be the ideal generated by all $\bar{a}$ $(a \in A)$ (indeed, $A_0 = \{\bar{a} \mid a \in A\}$). Then $A_0 = \langle \bar{a}_1,\bar{a}_2,\ldots,\bar{a}_s\rangle$ by (4.2.2). Let y be in A. Then $\bar{y} = \Sigma\, c_{0i}\bar{a}_i$, where c_{0i} is in $R[x]$. Since $\bar{a}_i$ and $\bar{y}$ are homogeneous of degree m_i and m, respectively, we may assume that the c_{0i} are also homogeneous of degree $m - m_i$. Hence $y_1 = y - \Sigma_i c_{0i}a_i \in (x_1,x_2,\ldots,x_n)^{m+1} \cap A$. Repeating this procedure, we have

$$y = \Sigma_i c_{0i}a_i + \Sigma_i c_{1i}a_i + \cdots + \Sigma_i c_{si}a_i + \cdots$$

where c_{si} is in $(x_1,\ldots,x_n)^{m+s-m_i}$. If we put $\Sigma\, c_{ki} = c_i \in R[[x]]$, then $y = \Sigma_i c_i a_i \in \langle a_1,a_2,\ldots,a_s\rangle$.

5.5.24 LEMMA Let $\{P_1,P_2,\ldots,P_n\}$ be a set of prime ideals in a commutative ring S. If an ideal A is not contained in any P_i, then A is not contained in the set union of the P_i.

Proof: We may assume that $P_i \not\supset P_j$ if $i \neq j$. Put $B_i = P_1 \cap P_2 \cap \overset{i}{\vee} \cdots \cap P_n$. Since P_i is prime, $P_i \not\supset B_i$ for all i. Let $b_i \in B_i - P_i$ and $a_i \in A - P_i$. We put $a = \Sigma_i a_i b_i$. Then $a \in A$. Further, noting that $b_j \in P_i$ for $i \neq j$, we know that if $a \in P_i$, then $a_i b_i \in P_i$, which is a contradiction.

5.5.25 THEOREM Let R be a semilocal noetherian ring and let $\{M_1,M_2,\ldots,M_s\}$ be the set of maximal ideals. Then $R[[x_1,x_2,\ldots,x_n]]$ is a semilocal noetherian ring and $\{M_i^*\} = \{\langle M_i,x_1,x_2,\ldots,x_n\rangle\}$ is the set of maximal ideals.

Proof: Since $R[[x]]/M_i^* \approx R/M_i$, M_i^* is maximal. Let A^* be an ideal of $R[[x]]$ not contained in any M_i^*. Then A^* contains an $a \notin M_i^*$ for all i by (5.5.24). Let $a = a_0 + b$, where a_0 is in R and b in $\langle x_1,x_2,\ldots,x_n\rangle$. Since $\langle x_1,x_2,\ldots,x_n\rangle \subset M_i^*$ for all i, $a_0 \notin M_i$ for all i. Hence a_0 is a unit in R. Therefore, a is a unit in $R[[x]]$ by the remark before (5.5.23), so $A^* = R[[x]]$, where $x = \{x_1,x_2,\ldots,x_n\}$.

5.5.26 THEOREM Let R be a semilocal noetherian ring and J = J(R). Then the completion $\hat{R}$ of R with respect to the J-adic topology is isomorphic to a factor ring of $R[[x_1,x_2,\ldots,x_n]]$. Hence $\hat{R}$ is a semilocal noetherian ring with $J(\hat{R}) = \hat{J}$.

Proof: Let $J = \langle a_1,a_2,\ldots,a_n \rangle$. Each element r in $\hat{R}$ is expressible as $r = r_0 + r_1 + \cdots + r_n + \cdots$; r_i is in J^i. Let $\{x_1,x_2,\ldots,x_n\}$ be a set of indeterminates. We define a ring homomorphism $\varphi : R[[x]] \to \hat{R}$ by setting $\varphi(a) = a$ for $a \in R$ and $\varphi(x_i) = a_i$. Then φ is an epimorphism from the above. Let M be a maximal ideal in $\hat{R}$. Then $\varphi^{-1}(M)$ is maximal in R[[x]]. Hence $\hat{R}$ is semilocal noetherian by (5.5.25). Furthermore, $\hat{R}/\hat{J} \approx R/J$ by (5.5.21). Hence $J(\hat{R}) \subset \hat{J}$. On the other hand, $\hat{R}/J(\hat{R})$ is semisimple, since $\hat{R}$ is semilocal and the number of simple components of $\hat{R}/J(\hat{R})$ is equal to or less than that for R[[x]]/J(R[[x]]), which is equal to the number for R/J by (5.5.25). Hence $\hat{J} = J(\hat{R})$.

Finally, we shall study the endomorphism ring of a quasi-injective module over a commutative noetherian ring.

5.5.27 PROPOSITION Let R be a commutative ring and M a quasi-injective module. If M is noetherian as an R-module, then $S = \mathrm{End}_R(M)$ is left and right artinian.

Proof: Let $M = \Sigma_{i=1}^{n} m_iR$ by (4.2.2). We can define an R-homomorphism $f : S \to \Sigma_{i=1}^{n} \oplus M$ by setting $f(s) = \Sigma_i sm_i$, since R is commutative. Then f is a monomorphism, so S is noetherian as an R-module. On the other hand, S is semiprimary by (5.5.8). Hence S is artinian.

5.5.28 PROPOSITION Let R and M be as in (5.5.27). Assume further that M is weakly distinguished. Put $S = \mathrm{End}_R(M)$. Then M is R-noetherian if and only if S is left artinian. In this case, M is R-artinian and S-injective, and R/A is artinian, where $A = r_R(M)$.

Proof: If M is R-noetherian, S is artinian by (5.5.27). Hence M is S-injective and M is R-artinian by (5.5.12), since S is noetherian. Further, R/A is an R-submodule of a finite direct sum of copies of M. Hence R/A is artinian. If S is (left) artinian, then M is R-noetherian by (5.5.10).

We shall determine the quasi-injective noetherian modules for some special cases.

5.5.29 LEMMA Let K be any ring and M a quasi-injective and weakly distinguished right K-module. Put $S = \mathrm{End}_K(M)$ and $T = \mathrm{End}_S(M)$. Then every K-submodule of M is a T-submodule of M.

Proof: Let N be a K-submodule of M. Then $r_M \ell_S(N) = N$ by (5.5.10). Hence N is a T-submodule.

Let R be a commutative noetherian ring and P a prime ideal of R. Let $E(R/P) = E$ be an injective hull of R/P. Then we obtained in (4.7.20) that $E = \cup_i A_i$, where $A_i = \ell_E(P^i)$ and we shall show in (5.5.34) that $\mathrm{End}_R(E)$ is a complete local noetherian ring.

5.5.30 LEMMA Let R be a commutative noetherian ring and $\{P_i\}$ a finite family of distinct maximal ideals in R. Then every R-submodule N of $\Sigma \oplus E(R/P_i)$ is weakly distinguished and quasi-injective.

Proof: We note that $E(R/P_i)$ is uniform. We first assume that N is an essential submodule of $E = \Sigma \oplus E_i$, where $E_i = E(R/P_i)$. Then for any x in N, we obtain $r_R(x) \supset \Pi_i P_i^{n_i}$ for some integers n_i. Let $N_1 \supset N_2$ be R-submodules of N such that N_1/N_2 is R-simple; then $N_1/N_2 \approx R/P_i$ for some P_i. Since $N \cap R/P_i \neq 0$, $\mathrm{Hom}_R(N_1/N_2, N) \neq 0$, which means that N is weakly distinguished. Hence E is R-weakly distinguished and injective. Moreover, if we put $S = \mathrm{End}_R(E)$, then $S = \mathrm{End}_S(E)$ [cf. (5.5.34)]. Hence every R-submodule N is an S-submodule by (5.5.29). In general, let E' be an injective hull of N contained in E. Then $E = E' \oplus E''$ and $E' \supset N$. $S' = \mathrm{End}_R(E')$ may be regarded as a subring of S. Hence N is also an S'-module. Therefore, N is R-quasi-injective by (5.4.2).

We are interested in a noetherian or an artinian quasi-injective module M and hence we may assume that M is directly indecomposable.

5.5.31 LEMMA Let T be an integral domain with K its quotient field. If K is a f.g. T-module, K = T.

Proof: If $K = \Sigma_{i=1}^n a_i T$, where $a_i = b_i c_i^{-1}$, then $cK \subset T$ for $c = \Pi_i c_i$. Hence $K = cK \subset T$.

5.5.32 THEOREM Let M be a directly indecomposable module over a commutative ring R. Then M is quasi-injective and noetherian if and only if there exist an ideal I such that R/I is noetherian and a maximal ideal P containing I such that M is contained in a submodule A_n of $E_{R/I}(R/P)$. In this case, M is R-artinian and hence R/I is artinian.

Proof: We assume that M is R-noetherian and quasi-injective. Put $I = r_R(M)$. Then $\bar{R} = R/I$ is noetherian as in the proof of (5.5.27). Hence we may assume that R is noetherian. Let E be an injective hull of M. Then $E = E_R(R/P)$ with P prime by (4.7.20). Put $S = \mathrm{End}_R(E)$. We know from (4.7.22) and its proof that $A_1 \approx Sa \approx K$ for any nonzero element a in A_1, where K is the quotient field of R/P. Since $M \cap A_1 \neq 0$ and M is quasi-injective, M contains a submodule that is isomorphic to K by (5.4.2). Hence P is a maximal ideal of R by (5.5.31), and M is contained in some A_n, since M is R-f.g. and each A_n has a composition length by (4.7.23). The remaining assertions are clear from the above and (5.5.30).

5.5.33 COROLLARY Let R be a commutative ring. Then there exists a noetherian injective R-module if and only if R contains a maximal ideal P such that R_P is artinian.

Proof: This is an immediate consequence of (5.5.32) and (4.7.23).

Finally, we shall consider quasi-injective artinian modules over a noetherian ring. Let R be a commutative noetherian ring. Let E be an indecomposable injective R-module. Then $S = \mathrm{End}_R(E) = \mathrm{End}_{R_P}(E)$ for some prime P by (4.7.21). Hence, when we study the structure of S, we may assume that R is a local ring.

5.5.34 THEOREM Let R be a commutative local noetherian ring with maximal ideal P and $E = E(R/P)$. Then $S = \mathrm{End}_R(E)$ is isomorphic to the completion $\hat{R}$ of R with respect to the P-adic topology. Hence S is noetherian.

Proof: Put $A_n = \ell_E(P^n)$. Then $E = \cup A_n$ by (4.7.20). Let $\hat{r}$ be in $\hat{R}$. We can express $\hat{r}$ as $\hat{r} = r_1 + r_2 + \cdots + r_n + \cdots$; $r_i \in P^i$. Let x be in E. Then $x \in A_n$ for some n. Define

$$\hat{r}x = (r_1 + r_2 + \cdots + r_{n-1})x$$

We can easily see that this operation is well defined (independent of the expression of $\hat{r}$ and A_n). If $\hat{r} \neq 0$, there exist some integers m and n such that $r_1 + r_2 + \cdots + r_m \notin P^n$, so $\hat{r}A_n \neq 0$ since $r_R(A_n) = P^n$ by (4.7.23). Hence $S \supset \hat{R}$. Let s be in S. Then $s(A_1) \subset A_1 = R/P$. Hence there exists an r_0 in R such that $r_E(s - r_0) \supset A_1$. Assume that we have already found r_0, $r_1, \ldots, r_{n-1}$ such that $r_i \in P^i$ and $r_E(s - (r_0 + r_1 + \cdots + r_{n-1})) \supset A_n$. Put $s_n = s - (r_0 + \cdots + r_{n-1})$. Then s_n induces an element in $\mathrm{End}_R(A_{n+1})$. Take an exact sequence

$$0 \to A_n \to A_{n+1} \to A_{n+1}/A_n \to 0$$

Then

$$0 \leftarrow \mathrm{Hom}_R(A_n,E) \leftarrow \mathrm{Hom}_R(A_{n+1},E) \leftarrow \mathrm{Hom}_R(A_{n+1}/A_n,E) \leftarrow 0$$

is exact. Since $A_iP^i = 0$, $\mathrm{Hom}_R(A_{n+1},E) = \mathrm{Hom}_R(A_{n+1},A_{n+1})$ and

$\mathrm{Hom}_R(A_{n+1}/A_n, E) = \mathrm{Hom}_R(A_{n+1}/A_n, R/P)$. Since $r_E(s_n) \supset A_n$, $s_n \mid A_{n+1}$ is given by an element in $\mathrm{Hom}_R(A_{n+1}/A_n, R/P) \approx P^n/P^{n+1}$ by (4.7.23). Thus there exists an r_n in $P^n - P^{n+1}$ such that $r_E(s_n - r_n) \supset A_{n+1}$ from the remark below (4.7.23). Hence $s = \lim(\Sigma_{i=1}^m r_i) \in \hat{R}$. Thus S is a local noetherian ring by (5.5.26).

5.5.35 LEMMA Let R be a commutative noetherian ring and P a maximal ideal in R. Then E = E(R/P) is artinian.

Proof: Put $S = \mathrm{End}_R(E)$. Then S is a commutative noetherian ring by (5.5.34) and E is weakly distinguished by (5.5.30). Hence E is R-artinian by (5.5.12).

5.5.36 THEOREM Let R be a commutative noetherian ring. An indecomposable module M is artinian and quasi-injective if and only if M is isomorphic to an R-submodule of E(R/P), where P is maximal.

Proof: Every R-submodule of E(R/P) is artinian and quasi-injective by (5.5.30) and (5.5.35). Let M be indecomposable and artinian quasi-injective. Then E = E(M) is indecomposable by (5.4.3), so E = E(R/P). Thus $M \supset R/P$ by the proof of (5.5.32). Since M is artinian, P is maximal.

5.6 DUALITIES

Let R and S be rings with identity. Let ${}_SM_R$ be a bimodule. Then the functor $h_M = \mathrm{Hom}_R(-,M)$ is a contravariant functor from $\underline{M}_R$ to $\underline{M}_S$. Since h_M is contravariant, $h_M(\Sigma_I \oplus N_\alpha) = \Pi_I\, h_M(N_\alpha)$. Hence the situation for h_M is different from that for h^M. We note that if M is f.g., $h^M(\Sigma_I \oplus N_\alpha) = \Sigma_I \oplus h^M(N_\alpha)$, since $f(M) \subset \Sigma_{I'} \oplus N$ for any f in $h^M(\Sigma_I \oplus N_\alpha)$, where I' is a finite subset of I.

Let $\underline{A}$ and $\underline{B}$ be categories. If there exist additive contravariant functors $X : \underline{A} \to \underline{B}$, $Y : \underline{B} \to \underline{A}$ such that XY (resp. YX) is naturally isomorphic to $1_{\underline{B}}$ (resp. $1_{\underline{A}}$), then we say that $\underline{A}$ and $\underline{B}$ are *dual*.

For equivalence we need the concept of generators. Hence we shall define a cogenerator, which is the dual concept to a generator. Let U be an object in a category $\underline{A}$. If U satisfies the condition that for $\alpha \neq \beta$ in [A,B] and A, B in $\underline{A}$, there exists $\gamma \in [B,U]$ such that $\gamma\alpha \neq \gamma\beta$, we say that U is a *cogenerator*.

5.6.1 LEMMA Let U_R be an R-module. Then if U is a cogenerator in $\underline{M}_R$, then every R-module is a submodule of a direct product Π_I U of I copies of U.

Proof: Let M be an R-module and put $T = \Pi_{[M,U]} U$, where $U_\alpha = U$ is indexed by $f_\alpha : M \to U$. Then we have a unique homomorphism $f : M \to T$ given by $\{f_\alpha : M \to U_\alpha\}$. Put ker f = K. Then $f_\alpha(k) = p_\alpha f(k) = 0$ for all f_α by the definition of product, where $p_\alpha : T \to U_\alpha$ is the projection and k is in K. Let $i : K \to M$ be the inclusion. If $i \neq 0$, there exists some $f_\alpha \in [M,U]$ with $f_\alpha i \neq 0$, which is a contradiction.

The following is dual to (5.2.1).

5.6.2 THEOREM Let U_R be an R-module. Then the following conditions are equivalent:

1. U is a cogenerator in $\underline{M}_R$.
2. Every R-module is a submodule of a direct product of copies of U.
3. U contains an injective hull of any simple R-module.

Proof: (1) ⇒ (2) is shown by (5.6.1).

(2) ⇒ (3). Let C be a simple module and E(C) = E. Then we have $E \subset \Pi_I U$ by (2). Since C is simple, $p_\alpha : C \to U$ is a monomorphism for some $\alpha \in I$ and hence $p_\alpha : E \to U$ is a monomorphism, since E is uniform.

(3) ⇒ (1). We assume that $E' \subset U$ for all $E' = E(C)$. Let $\alpha \neq \beta$ be in [T,S] and put $M = \text{im}(\alpha - \beta) \neq 0$. Then M contains submodules $M_1 \supset M_2$ with M_1/M_2 simple [see the proof of (5.5.10)]. $M_1 \to M_1/M_2 \to E(M_1/M_2) \neq 0$ can be extended to an element γ in $[M,E] \subset [S,E] \subset [S,U]$ and $\gamma\alpha \neq \gamma\beta$.

5.6.3 PROPOSITION Let U_R be a cogenerator and let X be a right ideal in R. Then $r_R \ell_U(X) = X$.

Proof: It is clear that $X \subset X' = r_R \ell_U(X)$. Assume that $X'/X \neq 0$. Then there exist right ideals $X' \supset X_1 \supset X_2 \supset X$ with X_1/X_2 simple. Since $U \supset E(X_1/X_2)$ by (5.6.2), there exists $\alpha \neq 0$ in $[R/X_2,U]$, which is an extension of $i : X_1/X_2 \to U$. Hence $\alpha f \neq 0$ is in [R/X,U], where $f : R/X \to R/X_2$ is the natural epimorphism. It is clear that $\alpha f(\bar{1})$ is in $\ell_U(X)$ [see the remark above (4.7.23)], so $\alpha f(X'/X) = \alpha f(\bar{1})X' = 0$. However, $\alpha f(X'/X) = \alpha(X'/X_2) \neq 0$, which is a contradiction.

The next theorem is dual to (5.3.5).

5.6.4 THEOREM Let R and R' be rings. By $\underline{M}_R^f$ (resp. ${}_{R'}\underline{M}^f$) we denote the category of all f.g. right R-modules (resp. left R'-modules). Assume that R is right artinian and R' is left artinian. If $\underline{M}_R^f$ and ${}_{R'}\underline{M}^f$ are dual, there exist R'-R bimodules U and V satisfying the following conditions:

1. U and V are injective cogenerators both in $\underline{M}_R$ and ${}_{R'}\underline{M}$.
2. $\text{End}_R(U) \approx R'$ and $\text{End}_{R'}(U) \approx R$.

3. $U \approx V$ as R'-R modules.
4. $h_U : \underline{M}_R^f \to {}_{R'}\underline{M}^f$ gives the duality.

Proof: Let $T : \underline{M}_R^f \to {}_{R'}\underline{M}^f$ and $S : {}_{R'}\underline{M}^f \to \underline{M}_R^f$ be contravariant functors with $ST = 1_{\underline{M}_R^f}$ and $TS = 1_{{}_{R'}\underline{M}^f}$. Put $U = T(R)$. Then ${}_{R'}U$ is a f.g. R'-module by the assumption. Since R is an R-projective generator, U is an R'-injective cogenerator. (We note that ${}_{R'}U$ is injective in ${}_{R'}\underline{M}^f$. Since R' is left noetherian, ${}_{R'}U$ is injective in ${}_{R'}\underline{M}$ by (4.4.2). Similarly, ${}_{R'}U$ is a cogenerator in ${}_{R'}\underline{M}$. For let $g \neq 0$ be in $[A,B]_{R'M}$. Then $Ra \to Rg(a)$ is in ${}_R\underline{M}^f$. Hence there exists γ in $[Rg(a),U]$ such that $\gamma(g(a)) \neq 0$ provided that $g(a) \neq 0$. Since U is injective in ${}_{R'}\underline{M}$, γ can be extended to an element in $[B,U]$.) Since $R = [R,R]_R$, $T(r) \in [U,U]$ for any $r \in R$. U is a left R'-module, so we shall write the element of ${}_{R'}[U,U]$ acting on the right side. T being a contravariant functor, U may be regarded as a right R-module via $T(r)$. Hence U is an R'-R bimodule and $R = {}_{R'}[U,U]$. Then U_R is injective by (5.5.12). Next we show that $h_U = T$. Since U is injective, h_U is an exact contravariant functor and $h_U(R) = U = T(R)$. Let A be a f.g. R-module and $0 \to A \to R^{(I_1)} \to R^{(I_2)}$ a projective resolution of A with I_1, I_2 finite. Then we have exact sequences

$$\begin{array}{ccccccc} 0 \longrightarrow TA & \longrightarrow & T(R^{(I_1)}) = U^{(I_1)} & \xrightarrow{\alpha} & T(R^{(I_2)}) = & U^{(I_2)} \\ & & \| & & & \| \\ 0 \longrightarrow [A,U] & \longrightarrow & U^{(I_1)} & \xrightarrow{\alpha} & & U^{(I_2)} \end{array}$$

Hence $TA \approx \ker \alpha = [A,U]$. We can check directly that T and h_U are naturally isomorphic via the isomorphisms above [cf. the proof of (5.3.5)]. Therefore, $[A,U] \neq 0$ if $A \neq 0$, so U_R is a cogenerator, since U_R is injective. Similarly, if we put $S(R') = V$, V satisfies condition (1). Since $ST = 1_{\underline{M}_R^f}$, $U \approx STU = [[U,U]_{R'},V] = V$ as R-modules. Similarly, we have an R'-isomorphism: $V \to [{}_R[V,V],U] = U$. Thus we have proved (5.6.4).

We note that if there exist U and V satisfying (1)-(3) in (5.6.4), then h_U and h_V give the duality between $\underline{M}_R^f$ and ${}_{R'}\underline{M}^f$ (since ${}_{R'}[[U,U]_{R'},V] = V$, $h_V h_U = 1_{\underline{M}_R^f}$ from the proof of (5.6.4)).

5.7 QUASI-FROBENIUS RINGS

Let K be a field and R an algebra over K. We assume that R is of finite dimension over K. Hence R is left and right artinian. Every R-module M_R may be regarded as a K-vector space. Set $M^* = \mathrm{Hom}_K(M,K)$. Then M^* is a left R-module. Since K is trivially K-injective, the functor $(\cdot)^*$ is exact from $\underline{M}_R$ to $_R\underline{M}$ (and also from $_R\underline{M}$ to $\underline{M}_R$). We consider M^{**}. There is a natural mapping $\nu : M \to M^{**}$ given by setting $\nu(m)(f) = f(m)$, where m is in M and f in M^*. Let $M = mK \oplus M'$ as K-modules. Then we obtain a g in M^* with $g(m) = 1$ and $g(M') = 0$. Hence $\nu(m)(g) = g(m) = 1 \neq 0$. Therefore, ν is a monomorphism for any M. Furthermore, the argument above shows that $[M : K] = [M^* : K]$ if $[M : K] < \infty$ $[M^* = \mathrm{Hom}_K(\Sigma_1^n \oplus K,K) = \Sigma_1^n \oplus \mathrm{Hom}_K(K,K) = \Sigma_1^n \oplus K]$. Hence, if M is a f.g. R-module, $M \approx M^{**}$ as R-modules. Thus we know that $\underline{M}_R^f$ and $_R\underline{M}^f$ are dual via $(\cdot)^*$. Therefore, R^* is an injective cogenerator in $\underline{M}_R$ and $_R\underline{M}$ by (5.6.4).

If $R^* \approx R$ as two sided R-modules, then we call R a *Frobenius algebra*.

5.7.1 THEOREM Let K be a field and G a finite group. Then the group algebra KG is a Frobenius algebra.

Proof: Let $R = KG = \Sigma_{g\in G} \oplus gK$ and $\theta : \Sigma_g gk_g \to k_1$ in R^*. Then $R^* = R\theta = \theta R$. We shall show that the mapping $\varphi : R \to R^*$ given by $\varphi(r) = \theta r = r\theta$ is the desired isomorphism. It is clear that φ is an R-homomorphism. Since $R^* = \theta R = R\theta$, φ is an epimorphism. Now $[R : K] = [R^* : K]$. Hence φ is an isomorphism.

If R is a Frobenius algebra, R is injective as a left and a right R-module. From this point of view we may generalize this property to a one-sided artinian ring. If R is a right artinian ring and R is injective as a right R-module, we call R a *quasi-Frobenius ring*. We abbreviate this by QF.

5.7.2 THEOREM Let R be a ring. Consider the following two groups of conditions:

(a-1) right artinian, (a-2) left artinian, (a-3) right noetherian, and (a-4) left noetherian

(b-1) injective as a right R-module and (b-2) injective as a left R-module

Then R is a QF-ring if and only if R satisfies each one of the conditions in the two groups.

Proof: We note that $R \approx \mathrm{Hom}_R(R,R)$ as a two-sided R-module. Assume that R is a QF-ring. Then R satisfies (a-1), (a-3) by (4.5.9) and (b-1). Then R is left noetherian by (5.5.6) and semiprimary by (5.5.8). Hence R is left artinian. Since R is artinian, R has the standard decomposition

$$R = \sum_{i=1}^{n} \sum_{j=1}^{\rho(i)} \oplus e_{ij}R \qquad [\text{see } \underline{(4.5.1)}]$$

where $\{e_{ij}\}$ is a set of mutually orthogonal primitive idempotents and $e_{ij}R \approx e_{i1}R$, $e_{i1}R \not\approx e_{j1}R$ if $i \neq j$. $e_{ij}R$ is indecomposable and injective, so it is uniform. Hence the socle $S(e_{i1}R)$ is not isomorphic to $S(e_{i'1}R)$ if $i \neq i'$ by (4.4.8). Therefore, R contains the injective envelope of all simple modules $\{S(e_{i1}R)\}_{i=1}^{n}$ and R is a cogenerator by (5.6.2). Accordingly, R is left R-injective by (5.5.12). Similarly, we can show the remaining assertion using Sec. 5.6.

From (5.7.2) we know that R is two-sided artinian if R is a QF-ring, and the definition of QF-ring is left- and right-sided symmetric. In the proof of (5.7.2) we showed that each indecomposable right ideal $e_{i1}R$ contains a unique minimal right ideal A_i and $A_i \not\approx A_j$ if and only if $i \neq j$. Since A_i is simple, $A_i \approx e_{i'}(R/J)$, where $e_{i'} = e_{i'1}$. The mapping $\pi : i \to i'$ is a permutation of $\{1,2,\ldots,n\}$. Thus we have shown that if R is QF, then:

Each e_iR contains a unique minimal right ideal A_i and $A_i \not\approx A_j$ if $i \neq j$. (5.7.1)

Each Re_i contains a unique minimal left ideal B_i. (5.7.2)

We shall show that if R satisfies (5.7.1) and (5.7.2), then R is a QF-ring.

5.7.3 THEOREM Let R be two-sided artinian. Assume that R satisfies (5.7.1) and (5.7.2). Then $B_{\pi(i)} \approx \bar{R}e_i$, where $\bar{R} = R/J$ and $J = J(R)$ and $r_R(J) = \ell_R(J)$ (= M). If we put $E_i = \Sigma_{j=1}^{\rho(i)} e_{ij}$, $E_iM = ME_{\pi(i)}$ is a minimal two-sided ideal.

Proof: First we note that $r_R(J) = S({}_RR)$. Let L be a simple left module. Then $JL = L$ or $JL = 0$. Since J is nilpotent, $JL = 0$. Hence $S({}_RR) \subset r_R(J)$. On the other hand, since $Jr_R(J) = 0$, $r_R(J)$ is an $\bar{R}$-module. Hence $r_R(J)$ is completely reducible by (4.5.4). Therefore, $r_R(J) \subset S({}_RR)$. Now $S(R_R) = \ell_R(J) = \Sigma_{i=1}^{n}\Sigma_{j=1}^{\rho(i)} \oplus S(e_{ij}R)$. Hence $A_i = e_i\ell_R(J)$ and $B_k = r_R(J)e_k$. Let $A_i = e_{\pi(i)}R$. Then $A_ie_{\pi(i)} \neq 0$, so $A_i \subset \ell_R(J)e_{\pi(i)}$ and $E_i\ell_R(J) = \Sigma_{i=1}^{n} \oplus e_i\ell_R(J) = \Sigma_{i=1}^{n} \oplus \ell_R(J)e_{\pi(i)}$ (note that π is a permutation).

Hence $E_i\ell_R(J) = \ell_R(J)E_{\pi(i)}$ is a two-sided ideal. Let $d \neq 0$ be an element in $E_i\ell_R(J)$ and $d = \Sigma_j e_{ij}d$. Then $e_{ij}d \neq 0$ for some j. Since $e_{ij}\ell_R(J)$ is minimal, $e_{ij}\ell_R(J) = e_{ij}dR$. Since $e_{ik}R \approx e_{ij}R$, $Re_{ij}dR = E_i\ell_R(J)$ by (3.1.7). Hence $E_i\ell_R(J)$ is a minimal two-sided ideal. Let C be a minimal left ideal in $E_i\ell_R(J)$. Then $E_i\ell_R(J) = CR = \Sigma_x Cx$ is semisimple as a left R-module by (4.1.2). Hence $E_i\ell_R(J) \subset r_R(J)$. Therefore, $\ell_R(J) = \Sigma_i E_i\ell_R(J) \subset r_R(J)$. $e_iB_{\pi(i)} = e_ir_R(J)e_{\pi(i)} \supset e_i\ell_R(J)e_{\pi(i)} = e_iRe_{\pi(i)} \neq 0$, so $B_{\pi(i)} \approx \overline{Re_i}$ (note that $e_j\bar{R}e_i = 0$ if $i \neq j$). Thus $B_i \not\approx B_j$ if $i \neq j$. By symmetry we obtain $r_R(J) \subset \ell_R(J)$.

5.7.4 LEMMA Let R be a ring and B a left ideal. If B' is a maximal left R-submodule in B, then $Br_R(B') \subset r_R(J)$. Let a be an element in B - B'. Then $r_R(B')/r_R(B) \approx ar_R(b')$.

Proof: Let x be in $r_R(B')$. Then $\ell_R(x) \supset B'$. Hence $B \supset \ell_R(x) \cap B \supset B'$. Since B' is maximal in B, $\ell_R(x) \cap B - B'$ or $B - \ell_R(x) \cap B$; i.e., $Bx = 0$. If $B' = \ell_R(x) \cap B$, $B' = \ker (b \to bx : b \in B)$, so Bx is minimal. In both cases $Bx \subset r_R(J)$. Since B' is maximal in B, B = B' + Ra. Thus $r_R(B) = r_R(B') \cap r_R(a)$. Consider the homomorphism: $y \in r_R(B') \to ay$. Then $r_R(B')/r_R(B) = r_R(B')/[r_R(B') \cap r_R(a)] \approx ar_R(B')$.

5.7.5 THEOREM Let R be two-sided artinian. If R satisfies (5.7.1) and (5.7.2), then

$r_R\ell_R(A) = A$ and $\ell_Rr_R(B) = B$ for each right ideal A and each left ideal B.

Proof: Let $B \supset B'$ be left ideals with B/B' simple. We assume that $B/B' \approx \overline{Re_i}$. Then $e_iB \not\subset B'$. Let e_ia be in $e_iB - B'$. Then $e_iar_R(B') \subset e_ir_R(J) = e_i\ell_R(J)$ by (5.7.3) and (5.7.4). Hence $e_iar_R(B') = 0$ or $e_iar_R(B')$ is simple. Therefore, $r_R(B) = r_R(B')$ or $r_R(B')/r_R(B)$ is simple by (5.7.4). Let

$$R = B_0 \supsetneq B_1 \supsetneq \cdots \supsetneq B_n = 0$$

be a composition series of $_RR$. Then

$$0 = r_R(B_0) \subset r_R(B_1) \subset \cdots \subset r_R(B_n) = R$$

and $r_R(B_i) = r_R(B_{i+1})$ or $r_R(B_i)/r_R(B_{i+1})$ is simple by the above. We obtain the same property for $\ell_R(\cdot)$. Noting that $\ell_Rr_R(B) \supset B$, we know that $\ell_Rr_R(B_i) = B_i$ for all i (use the length of the series).

5.7.6 THEOREM Let R be two-sided artinian. Then R is QF if and only if R satisfies (5.7.1) and (5.7.2).

Proof: If R satisfies (5.7.1) and (5.7.2), then $r_R(\cdot)$ and $\ell_R(\cdot)$ give anti-isomorphisms between one-sided ideals. Hence R satisfies (5.5.1). Then R is R-injective by the proof (1) ⇒ (3) in (5.5.12). The converse is shown above (5.7.3).

5.7.7 THEOREM Let R be right artinian. Then the following conditions are equivalent:

1. R is a QF-ring.
2. Every projective module is injective.
3. Every injective module is projective.

Proof: If R is artinian, every projective module is a direct sum of $\{e_iR\}$ by (7.6.10) below. Hence (1) ⇔ (2) is clear.

(1) ⇒ (3). Let E be injective. Then E = E(S(E)) by (4.7.19). Since R is a cogenerator, $E \approx \Sigma_{i=1}^{n} \oplus e_iR^{(I_i)}$ by (5.6.2).

(3) ⇒ (1). Let C be a simple module, say isomorphic to e_iR/e_iJ. Hence $E(e_iR/e_iJ) \approx e_{\pi(i)}R$ for some $\pi(i)$. It is clear that π is a permutation of $\{1,2,\ldots,n\}$. Therefore, R is R-injective.

We shall come back to the case of algebras R over a field K with $[R : K] < \infty$. Since the functor ( )* gives us the duality between $\underline{M}_R^f$ and ${}_R\underline{M}^f$, (eR)* is an indecomposable injective left R-module, where e is a primitive idempotent. Further, if $eR \not\approx e'R$, then $(eR)^* \not\approx (e'R)^*$ as left R-modules. Let $\{e_iR\}_{i=1}^{n}$ be a set of nonisomorphic representatives of indecomposable projective modules in $\underline{M}_R$. Then $\{(e_iR)^*\}_{i=1}^{n}$ is a set of indecomposable injective left R-modules. Let E be an injective left R-module. Then E is of the form $\Sigma \oplus E(S_i)$ by (4.7.19), where the S_i are simple left R-modules. Hence $\{(e_iR)^*\}_{i'=1}^{n}$ is a set of nonisomorphic representatives of injective left R-modules. Hence we have the following.

5.7.8 LEMMA Let R be an algebra of finite dimension over a field. Then $\{(Re_i)^*\}_{i=1}^{n}$ is a set of nonisomorphic representatives of indecomposable injective right R-modules. Hence every injective left R-module is a direct sum of these.

5.7.9 THEOREM Let R be an algebra of finite dimension over a field K. Then R is a QF-algebra if and only if R satisfies (5.5.1).

Proof: It is sufficient to show the "if" part. We shall show that R is self-injective. Hence we may assume that R is basic by (5.3.6).

Let $R = \Sigma_{i=1}^{n} \oplus e_iR$, where $\{e_i\}_{i=1}^{n}$ is a set of mutually orthogonal idempotents and $e_iR \not\approx e_jR$ if $i \neq j$. Since e_iR is uniform by the assumption, $E(e_iR)$ is indecomposable. Hence $E(e_iR) \approx (Re_{\pi(i)})^*$ by (5.7.8). Further, since $S(e_iR) \not\approx S(e_jR)$ for $i \neq j$, $E(e_iR) \not\approx E(e_jR)$, so π is a permutation of $\{1,2,\ldots,n\}$. Hence $[E(R) : K] = \Sigma_{i=1}^{n}[E(e_iR) : K] = \Sigma_{i=1}^{n}[Re_{\pi(i)} : K] = [R : K]$. Therefore, $E(R) = R$.

5.8 QF-1,2,3 RINGS

Following Thrall [215] we shall define some rings which generalize QF-rings. In this section we make use of the concept of perfect rings which are studied in Sec. 7.6. However, we may replace this by using artinian rings until Sec. 7.6.

First we give some properties for later use.

5.8.1 PROPOSITION Let M_R be an indecomposable module with $E(M)$ noetherian. Then $\mathrm{End}_R(M)$ is a local semiprimary ring.

Proof: We use the argument dual to the proof of (4.5.18). Put $S = \mathrm{End}_R(M)$ and $S_E = \mathrm{End}_R(E)$. Let s be in S. The series $r_M(s) \subset r_M(s^2) \subset \cdots \subset r_M(s^m) \subset \cdots$ terminates by the assumption. Hence $r_M(s^n) = r_M(s^{2n})$ for some n. Put $M_0 = s^n(M)$ and $s' = s^n \mid M_0 \in \mathrm{End}_R(M_0)$. Then $E(M_0)$ is also noetherian. Since $M_0 \cap r_M(s^n) = 0$, s' is a monomorphism. Hence s' can be extended to an automorphism h of $E(M_0)$, because $E(M_0)$ is noetherian. If s' is not an epimorphism, $h^{-1}(s'(M_0)) = M_0 \subsetneqq h^{-1}(M_0) \subsetneqq h^{-2}(M_0) \subsetneqq \cdots$, which is a contradiction. Thus s' is an isomorphism, so $M = M_0 \oplus r_M(s^n)$ by (1.7.5). Therefore, s is either nilpotent or an isomorphism by (1.2.7), and we obtain from the proof of (4.5.18) that S is a local ring whose radical consists of all nilpotent elements. Put $J' = J(S_E) \cap S$. Then S/J' is isomorphic to a subring of a semisimple ring $S_E/J(S_E)$, and J' is nilpotent by (5.5.8). We denote $S_E/J(S_E)$ by $\bar{S}_E$. Since $\bar{S}_E$ is semisimple, $\|\bar{S}_E\| = r < \infty$ as a right $\bar{S}_E$-module. We can regard $\bar{S}_E$ as $\mathrm{End}_{\bar{S}_E}(\bar{S}_E)$. Then the argument used in the last half of the proof of (4.5.18) shows that $\overline{J(S)}^r = 0$, since $\overline{J(S)}$ is nil. Therefore, $J(S)$ is nilpotent and $J(S) = J'$.

5.8.2 COROLLARY Let R be a right noetherian ring. If $E(R_R)$ is noetherian, then R is right artinian.

Proof: Since R is noetherian, $1 = \Sigma\, e_i$, where $\{e_i\}$ is a set of mutually orthogonal primitive idempotents by (3.1.2). Then $\mathrm{End}_R(e_iR) = e_iRe_i$ is a local semiprimary ring by (5.8.1). Hence $R = \mathrm{End}_R(R)$ is a semiprimary ring by the proof of (4.5.19). Since R is noetherian, R is right artinian.

Now we summarize the main properties of a QF-ring R. In general, we can define the functor $(\cdot)^*$ with $N^* = \mathrm{Hom}_S(N_S, S)$ for a ring S. Then there is a natural morphism φ from N to N^{**} by setting

$$\varphi(n)(f) = f(n) \qquad \text{where } n \text{ is in } N \text{ and } f \text{ in } N^*$$

If φ is a monomorphism, we say that N is *torsionless* and if φ is an isomorphism, we say that N is *reflexive*.

5.8.3 PROPOSITION Let N_S be an S-module. Then N is torsionless if and only if N is isomorphic to a submodule of a product of copies of S.

Proof: We can define $f : N \to \Pi_{[N,S]} S$ as in the proof of (5.6.2). If N is torsionless, there exists f in [N,S] with $f(n) \neq 0$ for an n in N. Hence f is a monomorphism. Conversely, assume that $N \subset \Pi_I S$. Let p_α be the projection of $\Pi_I S$ to the αth component. If n (= 0) $\in N$, $p_\alpha(n) \neq 0$ for some α and $p_\alpha|N$ is in [N,S].

If R is a QF-ring, R is an injective cogenerator in $\underline{M}_R$ and $_R\underline{M}$ and the functor ( )* from $\underline{M}_R^f$ to $_R\underline{M}^f$ gives a duality by (5.6.4). Hence we obtain that every f.g. R-module is reflexive.

Next, if R is a QF-ring, then eR and Re each contain a unique minimal submodule for a primitive idempotent e, and every indecomposable projective is of the form eR (Re). Hence we obtain that every indecomposable projective module contains a unique minimal submodule.

If every projective module in $_S\underline{M}$ and $\underline{M}_S$ for a ring S satisfies the condition above, then we say that S is a *QF-2 ring*. We can define similarly a one-sided QF-2 ring.

As the dual to the above, we obtain that every indecomposable injective module contains a unique maximal submodule, i.e., is cyclic hollow. In this case we say that S is a *QF-2* ring*.

If R is a QF-ring, R contains a minimal injective cogenerator $\Sigma \oplus e_iR = U$, where the e_i are primitive idempotents and $S(e_iR) \not\approx S(e_jR)$ if $i \neq j$.

5.8.4 PROPOSITION Let R and U be as above. Then every faithful module M_R contains a direct summand isomorphic to U.

Proof: Put $S_i = S(e_iR)$. Since M is faithful, $MS_i \neq 0$. Let $mS_i \neq 0$ for some m in M. Consider the mapping $\varphi : e_iR \to me_iR \subset M$; $e_ir \to me_ir$.

Then φ is a monomorphism, since $\varphi(S_i) \neq 0$. Further, since e_iR is injective, $M = M' \oplus M''$, $M' \approx e_iR$. e_iRS_j is a direct sum of copies of S_j, so $e_iRS_j = 0$ if $i \neq j$. Hence $M''S_j \neq 0$. Repeating the argument above, we have (5.8.4).

Thus, if R is a QF-ring, every faithful module contains a minimal faithful injective module. If every faithful module in ${}_S\underline{M}$ and $\underline{M}_S$ for a ring S satisfies the property above, then we say that S is a *QF-3 ring*. Similarly, we can define a one-sided QF-3 ring.

5.8.5 PROPOSITION Let R be a right and left artinian ring. Then the following are equivalent:

1. R is right QF-3.
2. $E(R_R)$ is projective.
3. R contains a faithful injective right ideal.

Proof: (1) ⇒ (2), (3). Assume that R is a right QF-3 ring. Since R is R-faithful, R contains a minimal faithful injective module E and $R = E \oplus K$. Hence E is projective and $E = \Sigma_{i=1}^{n} \oplus e_iR$, where the e_i are local idempotents, by (7.6.1). Further, $S_i = S(e_iR) \neq 0$ is simple. Let S be a simple submodule of R. Since E is faithful, there exists an x in E such that $xS\ (\neq 0) \approx S$. Hence $xS \approx S_i$ for some i, so $E(S) \approx e_iR$. Since $E(R) = E(S(R)) \approx \Sigma_{i=1}^{n} \oplus e_iR^{(n_i)}$, E(R) is projective.

(2) ⇒ (1). Assume that E(R) is projective. Then $E(R) \approx \Sigma_{i=1}^{n'} \oplus e_iR^{(I_i)}$ by (7.6.10). Since R is cyclic and R is essential in E(R), all I_i are finite. Further, $\Sigma_{i=1}^{n'} \oplus e_iR$ is faithful and injective, for E(R) is faithful. Then we can show from the proof of (5.8.4) that $\Sigma_{i=1}^{n'} \oplus e_iR$ is contained isomorphically in any faithful module.

(3) ⇒ (1). Use the same argument as above.

Finally, we define a QF-1 ring. Let M_S be an S-module and put $T = \mathrm{End}_S(M)$. Then M is a T-S bimodule. Let S' be the double centralizer of M, namely, $S' = \mathrm{End}_T(M)$. Then S' contains a ring homomorphic image S^o of S (Sec. 3.7). Further, M is S-faithful if and only if S^o is isomorphic to S by the natural mapping. In general, if $S' = S^o$, we say that M is *balanced*.

5.8.6 PROPOSITION Let M_S be an S-module. We assume that $M = \Sigma_{i=1}^{n}\Sigma_{j=1}^{\rho(i)} \oplus M_{ij}$, with $M_{i1} \approx M_{ik}$ for $k = 2, \ldots, \rho(i)$ and $M_{i1} \not\approx M_{j1}$ if $i \neq j$. We put $M_0 = \Sigma_{i=1}^{n} \oplus M_{i1}$. Then the double centralizer of M is isomorphic to that of M_0 via the natural correspondence.

Proof: Put $T = \mathrm{End}_R(M)$, $T_0 = \mathrm{End}_R(M_0)$, $S' = \mathrm{End}_T(M)$, and $S'_0 = \mathrm{End}_{T_0}(M_0)$. Let p_{ij} be the projection of M onto M_{ij}. Since $p_{ij} \in T$, $(M_{ij})s' = (p_{ij}(M))s' = p_{ij}(Ms') \subset M_{ij}$ for s' in S'. Let $a_{kl}^{(i)}$ be a given isomorphism from M_{ik} to M_{il}. Assuming that $a_{kl}^{(i)}(M_{pq}) = 0$ if $M_{pq} \neq M_{ik}$, we obtain that $a_{kl}^{(i)}$ is in T. Hence $a_{kl}^{(i)}(s'|M_{ik})(a_{kl}^{(i)})^{-1} = s'|M_{il}$. Therefore, we can establish (5.8.6) by the same argument as in (3.7.3).

5.8.7 COROLLARY Let R be a QF-ring and U a minimal injective cogenerator as given before (5.8.4). Then U is balanced.

5.8.8 PROPOSITION Let R be a QF-ring. Then every faithful module is balanced.

Proof: Let M be a faithful module. Then $M = U \oplus N$ by (5.8.4). Let R' be the double centralizer of M and r' in R'. Then $r'|U$ is given by right multiplication by an element r in R by (5.8.7) [cf. the proof of (5.8.6)]. We shall show that $nr' = nr$ for all n in N. On the contrary, assume that there exists an n in N with $nr - nr' \neq 0$. Then, since U is an injective cogenerator, there exists a g in $\mathrm{Hom}_R(N,U)$ such that $g(nr - nr') \neq 0$. We may assume that g is in $\mathrm{End}_R(M)$. Hence $g(nr) = g(n)r = g(n)r' = g(nr')$, for g(n) is in U and r' in R', which is a contradiction. Therefore, M is balanced.

If a ring S has the property that every faithful module is balanced in ${}_{S\text{-}}M$ and $M_{\text{-}S}$, then we say that S is a *QF-1 ring*.

6
Categories of Indecomposable Modules

We have studied the structure of semisimple modules and know that simple modules have a very simple structure; for example, the endomorphism ring of a simple module is a division ring. On the other hand, the endomorphism ring S of an indecomposable module M with finite length is a local ring; i.e., S/J(S) is a division ring, where J(S) is the Jacobson radical of S. From this point of view we can say that there might be some relationships between simple modules and indecomposable modules with finite length. Unfortunately, there are actually only a few known ring-theoretic relationships between them.

On the other hand, if we consider the above situation from the point of view of our study of category theory in Chap. 2, we might regard S/J(S) as a subset of morphisms [M,M]. Using this idea, we shall construct a new category as in the Krull-Remak-Schmidt-Azumaya theorem.

The main purpose of this chapter is to introduce the categorical machinery for the theorem above. We shall give some fundamental properties of the factor category related to a set of completely indecomposable modules $\{M_\alpha\}_I$. Further extended properties for direct sums $\Sigma_I \oplus M_\alpha$ will be given in the following chapters.

Using the results in Secs. 4.3 and 4.5, we study semiperfect modules in Sec. 6.1 and their categorical interpretations in Sec. 6.5. In Sec. 6.2 we give a ring-theoretical interpretation of Freyd's criterion for the abelian category. Section 6.3 is the principal section in this chapter. We investigate the factor category of the category induced from a set of completely indecomposable modules. We shall show that this factor category is Grothendieck. We will make great use of this fact in the following chapters. In Sec. 6.4 we investigate a factor category of the category induced from injective modules, which we shall utilize for the study of the Goldie dimension of modules in Chap. 7. Similarly, we study a factor category of the category induced from a set of projective modules in Sec. 6.5. In the final section, Sec. 6.6, we give a ring-theoretical proof of the main theorem (6.3.8). Its proof is somewhat simpler than the proof for (6.3.8). However, we will see later that the categorical method is much more useful and simpler than the ring-theoretical method.

6.1 SEMIPERFECT MODULES

We know that every R-module has an injective envelope (Sec. 4.4). We defined the concept dual to the above--projective cover--in Sec. 4.3. However, not all modules have projective covers. In this section we study modules with projective covers.

Let M be a projective module. If every factor module of any direct sum of copies of M has a projective cover, M is called *perfect*. If every factor module of any finite direct sum of copies of M has a projective cover, M is called *semiperfect*.

First we deal with general properties of endomorphism rings of projective modules.

6.1.1 LEMMA Let P_R be projective and $S = \mathrm{End}_R(P)$. For an R-submodule N of P, N is small in P if and only if $\mathrm{Hom}_R(P,N)_S$ is a right ideal contained in J(S), where J( ) means the Jacobson radical.

Proof: It is clear that $A = \mathrm{Hom}_R(P,N)$ is a right ideal in S. We assume that N is small in P and that $S = A + B$ for a right ideal B. Let $1 = a + b$; $a \in A$ and $b \in B$. Then $P = a(P) + b(P) = N + b(P) = b(P)$. Since P is projective, there exists an h in S such that $bh = 1_P$ by (4.3.2). Hence $B = S$. Therefore, $A \subset J(S)$ by (3.4.3). Conversely, we assume that $A \subset J$ and $P = N + T$ for some $T_R \subset P_R$. From the exact sequence $N \oplus T \to N + T \to 0$ we obtain $\mathrm{Hom}_R(P,N) + \mathrm{Hom}_R(P,T) = S$. Hence $\mathrm{Hom}_R(P,T)$ contains 1_P, which means that $T = P$.

6.1.2 COROLLARY Let P and S be as in (6.1.1). Then $J(S) = \{f \in S \mid f(P)$ is small in P$\}$.

Proof: Let f be in J(S). Put $N = f(P)$. Since $f : P \to N$ is an epimorphism, there exists an h in S such that $g = fh$ for any g in $\mathrm{Hom}_R(P,N)$. Hence $\mathrm{Hom}_R(P,N) \subset J$ and N is small in P. The converse is clear from (6.1.1).

Let F be a free R-module. Then $J(F) = FJ(R)$ by (3.3.10). Hence $J(P) = PJ(R)$ for a projective module P.

6.1.3 COROLLARY Let P be a nonzero R-projective module. Then $P \neq J(P)$ and hence P contains a maximal submodule.

Proof: $J(P) = \Sigma_I S_\alpha$ by (3.4.3), where the S_α are small submodules in P. We assume that $P = J(P)$. Then we have an epimorphism $f : \Sigma_I \oplus S_\alpha \to P$. Hence from the exact sequence $\mathrm{Hom}_R(P, \Sigma_I \oplus S_\alpha) \to S = \mathrm{End}_R(P) \to 0$, we obtain a g in $\mathrm{Hom}_R(P, \Sigma_I \oplus S_\alpha)$ such that $fg = 1_P$. We can express g as $g = \Sigma g_\alpha$, where the g_α are elements in $\mathrm{Hom}_R(P,S_\alpha)$ and $\{g_\alpha\}$ is summable. Hence $1 = \Sigma fg_\alpha$. Let $x \neq 0$ be in P. Then $fg_\alpha(x) = 0$ for all $\alpha \in I - I'$, where I' is finite. Hence $(1 - \Sigma_{I'} fg_\alpha)(x) = 0$. On the other hand, $\Sigma_{I'} fg_\alpha \in J(S)$ by (3.4.2) and (6.1.2), and hence $1 - \Sigma_{I'} fg_\alpha$ is a unit in S, which is a contradiction.

6.1.4 THEOREM Let P_R be projective. Then J(P) is small in P if and only if $J(S) = \mathrm{Hom}_R(P,J(P))$, where $S = \mathrm{End}_R(P)$. In this case we have $S/J(S) \approx \mathrm{End}_R(P/J(P))$.

Proof: We always have $J(S) \subset \mathrm{Hom}_R(P,J(P))$ by (6.1.2) and (3.4.3). If J(P) is small, $\mathrm{Hom}_R(P,J(P)) \subset J(S)$ by (6.1.2). Conversely, assume that $J(S) = \mathrm{Hom}_R(P,J(P))$. Let $P = J(P) + N$ for some $N_R \subset P$. Then we consider the diagram

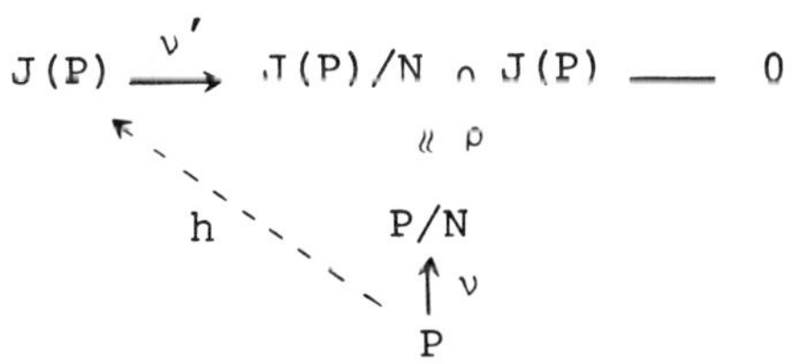

Since we have $h : P \to J(P)$ such that $\rho\nu = \nu'h$, $J(P) = h(P) + [N \cap J(P)]$ by (1.2.5). However, $h \in J(S)$, so h(P) is small in P by (6.1.2). Therefore, N = P and J(P) is small in P. Since P is projective, we have an exact sequence $0 \to \mathrm{Hom}_R(P,J(P)) \to S = \mathrm{Hom}_R(P,P) \to \mathrm{Hom}_R(P,P/J(P)) \to 0$. On the other hand, $\mathrm{Hom}_R(P,P/J(P)) = \mathrm{Hom}_R(P/J(P),P/J(P))$ by (3.3.9). Hence we have the last assertion.

6.1.5 COROLLARY Let e be idempotent. Then $J(eRe) = eJ(R)e$.

Proof: We have an isomorphism $\varphi : eRe \approx \mathrm{End}_R(eR)$ by (3.1.7), and eR is projective. Further $R = eR \oplus (1 - e)R$, so $J(R) = eJ(R) \oplus (1 - e)J(R)$. Hence $J(eR) = eJ(R)$ is small in eR by (3.3.12). Therefore, $J(eRe) \approx \mathrm{Hom}_R(eR,eJ(R))$ by (6.1.4). If we carefully check φ, we get $\mathrm{Hom}_R(eR, eJ(R)) \approx eJ(R)e$.

6.1.6 COROLLARY Let R be a ring. Then $J((R)_n) = (J(R))_n$.

Proof: Let $F = \Sigma_{i=1}^n \oplus u_i R$ be a free R-module; then $R_n \approx \text{End}_R(F)$ by (3.7.5). Since $J(F) = \Sigma_{i=1}^n \oplus u_i J(R)$ is small in F by (3.4.2), $J(\text{End}_R(F)) = \text{Hom}_R(F, \Sigma_{i=1}^n \oplus u_i J(R)) \approx (J(R))_n$.

We study next the structure of semiperfect modules.

6.1.7 LEMMA Let M be a semiperfect module and U a submodule of M. Let $\varphi_U : M \to M/U$ be the natural epimorphism. Then there exist projective submodules P and V of M and U, respectively, such that (1) $M = P \oplus V$, (2) $\varphi_U \mid P \to M/U$ is a projective cover of M/U, and (3) $U \cap P$ is small in P.

Proof: Take a diagram

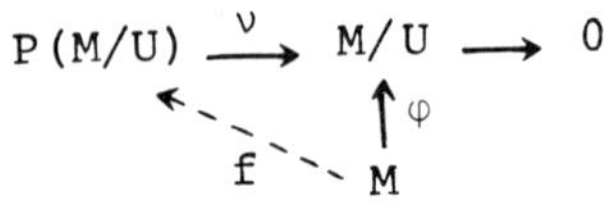

where $(P(M/U), \nu)$ is a projective cover of M/U. Then $M = P \oplus \ker f$ by (4.3.9), where $f : M \to P(M/U)$, $P \overset{f}{\approx} P(M/U)$ and $P \cap U$ is small in P. It is clear that $\ker f \subset U$.

6.1.8 COROLLARY Let M be semiperfect. Then for any submodule U of M, either U is small in M or there exists a nonzero direct summand V of M such that $U \supset V$.

Proof: If U is not small in M, $U \supsetneq U \cap P$ by (3.4.1) and (6.1.7). Hence $P \not\supset U$, so $V \neq 0$.

6.1.9 PROPOSITION Let M be a semiperfect module. Then S/J(S) is a regular ring in the sense of von Neumann, where $S = \text{End}_R(M)$.

Proof: Let $s \in S$. Then there exists a submodule P of M such that $M = \text{im } s + P$ and $P \cap \text{im } s$ is small in M by (6.1.7). We define an R-homomorphism $\phi : M/P \to M/s^{-1}(P)$ by setting $\phi(s(m) + P) = m + s^{-1}(P)$, which is clearly well defined. Now consider the diagram

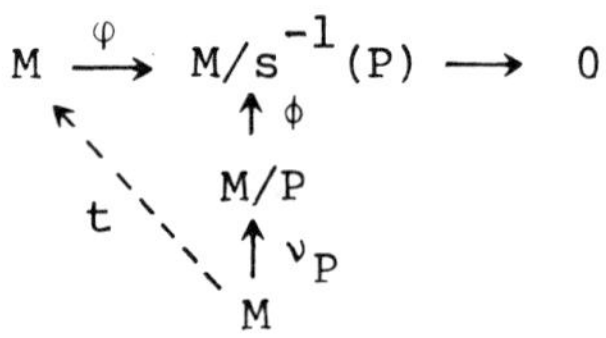

Then $ts(m) - m \in s^{-1}(P)$ and hence $s(ts(m) - m) \in P \cap \text{im } s$. Therefore, $s - sts \in J(S)$ by (6.1.2).

6.1.10 LEMMA If M is semiperfect, J(M) is small in M.

Proof: By (6.1.8), either J(M) is small in M or J(M) contains a nonzero submodule V such that $M = V \oplus V'$. If we had the latter, then $J(M) = J(V) \oplus J(V')$, and $J(V) \cap V = V$. Hence $V = 0$ by (6.1.3).

6.1.11 LEMMA Let M be semiperfect. Then M/J(M) is semisimple.

Proof: Put $\bar{M} = M/J(M)$ and $\bar{U} = U/J(M)$ for a submodule $U \supset J(M)$. By (6.1.7) there exist submodules P, V in M such that $M = P \oplus V$, $V \subset U$, and $U \cap P$ is small in M. Then $U \cap P \subset J(M)$. On the other hand, $[P + J(M)] \cap U = (P \cap U) + J(M) = J(M)$. Hence $\bar{M} = \bar{P} \oplus \bar{U}$. Therefore, $\bar{M}$ is semisimple by (4.1.2).

Let $M/J(M) = \Sigma_I \oplus A_\alpha$. If there exists a direct decomposition $M = \Sigma_I \oplus M_\alpha$ such that $\varphi(M_\alpha) = A_\alpha$ for each α, then we say that the decomposition $\bar{M} = \Sigma_I \oplus A_\alpha$ is *lifted to the decomposition* $M = \Sigma_I \oplus M_\alpha$, where $\varphi : M \to M/J(M)$ is the natural epimorphism.

6.1.12 LEMMA Let P be an R-projective module such that J(P) is small in P. Suppose that P/J(P) is a direct sum of submodules $\{\bar{P}'_\alpha\}_I$ as R/J(R)-modules, and that for each $\alpha \in I$, there exists a projective module Q_α such that $Q_\alpha/J(Q_\alpha) \approx \bar{P}'_\alpha$. Then the above decomposition of $\bar{P}$ is lifted to P and $P \approx \Sigma_I \oplus Q_\alpha$.

Proof: Set $Q = \Sigma_I \oplus Q_\alpha$. Since $P \to P/J(P)$ is a projective cover of P/J(P), $P/J(P) \overset{g}{\approx} Q/J(Q)$, and Q being projective, $Q = P \oplus Q'$ by (4.3.9) (see the diagram).

$$
\begin{array}{ccccccccccc}
0 & \longrightarrow & J(P) & \longrightarrow & P & \longrightarrow & P/J(P) & = & \Sigma_I \oplus \bar{P}_\alpha' & \longrightarrow & 0 \\
 & & & & \nwarrow & & \uparrow g^{-1} & & & & \\
 & & & & & \ddots & Q/J(Q) & = & \Sigma_I \oplus \bar{Q}_\alpha & & \\
 & & & & & & \uparrow & & & & \\
 & & & & & & Q & & & &
\end{array}
$$

We know that $Q' = J(Q')$ from the diagram. Hence $Q' = 0$.

6.1.13 COROLLARY Let M be semiperfect and $M/J(M) = \Sigma_I \oplus A_\alpha$ with A_α simple. Then there exists a decomposition $M = \Sigma \oplus M_\alpha$ of M which induces the above decomposition and $\mathrm{End}_R(M_\alpha)$ is a local ring.

Proof: We know from the proof of (6.1.11) that M satisfies the condition in (6.1.12). Hence we obtain the first assertion from (6.1.12). Since $\mathrm{End}_R(M_\alpha/J(M_\alpha)) = \mathrm{End}_R(M_\alpha)/J(\mathrm{End}_R(M_\alpha))$ is a division ring, $\mathrm{End}_R(M_\alpha)$ is a local ring.

6.1.14 THEOREM Let M be semiperfect. Then we obtain the following.

1. J(M) is small in M.
2. M/J(M) is semisimple.
3. Every decomposition of M/J(M) such that $M/J(M) = \bar{M}_1 \oplus \bar{M}_2$ is lifted to M.

Conversely, if a projective module M satisfies (1)-(3), then M is semiperfect. In this case $M = \Sigma_I \oplus M_\alpha$, where $End_R(M_\alpha)$ is local.

Proof: We have shown the first assertion. We assume that a projective module M satisfies (1)-(3). Let A be a submodule of M and put $\bar{M} = M/J(M)$ and $\bar{A} = [A + J(M)]/J(M)$. From (2) and (3), there exist submodules M_1' and M_2' such that $M = M_1' \oplus M_2'$ and $\bar{M}_1' = \bar{A}$. Then we have the diagram

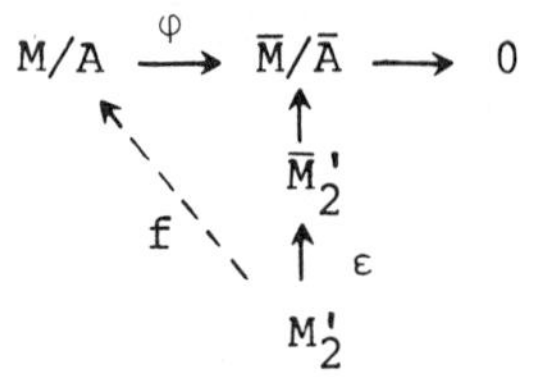

ker φ = [A + J(M)]/A is small in M/A by (1) and (3.4.1). Hence f is an epimorphism by (3.4.4). On the other hand, ker f ⊂ ker $\varepsilon = J(M_2')$, which is small in M_2' by (1). Therefore, (M_2', f) is a projective cover of M/A.

6.1.15 COROLLARY Let $\{P_i\}_{i=1}^n$ be a finite set of R-projective modules. Then $P = \Sigma_{i=1}^n \oplus P_i$ is semiperfect if and only if each P_i is.

Proof: This is clear from (3.4.2) and (6.1.14), since $\bar{P} = \Sigma \oplus \bar{P}_i$ is semisimple if and only if each $\bar{P}_i$ is.

We note that a semiperfect module P is perfect if and only if any direct sum $P^{(I)}$ of $|I|$ copies of P is semiperfect.

Similarly we obtain the following.

6.1.16 COROLLARY Let $\{P_\alpha\}_I$ be a set of semiperfect R-modules. Then $P = \Sigma_I \oplus P_\alpha$ is semiperfect if and only if J(P) is small in P.

If R is perfect as a right R-module, we say that R is *right perfect*. Similarly, we can define a *right semiperfect ring* and a *left perfect* ring. It is clear that R is right perfect if and only if every right R-module has a projective cover.

6.1.17 THEOREM Let P be a projective R-module. Then the following conditions are equivalent:

1. $S = End_R(P)$ is a local ring; i.e., S/J(S) is a division ring.
2. Every proper submodule of P is small in P.

3. P is semiperfect and directly indecomposable.

Proof: (1) ⇒ (2). Since S is local, J(S) consists of all nonisomorphisms by (3.3.14). Let N be a proper submodule of P and assume that P = T + N. Since $P/T \overset{\varphi}{\approx} N/N \cap T$, we have the diagram

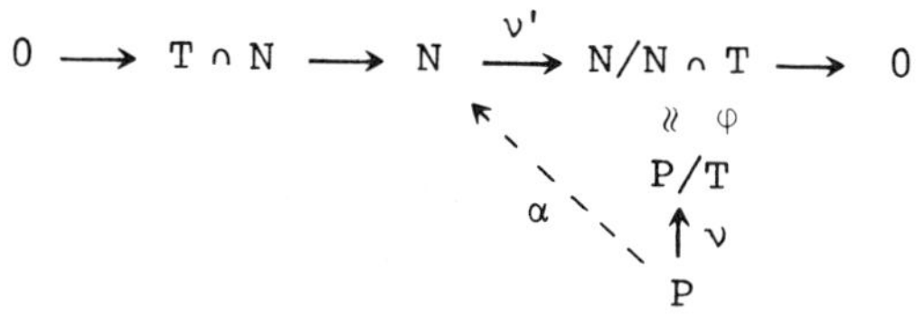

where ν and ν' are the natural epimorphisms. Since P is projective, we obtain an α in $\mathrm{Hom}_R(P,N) \subset S$ such that $\nu'\alpha = \varphi\nu$. Since $N \neq P$, $\alpha \in J(S)$. Hence $N = \mathrm{im}\,\alpha + T \cap N$, and $P = \mathrm{im}\,\alpha + T$. Therefore, P = T by (6.1.2).

(2) ⇒ (1). Assume f ($\in S$) is not an isomorphism. If im f = P, P = $P_0 \oplus \ker f$. Since ker f is proper, ker f is small in P, which is a contradiction. Hence im f ≠ P. Let g be another nonisomorphism. Since im f and im g are small in P, $P \neq \mathrm{im}\, f + \mathrm{im}\, g \supset \mathrm{im}\,(f + g)$. Hence S is a local ring.

(2) ⇒ (3). This is clear from the definition.

(3) ⇒ (2). Let T be a proper submodule of P, and $P \to P/T \to 0$ a projective cover of P/T. Since P is indecomposable, $P \approx P'$. Hence T is small in P.

6.1.18 PROPOSITION Let P be semiperfect and P_0 a projective R-submodule of P. Then P_0 is a direct summand of P if and only if $J(P) \cap P_0 = J(P_0)$.

Proof: Suppose that $J(P_0) = J(P) \cap P_0$; then $P_0/J(P_0) \subset P/J(P)$. Hence by (6.1.14) there exists a direct summand P_1 of P such that $P_1/J(P_1) \oplus P_0/J(P_0) = P/J(P)$. On the other hand, the formal direct sum $P_1 \oplus P_0$ is isomorphic to P by (6.1.12). Hence $J(P_0)$ is small in P_0. Consider the diagram

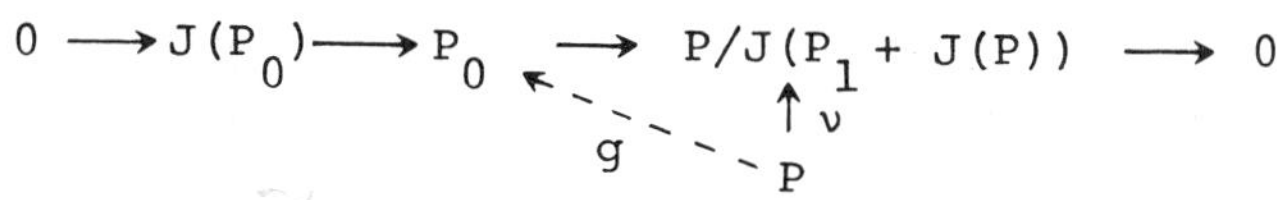

where $\nu : P \to P/J(P) \to P/J(P_1 + J(P)) \approx P_0/J(P_0)$. Then $(1_{P_0} - g_i)(P_0) \subset J(P_0)$ and hence $1_{P_0} - gi \in J(S_{P_0})$ by (6.1.4), where i is the inclusion and $S_{P_0} = \mathrm{End}_R(P_0)$. Therefore, gi is an isomorphism, which means that P_0 is a direct summand of P. The converse is clear.

6.2 FACTOR CATEGORIES

As usual, we shall assume that categories $\underline{A}$ in this section are additive unless otherwise stated. We shall use the following notation.

$\underline{M}_R$ is the category of all right R-modules, where R is a ring.
$\underline{A}_m$ is the class of all morphisms in $\underline{A}$.

For α, β in $\underline{A}_m$, "$\alpha\beta$ is defined" means that codomain of β = domain of α, and "$\alpha \pm \beta$ is defined" means that domain of α = domain of β and codomain of α = codomain of β. Since $\underline{A}$ is additive, $[A,A]$ is a ring for any $A \in \underline{A}$ and $[A,B]$ is a right $[A,A]$- and a left $[B,B]$-bimodule. We shall define ideals in an additive category $\underline{A}$. Let $\underline{C}$ be a subclass of $\underline{A}_m$. If $\underline{C}$ satisfies the following conditions, $\underline{C}$ is called a *left ideal* of $\underline{A}$.

1. For any $\alpha \in \underline{A}_m$ and $\beta \in \underline{C}$, if $\alpha\beta$ is defined, $\alpha\beta \in \underline{C}$. (6.2.1)
2. For any $\gamma,\delta \in \underline{C}$, if $\gamma \pm \delta$ is defined, $\gamma \pm \delta \in \underline{C}$.

We can define similarly *right* or *two-sided ideals* in $\underline{A}$. Let $\underline{C}$ be a two-sided ideal in $\underline{A}$. If $[A,A] \cap \underline{C}$ is the Jacobson radical of $[A,A]$ for all $A \in \underline{A}$, $\underline{C}$ is called a *Jacobson radical* of $\underline{A}$. Now, we further assume that $\underline{A}$ has finite coproducts. Let $\underline{C}$ be a Jacobson radical of $\underline{A}$. We take two objects A_1, A_2 in $\underline{A}$, and put $A = A_1 \oplus A_2$. Then

$$[A,A] = \begin{bmatrix} [A_1,A_1] & [A_2,A_1] \\ [A_1,A_2] & [A_2,A_2] \end{bmatrix}$$

and $[A_{k_1},A_{k_2}] = p_{k_2}[A,A]i_{k_1}$, where $i_k : A_k \to A$ is the injection and $p_k : A \to A_k$ is the projection. If we put $J([A_i,A_j]) = [A_i,A_j] \cap \underline{C}$, $p_{k_2}J([A,A])i_{k_1} \subset J([A_{k_1},A_{k_2}])$. Conversely, $i_{k_2}J([A_{k_1},A_{k_2}])p_{k_1} \subset [A,A] \cap \underline{C} = J([A,A])$. Therefore, $J([A_{k_1},A_{k_2}]) = p_{k_2}J([A,A])i_{k_1}$, so $\underline{C}$ is uniquely determined. We call $\underline{C}$ the *Jacobson radical* of $\underline{A}$.

The following concept is essential in this section. Let $\underline{A}$ be an additive category and $\underline{C}$ a two-sided ideal in $\underline{A}$. We define the *factor category* $\underline{A}/\underline{C}$ with respect to $\underline{C}$ as follows.

1. The objects in $\underline{A}/\underline{C}$ coincide with objects in $\underline{A}$ ($\bar{A}$ means that A is considered as in $\underline{A}/\underline{C}$). (6.2.2)
2. For $\bar{A}$, $\bar{B} \in \underline{A}/\underline{C}$, $[\bar{A},\bar{B}] = [A,B]/[A,B] \cap \underline{C}$ ($\bar{f}$ means the residue class of f in $[A,B]/[A,B] \cap \underline{C}$).

The identity mapping $\underline{A} \to \underline{A}/\underline{C}$ on objects of $\underline{A}$ and the natural mapping $f \to \bar{f}$ give us a covariant functor from $\underline{A}$ to $\underline{A}/\underline{C}$, where $\underline{C}$ is a two-sided ideal.

Remarks 1. It is clear that $\underline{A}/\underline{C}$ is also an additive category. In general, even if $\underline{A}$ is abelian, $\underline{A}/\underline{C}$ is not abelian. If we want to use properties of the factor category, we should find good ideals $\underline{C}$ such that $\underline{A}/\underline{C}$ becomes a good category.

2. Let $A = \Sigma_{i=1}^{n} \oplus A_i$ in $\underline{A}$. Then there exists inclusions i_k and projections p_k such that $1_A = \Sigma_{i=1}^{n} i_k p_k$; $p_k i_k = 1_{A_k}$ and $i_j p_k = 0$, if $j \neq k$. These relations are preserved in $\underline{A}/\underline{C}$; i.e., $1_{\bar{A}} = \Sigma\, \bar{i}_k \bar{p}_k$, $\bar{p}_k \bar{i}_k = 1_{\bar{A}_k}$ and $\bar{i}_j \bar{p}_k = 0$ if $j \neq k$. Hence $\bar{A} = \Sigma_{i=1}^{n} \oplus \bar{A}_i$ in $\underline{A}/\underline{C}$ provided that $\underline{A}/\underline{C}$ is exact. This is not true for infinite coproducts.

3. If A, B are isomorphic to one another in $\underline{A}$, then there exist morphisms $\alpha : A \to B$ and $\beta : B \to A$ such that $\alpha\beta = 1_B$ and $\beta\alpha = 1_A$. Hence $\bar{A}$, $\bar{B}$ are isomorphic to one another in $\underline{A}/\underline{C}$. However, the converse is not true in general. If $\underline{C}$ is the Jacobson radical, the converse is also true; because, if $\bar{A}$, $\bar{B}$ are isomorphic to one another, there exist $\alpha' : A \to B$, $\beta' : B \to A$ such that $\bar{\alpha}'\bar{\beta}' = 1_{\bar{B}}$ and $\bar{\beta}'\bar{\alpha}' = 1_{\bar{A}}$. Hence $1_A - \beta'\alpha'$ is in the radical of $[A,A]$. Therefore, $\beta'\alpha'$ is a unit in $[A,A]$. Similarly, $\alpha'\beta'$ is a unit in $[B,B]$. Hence α, β are isomorphisms.

4. The concept of the factor category is quite different from that of the quotient category defined by J. P. Serre and P. Gabriel.

6.2.1 PROPOSITION Let $\underline{A}$, $\underline{B}$ be additive categories and $T : \underline{A} \to \underline{B}$ an additive covariant functor. Then $\underline{C} = \{\alpha \in \underline{A}_m \mid T\alpha = 0\}$ is a two-sided ideal in $\underline{A}$ (in general, this is not a set, but we always use set notation in this book), and $T = \bar{T} \circ \psi$, where $\psi : \underline{A} \to \underline{A}/\underline{C}$ is the natural functor, and $\bar{T} : \underline{A}/\underline{C} \to \underline{B}$ is naturally induced from T.

Proof: This is clear from Sec. 2.3.

We recall the definition of an abelian category. Let $\underline{A}$ be an additive category. $\underline{A}$ is abelian if $\underline{A}$ satisfies the following:

(i) For any two objects A, B in $\underline{A}$, the coproduct $A \oplus B$ of A and B is defined and belongs to $\underline{A}$.

(ii) $\underline{A}$ contains a zero object.

(iii) For each morphism f in $\underline{A}$, ker f and coker f exist in $\underline{A}$.

(iv) (normal) For each monomorphism f in $\underline{A}$, f is a kernel of some morphism in $\underline{A}$.

(iv') (conormal) For each epimorphism f in $\underline{A}$, f is a cokernel of some morphism in $\underline{A}$ [see (2.2.4)].

In this section we rewrite the above definition of an abelian category in another notation, which is very familiar to us in ring theory.

Let $\underline{A}$ be an additive category and S a subclass of $\underline{A}_m$. Put $(S : \alpha)_r = \{\beta \in \underline{A}_m \mid \alpha\beta \text{ is defined and } \alpha\beta \in S\}$, $(S : \alpha)_\ell = \{\beta \in \underline{A}_m \mid \beta\alpha \text{ is defined and } \beta\alpha \in S\}$. If $(0 : \alpha)_r \neq 0$ for some $\alpha \in \underline{A}_m$, α is called a left zero divisor. Similarly, we define a right zero divisor. By the definition, we know that α is a monomorphism (an epimorphism) if and only if α is not a left (right) zero divisor. Let $C \xrightarrow{\alpha'} A \xrightarrow{\alpha} B$ be a sequence. Then α' is the kernel of α if and only if $(0 : \alpha')_r = 0$ and $(0 : \alpha)_r = \alpha'\underline{A}_m$, where $\alpha'\underline{A}_m = \{\alpha'\gamma \mid r \in \underline{A}_m, \alpha'\gamma \text{ is defined}\}$. α is the cokernel of α' if and only if $(0 : \alpha)_\ell = 0$ and $(0 : \alpha')_\ell = \underline{A}_m\alpha$.

6.2.2 PROPOSITION Let $\underline{A}$ be an additive category with finite coproducts. Then $\underline{A}$ is abelian if and only if $\underline{A}$ satisfies the following conditions.

1. For each $\alpha \in \underline{A}_m$, there exists $\beta \in \underline{A}_m$ such that $(0 : \beta)_r = 0$ and $(0 : \alpha)_r = \beta\underline{A}_m$.
2. For each $\alpha \in \underline{A}_m$, there exists β' such that $(0 : \beta')_\ell = 0$ and $(0 : \alpha)_\ell = \underline{A}_m\beta'$.
3. For each $\gamma \in \underline{A}_m$ such that $(0 : \gamma)_r = 0$, $(0 : (0 : \gamma)_\ell)_r = \gamma\underline{A}_m$.
4. For each $\gamma' \in \underline{A}_m$ such that $(0 : \gamma')_\ell = 0$, $(0 : (0 : \gamma')_r)_\ell = \underline{A}_m\gamma'$.

Proof: By the assumption, $\underline{A}_m$ satisfies (i), (ii) in the above definition, and (iii) corresponds to conditions 1 and 2. We assume that $\underline{A}$ is abelian. Let γ be as in 3. Then there exists a cokernel β of $\gamma : 0 \to A \xrightarrow{\gamma} B \xrightarrow{\beta} C \to 0$ (exact), and $\gamma = \ker \beta$ and $\beta = \operatorname{coker} \gamma$. Hence $(0 : \beta)_r = \gamma\underline{A}_m$, and $(0 : \gamma)_\ell = \underline{A}_m\beta$ by the above remark. Therefore $(0 : (0 : \gamma)_\ell)_r = (0 : \underline{A}_m\beta)_r = (0 : \beta)_r = \gamma\underline{A}_m$. Condition 4 is dual to 3.

Conversely, assume that $\underline{A}_m$ satisfies 1-4. We know from the conditions that 1 and 2 guarantee the existence of a kernel and a cokernel for any α in $\underline{A}_m$. Let $\gamma : A \to B$ be a monomorphism. Then there exists a β in $\underline{A}_m$ such that β is an epimorphism and $(0 : \gamma)_\ell = \underline{A}_m\beta$ from 2. Furthermore, $(0 : (0 : \gamma)_\ell)_r = (0 : \underline{A}_m\beta)_r = (0 : \beta)_r = \gamma\underline{A}_m$ by 3. Hence $\gamma = \ker \beta$, and we have (iv). (iv') is dual to (iv). Therefore, $\underline{A}$ is abelian.

We shall define some special categories which we shall use later.

Let $\underline{A}$ be an additive category. $\underline{A}$ is called *regular* if [A,A] is a regular ring in the sense of von Neumann for all $A \in \underline{A}$. $\underline{A}$ is called *amenable* if $\underline{A}$ has finite coproducts and for any idempotent e in [A,A], im e exists and e : A → im e splits; i.e., A = im e ⊕ ker e for all A ∈ $\underline{A}$. $\underline{A}$ is called *spectral* if each f ∈ $\underline{A}_m$ splits.

6.2.3 PROPOSITION Let $\underline{A}$ be an (additive) amenable and regular category. Then $\underline{A}$ is abelian.

Proof: We shall show that $\underline{A}$ satisfies 1-4 in (6.2.2). Let $\alpha : A_1 \to A_2$ be a monomorphism. Put $\alpha' = \begin{bmatrix} 0 & 0 \\ \alpha & 0 \end{bmatrix} : A_1 \oplus A_2 \to A_1 \oplus A_2$. Since $\underline{A}$ is regular, there exists $x = (x_{ij})$ in $[A_1 \oplus A_2, A_1 \oplus A_2]$ such that $\alpha' x \alpha' = \alpha'$, where $x_{ij} \in [A_j, A_i]$. Hence $\alpha = \alpha x_{12}\alpha$. Set $e = x_{12}\alpha$; then $e = e^2$ and $\alpha e = \alpha$. Hence $\underline{A}_m\alpha = \underline{A}_m e$. Since $\underline{A}$ is amenable, $e = i_e e'$, where $e' : A \to$ im e is an epimorphism and i_e : im e → A is the inclusion. Then we have $(0 : \alpha)_r = (0 : \underline{A}_m\alpha)_r = (0 : \underline{A}_m e)_r = (0 : e)_r = (1_A - e)\underline{A}_m \subset i_{(1-e)}\underline{A}_m \subset (0 : \alpha)_r$. Hence $(0 : \alpha)_r = i_{(1-e)}\underline{A}_m$ and $(0 : (0 : \alpha)_r)_\ell = (0 : i_{(1-e)})_\ell = \underline{A}_m e = \underline{A}_m\alpha$, which gives 2 and 4 in (6.2.2). From the dual, we obtain 1 and 3. Therefore, $\underline{A}$ is abelian.

We can easily see from the proof above that im e = im α. Thus we have the following.

6.2.4 COROLLARY Let $\underline{A}$ be an additive and amenable category. Then $\underline{A}$ is abelian and spectral if and only if $\underline{A}$ is abelian and regular.

6.3 CATEGORIES OF INDECOMPOSABLE MODULES

We shall often use some results in Chap. 2, and the main theorem of this section will be used very often in the following chapters. Hence in Sec. 6.6 we shall give a ring-theoretical proof of the principal theorem for the reader who is not familiar with category theory. However, its method is much more complicated than that in this section.

Let R be a ring with identity and M_R an R-module. If $\mathrm{End}_R(M)$ is a local ring, we call M *completely indecomposable* (c.ind.). If M is completely indecomposable, M is directly indecomposable. However, the converse is not true. If M is indecomposable and is one of the following: (1) $\|M\| < \infty$, (2) quasi-injective, and (3) semiperfect, then M is completely indecomposable.

6.3.1 LEMMA Let M_i be c.ind. and let $f_i : M_i \to M_{i+1}$ be R-homomorphisms for i = 1, 2. If $f_2 f_1$ is an isomorphism, then the f_i are isomorphisms.

Proof: Since $f_2 f_1$ is an isomorphism, f_1 is a monomorphism and f_2 is an epimorphism. Furthermore, $M_2 = \mathrm{im}\, f_1 \oplus \ker f_2$ by (1.7.5). Hence $\ker f_2 = 0$ and $\mathrm{im}\, f_1 = M_2$.

Let $\{M_\alpha\}_I$, $\{N_\beta\}_J$ be sets of modules, and put $M = \Sigma_I \oplus M_\alpha$ and $N = \Sigma_J \oplus N_\beta$. We can describe $\mathrm{Hom}_R(M,N)$ as the set of matrices with entries $\alpha_{ij} : M_j \to N_i$ [see (3.7.4)]. Let $T = \Sigma_K \oplus T_\delta$ be another module and let f be an element in $\mathrm{Hom}_R(M,N)$ and g in $\mathrm{Hom}_R(N,T)$. We assume that $f = (\alpha_{ij})$ and $g = (\beta_{pq})$ as above. Then we can easily show that $gf = (\beta_{pq})(\alpha_{ij})$ (product of matrices). Thus, if $M = N = T$, $\mathrm{End}_R(M)$ is isomorphic to the ring of all column-summable matrices (α_{ij}).

Now, we shall assume that all M_α, N_β, and T_δ are c.ind. We define a particular subset

$$J'^{(\beta,\alpha)} = \{(\alpha_{ij}) \in \mathrm{Hom}_R(M,N) \mid \text{no } \alpha_{ij} \text{ is an isomorphism}\} \qquad \underline{(6.3.1)}$$

$J'^{(\beta,\alpha)}$ may depend on decompositions of M and N.

6.3.2 LEMMA Let $M = \Sigma_I \oplus M_\alpha$, $N = \Sigma_J \oplus N_\sigma$, and $T = \Sigma_K \oplus T_\rho$, where the M_α, the N_σ, and the T_ρ are c.ind. modules. Then $\mathrm{Hom}_R(N,T)J'^{(\sigma,\alpha)} \subset J'^{(\beta,\alpha)}$ and $J'^{(\rho,\sigma)}\mathrm{Hom}_R(M,N) \subset J'^{(\rho,\alpha)}$.

Proof: Let $f = (a_{ij})$ be in $J'^{(\sigma,\alpha)}$ and $h = (b_{jk})$ in $\mathrm{Hom}_R(N,T)$. We put $hf = (x_{ts})$, where $x_{ts} = \Sigma_k b_{tk}a_{ks}$. If $M_s \not\approx T_t$, x_{ts} is not an isomorphism. Suppose that $M_s \approx T_t$. Let $m \neq 0$ be in M_s. Since (a_{ij}) is column summable, there exists a finite subset J' of J such that $a_{ks}(m) = 0$ if $k \in J - J'$. Put $x_{ts} = \Sigma_{k_i \in J'} b_{tk_i}a_{k_is} + \Sigma_{J-J'} b_{tk}a_{ks}$. Then neither the the latter nor the former term is an isomorphism by the definition of J' and (6.3.1). Thus x_{ts} is not an isomorphism since the M_α are c.ind. and $M_s \approx T_t$. Hence $hf \in J'^{(\rho,\alpha)}$. Similarly, we have the last assertion.

6.3.3 PROPOSITION Module $J'^{(\sigma,\alpha)}$ above does not depend on the decomposition of M and N. If, in particular, $M = N$, J' is a two-sided ideal in $\mathrm{End}_R(M)$.

Proof: Let $M = \Sigma_I \oplus M_\alpha$ and $N = \Sigma_J \oplus N_\sigma = \Sigma_{J'} \oplus N'_\sigma$. Put $T = N = \Sigma_{J'} \oplus N'_\sigma$ in (6.3.2). Then $f = 1_N f \in J'^{(\sigma,\alpha)}$ for any $f \in J'^{(\sigma,\alpha)}$. Therefore, $J'^{(\sigma,\alpha)} \subset J'^{(\sigma',\alpha)}$. Similarly, we obtain $J'^{(\sigma',\alpha)} \subset J'^{(\sigma,\alpha)}$, and hence $J'^{(\sigma',\alpha)} = J'^{(\sigma,\alpha)}$. We have analogously $J'^{(\sigma,\alpha)} = J'^{(\sigma,\alpha')}$.

We shall give here two elementary properties of a ring.

6.3.4 LEMMA Let R be a ring and let e and f be idempotents such that $eR \overset{\phi_1}{\approx} fR$ and $(1 - e)R \overset{\phi_2}{\approx} (1 - f)R$. Then there exists a regular element a in R such that $f = a^{-1}ea$.

Proof: $R = eR \oplus (1 - e)R = fR \oplus (1 - f)R$ by (3.1.1). Put $\phi = \phi_1 + \phi_2 \in \mathrm{End}_R(R)$. Since $\mathrm{End}_R(R) = R$, ϕ is given by the left-sided multiplication of an element a in R. Since ϕ is an isomorphism, a is a unit [note that e and f are considered in $R = \mathrm{End}_R(R)$].

We shall use the following.

6.3.5 COROLLARY Let P be a vector space over a division ring Δ, say $P = \Sigma_I \oplus u_\sigma\Delta$. Let $S = \mathrm{End}_\Delta(P)$ and e an idempotent in S. Then there exist a subset J of I and a regular element a in S such that $e = a^{-1}fa$ for the projection $f : P \to \Sigma_J \oplus v_\gamma\Delta$.

Proof: Let $eP = \Sigma_J \oplus v_\gamma\Delta$ and we may assume that $P = \Sigma_J \oplus u_\rho\Delta \oplus \Sigma_{I-J} \oplus u_\sigma\Delta$. Since $eS \approx \mathrm{Hom}_\Delta(P,eP) \approx \mathrm{Hom}_\Delta(P,fP)$, we have (6.3.5) from (6.3.4).

Now we shall enter into the main part of this section. Let $\{M_\alpha\}_I$ be a set of c.ind. modules. By $\underline{A}$ ($\underline{A}_f$) we shall denote the full subadditive category in $\underline{M}_R$, whose objects consist of all kinds of (finite) direct sums $\Sigma_K \oplus T_\gamma$ such that the T_γ are isomorphic to some M_β in $\{M_\alpha\}_I$. We call $\underline{A}$ ($\underline{A}_f$) the *category (finitely) induced from* $\{M_\alpha\}_I$ (we shall use the same terminology even if $\{M_\alpha\}_I$ are not c.ind.; we simply refer to the category induced from certain modules).

Let $\underline{B}$ be an additive category. If $\underline{B}$ satisfies the following properties, then $\underline{B}$ is called a Grothendieck category.

1. $\underline{B}$ is abelian and cocomplete.
2. $\underline{B}$ has a generator.
3. Let B be in $\underline{B}$ and let $\{B_\alpha\}$, C be subobjects of B such that $\{B_\alpha\}$ is a directed set. Then

$$(\cup B_\alpha) \cap C = \cup (B_\alpha \cap C) \qquad \text{(see Sec. 2.4)}$$

If every object in $\underline{B}$ is artinian (noetherian) with respect to subobjects, $\underline{B}$ is called *artinian* (*noetherian*). If every object in $\underline{B}$ is a coproduct of minimal objects, $\underline{B}$ is called *completely reducible*. If the Jacobson radical of $\underline{B}$ is zero, $\underline{B}$ is called *semisimple*.

6.3.6 LEMMA Let $\underline{A}$ be a semisimple category with finite coproducts. If $\alpha \neq 0 \in [M,N]$, there exist $\beta,\beta' \in [N,M]$ such that $\beta\alpha \neq 0$ and $\alpha\beta' \neq 0$.

Proof: Set $P = M \oplus N$, $S = [P,P]$, and $\alpha^* = \begin{bmatrix} 0 & 0 \\ \alpha & 0 \end{bmatrix}$. If $[N,M]\alpha = 0$, $S\alpha^*$ is nilpotent, which is a contradiction. Similarly, we have $\alpha[N,M] \neq 0$.

6.3.7 COROLLARY Let $\underline{A}$ be as in (6.3.6). If [M,M] is a division ring, M is a minimal object.

Proof: Let $M \supsetneqq N$. Then [M,N] = 0. Hence [N,M] = 0 by (6.3.6), so the inclusion map $N \to M$ is zero.

From now on, we shall denote $\mathrm{Hom}_R(M,N)$ by [M,N] for R-modules M and N.

6.3.8 PRINCIPAL THEOREM Let $\{M_\alpha\}_I$ be a set of completely indecomposable modules and let $\underline{A}$, $\underline{A}_f$ be the category induced from $\{M_\alpha\}_I$ and finitely induced from $\{M_\alpha\}_I$, respectively. Let $\underline{J}'$ be the ideal in $\underline{A}$ defined by (6.3.1). Then $\underline{A}/\underline{J}'$ (resp. $\underline{A}_f/\underline{J}'$) is a Grothendieck and completely reducible (resp. completely reducible) category.

Proof: We put $\bar{A} = \underline{A}/\underline{J}'$ ($\bar{A}_f = \underline{A}_f/\underline{J}'$). By the definition of coproducts and (6.3.3) we can easily show that $\overline{\Sigma_J \oplus M_\gamma} = \Sigma_J \oplus \bar{M}_\gamma$. Put $\bar{S}_M = [M,M]/[M,M] \cap J'$ for an object $M = \Sigma_J \oplus M_\gamma$ in $\underline{A}$. Then $\bar{S}_M = \{(\bar{a}_{\sigma\tau})$, column-finite matrices$\}$, since $\bar{a}_{\sigma\tau} = 0$ for almost all σ. We rearrange $\{M_\alpha\}_I$ as follows. $M = \Sigma_\alpha \Sigma_{I_\alpha \ni \beta} \oplus M_{\alpha\beta}$, $M_{\alpha\rho} \approx M_{\alpha\rho'}$, and $M_{\alpha\rho} \not\approx M_{\alpha'\rho'}$ if $\alpha \neq \alpha'$. Then $[\bar{M},\bar{M}] = \Pi_\alpha[\Sigma_{I_\alpha} \oplus \bar{M}_{\alpha\beta}, \Sigma_{I_\alpha} \oplus \bar{M}_{\alpha\beta}]$. $[\Sigma_{I_\alpha} \oplus \bar{M}_{\alpha\beta}, \Sigma_{I_\alpha} \oplus \bar{M}_{\alpha\beta}] \approx \{(x_{\alpha\beta}) \mid$ column finite, and $x_{\alpha\beta} \in [\bar{M}_\alpha, \bar{M}_\alpha] = \Delta_\alpha\}$, where Δ_α is a division ring. Therefore, $\underline{\bar{A}}$ and $\underline{\bar{A}}_f$ are regular and semisimple by (5.4.5). Next we show that they are amenable. Set $\bar{S}_\alpha = [\Sigma_{I_\alpha} \oplus \bar{M}_{\alpha\beta}, \Sigma_{I_\alpha} \oplus \bar{M}_{\alpha\beta}]$; then $\bar{S}_M = \Pi_\alpha \bar{S}_\alpha$. Let $\bar{e}$ be an idempotent in $\bar{S}_M$ and $\bar{e} = \Pi \bar{e}_\alpha$, where $\bar{e}_\alpha \in \bar{S}_\alpha$ and $\bar{e}_\alpha^2 = \bar{e}_\alpha$. Then there exist a regular element $\bar{a}_\alpha \in \bar{S}_\alpha$ and a projection $f_\alpha : \Sigma_{I_\alpha} \oplus M_{\alpha\beta} \to \Sigma_{J_\alpha} \oplus M_{\alpha\beta}$, in $\underline{M}_R$ such that $\bar{e}_\alpha = \bar{a}_\alpha^{-1}\bar{f}_\alpha\bar{a}_\alpha$ by (6.3.5) (note that $\bar{S}_\alpha$ may be regarded as the endomorphism ring of a vector space). Since f_α is the projection in $\underline{M}_R$, $\bar{f}_\alpha$ splits in $\underline{\bar{A}}$. Hence, $\bar{e}$ does, since $\bar{a}_\alpha$ is regular. Therefore, $\bar{e}$ splits, which implies that $\underline{A}$ ($\underline{A}_f$) is amenable. Thus $\underline{A}$ ($\underline{A}_f$) is abelian and spectral by (6.2.3) and (6.2.4). On the other hand, $\bar{M}_\alpha$ is a minimal object by (6.3.7). Hence $\underline{\bar{A}}$ is completely reducible. Finally, we show that $\underline{\bar{A}}$ satisfies condition 3 in the definition of Grothendieck category. Let $\{\bar{A}_\alpha\}_K$ be a directed set of subobjects in an object $\bar{F}$ and let $\bar{B}$ a subobject in $\bar{F}$. Set $\bar{C} = \cup_K(\bar{A}_\alpha \cap \bar{B})$; then $\bar{B} = \bar{C} \oplus \bar{B}_0$, since $\underline{\bar{A}}$ is spectral. $(\cup \bar{A}_\alpha) \cap \bar{B} = (\cup_K \bar{A}_\alpha) \cap (\bar{C} \cup \bar{B}_0) = \bar{C} \cup [(\cap_K \bar{A}) \cap \bar{B}_0]$ by (2.2.5), for $\bar{C} \subset \cup_K \bar{A}_\alpha$. We assume that $(\cup_K \bar{A}_\alpha) \cap \bar{B}_0 = \bar{D} \neq 0$. From the exact sequence $\Sigma_K \oplus \bar{A}_\alpha \xrightarrow{\bar{f}} \cup_K \bar{A}_\alpha \to 0$ [see (2.4.3)], we obtain a monomorphism $\bar{g} : \bar{D} \to \Sigma \oplus \bar{A}_\alpha$ such that $\bar{f}\bar{g} = 1_{\bar{D}}$, because $\underline{\bar{A}}$ is spectral. Let $\bar{D}_0$ be a minimal subobject in $\bar{D}$. Then $\bar{g}|\bar{D}_0$ is a column-finite matrix from the first observation. Hence $\mathrm{im}(\bar{g}|\bar{D}_0) \subset \Sigma_{i=1}^n \oplus \bar{A}_{\alpha_i}$, so $\bar{D}_0 \subset \cup_{i=1}^n \bar{A}_{\alpha_i} \subset \bar{A}_\beta$ for some $\beta \in K$ such that $\beta \geq \alpha_i$. Thus

$\bar{D}_0 \subset \bar{A}_\beta \cap \bar{B} \subset \bar{C}$, and $\bar{D}_0 \subset \bar{B}_0$, which is a contradiction. Therefore, $(\cup_K \bar{A}_\alpha) \cap \bar{B} = \cup_K(\bar{A}_\alpha \cap \bar{B})$.

6.4 CATEGORIES OF INJECTIVE MODULES

We shall give here an application of (6.2.3) to injective modules. Let M be an R-injective module. We define a full subadditive category $\underline{C}(M)$ in $\underline{M}_R$ as follows (cf. the category induced from modules in Sec. 6.3). The objects in $\underline{C}(M)$ consist of all direct summands of products $\Pi_I M_\alpha$; $M_\alpha \approx M$. If M is an injective cogenerator in $\underline{M}_R$, then $\underline{C}(M)$ is the category of all injective modules. We call $\underline{C}(M)$ the *category of injective modules* induced from M. Let $\underline{J}$ be the radical of $\underline{C}(M)$. $(J \cap [A,B] = \{f \in [A,B] \mid f^{-1}(0)$ is essential in A$\}$ by (3.4.5) and (5.4.4).)

6.4.1 THEOREM Let M_R, $\underline{C}(M)$, and $\underline{J}$ be an injective module, the category of injective modules induced from M, and the Jacobson radical of $\underline{C}(M)$, respectively. Then $\underline{C}(M)/\underline{J}$ is a Grothendieck and spectral category.

Proof: We shall denote $\underline{C}(M)/\underline{J}$ by $\bar{\underline{C}}(M)$. Then $\bar{\underline{C}}(M)$ has finite coproducts from the definition and Remark 2 in Sec. 6.2 (see the proof below), and $\bar{\underline{C}}(M)$ is a regular category by (5.4.4). Furthermore, (5.4.4) shows that $\bar{\underline{C}}(M)$ is amenable. Hence $\bar{\underline{C}}(M)$ is an abelian spectral category by (6.2.3) and (6.2.4). We shall show that $\bar{\underline{C}}(M)$ has infinite coproducts. Let $\{\bar{A}_\alpha\}_I$ be a set of objects in $\bar{\underline{C}}(M)$. Since A_α is a direct summand of some $\Pi_{I_\alpha} M$, $\Sigma_I \oplus A_\alpha$ is a direct summand of $\Pi_I \Pi_{I_\alpha} M$. Then $E = E(\Sigma_I \oplus A_\alpha)$ is an object in $\underline{C}(M)$, for $\Pi_I \Pi_{I_\alpha} M$ is injective. We show that $\bar{E} = \Sigma_I \oplus \bar{A}_\alpha$. Let N be any object in $\underline{C}(M)$ and $\{\bar{f}_\alpha : \bar{A}_\alpha \to \bar{N}\}$ a set of morphisms, where $f_\alpha : A_\alpha \to N$ is a representative of $\bar{f}_\alpha$. Then there exists an $f : \Sigma_I \oplus A_\alpha \to N$ in $\underline{M}_R$ making the following diagram commute.

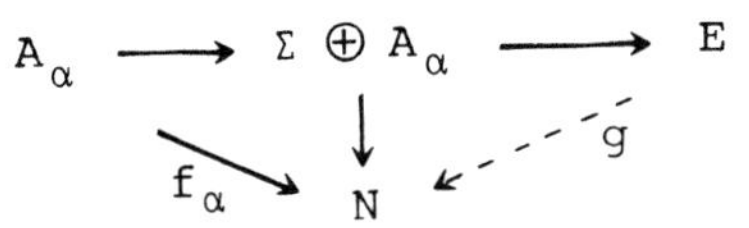

Since N is injective, there exists a $g : E \to N$ making the diagram above commute on the right side. We can easily show from (5.4.4) that g does not depend on the choice of representatives $\{f_\alpha\}$ and that $\bar{g}$ is uniquely determined. Similarly, we can show that for a given $\bar{g} : \bar{E} \to \bar{N}$, there exists a unique set $\{\bar{f}_\alpha : \bar{A}_\alpha \to \bar{N}\}$ such that $\bar{g} = \Pi_\alpha \bar{f}_\alpha$. Hence $\bar{E} = \Sigma \oplus \bar{A}_\alpha$.

Next we show that $\bar{\underline{C}}(M)$ has a generator. Let $\underline{S}$ be the set of right ideals K in R such that $E_K = E(R/K) \in \underline{C}(M)$. Put $\bar{U} = \Sigma_{K \in \underline{S}} \oplus \bar{E}_K$. Let T be an object in $\underline{C}(M)$ and let $t \neq 0$ in T. Then $T \supset tR \approx R/r_R(t)$, and $E(R/r_R(t)) \in \underline{C}(M)$, since T is an injective and is in $\underline{C}(M)$. Therefore, $E_{r_R(t)}$ is isomorphic to a direct summand of T, which implies that $[\bar{U},\bar{T}] \neq 0$.

Finally, we show, similarly to the proof of (6.3.8), that $(\cup_K \bar{A}_\alpha) \cap \bar{B} = \cup_K(\bar{A}_\alpha \cap \bar{B})$ for a subobject $\bar{B}$ and a directed set of subobjects $\{\bar{A}_\alpha\}_K$ in a given object $\bar{F}$. Set $\bar{C} = \cup_K(\bar{A}_\alpha \cap \bar{B})$; then $\bar{B} = \bar{C} \oplus \bar{B}_0$, and $(\cup_K \bar{A}_\alpha) \cap \bar{B} = \bar{C} \cup [(\cup \bar{A}_\alpha) \cap \bar{B}_0]$. Set $\bar{D} = (\cup_K \bar{A}_\alpha) \cap \bar{B}_0$ and assume that $\bar{D} \neq 0$. From the exact sequence $\Sigma_K \oplus \bar{A}_\alpha \xrightarrow{\bar{f}} \cup_K \bar{A}_\alpha \to 0$, we obtain a monomorphism $\bar{g} : \bar{D} \to \Sigma_K \oplus \bar{A}_\alpha$ such that $\bar{f}\bar{g} = 1_{\bar{D}}$. We note that g is an R-monomorphism, since $\underline{J}$ is the Jacobson radical, and that $\Sigma_K \oplus \bar{A}_\alpha = \overline{E(\Sigma_I \oplus A_\alpha)}$. Put D' = im g in $\underline{M}_R$. Then $\bar{D}' = \text{im } \bar{g}$. Since $\bar{D}' \neq 0$, $D' \cap \Sigma \oplus A_\alpha \neq 0$ in $\underline{M}_R$. Let $x \neq 0$ be an element in $D' \cap \Sigma_K \oplus A_\alpha$ and let $E(xR)$, $E_1(xR)$ be injective hulls of xR in D' and $\Sigma_{i=1}^n \oplus A_{\alpha_i}$, respectively, where $x \in \Sigma_{i=1}^n \oplus A_{\alpha_i}$. Then $\overline{E(xR)} = \overline{E_1(xR)} \subset \Sigma_{i=1}^n \oplus \bar{A}_{\alpha_i}$ (cf. Remark 2 below). Hence $\bar{E}(g^{-1}(x)R) \subset \cup \bar{A}_{\alpha_i} \subset \bar{A}_\beta$ for some β such that $\beta \geq \alpha_i$ and $\bar{E}(g^{-1}(x)R) \subset \bar{D}$, which is a contradiction.

6.4.2 THEOREM Let R be a ring. If every right injective R-module is a direct sum of indecomposable modules, then R is right noetherian.

Proof: It is well known by (4.7.18) that R is noetherian if and only if any direct sum of injective modules is also injective. Let $\underline{A}$ be the full subcategory of all injective R-modules in $\underline{M}_R$ and $\underline{J}$ the Jacobson radical of $\underline{A}$. Then $\underline{A}/\underline{J}$ is a completely reducible Grothendieck category by the assumption and the proof of (6.4.1). Let $\{Q_i\}_1^\infty$ be a family of injective modules (we may assume that the Q_i are c.ind.) and E an injective hull of $\Sigma \oplus Q_i$. From the assumption, $E = \Sigma \oplus E_i$, where the E_i are (completely) indecomposable. Hence $\Sigma \oplus \bar{E}_j = \Sigma \oplus \bar{Q}_i$ in $\underline{A}/\underline{J}$ by (6.4.1). Therefore, since $\bar{E}_i \approx \bar{Q}_j$ implies that $E_i \approx Q_j$ by Remark 3 in Sec. 6.2, $E \approx \Sigma \oplus Q_i$, which means that $\Sigma \oplus Q_i$ is injective. Hence R is noetherian.

Remarks 1. We noted, in the proof of (6.3.8), that $\overline{\Sigma_I \oplus M_\alpha} = \Sigma_I \oplus \bar{M}_\alpha$ in the factor category of c.ind. modules. However, $\Sigma_I \oplus A_\alpha$ is not an object in $\underline{C}(M)$, in general, and $\Sigma_I \oplus \bar{A}_\alpha$ means that $\bar{E}(\Sigma_I \oplus A_\alpha)$.

2. Let E, E' be injective and let $f : E \to E'$ be an R-homomorphism. We can directly find ker $\bar{f}$ and im $\bar{f}$ in $\bar{\underline{C}}(M)$. Let K = ker f in $\underline{M}_R$ and E'' = E(K) in E. Then $E = E'' \oplus E_1$. We define an $f' \in [E,E']$ by setting $f' = 0 + f$. Then $\ker(f - f') = K \oplus E_1 \subset E$. Hence $\bar{f} = \bar{f}'$. Therefore, ker $\bar{f}$ = ker $\bar{f}'$ = $\bar{E}''$, and im $\bar{f}$ = im $\bar{f}'$ = $\overline{f(E_1)}$. This argument shows that ker $\bar{f}$ (im $\bar{f}$) does

not depend on the choice of injective hulls of K in E, and that we can give direct proofs of many results in the following without the factor category. However, if we use the factor category, the proof is simple and natural in some sense.

3. If $\underline{J} \neq 0$, in general $\overline{\Pi A_\alpha} \neq \Pi \bar{A}_\alpha$ in $\bar{\underline{C}}(M)$ for $A_\alpha \in \underline{C}(M)$.

6.5 CATEGORIES OF PROJECTIVE MODULES

We have studied the category of injective modules and the factor category with respect to the Jacobson radical in Sec. 6.4. We consider the category of projective modules in this section.

Let R be a ring and $\underline{A}$ a full additive subcategory of $\underline{M}_R$. We put

$$\underline{C} = \{f \in \underline{A}_m \mid f : A \to A',\ \text{im } f \text{ (in } \underline{M}_R) \text{ is small in } A'\} \qquad \underline{(6.5.1)}$$

Then we have, by (3.4.1), the following.

6.5.1 PROPOSITION $\underline{C}$ is an ideal in $\underline{A}$.

Let $\{P_\alpha\}_I$ be a set of R-projective modules. Let $\underline{A}$ and $\underline{A}_f$ be the category induced from $\{P_\alpha\}_I$ in $\underline{M}_R$ and finitely induced from $\{P_\alpha\}_I$, respectively. Then every object P (resp. P') in $\underline{A}$ (resp. in $\underline{A}_f$) is a direct sum of some (finite) projective modules which are isomorphic to some P_α in $\{P_\alpha\}_I$. We denote $\text{Hom}_R(P,P')$ by $[P,P']$.

6.5.2 PROPOSITION Let $\underline{A}$ and $\underline{A}_f$ be as above, and let $\underline{C}$ be as in $\underline{(6.5.1)}$. Then $\underline{C}$ is the Jacobson radical of $\underline{A}$ and $\underline{A}_f$, respectively.

Proof: This is clear by (6.1.2).

We shall study the above $\underline{A}$ in this section, so we shall denote $\underline{C}$ by $\underline{J} = J(\underline{A})$.

6.5.3 LEMMA If R has a family of mutually orthogonal nonzero idempotents $\{e_i\}_{i=1}^\infty$, then R_I is not regular in the sense of von Neumann for any infinite set I.

Proof: We may assume that $|I| = \aleph_0$. We denote a family of matrix units in R_I by $\{e_{ij}\}$. Set $B = \Sigma_{i=1}^\infty e_i e_{ij}$. If R_I is a regular ring, then there exists an A in R_I such that BAB = B, say $A = \Sigma\, a_{ij} e_{ij}$. We assume that $a_{i1} = 0$ if i > t for a large t. Then BAB = B implies that $\Sigma_{i=1}^t e_i a_{i1} e_j = e_j$ for all j. If j > t, $e_j = e_j^2 = \Sigma_{i=1}^t e_j a_{ij} e_j = 0$, which is a contradiction.

6.5.4 COROLLARY Let R be a regular ring. Then R_I is regular for any set I if and only if R is artinian.

Proof: If R is artinian, then it is clear that R_I is regular by (5.4.5). We assume that there exists an infinite series of principal ideals of R: $Ra_1 \supsetneq Ra_2 \supsetneq \cdots \supsetneq Ra_n \supsetneq \cdots$. Since R is regular, $Ra_n = Re'_n$ for some idempotent e'_n. Hence R contains an infinite set of nonzero mutually orthogonal idempotents $\{e_i\}$ by (3.1.5), which is a contradiction to (6.5.3). Therefore, R has the nonzero artinian socle, and hence R is artinian, since R is equal to the socle.

6.5.5 PROPOSITION Let $\underline{A}$ be as in (6.5.2). Then we have the following.

1. If $\underline{A}/\underline{J}$ is spectral abelian, J(P) is small in P for any $P \in A$.

Further, we assume that $\underline{A}/\underline{J}$ is a Grothendieck category.

2. If P is a direct sum (in $\underline{M}_R$) of subobjects $\{P_\alpha\}_I$ in $\underline{A}$, then $\bar{P} = \Sigma_I \oplus \bar{P}_\alpha$ in $\underline{A}/\underline{J}$.
3. If $\bar{P}$ is a direct sum of minimal objects in $\underline{A}/\underline{J}$, then P is semiperfect in $\underline{M}_R$.
4. If Q in $\underline{A}$ is a finitely generated R-module, then Q is perfect in $\underline{M}_R$.

Proof: 1. Put $S = [P,P]$ and $J'(S) = [P,J(P)]$. We assume that $J'(S) \neq J(S)$. Since $S/J(S)$ is a regular ring by (6.2.4) and $J'(S)$ is a two-sided ideal in S, there exists a nonzero element e' in $J'(S)/J(S)$ with $e'^2 = e'$. Hence $\bar{P} = \bar{P}_1 \oplus \bar{P}_2$ and $e' \equiv ip \pmod{J(S)}$, where $p : P \to P_1$ and $i : P_1 \to P$. Since $pi \equiv 1_{P_1} \pmod J$, $\alpha = pi$ is an R-isomorphism in $\underline{M}_R$ by Remark 3 in Sec. 6.2. Then $1_{P_1} = (\alpha^{-1}p)i$ and $p = 1_{P_1}p = (\alpha^{-1}p)ip = \alpha^{-1}pe'$ (mod J). Thus P_1 is a direct summand of P, $\alpha^{-1}pe' \in J'(S)$, and hence $p \in J'(S)$. However, p is an epimorphism in $\underline{M}_R$, so $p(P) = P_1 \subset J(P) \cap P_1 = J(P_1)$, which is a contradiction [cf. (6.1.1)].

2. We shall show that $\Sigma_I \oplus \bar{P}_\alpha = \overline{\Sigma_I \oplus P_\alpha}$ in $\underline{A}/\underline{J}$. Let J be a finite subset of I; then $P_J = \Sigma_J \oplus P_\alpha$ is a direct summand of $P = P_I$. Hence $\Sigma_J \oplus \bar{P}_\alpha = \bar{P}_J$ is a direct summand of $\bar{P}$. Therefore, $\cup \bar{P}_J = \Sigma_I \oplus \bar{P}_\alpha$ is a subobject of $\bar{P}$ by (2.4.2) and (2.4.5). Put $\bar{P} = \Sigma_I \oplus \bar{P}_\alpha \oplus \bar{Q}$ and let f be a projection of $\bar{P}$ to $\bar{Q}$. Then $fg \equiv 1_Q \pmod{\underline{J}}$ for some $g \in [Q,P]$. Since J is the radical, fg is an R-isomorphism. $f(\Sigma_I \oplus \bar{P}_\alpha) = 0$ implies that $f(\Sigma_I \oplus P_\alpha) \subset J(Q)$. Hence $J(Q) \supset f(P) \supset fg(Q) = Q$. Therefore, $Q = 0$.

3. We assume that $\bar{P} = \Sigma_I \oplus \bar{P}'_\alpha$ with $\bar{P}'_\alpha$ minimal. Set $P' = \Sigma_I \oplus P'_\alpha$. Then $\bar{P} \approx \bar{P}'$ by (2). Therefore, $P \approx P'$ as R-modules, since J is the radical. Furthermore, P'_α is semiperfect and so is P by (6.1.16) and (6.1.17).

4. Let Q be a finitely generated projective R-module in $\underline{A}$ and let $S_Q = [Q,Q]$. Set $Q^* = \Sigma_{i=1}^{\infty} \oplus Q_i$; $Q_i \approx Q$ for all i. Since Q is f.g., S_{Q^*}

is the ring $(S_Q)_\infty$ of column-finite matrices with entries in S_Q. From the assumption, $S_{Q^*}/J(S_{Q^*})$ is regular. On the other hand, $(J(S_Q))_\infty \supset J((S_Q)_\infty)$ by (6.1.2). Being a homomorphic image of $(S_Q)_\infty/J((S_Q)_\infty)$, $(S_Q/J(S_Q))_\infty$ is regular. Therefore, $S_Q/J(S_Q)$ is an artinian ring by (6.5.4). Let $\bar{1} = \Sigma_{i=1}^n \bar{e}_i$ in $S_Q/J(S_Q)$, where $\{\bar{e}_i\}_1^n$ is a complete set of mutually orthogonal primitive idempotents. Then $\bar{Q} = \Sigma_{i=1}^r \oplus \operatorname{im} \bar{e}_i$. Put $\operatorname{im} \bar{e}_i = \bar{Q}_i$. Then the $\bar{Q}_i$ are minimal by (6.3.7). Hence $Q \approx \Sigma_{i=1}^r \oplus Q_i$ and the Q_i are c.ind. It is clear from the first observation that Q is perfect by (6.1.16).

6.5.6 THEOREM Let $\underline{A}$ be the full subadditive category of all projectives in $\underline{M}_R$ and $\underline{J}$ the radical of $\underline{A}$. Then the following conditions are equivalent:

1. $\underline{A}/\underline{J}$ is a Grothendieck and completely reducible category.
2. $\underline{A}/\underline{J}$ is a Grothendieck and spectral category.
3. R is a right perfect ring.

Proof: (1) ⇒ (2). This is clear.

(2) ⇒ (3). Since R is a finitely generated R-module, R is right perfect by (6.5.5).

(3) ⇒ (1). If R is right perfect, then every object P in $\underline{A}$ is perfect and hence P is a direct sum of c.ind. modules. Furthermore, $\underline{J}$ $[P,P] = [P,J(P)]$ is equal to the ideal defined by (6.3.1). Hence $\underline{A}/\underline{J}$ is a Grothendieck, reducible abelian category by (6.3.8).

Remark If P is semiperfect, $J(\operatorname{End}_R(P)) = \operatorname{Hom}_R(P,J(P))$ by (6.1.4), and $\operatorname{End}_R(P)/J(\operatorname{End}_R(P)) \approx \operatorname{End}_R(P/J(P))$. Hence in this case the investigation on $\underline{A}/\underline{J}$ is equivalent to that on $\operatorname{End}_R(P/J(P))$. We shall use some properties of A/J in Chap. 7. However, this argument can easily be translated back to ring theory in case of projectives from the remark above.

6.6 RING-THEORETICAL PROOF

In this section we give a proof of (6.3.8) using a less categorical method. Let $\{A_\alpha\}_I$ be a set of c.ind. modules such that $A_\alpha \not\approx A_\beta$ if $\alpha \neq \beta$. Put $\Delta_\alpha = \operatorname{End}_R(A_\alpha)/J(\operatorname{End}_R(A_\alpha))$, a division ring. Then $\Delta = \Pi_I \Delta_\alpha$ is a regular ring and $\{\Delta_\alpha\}_I$ is a set of irreducible Δ-modules. By $\underline{S}$ we denote the category induced from $\{\Delta_\alpha\}$ in $\underline{M}_\Delta$. It is clear that $\underline{S}$ is a completely irreducible and Grothendieck subcategory in $\underline{M}_\Delta$. We note that for any $f \in \underline{S}_m$, im f and ker f in $\underline{S}$ coincide with those in $\underline{M}_\Delta$.

Let $\underline{A}$ be the category induced from $\{A_\alpha\}$, and $\underline{J}'$ the ideal in (6.3.1). Then (6.3.8) is given by the following theorem.

6.6.1 THEOREM $\underline{A}/\underline{J}'$ is equivalent to $\underline{S}$.

We need a lemma to prove this.

Let $\{T_\alpha\}_I$ be a set of R-modules and $\underline{T}$ the category induced from $\{T_\alpha\}_I$. We shall define a new category T*. Let $M = \Sigma_K \oplus M_\gamma$ be an object in $\underline{T}$, where $M_\gamma \approx T_\gamma$. We consider a pair $(M, \Sigma_K \oplus M_\gamma)$. We specify that $(M, \Sigma_K \oplus M_\gamma) \equiv (N, \Sigma_{K'} \oplus N_{\gamma'})$ if and only if $M = N$ and $\Sigma_K \oplus M_\gamma$, $\Sigma_{K'} \oplus N_{\gamma'}$ are the same decomposition of M. We denote the set of morphisms $[(M, \Sigma_K \oplus M_\gamma),(N, \Sigma_{K'} \oplus N_{\gamma'})]$ by $\mathrm{Hom}_R(M,N)$ (we should distinguish $[(M, \Sigma_K \oplus M_\gamma), (M', \Sigma_{K'} \oplus M'_{\gamma'})]$ and $[(M, \Sigma_{K'} \oplus M''_{\gamma''}),(M', \Sigma_{K'''} \oplus M'''_{\gamma'''})]$ if the decompositions are different, even though they are equal to $\mathrm{Hom}_R(M,M')$). Then we have an additive category $\underline{T}^* = \{(M, \Sigma_K \oplus M_\gamma)\}$.

6.6.2 LEMMA $\underline{T}$ is equivalent to $\underline{T}^*$.

Proof: We shall define functors $P : \underline{T} \to \underline{T}^*$ and $U : \underline{T}^* \to \underline{T}$ such that PU (resp. UP) is equivalent to $1_{\underline{T}^*}$ (resp. $1_{\underline{T}}$). For each M in $\underline{T}$, we fix one decomposition: $M = \Sigma \oplus M_\alpha$, $M_\alpha \approx T_\alpha$, and we define

$$P(M) = (M, \Sigma_K \oplus M_\gamma)$$

Conversely, let $M' = (M', \Sigma \oplus M'_\gamma)$ be in $\underline{T}^*$. Put

$$U(M') = \Sigma \oplus M'_\gamma = M'$$

It is clear that P and U are functors and $UP = 1_{\underline{T}}$. Let $M' = U(M') = \Sigma_L \oplus M^*_\varepsilon$ be the fixed decomposition of M'. Then $PU(M') = (M', \Sigma \oplus M^*_\varepsilon)$. Put $\varphi_{M'} = 1_{M'}$ $(\in [M', \Sigma \oplus M'),(M', \Sigma \oplus M^*_\varepsilon])$. Then it is clear the PU and $1_{\underline{T}^*}$ are naturally equivalent through $\{\varphi_{M'}\}$.

Let $\underline{A}$ be the category induced from $\{A_\alpha\}_I$ in $\underline{M}_R$. We can define the ideal $\underline{J}'$ as in (6.3.1). Since $\underline{J}'$ does not depend on decompositions of M by (6.3.3), we have the following.

6.6.3 COROLLARY $\underline{A}/\underline{J}'$ is equivalent to $\underline{A}^*/\underline{J}'$.

Proof of (6.6.1): We shall show an equivalence between $\underline{A}^*/\underline{J}'$ and S*. Let $M = (M, \Sigma_K \oplus M_\gamma)$ be an object in $\underline{A}^*/\underline{J}'$, where $M_\gamma \approx A_\gamma \in \{A_\alpha\}_I$. We put $M = \Sigma \oplus M_\gamma = \Sigma_I \Sigma_{I_\alpha} \oplus M_{\alpha\beta}$; $M_{\alpha\beta} \approx A_\alpha$. Let $N = (N, \Sigma_I \Sigma_{J_\alpha} \oplus N_{\alpha\beta})$ be another object. We divide matrices in $\mathrm{Hom}_R(M,N)$ into $|I| \times |I|$ blocks. Then it is clear from the decomposition above that every entry of a matrix in every block except the diagonal consist of nonisomorphisms, and hence these blocks belong to $\underline{J}'$. We put $M(\alpha) = \Sigma_{J_\alpha} \oplus M_{\alpha\beta}$ and $S(\alpha) = \Sigma_{J_\alpha} \oplus \Delta_\alpha = \Delta^{(J_\alpha)}$. Then we have a natural isomorphism

$$\mathrm{End}_R(M(\alpha))/\underline{J}' \cap \mathrm{End}_R(M(\alpha)) \approx \mathrm{End}(S(\alpha)) \tag{6.6.1}$$

We define $T : \underline{A}^*/\underline{J}' \to S^*$ by setting $T(M) = \Sigma_I \oplus \Delta_\alpha^{(J_\alpha)}$ and $T(\mathrm{Hom}_R(M,N)/\underline{J}' \cap \mathrm{Hom}_R(M,N)) = \mathrm{Hom}_\Delta(T(M),T(N))$ via (6.6.1). Then T is a functor. Conversely, let $S = (S, \Sigma_I\Sigma_{I_\alpha} \oplus \Delta_{\alpha\beta})$ be in S^*, where $\Delta_{\alpha\beta} \approx \Delta_\alpha$. We define

$$V(S) = (M, \Sigma_I \oplus A_\alpha^{(I_\alpha)})$$

Since $\Delta_\alpha \approx \mathrm{End}_\Delta(\Delta_{\alpha\beta}) \approx \mathrm{End}_R(A_\alpha)/J(\mathrm{End}_R(A_\alpha))$, and $J(\mathrm{End}_R(A_\alpha))$ consists of all nonisomorphisms, we define

$$V(\mathrm{Hom}_\Delta(S,S')) = \mathrm{Hom}_R(V(S),V(S'))/\underline{J}' \cap \mathrm{Hom}_R(V(S),V(S'))$$

V is also a functor $\underline{S}^* \to \underline{A}^*/\underline{J}'$. We have a natural isomorphism $\varphi_M : \Sigma_I\Sigma_{J_\alpha} \oplus M_{\alpha\beta} \to \Sigma_I \oplus A_\alpha^{(J_\alpha)}$ [and $\psi_S : \Sigma_I\Sigma_{I_\alpha} \oplus \Delta_{\alpha\beta} \to \Sigma_I \oplus \Delta_\alpha^{(I_\alpha)}$]. Hence TV and VT are natural equivalents to $1_{\underline{S}^*}$ and $1_{\underline{A}^*/\underline{J}'}$, respectively. Let $P_{\underline{A}/\underline{J}'}$ and $U_{\underline{A}/\underline{J}'}$ (resp. $P_{\underline{S}}$ and $U_{\underline{S}}$) be functors as in the proof of (6.6.2). Then $X = U_{\underline{S}}TP_{\underline{A}/\underline{J}'}$ and $Y = U_{\underline{A}/\underline{J}'}VP_{\underline{S}}$ give the desired equivalence:

$$\begin{array}{ccc}
\underline{A}/\underline{J}' & \underset{Y}{\overset{X}{\rightleftarrows}} & \underline{S} \\
P_{\underline{A}/\underline{J}'} \downarrow \uparrow U_{\underline{A}/\underline{J}'} & & P_{\underline{S}} \downarrow \uparrow U_{\underline{S}} \\
\underline{A}^*/\underline{J}' & \underset{V}{\overset{T}{\rightleftarrows}} & \underline{S}^*
\end{array}$$

Remark The functor Y above does not depend on the choice of decompositions of M in $\underline{A}$. However, X does. If we fix another decomposition of M: $M = \Sigma_{K'} \oplus M'_{\alpha'}$, and define $P'(M) = (M, \Sigma_{K'} \oplus M'_{\alpha'})$ in the proof of (6.2.2), then $1_{\underline{A}/\underline{J}'} \sim YX \sim YX'$, where $X' = UTP'$. Hence $X \sim X'$.

6.6.4 COROLLARY Let $M = \Sigma_I\Sigma_{J_\alpha} \oplus M_{\alpha\beta} = \Sigma_{I'}\Sigma_{J'_{\alpha'}} \oplus M'_{\alpha'\beta'}$ be two decompositions of M by c.ind. modules M_α and $M'_{\alpha'}$. Then there exists a one-to-one mapping φ of K onto K' such that $M_\alpha \approx M'_{\varphi(\alpha)}$ for all $\alpha \in K$.

Proof: Let $M = \Sigma_I\Sigma_{J_\alpha} \oplus M_{\alpha\beta} = \Sigma_{I'}\Sigma_{J'_{\alpha'}} \oplus M'_{\alpha'\beta'}$. We may assume that $I = I'$, since we can assume either $J_\alpha = \emptyset$ or $J'_\alpha = \emptyset$ if necessary. Then $X(M) = \Sigma \oplus \Delta_\alpha^{(J_\alpha)} \approx X'(M) = \Sigma_I \oplus \Delta_\alpha^{(J'_\alpha)}$ from the remark above. Hence $|J_\alpha| = |J'_\alpha|$ by (4.1.4).

7
Applications I: Krull-Remak-Schmidt-Azumaya Theorem

We investigated the factor category of completely indecomposable modules, injective modules, and projective modules in Chap. 6. These results will now be applied to prove the Krull-Remak-Schmidt-Azumaya theorem and to study the structure of a perfect module and the Goldie dimension of a module.

As mentioned in Chap. 6, we may consider injective envelopes instead of the factor category induced from injective modules, and for projective modules P we may use the structure of P/J(P) and its endomorphism ring instead of the factor category induced from projective modules. However, up to now, no ring-theoretical interpretation of the factor category induced from completely indecomposable modules has been given. In order to get such an interpretation, we need new concepts in ring theory. With these, we are able to use the categorical method in ring theory, which is a very useful tool although initially difficult to understand.

In this chapter we make free use of the results in Chap. 6. Section 7.1 is the main section of this chapter. A categorical proof of the Krull-Remak-Schmidt-Azumaya theorem is given by making use of the factor category. In order to extend the theorem above, the concept of semi-T-nilpotency is introduced in Sec. 7.2, which we shall use frequently in the sequel. In Sec. 7.3, a new submodule--the full submodule--is defined with reference to the factor category, and a partial answer to the problem of Azumaya and Matlis is given using full submodules. We are able to translate back some statements for the factor category to those in ring theory with the aid of this submodule. As an application of Sec. 7.2, a small submodule in a projective module and a perfect module are studied in Secs. 7.4 and 7.5, respectively. We give the structure of a perfect ring in Sec. 7.6 by specializing Secs. 7.4 and 7.5. In Sec. 7.7, the Goldie dimension of a module is investigated from the categorical point of view. The final section, Sec. 7.7, gives a ring-theoretical proof of (7.2.6) and (7.3.14).

7.1 KRULL-REMAK-SCHMIDT-AZUMAYA THEOREM

In this section we prove the general Krull-Remak-Schmidt-Azumaya theorem as an application of (6.3.8). We proved some of this in (6.6.4). We denote the category of all right R-modules by $\underline{M}_R$.

7.1.1 LEMMA Let $M = \Sigma_{i=1}^n \oplus N_i$, where the N_i are c.ind. modules, i.e., $End_R(N_i)$ is local. Then J' in (6.3.1) is the Jacobson radical of $S_M = End_R(M)$.

Proof: $S_M = \{(x_{ij}) \mid x_{ij} \in Hom_R(N_j, N_i)\}$ by (3.7.4). Let $x = (x_{ij})$ be in J'. Then we note that $1 - x_{ii}$ is regular in S_{N_i} and a sum of non-isomorphisms of S_{N_i} is not an isomorphism. Since $1 - x_{11}$ is a unit, we can find regular matrices P_1 and Q_1 such that $P_1(1 - x_{11})Q_1 = \begin{bmatrix} 1 & 0 \\ 0 & x'_{ij} \end{bmatrix}$ by standard arguments of matrix theory. Since the x_{ij} are in J', the x'_{ij} are also in J'. Repeating this argument, we can find regular matrices P, Q in S_M such that $P(1 - x)Q - 1_M$. Hence x is quasi-regular.

7.1.2 THEOREM (Krull-Remak-Schmidt-Azumaya) Let $\{M_\alpha\}_I$ and $\{N_\beta\}_J$ be sets of completely indecomposable modules such that $M = \Sigma_I \oplus M_\alpha = \Sigma_J \oplus N_\beta$. Then we have the following.

1. There exists a one-to-one mapping ϕ of I onto J such that $M_\alpha \approx N_{\phi(\alpha)}$ for all $\alpha \in I$, and hence $|I| = |J|$, where $|I|$ is the cardinal number of I.
2. We assume that $M = \Sigma_{I'} \oplus M'_{\alpha_i} \oplus M^*$ and $M'_{\alpha_i} \approx M_{\alpha_i}$ for any finite subset I' of I. Then there exists a one-to-one mapping ψ of I' into J such that $M'_{\alpha_i} \approx N_{\psi(\alpha_i)}$ for all $\alpha_i \in I'$ and $M = \Sigma_{I'} \oplus N_{\psi(\alpha_i)} \oplus \Sigma_{J-I'} \oplus M_{\alpha'}$.

2'. Let $M = \Sigma_{I'} \oplus M'_{\alpha_i} \oplus M^*$ be as in (2). Then there exists a one-to-one mapping ψ of I' into J such that $M'_{\alpha_i} \approx N_{\psi(\alpha_i)}$ for all $\alpha_i \in I'$ and $M = \Sigma_{I'} \oplus M'_{\alpha_i} \oplus \Sigma_{J-\psi(I')} \oplus N_{\beta'}$.

3. Let M' be a direct summand of M; then M' is either isomorphic to some $\Sigma_{i=1}^m \oplus M_{\alpha_i}$ for some $m < \infty$, or M' contains a direct summand isomorphic to $\Sigma_{i=1}^m \oplus M_{\alpha_i}$ for every m.

Proof: 1. Let $\underline{A}$ be the category induced from $\{M_\alpha, N_\beta\}_{I,J}$ and $\underline{J}'$ the ideal in $\underline{A}$ defined in (6.3.1). Then $\underline{A}/\underline{J}' = \bar{\underline{A}}$ is a Grothendieck completely reducible category by (6.3.8). Furthermore, we know from its proof that $\bar{M} = \Sigma_I \oplus \bar{M}_\alpha = \Sigma_J \oplus \bar{N}_\beta$. Since $\bar{M}_\alpha$ and $\bar{N}_\beta$ are minimal objects, there exists a one-to-one mapping ϕ of I onto J such that $\bar{M}_\alpha \approx \bar{N}_{\phi(\alpha)}$ by (4.1.4).

(Note that we may use in $\underline{A}$ an argument similar to the ring-theoretical one, since $\underline{A}$ is a good category.) On the other hand, $S_{M_\alpha} \cap J'$ is equal to the Jacobson radical. Hence the fact that $\bar{M}_\alpha \approx \bar{N}_{\phi(\alpha)}$ implies that $M_\alpha \approx N_{\phi(\alpha)}$ as R-modules by Remark 3 in Sec. 6.2.

2. Put $M_0 = \Sigma_{I'} \oplus M_{\alpha_i}$ and let $p : M \to M_0$ be the projection. Then $\bar{M} = \ker \bar{p} \oplus \Sigma_{I'} \oplus \bar{N}_{\psi(\alpha_i)}$, since $\bar{\underline{A}}$ is completely reducible, where $\bar{M}_{\alpha_i} \approx \bar{N}_{\psi(\alpha_i)}$. It is clear that $\operatorname{im} \bar{p} = \Sigma_{I'} \oplus \bar{M}_{\alpha_i}$. Put $N_0 = \Sigma_{I'} \oplus N_{\psi(\alpha_i)}$ and let $i : N_0 \to M$ be the inclusion. Then $\overline{pi}$ is an isomorphism in $\bar{\underline{A}}$. Since $N_0 \in \underline{A}_f$, $\underline{J}' \cap [N_0,N_0]$ is equal to the radical of $[N_0,N_0]$ by (7.1.1). Hence pi is an isomorphism in $\underline{M}_R$ by Remark 3 in Sec. 6.2. Therefore, $M = N_0 \oplus \ker p$ in $\underline{M}_R$, so $M = N_0 \oplus \Sigma_{J-I'} \oplus M_{\alpha'}$. It is clear that $M_{\alpha_i} \approx N_{\psi(\alpha_i)}$ in $\underline{M}_R$.

2'. The following argument is dual to that above. Set $M_0' = \Sigma_{I'} \oplus M'_{\alpha_i}$. Since $\bar{\underline{A}}$ is completely reducible and $M = M_0' \oplus M^*$,

$$\bar{M} = \bar{M}_0' \oplus \Sigma_{J-\psi(I')} \oplus \bar{N}_\beta \qquad \text{where } \psi : I' \to J \text{ and } \bar{M}'_{\alpha_i} \approx \bar{N}_{\psi(\alpha_i)}$$

Let p' be the projection of M to $N_0 = \Sigma_{I'} \oplus N_{\psi(\alpha_i)}$. It is clear that $\ker \bar{p}' = \Sigma_{J-\psi(I')} \oplus \bar{N}_{\beta'}$, $\operatorname{im} \bar{p} = \Sigma_{I'} \oplus \bar{N}_{\psi(\alpha_i)}$, and $\bar{p}|M_0'$ is an isomorphism by (a). Let $i : M_0' \to M$ be the inclusion. Then $\overline{p'i}$ is an isomorphism. Since M_0' is in $\underline{A}_f$, p'i is an isomorphism in $\underline{M}_R$. Therefore, $M = M_0' \oplus \ker p'$ in $\underline{M}_R$ and $M = M_0' \oplus \Sigma_{J-\psi(I')} \oplus N_{\beta'}$.

3. Let e be the projection of M to M'. Since $\bar{\underline{A}}$ is completely reducible, $\operatorname{im} \bar{e} = \Sigma_{I'} \oplus \bar{M}'_\alpha$, where the M'_α are isomorphic to some M_β in $\{M_\alpha\}_I$. Put $M_0 = \Sigma_{I'} \oplus M'_\alpha$ and $M_0' = \Sigma_{i=1}^n \oplus M'_{\alpha_i}$ for some n (external direct sum) in $\underline{M}_R$. Then by the definition of $\bar{\underline{A}}$, we have the following R-homomorphisms: $i : M_0' \to M_0 \xrightarrow{i'} M$ and $p : M \xrightarrow{e'} M_0 \xrightarrow{i'} M$ such that i is the inclusion $\bar{M}_0' \to \bar{M}$, $\bar{p} : \bar{M} \to \bar{M}_0'$ is the projection, and $\bar{i}'\bar{e} = \bar{e}$. Since $M_0' \in \underline{A}_f$ and $\overline{pei}$ is an isomorphism in $\bar{\underline{A}}$, so is pei in $\underline{M}_R$:

$$\text{(b)} \quad M_0' \xrightarrow{i} M \xrightarrow{e} M \xrightarrow{p} M_0'$$

Hence im e in $\underline{M}_R$ contains im ei, which is a direct summand of M and isomorphic to $\Sigma_{i=1}^n \oplus M'_{\alpha_i'}$. If I' is infinite, the above argument gives the last assertion in (3). Assume that I' is finite. In this case, we can take $M_0' = M_0$. Hence M' = im e contains im ei as an R-direct summand from (b); i.e., $M' = M_0' \oplus M''$. If $M'' \neq 0$, we have a nonzero direct summand of M'', which is a contradiction, since $\operatorname{im} \bar{e} = \operatorname{im} \overline{ei}$.

Remark In this proof, we used only the assumption "I' is finite" to obtain that $\underline{J}' \cap [M_0,M_0]$ is equal to the radical of $[M_0,M_0]$ for some module M_0 in $\underline{A}$. Hence if we can obtain the property above from another assumption, then the proofs given above are still valid. We shall later make use of this fact.

We shall give, in the following section, conditions for $\underline{J}'$ to be the Jacobson radical of $\underline{A}$. In general, we have the following.

7.1.3 PROPOSITION $\underline{J}'$ contains the Jacobson radical of $\underline{A}$.

Proof: Since $\underline{A}/\underline{J}'$ is a regular category, $\underline{J}'$ contains the Jacobson radical.

7.2 SEMI-T-NILPOTENCY

In the preceding section we proved the Krull-Remak-Schmidt-Azumaya theorem with the structure of $\underline{A}/\underline{J}'$. Even though $\underline{A}/\underline{J}'$ is a good category, we can get little ring-theoretical information from this. As is known from Sec. 7.1, if $\underline{J}'$ is the Jacobson radical of $\underline{A}$, then we can get further information for modules from the structure of $\underline{A}/\underline{J}'$. We shall give a necessary and sufficient condition for $\underline{J}'$ to be the Jacobson radical, and show, in the later sections, that almost all of the known results concerning the Krull-Remak Schmidt Azumaya theorem are contained in the above result.

Let $\{M_\alpha\}_I$ be a set of modules (not necessarily c.ind.). Let $\underline{A}$ be the category induced from $\{M_\alpha\}_I$ and $\underline{C}$ an ideal in $\underline{A}$. Take any countably infinite subset $\{M_{\alpha_i}\}_{i=1}^{\infty}$ of $\{M_\alpha\}_I$ and a set of morphisms $\{f_i : M_{\alpha_i} \to M_{\alpha_i+1} \mid f_i \in \underline{C}\}$. If for any set above and any element m in M_{α_1}, there exists a natural number n (depending on the sets and m) such that $f_n f_{n-1} \cdots f_1(m) = 0$, $\{M_\alpha\}_I$ is called a locally *semi-T-nilpotent* set with respect to $\underline{C}$. Let $\{M_i\}_1^{\infty}$ be a countable set of modules M_i such that the M_i are isomorphic to some modules in $\{M_\alpha\}_I$ (the case $M_i = M_{i+1}$ may occur). If any set above and any set of morphisms f_i satisfy the condition above, we call $\{M_\alpha\}_I$ a *locally T-nilpotent* set. If I is finite, we say by definition that $\{M_\alpha\}_I$ is a *locally semi-T-nilpotent* set. If the above n does not depend on any element m in M_{α_1}, we omit the word "locally." If every M_α is finitely generated, we have this case.

Let S be a partially ordered set. If a subset $T = \{s(1,i_2',\ldots,i_n') \mid 1 \le i_j' \le i_n'\}$ of S satisfies $s(1) = s$, $s(1,i_2) > s$, $s(1,i_2,i_3) > s(1,i_2)$, $\cdots$, $s(1,i_2,\ldots,i_{t-1},i_t) > s(1,i_2,\ldots,i_{t-1})$, then we call T a *tree*.

7.2.1 LEMMA (Koenig Graph Theorem) Let T be a tree. If each path $s_1 < s(1,i_2) < \cdots < s(1,i_2,\ldots,i_t)$ terminates (has finite length), then there exists an integer N such that each path has length less than N.

Proof: We note that every point in T ramifies a finite number of paths. Assume that there is no bound as in (7.2.1). Start from s_1. Then there is a path from s_1 to $s(1,i_2)$ connected to a path that is rising infinitely. Next, we can again find a path from $s(1,i_2)$ to $s(1,i_2,i_3)$ similar to the above. Continuing this argument, we can get an infinite path s_1, $s(1,i_2)$, ..., $s(1,i_2,\ldots,i_n)$, ... which is a contradiction.

We shall come back to the category $\underline{A}$.

Let $M = \Sigma_I \oplus M_\alpha$, and represent $\mathrm{End}_R(M) = S_M$ as the ring of column-summable matrices. We may assume that I is well ordered. Let $\alpha_1 < \alpha_2 < \cdots < \alpha_n$ (or $\alpha_1 > \alpha_2 > \cdots > \alpha_n$) be in I, and $b_{\alpha_i\alpha_{i-1}}$ in $[M_{\alpha_{i-1}}, M_{\alpha_i}]$. Then for simplicity we denote $b_{\alpha_n\alpha_{n-1}} b_{\alpha_{n-1}\alpha_{n-2}} \cdots b_{\alpha_2\alpha_1}$ by $b(\alpha_n,\alpha_{n-1},\ldots,\alpha_1)$, where $[-,-] = \mathrm{Hom}_R(-,-)$.

7.2.2 LEMMA Let M and $\{M_\alpha\}_I$ be as above. We take an ideal $\underline{C}$ in the category $\underline{A}$ induced from $\{M_\alpha\}$. Let $f = (b_{\sigma\tau})$ be in $S_M \cap \underline{C}$. Set $\underline{F}_\tau = \{b(\alpha_n, \alpha_{n-1},\ldots,\alpha_2, \alpha_1 = \tau) \mid n \geq 2\}$. We assume that $\{M_\alpha\}_I$ is a locally semi-T-nilpotent set with respect to $\underline{C}$. Then for any element x_τ in M_τ, $b(\alpha_n,\alpha_{n-1},\ldots,\alpha_1)(x_\tau) = 0$ for almost all b in $\underline{F}_\tau$, where $S_M = \mathrm{End}_R(M)$.

Proof: Since $(b_{\sigma\tau})$ is column summable, there exists a finite subset T_1 of I such that $b_{\sigma\tau}(x_\tau) = 0$ for all $\sigma \in I - T_1$. Let β be in T_1. Then the subset $T_2 = \{\gamma \mid b(\gamma,\beta,\tau)(x_\tau) \neq 0\}$ of I is also finite. On the other hand, $\{M_\alpha\}_I$ is locally semi-T-nilpotent, and $b_{\sigma\tau} \in \underline{C}$, since $\underline{C}$ is an ideal. Hence (7.2.2) is clear from (7.2.1).

Remark Let $b(\alpha_n,\alpha_{n-1},\ldots,\alpha_1)$ be as above. Then for $\tau < \sigma$, $\Sigma_{\alpha_i} b(\sigma, \alpha_t,\ldots,\alpha_2,\tau)$ is an element in $[M_\tau, M_\sigma]$.

7.2.3 LEMMA Let $\{M_\alpha\}_I$, M and $\underline{C}$ be as in (7.2.2). We assume that $\{M_\alpha\}_I$ is locally semi-T-nilpotent with respect to $\underline{C}$. Let $(b_{\sigma\tau})$ be in $S_M \cap \underline{C}$ such that $b_{\sigma\tau} = 0$ if $\sigma \leq \tau$; then $(b_{\sigma\tau})$ is quasi-regular.

Proof: Put $B = (b_{\sigma\tau})$. Then each entry of the column of B^n consists of some elements in $\underline{F}_\tau$. Hence $\Sigma_{n=1}^\infty B^n$ is defined and is an element in S_M by (7.2.2) and the remark above. Put $A = \Sigma_{n=1}^\infty B^n$. Then $AB + B = A = BA + B$. Hence B is quasi-regular.

7.2.4 LEMMA Let $\{M_\alpha\}_I$ be a set of modules (not necessarily c.ind.) and $\underline{C}$ an ideal in the category induced from $\{M_\alpha\}_I$. By S_α we denote $\mathrm{End}_R(M_\alpha)$. Suppose the following:

1. $\underline{C} \cap S_\alpha \subset J(S_\alpha)$ for $\alpha \in I$.
2. If $\{a_\alpha\}_{I'}$ is a set of morphisms in $\underline{C} \cap [M_\alpha, M_\tau]$ such that $\{a_\alpha\}_{I'}$ is summable, then $\Sigma_{I'} a_\sigma \in \underline{C} \cap [M_\alpha, M_\tau]$.
3. $\{M_\alpha\}_I$ is a locally semi-T-nilpotent set with respect to $\underline{C}$. Then $\underline{C} \cap S_M \subset J(S_M)$.

Proof: Let $A' = (a'_{\sigma\tau})$ be in $\underline{C} \cap S_M$ and put $A = (a_{\sigma\tau}) = E - A'$, where E is the identity matrix. We shall show that A is regular in S_M by an argument similar to (7.1.1). Since $a'_{\sigma\sigma}$ is in $J(S_\sigma)$ by (1), $a_{\sigma\sigma}$ is regular in S_σ. Put $b_{\sigma 1} = -a_{\sigma 1}a_{11}^{-1}$ for $\sigma > 1$; then $\{b_{\sigma 1}\}_\sigma$ is summable, and $b_{\sigma 1} \in \underline{C}$. We shall define $b_{\sigma\tau}$ for $\sigma > \tau$ with the following properties:

(i) $\{b_{\sigma\tau}\}_\sigma$ is summable and $b_{\sigma\tau} \in \underline{C}$.

(ii) $b_{\sigma\tau} = -y_{\sigma\tau}y_{\tau\tau}^{-1}$, where

$$\text{(a)}\quad y_{\sigma\tau} = a_{\sigma\tau} + \Sigma_{\tau>\alpha_\tau} b(\sigma, \sigma_t, \sigma_{t-1}, \ldots, \sigma_1)a_{\sigma_1\tau}$$

where the sum runs through all finite chains $(\tau >)\ \alpha_t > \cdots > \alpha_1$. We define $\{b_{\sigma 1}\}$ with (i) and (ii). Suppose that we have defined $\{b_{\sigma\rho}\}$ for $\rho < \tau$. Then since every term in (a) is defined, we can define $y_{\sigma\tau}$ by (a). Since $\Sigma_{\tau>\alpha_t} b(\tau, \alpha_t, \ldots, \alpha_1)a_{\alpha_k\tau} \in \underline{C} \cap S_\tau \subset J(S_\tau)$ by (1), (2), and (7.2.2), $y_{\tau\tau}$ is regular in S_τ. Hence we can define $b_{\sigma\tau}$ by (ii). It is clear by (7.2.2) and (2) that $\{b_{\sigma\tau}\}$ is summable and $b_{\sigma\tau} \in \underline{C}$. Now define $C = (c_{\sigma\tau})$ by setting $c_{\sigma\sigma} = 1$, $c_{\sigma\tau} = 0$ for $\alpha < \tau$ and $c_{\sigma\tau} = \Sigma_{\alpha_i} b(\sigma, \alpha_t, \ldots, \alpha_2, \tau) \in \underline{C} \cap [M_\tau, M_\sigma]$ for $\sigma > \tau$. Then C is column summable, and hence $C \in S_M$. Put $CA = (d_{\sigma\tau})$. First we show $d_{\sigma\tau} = 0$ for $\sigma > \tau$.

$$
\begin{aligned}
d_{\sigma\tau} &= \Sigma_\rho c_{\sigma\rho}a_{\rho\tau} = a_{\rho\tau} + \Sigma_{\rho<\sigma} c_{\sigma\rho}a_{\rho\tau} \\
&= a_{\sigma\tau} + \Sigma_{\rho<\sigma}\Sigma_{\alpha_i} b(\sigma, \alpha_t, \ldots, \alpha_2, \rho)a_{\rho\tau} \\
&= a_{\sigma\tau} + \Sigma_{\tau>\alpha_t} b(\sigma, \alpha_t, \ldots, \alpha_1)a_{\alpha_1\tau} \\
\text{(b)}\qquad &\quad + b_{\sigma\tau}[\Sigma_{\alpha'_i} b(\tau, \alpha'_t, \ldots, \alpha'_1)a_{\alpha'_1\tau} + a_{\tau\tau}] \\
&\quad + \Sigma_{\sigma>\alpha''_{t''}>\tau} b_{\sigma\alpha''_{t''}}[\Sigma_{\alpha''_i} b(\alpha''_{t''}, \ldots, \alpha''_1)a_{\alpha''_1\tau} + a_{\alpha''_1\tau}] \\
&= y_{\sigma\tau} + b_{\sigma\tau}y_{\tau\tau} + \Sigma_{\sigma>\alpha''_t>\tau} b_{\sigma\alpha''_{t''}}d_{\alpha''_{t''}\tau} \\
&= \Sigma_{\sigma>\alpha''_{t''}>\tau} b_{\sigma\alpha''_{t''}}d_{\alpha''_{t''}\tau} \qquad \text{[by (a)]}
\end{aligned}
$$

It is clear by (b) that $d_{\tau+1\tau} = 0$ for all τ. If we use transfinite induction on σ, τ, we can show that $d_{\sigma\tau} = 0$ if $\sigma > \tau$ by (b). Furthermore, $d_{\sigma\sigma} = \Sigma\, b(\sigma,\alpha_t,\ldots,\alpha_1)a_{\alpha_1\sigma} + a_{\sigma\sigma}$ is regular in S_σ. Set $C_1 = \mathrm{diag}[d_{11}^{-1},\ldots, d_{\sigma\sigma}^{-1},\ldots]$ and $K = E - C_1CA = E - C_1D$. Then the entries of K in the diagonal and under the diagonal are all zero and the entries above the diagonal belong to $\underline{C}$ by (ii) and (2). Hence K is quasi-regular by (3) and (7.2.3) (consider the case $\alpha_1 > \alpha_2 \cdots > \alpha_n$). Therefore, C_1CA is regular in S_M. Again using (7.2.3), we know that C is regular in S_M. Thus A is regular, and $\underline{C} \cap S_M \subset J(S_M)$.

7.2.5 LEMMA Let $\{M_\alpha\}_I$ be a set of c.ind. modules and $M = \Sigma_I \oplus M_\alpha$. Suppose that M satisfies condition (2) in (7.1.2) for any subset I' with $|I'| \leq \aleph_0$. Then $\{M_\alpha\}_I$ is a locally semi-T-nilpotent set (with respect to $\underline{J}'$).

Proof: Let $\{M_i\}_1^\infty$ be a subset of $\{M_\alpha\}_I$ and $\{f_i : M_i \to M_{i+1}\}$ a given set of morphisms. First we show that some of the $\{f_i\}$ are not monomorphisms. Set

$$M_i(f_i) = (M_i' =) \{m_i + f_i(m_i) \mid m_i \in M_i\} \subset M_i \oplus M_{i+1} \tag{7.2.1}$$

and $M_0 = \Sigma_{I-I_0} \oplus M_\alpha$, where $I_0 = (1,2,\ldots,n,\ldots)$. Then it is clear that

$$\text{(c)}\quad \begin{aligned} M &= M_1 \oplus M_2' \oplus M_3 \oplus M_4' \oplus \cdots \oplus M_0 \\ &= M_1' \oplus M_2 \oplus M_3' \oplus M_4 \oplus \cdots \oplus M_0 \end{aligned}$$

Assume that all f_i are monomorphisms and use the condition (2) in (7.1.2) for the decomposition (c). We take a subset $I' = (2,4,\ldots,2n,\ldots)$. Then we obtain from the above property that

$$\text{(d)}\quad M = M_1' \oplus M_3' \oplus \cdots \oplus M_0 \oplus \psi_2(M_2) \oplus \psi_4(M_4) \oplus \cdots \oplus \psi_{2n}(M_{2n}) \oplus \cdots$$

where $\psi_{2n}(M_{2n})$ is equal to some one of the modules in the first decomposition except modules in M_0. From the above assumption no one of $\{f_i\}$ is an epimorphism. Hence every M_{2n}' has to be equal to some $\psi_{2m}(M_{2m})$. Therefore, $\Sigma_{I'} \oplus \psi_{2n}(M_{2n}) \supset \Sigma_{I'} \oplus M_{2m}'$. We shall show that $\Sigma_{I'} \oplus \psi_{2n}(M_{2n}) = \Sigma_{I'} \oplus M_{2m}'$. If $\Sigma_{I'} \oplus \psi_{2n}(M_{2n}) \neq \Sigma_{I'} \oplus M_{2m}'$, we have some 2i such that $\psi_{2i}(M_{2i})$ is equal to some M_{2k+1}. First we assume that $\psi_{2n}(M_{2n}) = M_{2i+1}$ and $\psi_{2m}(M_{2m}) = M_{2j+1}$ for $i < j$. Then since M_{2k}' is equal to some $\psi_{2p}(M_{2p})$,

$$\text{(e)}\quad \{M_{2i+1}, M_{2i+1}', M_{2i+2}', \ldots, M_{2j}', M_{2j+1}\}$$

are independent from (d). We shall denote

$$f_p f_{p-1} \cdots f_q \quad \text{by } \theta(p,q) \text{ for } p > q \tag{7.2.2}$$

Let $x \neq 0$ be in M_{2i+1}; then

$$\begin{aligned} x = x + f_{2i+1}(x) & \quad \in M'_{2i+1} \\ - f_{2i+1}(x) - f_{2i+2}f_{2i+1}(x) & \quad \in M'_{2i+2} \\ \vdots & \\ \pm [\theta(2j-1,2i+1)(x) + \theta(2j,2i+1)(x)] & \quad \in M'_{2j} \\ \mp \theta(2j,2i+1)(x) & \quad \in M_{2j+1} \end{aligned} \tag{f}$$

which is a contradiction to (e). Therefore, if $\Sigma \oplus \psi_{2n}(M_{2n}) \neq \Sigma \oplus M'_{2n}$, we should have only one $\psi_{2k}(M_{2k})$ equal to some M'_{2i+1}. Thus $M = \Sigma_{p=1}^{2i} \oplus M'_p \oplus M_{2i+1} \oplus M'_{2i+1} \oplus \Sigma_{k>2i+1} \oplus M'_k \oplus M_0 = \Sigma_{q=1}^{2i+1} \oplus M_q \oplus (\text{im } f_{2i+1}) \oplus \Sigma_{k>2i+1} \oplus M'_k \oplus M_0$. Since f_{2i+1} is not an epimorphism, we can show by the same argument applied to (f) that $M_{2i+2} \not\subset M$. Therefore, some of $\{f_i\}$ are not monomorphisms. From this argument, we may assume that there are infinitely many nonmonomorphisms f_j among $\{f_i\}$. Let $\{f_{i_1}, f_{i_2}, \ldots, f_{i_n}, \ldots\}$ be such a set. Put $\theta(i_{k+1} - 1, i_k) = g_k$. Then all g_k are not monomorphisms. In order to show that $\{f_i\}$ is a locally semi-T-nilpotent set, it is sufficient to show the same for $\{g_k\}$. We set $M_k^* = M_{i_k}$. Let $x \neq 0 \in \ker g_i$; then $x \in M_i^* \cap M_i^{*\prime}$. When we use the above argument for $\{M_k^*\}$, we know from (d) that $\psi_{2n}(M_{2n}^*)$ is not equal to any M_{2m+1}^*. Therefore, $\psi_{2n}(M_{2n}^*)$ is equal to some $M_{2m}^{*\prime}$, and $M = M_1^{*\prime} \oplus M_2^{*\prime} \oplus \cdots \oplus M_0$ (it is possible that some $M_{2m}^{*\prime}$ may not appear in this decomposition). Take $x \neq 0$ in M_1^* and use formula (f). Then we know that there exists some t such that $\theta(t,1)(x) = 0$. Therefore, $\{f_i\}$ is a locally semi-T-nilpotent set.

7.2.6 THEOREM Let $\{M_\alpha\}_I$ be a set of completely indecomposable modules and $M = \Sigma_I \oplus M_\alpha$. Then the following conditions are equivalent.

1. M satisfies condition (2) in (7.1.2) for any I' and any other decomposition.
2. $\{M_\alpha\}_I$ is locally semi-T-nilpotent with respect to $\underline{J}'$ defined by (6.3.1).
3. $\underline{J}' \cap \text{End}_R(M)$ is equal to the Jacobson radical of $\text{End}_R(M)$.

Proof: (1) ⇒ (2). This is clear from (7.2.5).

(2) ⇒ (3). $S_M \cap \underline{J}' \supset J(S_M)$ by (7.1.3), where $S_M = \text{End}_R(M)$. We shall prove the opposite inclusion from (7.2.4). The first condition in (7.2.4)

is clear for S_α. Let $\{a_i\}$ be a set of elements in $\underline{J}' \cap [M_\sigma, M_\tau]$ such that $\{a_i\}$ is summable. Set $a = \Sigma\, a_i$. If $M_\sigma \not\approx M_\tau$, then $a \in \underline{J}' \cap [M_\sigma, M_\tau]$. If $M_\sigma \approx M_\tau$, we can show by the argument in the proof of (6.3.2) that a is not an isomorphism. Hence $a \in \underline{J}' \cap [M_\sigma, M_\tau]$, which is the second condition in (7.2.4). The third one is the same as (2). Hence $\underline{J}' \cap S_M \subset J(S_M)$ by (7.2.4).

(3) ⇒ (1). Let $M' = \Sigma_{I'} \oplus M_\gamma$ and e the projection of M to M'. It is clear by (6.3.3) that $(\underline{J}' \cap S_M) \cap S_{M'} = \underline{J}' \cap S_{M'}$. On the other hand, $eS_Me = S_{M'}$, and $J(S_{M'}) = eJ(S_M)e$ by (3.1.8) and (6.1.5). Hence $J(S_{M'}) = \underline{J}' \cap S_{M'}$, which guarantees (1) by the remark in Sec. 7.1.

As we frequently use the method in the proof of (7.2.5), we record it as a proposition.

7.2.7 PROPOSITION Let $\{M_i\}_{i=1}^\infty$ be a set of R-modules and let $\{f_i : M_i \to M_{i+1}\}$ be a set of homomorphisms. We put $M = \Sigma_{i=1}^\infty \oplus M_i$ and $M' = \Sigma_{i=1}^\infty \oplus M_i(f_i)$. If $M = M'$, $\{f_i\}$ is locally T-nilpotent. Further we assume that each M_i is c.ind. and each f_i is not an isomorphism. If M' is a direct summand of M, M = M'.

Proof: Let i_{M_j} and $i_{M_j(f_j)}$ be the inclusions of M_j and $M_j(f_j)$. Since $i_{M(f_i)}(1_{M_i} - f_i) \equiv i_{M_i} \pmod{\underline{J}'}$, $\bar{M} = \bar{M}'$. We assume that $M = M' \oplus M^*$. Since $\bar{M} = \bar{M}'$, M* does not contain any c.ind. direct summand by (7.1.2). Hence M* = 0 by (7.1.2).

7.3 FULL SUBMODULES

Let $\{M_\alpha\}_I$ be a set of c.ind. modules. We set $M = \Sigma_I \oplus M_\alpha$. Let N be a direct summand of M. Then N contains a direct summand N' of M which is a finite direct sum of N_i, where the N_i are isomorphic to some M_α in $\{M_\alpha\}_I$ by (7.1.2). The problem (called Azumaya and Matlis's problem) of whether N is also a direct sum of c.ind. modules has not been solved up to now. We shall study this problem in the later sections in particular cases. We give here some general results.

First we state a useful theorem.

7.3.1 THEOREM Let $\{K_\alpha\}_I$ be a set of countably generated R-modules and $K = \Sigma_I \oplus K_\alpha$. Then for any direct summand P of K, $P = \Sigma_I \oplus P_i$ with P_i countably generated.

Proof: Let $K = P \oplus Q$. First we construct a sequence of submodules S_α of K such that $K = \cup\, S_\alpha$ and

(i) $S_\alpha \subset S_{\alpha+1}$ and if α is a limit ordinal, $S_\alpha = \bigcup_{\alpha>\beta} S_\beta$.

(ii) $S_{\alpha+1}/S_\alpha$ is countably generated.

(iii) $S_\alpha = \Sigma_{I_\alpha} \oplus K_i$ for a subset I_α of I.

(iv) $S_\alpha = P_\alpha \oplus Q_\alpha$; $P_\alpha = S_\alpha \cap P$ and $Q_\alpha = S_\alpha \cap Q$.

If we construct the set $\{S_\alpha\}$, $S_\alpha = P_\alpha \oplus Q_\alpha$ and S_α is a direct summand of K by (iii). Hence P_α is a direct summand of $P_{\alpha+1}$ by (1.7.4), say $P_{\alpha+1} = P_\alpha \oplus P'_\alpha$. Then P'_α is countably generated by (ii), and $P' = \Sigma_J P'_\alpha = \Sigma_J \oplus P'_\alpha$, where $P_0 = 0$. Similarly, we obtain Q' and Q'. We note that J does not contain any limit ordinal. Then $P \supset P'$, $Q \supset Q'$, and $K = P \oplus Q \supset P' \oplus Q' \supset \bigcup S_\alpha = K$. Hence $P = \Sigma \oplus P'_\alpha$ and $Q = \Sigma \oplus Q'_\alpha$. Now, we shall construct S_α. Put $S_0 = 0$. Assume that the S_β are already constructed for all $\beta < \alpha$. If α is a limit ordinal, put $S_\alpha = \bigcup_{\alpha>\beta} S_\beta$. Hence we may assume that α is not a limit ordinal, and we define $S_{\alpha+1}$. If $S_\alpha \neq K$, there exists K_i not contained in S_α. We shall denote the projection of K to a direct summand N by p_N. Let $K_i = \langle x_{11}, x_{12}, x_{13}, \dots \rangle$ by the assumption. First we take x_{11} and $x_{11} = p_P(x_{11}) + p_Q(x_{11})$. Then $p_P(x_{11})$ and $p_Q(x_{11})$ are contained in some finite sum $\Sigma_{i=1}^{n_1} \oplus K_{i(1)}$ of $\{K_i\}$. Let $\Sigma_{i=1}^{n_1} \oplus K_{i(1)} = \langle x_{21}, x_{22}, x_{23}, \dots \rangle$ containing $p_P(x_{11})$ and $p_Q(x_{11})$. Repeat this argument for x_{12} and let $\Sigma \oplus K_{i(2)} = \langle x_{31}, x_{32}, x_{33}, \dots \rangle$ containing $p_P(x_{12})$ and $p_Q(x_{12})$. Again use the argument for x_{21}, and put $\Sigma K_{i(3)} = \langle x_{41}, x_{42}, x_{43}, \dots \rangle$ containing $p_P(x_{21})$, $p_Q(x_{21})$. Repeat this argument successively for x_{11}, x_{12}, x_{21}, x_{31}, x_{22}, ... We get a matrix (x_{ij}) of infinite degree. Set $S_{\alpha+1} = S_\alpha + \Sigma_{i=1}^{\infty} \Sigma_{k=1}^{n_i} K_{i(k)} = \Sigma_{I'} \oplus K_j$, where $I' \subset I$. Then $S_{\alpha+1}/S_\alpha$ is countably generated. Let y be any element in $K_{i(k)}$ and $y = \Sigma\, x_{kj} r_j(k)$, where $r_j(k) \in R$. Then $p_P(y)$ and $p_Q(y)$ are in $S_{\alpha+1}$ from the construction. Hence $S_{\alpha+1} = P_{\alpha+1} \oplus Q_{\alpha+1}$.

Now, in this section we fix a set of c.ind. modules $\{M_\alpha\}_I$. Set $M = \Sigma_I \oplus M_\alpha$ and by $\underline{A}$ we denote the category induced from $\{M_\alpha\}_I$. We need (8.1.4) and (8.4.5), which are independent of the contents of this chapter.

7.3.2 LEMMA Let $M = \Sigma_I \oplus M_\alpha$ be as before. We assume that $M = N_1 \oplus N_2$. Then for any x in N_1 there exists a direct summand N_0 of N_1 such that $x \in N_0$ and $N_0 \in \underline{A}$.

Proof: It is clear that there exists a finite subset J of I such that $x \in M_J = \Sigma_J \oplus M_\alpha$. Since M_J has the exchange property in M by (8.1.4) and (8.4.5), $M = M_J \oplus N'_1 \oplus N'_2$, where $N'_1 \subset N_i$. Set $N''_i = N_i \cap (M_J \oplus N'_j)$, $i \neq j$. Then $x \in N''_1$ and $M = \Sigma_{i=1}^{2} \oplus (N'_i \oplus N''_i)$. Hence $M_J \approx \Sigma_{i=1}^{2} \oplus N''_i$, so $N''_i \in \underline{A}$ by (7.1.2).

7.3.3 COROLLARY Let $M = N_1 \oplus N_2$ be as in (7.3.2). If N_1 is countably generated, then N_1 is a direct sum of c.ind. modules.

Proof: Let $N_1 = \langle n_1, n_2, \ldots, n_t, \ldots \rangle$. We assume that there exists a direct summand N_{1t} ($\in \underline{A}$) of N_1 containing $n_1, n_2, \ldots, n_t$. Let $N_1 = N_{1t} \oplus N'_{1t}$ and $n_{t+1} = a + a'$; $a \in N_{1t}$, $a' \in N'_{1t}$. Then there exists a direct summand N'_{1t+1} ($\in \underline{A}$) of N'_{1t} containing a' by (7.3.2). Hence $N_{1t} \oplus N'_{1t+1}$ contains n_j for $j \le t + 1$. Use induction.

7.3.4 THEOREM Let $M = \Sigma \oplus M_\alpha$ and let all M_α be countably generated and c.ind. modules. Then every summand of M is a direct sum of c.ind. submodules.

Proof: This is clear from (7.3.1) and (7.3.3).

7.3.5 THEOREM Let $\{M_\alpha\}$ and M be as in (7.3.2). Then the following statements are equivalent:

1. Every direct summand of M is a direct sum of c.ind. modules M'_α such that the M'_α are isomorphic to some M_γ in $\{M_\alpha\}$.
2. For any idempotents e and f in S_M, we have $eS_M \approx fS_M$ if and only if $eS_M/e(\underline{J}' \cap S_M) \approx fS_M/f(\underline{J}' \cap S_M)$.

Proof: (1) $\Rightarrow$ (2). Put $\bar{S}_M = S_M/\underline{J}' \cap S_M$, $eM = \Sigma \oplus M'_{\alpha'}$, and $fM = \Sigma \oplus M''_{\alpha''}$. Assume that $\overline{eS_M} \approx \overline{fS_M}$. Then $\operatorname{im} \bar{e} \approx \operatorname{im} \bar{f}$ in $\underline{\bar{A}}$ by (3.1.10). Hence, since $\operatorname{im} \bar{e} = \Sigma \oplus \bar{M}'_{\alpha'}$ and $\operatorname{im} \bar{f} = \Sigma \oplus \bar{M}''_{\alpha''}$, $M'_{\alpha'}$ is isomorphic to some $M''_{\alpha''}$ by Remark 3 in Sec. 6.2. Therefore, $eM \approx fM$, which implies that $eS_M \approx fS_M$ by (3.1.10).

(2) $\Rightarrow$ (1). Let M' be a direct summand of M and e the projection. We showed in the proof of (7.1.2) that there exists an idempotent f in S_M such that $fM = \Sigma_{I'} \oplus M_\delta$, where $I' \subset I$ and $\overline{eS_M} \approx \overline{fS_M}$. Hence $eS_M \approx fS_M$ implies that $eM \approx fM$ by (3.1.10).

7.3.6 COROLLARY Let $\{M_\alpha\}$ be as in (7.3.2). If one of the conditions in (7.2.6) is satisfied, then every direct summand of M is a direct sum of c.ind. modules [see (4.5.15)].

Remarks 1. Let Z be the ring of integers and p a prime. Then $\{Z/p^i\}_{i=1}^{\infty}$ is not a semi-T-nilpotent set. Hence $M = \Sigma_{i=1}^{\infty} \oplus Z/p^i$ does not satisfy any conditions in (7.2.6). However, every direct summand of M is a direct sum of c.ind. modules.

2. Let K be a field and let R be the ring of lower triangular matrices with infinite degree. Set $M = \Sigma \oplus e_{ii}R$, where the e_{ii} are matrix units in R. Then $\{e_{ii}R\}$ is not a semi-T-nilpotent set, but for M we have the same conclusions as in Remark 1.

3. Let R be the ring of upper triangular matrices. Then $\{e_{ii}R\}$ is a T-nilpotent set.

Next we shall try to translate back some results for $\underline{A}/\underline{J}'$ to properties of modules. Let M and N be in $\underline{A}$ and N a submodule of M with $i : N \to M$ the inclusion. If $\bar{i}$ is an isomorphism in $\bar{\underline{A}}$ (i.e., $\bar{N} = \bar{M}$), then N is called a *full submodule* in M.

Notation Let e be an idempotent in $S_M = \mathrm{End}_R(M)$. Then $M = eM \oplus (1 - e)M$ in $\underline{M}_R$. We do not know whether $eM \in \underline{A}$; however, we shall denote im $\bar{e}$ in $\bar{\underline{A}}$ by $\overline{eM}$ for convenience. It is clear that if $eM \in \underline{A}$, im $\bar{e} = \overline{eM}$ in $\bar{\underline{A}}$. We note that even if f(M) is in $\underline{A}$ for some $f \in S_M$, im $\bar{f}$ is not equal to $\overline{f(M)}$ in general.

7.3.7 PROPOSITION Every full submodule of M is isomorphic to M.

Proof: Since $\bar{M} = \Sigma_I \oplus \bar{M}_\alpha = \bar{N} = \Sigma_J \oplus \bar{N}_\gamma$, $M \approx N$ as R-modules by Remark 3 in Sec. 6.2, where the N_γ are c.ind. modules.

7.3.8 PROPOSITION Let M and P be in $\underline{A}$. We assume that $\bar{M} \supset \bar{P}$ in $\bar{\underline{A}}$. Then there exists a submodule P_0 of M which satisfies the following:

1. P_0 is in $\underline{A}$; that is, $P_0 = \Sigma_J \oplus M'_{\alpha'}$.
2. For any finite subset J' of J, $\Sigma_{J'} \oplus M'_{\alpha'}$ is a direct summand of M. If $\{M'_\alpha\}_J$ is a locally semi-T-nilpotent set with respect to $\underline{J}'$, then P_0 is a direct summand of M.
3. $P_0 \approx P$ as R-modules.

Furthermore, if e is an idempotent in S_M and $\bar{P}$ = im $\bar{e}$, then we can find such P_0 in im e in $\underline{M}_R$.

Proof: Since $\bar{\underline{A}}$ is completely reducible by (6.3.8), there exist R-homomorphisms $i : P \to M$ and $p : M \to P$ such that $\overline{pi} = 1_{\bar{P}}$. Let $P = \Sigma_K \oplus P_\gamma$: the P_γ are c.ind. For a subset K' of K, we denote the injection $P_{K'} = \Sigma_{K'} \oplus P_\gamma \to P$ and the projection $P \to P_{K'}$ by $i_{K'}$ and $p_{K'}$, respectively:

$$P_{K'} \underset{p_{K'}}{\overset{i_{K'}}{\rightleftarrows}} P \underset{p}{\overset{i}{\rightleftarrows}} M$$

Then $\overline{p_{K'}pii_{K'}} = 1_{\bar{P}_{K'}}$. If either K' is finite or $\{P_\gamma\}_{K'}$ is semi-T-nilpotent, $S_{P_{K'}} \cap \underline{J}' = J(S_{P_{K'}})$ by (7.1.1) and (7.2.6). Hence $p_{K'}pii_{K'}$ is an R-isomorphism. Therefore, $ii_{K'}$ is a monomorphism in $\underline{M}_R$ for every finite subset K' of K, which means that i is a monomorphism in $\underline{M}_R$. Put P_0 = im i in $\underline{M}_R$. Then P_0 satisfies (1)-(3). Suppose that im $\bar{e} = \bar{P}$. Then there exist $i' : P \to M$ and $p' : M \to P$ such that $\bar{e} = \overline{i'p'} = \overline{ei'p'}$ and $\overline{p'i'} = 1_{\bar{P}}$. Put $P_0 = \text{im } ei'$ in M_R. From the argument above, we know that P_0 satisfies the requirements in (7.3.8).

7.3.9 REMARK Let $N = \Sigma_{i=1}^{n} \oplus M_i$ be a submodule of M via the inclusion i_N. Then we know by the above proof that $\bar{i}_N$ is a monomorphism in $\bar{A}$ if and only if N is a direct summand of M. If n is infinite and $\{M_i\}$ is locally semi-T-nilpotent, then we have the same.

7.3.10 LEMMA Let $\{M_\alpha\}_I$ be a set of c.ind. modules and $M = \Sigma_I \oplus M_\alpha$. Let $S = \text{End}_R(M)$ and let a be any element in S. Then, for any finite subset $\{M_{\alpha_i}\}_{i=1}^{n}$ of $\{M_\alpha\}_I$, there exists a set $\{M_i'\}_{i=1}^{n}$ of direct summands of M such that $M = \Sigma_{i=1}^{n} \oplus M_i' \oplus \Sigma_{\alpha \notin \{\alpha_i\}} \oplus M_\alpha$ and each M_{α_i} is isomorphic to M_i' via a or 1 - a.

Proof: Let e_1 be the projection of M to M_{α_1}. Then $e_1ae_1 \mid M_{\alpha_1}$ and $e_1(1 - a)e_1 \mid M_{\alpha_1}$ are in $[M_{\alpha_1}, M_{\alpha_1}]$ and $1_{M_{\alpha_1}} = [e_1ae_1 + e_1(1 - a)e_1] \mid M_{\alpha_1}$. Since M_{α_1} is c.ind., either $e_1ae_1 \mid M_{\alpha_1}$ or $e_1(1 - a)e_1 \mid M_{\alpha_1}$ is an isomorphism: $M_{\alpha_1} \xrightarrow{b} b(M_{\alpha_1}) \xrightarrow{e_1} M_{\alpha_1}$, where b = a or 1 - a. Hence $M = b(M_{\alpha_1}) \oplus \ker e_1 = b(M_{\alpha_1}) \oplus \Sigma_{\alpha \neq \alpha_1} \oplus M_\alpha$. Repeating this argument on the last decomposition, we obtain (7.3.10).

7.3.11 LEMMA Let $M = \Sigma_I \oplus M_\alpha$ be as in (7.3.10) and $S = \text{End}_R(M)$. Let $J' = \underline{J}' \cap S$ be the ideal in (6.3.1). Then J' does not contain a nonzero idempotent.

Proof: Let e be a nonzero idempotent in S. Then there exists a finite subset $\{M_{\alpha_i}\}_1^n$ of $\{M_\alpha\}_I$ such that $eM \cap \Sigma_{i=1}^{n} \oplus M_{\alpha_i} \neq 0$. We apply (7.3.10) to e and $\{M_{\alpha_i}\}_1^n$. Then we can find a direct summand $\Sigma_{i=1}^{n} \oplus M_i$ of M such that $M_i = b_i(M_{\alpha_i})$, where b_i = e or 1 - e. It is impossible that all b_i are equal to 1 - e. Hence $e_iee_{\alpha_i}$ is an isomorphism for some i, where $e_{\alpha_i} : M \to M_{\alpha_i}$, $e_i : M \to M_i$ are projections. Therefore, $e \notin J'$ by (6.3.3).

7.3.12 LEMMA Let M and $\underline{J}'$ be as in (7.3.11). Then for any $f \in \underline{J}' \cap S_M$, $1_M - f$ is a monomorphism.

Proof: Suppose that ker (1 - f) ≠ 0. Then there exists a finite subset I' of I such that $\ker(1 - f) \cap \Sigma_{I'} \oplus M_\alpha \neq 0$. By (7.3.10) we obtain a set of direct summands $\{M'_{\phi(\alpha)}\}_{I'}$ such that $M = \Sigma_{I'} \oplus M'_{\phi(\alpha)} \oplus \Sigma_{I-I'} \oplus N_\alpha$ and $M_{\alpha'} \approx M_{\phi(\alpha)}$ for each $a \in I'$ via either f or 1 - f.

However, f is in $\underline{J}'$, and hence we must obtain these isomorphisms by 1 - f, which is a contradiction. Therefore, ker (1 - f) = 0.

We shall give some criteria for a submodule to be a full submodule.

7.3.13 THEOREM Let $\{M_\alpha\}_I$, $\underline{A}$, and $\underline{J}'$ be a set of c.ind. modules, the category induced from $\{M_\alpha\}$, and the ideal in $\underline{A}$ defined by (6.3.1), respectively. Let N be in $\underline{A}$; i.e., $N = \Sigma_K \oplus N_\gamma$, which is a submodule of M via the inclusion $i_N : N \to M$. Then the following are equivalent:

1. N is a full submodule of M.
2. $\bar{i}_N$ is a monomorphism in $\underline{A}/\underline{J}'$, and for any direct summand P of M, there exists a finite subset K' of K such that $P \cap N_{K'} \neq 0$ or $P \oplus N_{K'}$ is not a direct summand of M, where $N_{K'} = \Sigma_{K'} \oplus N_\gamma$.
3. $\bar{i}_N$ is a monomorphism and N contains im (1 - f) in $\underline{M}_R$ for some $f \in \underline{J}'$.

Hence im (1 - f) is a full submodule in M for all $f \in \underline{J}'$. Furthermore, the above $N_{K'}$ is a direct summand of M if either K' is finite or $\{N_\gamma\}_{K'}$ is semi-T-nilpotent.

Proof: (1) ⇒ (2). Since P contains a c.ind. direct summand of M, we may assume that P is c.ind. Furthermore, since $\bar{\underline{A}}$ is a Grothendieck category and $\bar{P}$ is minimal in $\bar{\underline{A}}$, $\bar{P} \subset \Sigma_{K'} \oplus \bar{N}_\gamma = \bar{N}_{K'}$ for some finite subset K' of K by (2.4.2). Suppose that $P \cap N_{K'} = 0$ and $P \oplus N_{K'}$ is a direct summand of M; $M = P \oplus N_{K'} \oplus M_0$. Let $i : P \oplus N_{K'} \to M$ be the inclusion. Then im $\bar{i}$ = $\bar{P} \oplus \bar{N}_{K'}$, which is a contradiction. Hence $P \oplus N_{K'}$ is not a direct summand of M.

(2) ⇒ (1). We assume that i_N is a monomorphism and that $\bar{M} \neq \bar{N}$. Then there exists a minimal object $\bar{M}_\alpha$ such that $\bar{M}_\alpha \cap \bar{N} = 0$. Hence, for any finite subset K' of K, $\bar{M}_\alpha \cap \bar{N}_{K'} = 0$. Take a formal direct sum $M_\alpha \oplus N_{K'}$ and consider a natural mapping from $M_\alpha \oplus N_{K'}$ to $M_\alpha \cup N_{K'} \subset M$. Then $M_\alpha \oplus N_{K'}$ is a direct summand of M by (7.3.9).

(1) ⇒ (3). Since $\bar{i}_N$ is an isomorphism, there exists an R-homomorphism $j \in [M,N]$ such that $\bar{i}_N\bar{j} = \bar{1}_M$. Then $f = 1 - i_N j \in \underline{J}'$ and im (1 - f) in $\underline{M}_R$ is contained in im i_N = N.

(3) ⇒ (1). Since 1 - f is a monomorphism by (7.3.12), N' = im (1 - f) in $\underline{M}_R$ is in $\underline{A}$. We factor 1 - f as $1 - f : M \xrightarrow{(1-f)'} N \xrightarrow{i} M$. Then $\bar{1}_M = 1 - f = \bar{i}(1 - f)'$. Hence $\bar{i}$ is an isomorphism in $\bar{\underline{A}}$, since (1 - f)' is an isomorphism in $\underline{M}_R$. Therefore, im (1 - f)' is a full submodule of M. Since $\bar{i}_N$ is a monomorphism and N ⊃ im (1 - f), N is also a full submodule.

The remaining assertion is clear from (7.2.6) and (7.3.9).

In the above we showed that if K' is a finite set, then $N_{K'}$ is a direct summand of M for a full submodule N. We generalize this property as follows:

Let $A \supset B$ be R-modules and $B = \Sigma_K \oplus B_\gamma$. If for any finite subset K' of K, $\Sigma_{K'} \oplus B_\gamma$ is a direct summand of A, then we call B a *locally direct summand* of A (with respect to the decomposition $B = \Sigma_K \oplus B_\gamma$). We note that if all B_γ are injective, B is always a locally direct summand of A. We shall use this fact in the next section. In general, $B = \Sigma_I \oplus B_\gamma$ is a locally direct summand of $\Pi_I B_\gamma$.

The following theorem is a generalization of (7.3.13) and is a particular solution of the Azumaya and Matlis problem.

7.3.14 THEOREM Let $\{M_\alpha\}_I$, M, and $\underline{A}$ be as in (7.3.13). Let $M = \Sigma_K \oplus N_\gamma$, where the N_γ need not be in $\underline{A}$. Then there exists a set of submodules $\{P_\gamma\}_K$ of M as follows.

1. $N_\gamma \supset P_\gamma$ and $P_\gamma \in \underline{A}$.
2. $\Sigma_K \oplus P_\gamma$ is a full submodule of M.

Proof: Let π_γ be the projection of M to N_γ (note that π_γ is regarded as an element in [M,M]). It is clear that $\{\pi_\gamma\}$ is a summable set and $1_M = \Sigma_J \pi_\gamma$. Let M_1 be a member of $\{M_\alpha\}$. For any nonzero element m_1 in M_1 we have $\pi_\gamma(m_1) = 0$ for all $\gamma \in K - K'$, where K' is a finite subset of K. Hence $\pi_\gamma \mid M_1 \in \underline{J}'$ for all $\gamma \in K - K'$. We shall express the π_γ as matrices $(x^\gamma_{\alpha\beta})$. Since $\{\pi_\gamma\}$ is summable, so is $\{x^\gamma_{\alpha\beta}\}_\gamma$ for any α and β. It is clear that $\pi_\gamma \mid M_1 = (x^\gamma_{\alpha 1})_{\alpha\in I}$. Therefore, $\Sigma_{K-K'} \pi_\gamma \mid M_1 \in \underline{J}'$ [see the proof of (6.3.2)]. Then $\bar{M}_1 = \text{im } \bar{1}_M \mid \bar{M}_1 \subset \text{im } (\Sigma_{K'} \bar{\pi}_\gamma \mid \bar{M}_1 + (\Sigma_{K-K'} \bar{\pi}_\gamma \mid \bar{M}_1)) = \text{im } (\Sigma_{K'} \bar{\pi}_\gamma \mid \bar{M}_1) \subset \Sigma_{K'} \text{ im } \bar{\pi}_\gamma$. Hence $\bar{M} = \Sigma_K \text{ im } \bar{\pi}_\gamma$ [see (2.4.1)]. On the other hand, there exists a set $\{P_\gamma\}_K$ of submodules P_γ in N_γ such that $P_\gamma \in \underline{A}$ and $\bar{P}_\gamma = \text{im } \bar{\pi}_\gamma$ by (7.3.8). It is clear that $\cup_{\delta\in K''} \text{im } \bar{\pi}_\delta = \Sigma_{\delta\in K''} \oplus \text{im } \bar{\pi}_\delta$ for any finite subset K'' of K, so $\bar{M} = \Sigma_K \oplus \text{im } \bar{\pi}_\gamma = \Sigma_K \oplus \bar{P}_\gamma$.

We shall call such P_γ a *full submodule* of N_γ.

7.3.15 THEOREM Let $\{M_\alpha\}_I$ be a set of c.ind. modules and $M = \Sigma_I \oplus M_\alpha$. Then the following statements are equivalent:

1. $\{M_\alpha\}_I$ is a locally semi-T-nilpotent set.
2. Every full submodule coincides with M.
3. Every locally direct summand M' of M with respect to $M' = \Sigma_K \oplus T_\alpha$ with any cardinal $|K|$ is a direct summand of M.
4. (3) is true for decompositions with $|K| \leq \aleph_0$.
5. (4) is true whenever all T_α are c.ind. modules.

6. $S_M/J(S_M)$ is a regular ring in the sense of von Neumann and every idempotent in $S_M/J(S_M)$ is lifted to S_M.

Proof: (1) ⇒ (2). Every full submodule N of M is a direct summand of M by the last assertion of (7.3.13). Hence N = M by (7.3.11).

(2) ⇒ (1). Since im (1 - f) is full in M for $f \in \underline{J}' \cap S_M$, 1 - f is an isomorphism by (2). Hence $\underline{J}' \cap S_M \subset J(S_M)$, which implies (1) from (7.2.6).

(1) ⇒ (3). Every direct summand of M is a direct sum of c.ind. modules by (7.3.6). The assumption of locally direct summand implies that $\Sigma_K \oplus \bar{T}_\alpha$ is a subobject of $\bar{M}$ via $\bar{i}_{M'}$, where $i_{M'} : M' \to M$ is the inclusion. Hence M' is a direct summand by (7.3.9).

(3) ⇒ (4) ⇒ (5). This is clear.

(5) ⇒ (1). We shall recall the proof of (7.2.5). Let $\{M_i\}_1^\infty$ be a countable subset of $\{M_\alpha\}$ and let $\{f_i : M_i \to M_{i+1}\}$ be a given set of homomorphisms in $\underline{J}'$. We defined the submodule $M' = M_1' \oplus M_2' \oplus \cdots$ in M. Since $M_1' \oplus \cdots \oplus M_n' \oplus M_{n+1} = \Sigma_{i=1}^{n+1} \oplus M_i$ for any n, M' is a locally direct summand of M. Hence M' is a direct summand of M, and hence M' is a direct summand of $M_0 = \Sigma_{i=1}^\infty \oplus M_i$. Therefore, $M' = M_0$ by (7.2.7), and hence $\{f_i\}$ is locally T-nilpotent by (7.2.7).

(1) ⇒ (6). Since $\underline{J}' \cap S_M = J(S_M)$ by (7.2.6) and $\underline{A}/\underline{J}'$ is a regular ring by (6.3.8), so is $S_M/J(S_M)$. Let f be in S_M such that $\bar{f}^2 = \bar{f}$. Then there exists a direct summand M_1 of M such that $\bar{M}_1 = \text{im } \bar{f}$ by (7.3.8). Let e be the projection of M to M_1. Since $\text{im } \bar{f} = \text{im } \bar{e}$, $\text{im } (\overline{1 - f}) \approx \text{im } (\overline{1 - e})$ in $\bar{\underline{A}}$. Hence there exists a regular element $\bar{a}$ in $\bar{S}_M$ such that $\bar{f} = \bar{a}^{-1}\bar{e}\bar{a}$ by (6.3.4). Since $\underline{J}' \cap S_M = J(S_M)$, a is a unit in S_M and hence $a^{-1}ea$ is an idempotent.

(6) ⇒ (1). $\underline{J}' \cap S_M \supset J(S_M)$ by (7.1.3). Since $S_M/J(S_M)$ is regular, $(\underline{J}' \cap S_M)/J(S_M)$ contains a nonzero idempotent if $(\underline{J}' \cap S_M)/J(S_M) \neq 0$. Then this idempotent is lifted to S_M by (6), and hence it is in $\underline{J}' \cap S_M$, which contradicts (7.3.11).

7.4 SMALL SUBMODULES IN PROJECTIVE MODULES

If a projective R-module P is semiperfect, J(P) is small in P by (6.1.14). We shall generalize this fact to arbitrary projective modules.

Let $F = \Sigma \oplus u_\alpha R$ be a free R-module and S a small submodule in F. Then $S \subset J(F) = \Sigma \oplus u_\alpha J(R)$. In order to decide the form of S, we shall define a set of right semi-T-nilpotent right ideals. Let $\{A_\alpha\}_K$ be a set of right ideals in R and K an infinite set. If $\{A_\alpha\}_K$ satisfies the following condition (7.4.1), we call $\{A_\alpha\}_K$ a *right semi-T-nilpotent* set or a *vanishing* set of ideals.

$$\text{For any countable subset } \{A_{n_i}\}_1^\infty \text{ of } \{A_\alpha\}_K \text{ and } \{a_i \in A_{n_i}\}, \text{ there exists n, depending on } \{a_i\}, \text{ such that } a_n a_{n-1} \cdots a_1 = 0. \qquad (7.4.1)$$

Let $\{b_\sigma\}_K$ be a set of elements in R. If $\{b_\sigma R\}$ is a semi-T-nilpotent set, we call $\{b_\sigma\}$ a *right semi-T-nilpotent* set. If we allow $\alpha_i = \alpha_j$ for $i \neq j$, we call $\{A_\alpha\}$ and $\{b_\sigma\}$ *right T-nilpotent sets*. If $A_\alpha = A$ for all α, then we call A right T-nilpotent. If K is a finite set, we understand that $A_\alpha = 0$ for almost all α; then $\{A_\alpha\}$ is always a semi-T-nilpotent set, but not a T-nilpotent set.

First we generalize the concept of a right semi-T-nilpotent set of right ideals. Let $\{Q_\alpha\}_I$ be an infinite set of R-modules and let $\{S_\beta \mid \subset Q_\beta\}_K$ be an infinite set of R-submodules. We take a countable subset $\{Q_{\alpha_i}\}$ of $\{Q_\alpha\}_K$ and a set of homomorphisms $\{f_i : Q_{\alpha_i} \to Q_{\alpha_{i+1}}\}$ such that $f_i(Q_{\alpha_i}) \subset S_{\alpha_{i+1}}$. If for any element t in Q_{α_1}, there exists n, depending on t, and $\{f_i\}$ such that $f_n f_{n-1} \cdots f_1(t) = 0$, then we call $\{f_i\}$ a *locally T-nilpotent* set of homomorphisms. If for any countable subset $\{Q_{\alpha_i}\}$ and any set of homomorphisms f_i as above, $\{f_i\}$ is always locally T-nilpotent, then we call $\{S_\alpha\}_K$ a *locally (right) semi-T-nilpotent* set of submodules. In the above, if we set $Q_\alpha = R$ for all α and $f_i : Q_i \to Q_{i+1}$ is left-sided multiplication by an element in R, the above concept coincides with that of right semi-T-nilpotent.

7.4.1 THEOREM Let $\{M_\alpha\}_I$ be a set of f.g. R-modules and let S be a small submodule of $M = \Sigma_I \oplus M_\alpha$. Then $\{S_\alpha = p_\alpha(S)\}_I$ is a (right) semi-T-nilpotent set of submodules, where the $p_\alpha : M \to M_\alpha$ are the projections.

Proof: We may assume that I is well ordered and infinite. Further, from the definition and (3.4.1), we may assume that $I = \{1,2,\ldots,n,\ldots\}$ and $f_j : M_j \to S_{j+1}$. Let $M_n = \langle m_1^{(n)}, m_2^{(n)}, \ldots, m_{k_n}^{(m)} \rangle$. Then there exists $s_{(1,i)}$ in S for any $m_i^{(1)}$ such that

$$\text{(a)} \quad s_{(1,i)} = s_{(1,i)}^{(1)} + f_1(m_i^{(1)}) + s_{(1,i)}^{(3)} + \cdots + s_{(1,i)}^{(p_1)}$$
$$i = 1, 2, \ldots, k_1$$

where $s_{(1,i)}^{(t)} \in S_t$. We take an $n_3 > p_1$ and put $g_1 = f_1$, $g_2 = f_{n_3-1} \cdots f_2$. Then $g_2 \in \mathrm{Hom}_R(M_s, S_{n_3})$. Consider the same on g_2, instead of f_1. Then

(b) $s_{(2,i)} = s^{(1)}_{(2,i)} + \cdots + s^{(n_3-1)}_{(2,i)} + g_2(m^{(2)}_i) + \cdots + s^{(p_2)}_{(2,i)}$

$(i = 1, 2, \ldots, k_2)$

Again take an $n_4 > p_2$, and put $g_3 = f_{n_4-1} \cdots f_{n_3}$. Repeating this argument, we can get a series $\{g_i : M_{n_i} \to S_{n_i+1}; \mid n_1 = 1, n_2 = 2\}$ such that $p_i < n_{i+1}$ [see (a), (b)]. It is sufficient to show that $\{g_i\}$ is locally T-nilpotent. We put $M'_{n_i} = M_{n_i}(g_i) = \{m + g_i(m) \mid m \in M_{n_i}\}$ and $M' = \Sigma^{\infty}_{i=1} M'_{n_i} + \Sigma_{i \notin \{n_i\}} M_i + S$. We show $M = M'$. Since $M'_1 + S$ $(\subset M')$ contains $m^{(1)}_i + g_1(m^{(1)}_i) - s_{(1,i)} = m^{(1)}_i - s^{(1)}_{(1,i)} - s^{(3)}_{(1,i)} - \cdots - s^{(p_1)}_{(1,i)}$, $m^{(1)}_i \equiv s^{(1)}_{(1,i)}$ (mod M'), $(i = 1, 2, \ldots, k_1)$. Hence $(M_1 + M')/M' = (S_1 + M')/M'$. Since S_1 is small in M_1 by (3.4.1), $(S_1 + M')/M'$ is small in $(M_1 + M')/M'$. Hence $M_1 \subset M'$. We can easily see, by induction and the construction of M', that $M' \supset M_n$ for all n. Therefore, $M = M'$, so $\{g_i\}$ is locally T-nilpotent by (7.2.7).

7.4.2 COROLLARY Let $\{N_\alpha\}_I$ be a set of R-modules and $\{T_\alpha\}_I$, $T_\alpha \subset N_\alpha$ a set of submodules. If $\Sigma_I \oplus T_\alpha$ is a small submodule of $\Sigma_I \oplus N_\alpha$, then $\{T_\alpha\}_I$ is a locally (right) semi-T-nilpotent set of submodules.

Proof: This is clear from the above proof.

7.4.3 THEOREM Let $\{N_\alpha\}_I$ be a set of R-modules and let $\{T_\alpha\}_I$, $T_\alpha \subset N_\alpha$, be a set of submodules. Put $N = \Sigma_I \oplus N_\alpha$, $T = \Sigma_I \oplus T_\alpha$ and $S_N = \mathrm{End}_R(N)$. Then $J(S_N) \supset \mathrm{Hom}_R(N,T)$ if and only if $\{T_\alpha\}_I$ is a locally (right) semi-T-nilpotent set of submodules and $\mathrm{Hom}_R(N_\alpha,T_\alpha) \subset J(S_{N_\alpha})$.

Proof: We assume that $J(S_N) \supset \mathrm{Hom}_R(N,T)$. Let $\{N_{\alpha_i}\}$ and $\{f_i : N_{\alpha_i} \to T_{\alpha_{i+1}}\}$ be given sets. We assume that $\alpha_j = j$ for all j. Then

$$F = \begin{bmatrix} 0 & & & & 0 \\ f_1 & 0 & & & \\ & f_2 & & & \\ & & \cdots & 0 & \\ & 0 & & f_n & \\ & & & & \cdots \end{bmatrix}$$

is in $J(S_N)$ by the assumption. Hence $1 - F$ is regular in S_N. Taking the inverse of $1 - F$, we know that $\{f_i\}$ is locally right T-nilpotent (note that the inverse is a summable matrix). It is clear by (3.1.8) and (6.1.5) that $J(S_{N_\alpha}) = e_\alpha J(S_N)e_\alpha \supset e_\alpha \mathrm{Hom}_R(N,T)e_\alpha = \mathrm{Hom}_R(N_\alpha,T_\alpha)$, where the e_α are the projections of N onto N_α. Conversely, we note that the $C_{\sigma\tau} = \mathrm{Hom}_R(N_\tau,T_\sigma)$ satisfy conditions (1)-(3) in (7.2.4) and that we used only the fact that $C_{\sigma\tau}\mathrm{Hom}_R(N_\varepsilon,N_\tau) \subset C_{\sigma\varepsilon}$ in the proof of (7.2.4). Hence $J(S_N) \supset \mathrm{Hom}_R(N,T)$ by (7.2.4).

7.4.4 COROLLARY Let $\{P_\alpha\}_I$ be a set of R-projectives and let $\{S_\alpha \subset P_\alpha\}$ be a set of R-submodules. Then $\Sigma_I \oplus S_\alpha$ is small in $\Sigma_I \oplus P_\alpha$ if and only if $\{S_\alpha\}_I$ is a locally (right) semi-T-nilpotent set of small submodules S_α in P_α.

Proof: This is clear from (7.4.3) and (6.1.4).

7.4.5 COROLLARY Let $F = \Sigma_I \oplus u_\alpha R$. Then J(F) is small in F if and only if J(R) is right T-nilpotent. Hence we obtain that $J(R_I) = (J(R))_I$ for an infinite set I if and only if J(R) is right T-nilpotent.

Proof: We obtain the last assertion by (6.1.4). The first one is clear.

7.4.6 COROLLARY Let M be an R-module. If $\{A_m\}_{m\in X}$ is a right semi-T-nilpotent set of right ideals in J(R), then $\Sigma_X\, mA_m$ is small in M. Conversely, if M is projective and S is a small submodule in M, then there exists a semi-T-nilpotent set $\{A_m\}_X$ of right ideals in J(R) such that $S \subset \Sigma_X\, mA_m$, where X is any set of generators of M.

Proof: Consider the epimorphism $\varphi : F = \Sigma_X \oplus u_m R \to M$ given by setting $\varphi(u_m) = m$. Since $\Sigma_X \oplus u_m A_m$ is small in F from (7.4.4), $\varphi(\Sigma_X \oplus u_m A_m) = \Sigma_X\, mA_m$ is small in M. Conversely, we assume that M is projective. M is a direct summand of F, so $\varphi i = 1_M$ for a monomorphism $i : M \to F$. i(S) is also small in F and hence there exists a right semi-T-nilpotent set $\{A_m\}_X$ of right ideals in J(R) such that $i(S) \subset \Sigma_X \oplus u_m A_m$ from (7.4.1). Therefore, $S = \varphi i(S) \subset \Sigma_X\, mA_m$.

7.4.7 THEOREM Let $\{P_\alpha\}_I$ be a set of R-projectives. Let $M = \Sigma_I \oplus P$ and S a submodule of M. Then the following statements are equivalent:

1. S is small in M.
2. Let $p_\alpha : M \to P_\alpha$ be the projection of M onto P_α and $S_\alpha = p_\alpha(S)$. Then $\{S_\alpha\}_I$ is a locally (right) semi-T-nilpotent set of small submodules.
3. There exists a locally semi-T-nilpotent set $\{S_\alpha\}_I$ of small submodules S_α in P_α such that $S \subset \Sigma_I \oplus S_\alpha$.

Proof: (2) ⇒ (3) ⇒ (1). This is clear from (7.4.4).

(1) ⇒ (2). We shall prove this in a general form.

7.4.8 LEMMA In (7.4.1), we assume that every M_α is a direct summand of a direct sum Q_α of finitely generated R-modules $M_{\alpha\beta}$. Then the statement in (7.4.1) is valid.

Proof: Put $Q_\alpha = \Sigma_{\beta\in I_\alpha} \oplus M_{\alpha\beta}$ and $M^* = \Sigma_I \oplus Q_\alpha$. Then M is a direct summand of M*. Let i be the injection of M into M*. Then i(S) is small in M*. By $p_{\alpha\beta}$ we denote the projection of M* onto $M_{\alpha\beta}$. Then $\Sigma_I\Sigma_{I_\alpha} \oplus p_{\alpha\beta}(i(S)) \supset \Sigma_I \oplus i(S_\alpha)$ and $i(S_\alpha) \subset \Sigma_{I_\alpha} \oplus p_{\alpha\beta}(i(S))$. Now, $\{p_{\alpha\beta}(i(S))\}_{I,I_\alpha}$ is a locally semi-T-nilpotent set from (7.4.1). Let $\{M_{\alpha_i}\}_1^\infty$ and $\{f_i : M_{\alpha_i} \to S_{\alpha_{i+1}}\}$ be given sets. Then we can extend f_i to $f_i' : Q_{\alpha_i} \to i(S_{\alpha_{i+1}})$ by sending a direct complement to zero. We shall denote f_i' by a column-finite matrix $(a_{\rho\tau}^{(i)})$ [see (3.7.4)], where $a_{\rho\tau}^{(i)} \in \mathrm{Hom}_R(M_{\alpha_i\tau}, P_{\alpha_{i+1}\rho}(i(S)))$. Let m be in M_{α_1} and $i(m) = \Sigma_{j=1}^t m_{\alpha_1\beta_j}$; $m_{\alpha_1\beta_j} \in M_{\alpha_1\beta_j}$. Then $f_1(m) = f_1'(i(m)) = \Sigma_{j=1}^t \Sigma_k a_{\sigma_k\beta_j}^{(1)}(m_{\alpha_1\beta_j})$, where $a_{\sigma_k\beta_j}^{(1)} = 0$ for almost all k.

$$f_2f_1(m) = f_2'f_1'(i(m)) = \Sigma_{j=1}^t \Sigma_{k'}\Sigma_k a_{k'\sigma_k}^{(2)} a_{\sigma_k\beta_j}^{(1)}(m_{\alpha_1\beta_j})$$

Since $p_{\alpha\beta}(i(S))$ is locally semi-T-nilpotent, we can obtain n such that $f_n f_{n-1} \cdots f_1(m) = 0$ from (7.2.1).

7.5 PERFECT MODULES

We defined a perfect module in Sec. 6.1. We now give some relationships between (semi-) perfect module and (semi-) T-nilpotency.

7.5.1 THEOREM Let $\{P_\alpha\}_I$ be a set of projective modules and $P = \Sigma_I \oplus P_\alpha$. Then P is semiperfect if and only if each P_α is semiperfect and $\{J(P_\alpha)\}_I$ is a locally semi-T-nilpotent set of small submodules. Furthermore, P is perfect if and only if $\{J(P_\alpha)\}$ is locally T-nilpotent.

Proof: We assume that P is semiperfect. Then each P_α is semiperfect, and J(P) is small in P by (6.1.4). Hence $\{J(P_\alpha)\}_I$ is locally semi-T-nilpotent by (7.4.7). If P is perfect, consider any direct sum of copies of P; then the argument above shows that $\{J(P_\alpha)\}_I$ is locally semi-T-nilpotent. Conversely, we assume that each P_α is semiperfect. Then by (6.1.4) P/J(P) is a semisimple module and $P = \Sigma_J \oplus P_\beta$, where the P_β are c.ind. Since $\{J(P_\alpha)\}_I$ is a locally semi-T-nilpotent set of small submodules, J(P) is small in P by (7.4.4). Therefore, P is semiperfect by (6.1.14). If $\{J(P_\alpha)\}_I$ is locally T-nilpotent, we can use the argument above on any direct sum of copies of P. Hence P is perfect.

7.5.2 PROPOSITION Let P be a semiperfect module. Then there exists a maximal member among perfect direct summands of P and all these modules are isomorphic to one another.

Proof: Let $P = \Sigma_I \oplus P_\alpha$ and the P_α c.ind. Let $\underline{S}$ be the set of subsets $\{P_\gamma\}_J$ of $\{P_\alpha\}_I$ such that $\{P_\gamma\}_J$ is locally T-nilpotent. We can find a maximal member in $\underline{S}$ by Zorn's lemma, since $\{P_\alpha\}_I$ is semi-T-nilpotent. Put $P_0 = \Sigma_{J'} \oplus P_{\gamma'}$; then P_0 is a desired perfect summand of P by (7.5.1). Put $P = \Sigma_J \oplus P_\gamma \oplus \Sigma_K \oplus P_\delta = \Sigma_{J'} \oplus P'_{\gamma'} \oplus \Sigma_{K'} \oplus P'_{\delta'}$, where $\Sigma_J \oplus P_\gamma$ and $\Sigma_{J'} \oplus P'_{\gamma'}$ are maximal submodules as in (7.5.2). Then P_γ and $P'_{\gamma'}$ are themselves T-nilpotent. Hence, if P_γ is isomorphic to some $P'_{\delta'}$ in $\{P'_{\sigma'}\}_{K'}$, $\{P'_{\gamma'}\}_{J'} \cup \{P_\gamma\}$ is locally T-nilpotent, which contradicts the maximality of $\Sigma_{J'} \oplus P'_{\gamma'}$. Therefore, $\Sigma_J \oplus P_\gamma \approx \Sigma_{J'} \oplus P'_{\gamma'}$ by (7.1.2).

7.5.3 PROPOSITION Let P be a finitely generated and projective R-module and $M = \Sigma_{i=1}^{\infty} \oplus P$. Then P is perfect if and only if $S_M/J(S_M)$ is a regular ring in the sense of von Neumann and every idempotent in $S_M/J(S_M)$ is lifted to S_M.

Proof: If P is perfect, the statements are obtained by (7.3.15). Conversely, let $S_P = \mathrm{End}_R(P)$. Then $S_M = (S_P)_I$. Let e be an idempotent in $(J(S_P))_I/J(S_M)$. We may assume that e is idempotent in $(J(S_P))_I$ from the assumption. Since $J(S_P) = \mathrm{Hom}_R(P,J(P))$ from (6.1.4), $e(M) \subset \Sigma \oplus J(P) = \Sigma \oplus PJ(R) = MJ(R)$. Hence $e(M) = e(M)J(R)$. Therefore, $e = 0$ by (6.1.3). On the other hand, $S_M/J(S_M)$ is regular, so $J(S_M) = (J(S_P))_I$ by the above fact. Accordingly, $J(S_P)$ is right T-nilpotent by (7.4.5) and $S_P/J(S_P)$ is semisimple artinian from (6.5.4). On the other hand, for every direct summand K of M/J(M), there exists a direct summand N [$\supset$ J(M)] of M such that K = N/J(M) by the assumption. Hence $P = \Sigma \oplus P_\alpha$, and $\mathrm{End}_S(P_\alpha)$ is a local ring by (6.1.12), which implies that P is perfect by (7.5.1).

The following lemma is a generalization of (4.5.1).

7.5.4 LEMMA Let M be artinian and projective. If $AM = A^2M \neq 0$ for a right ideal A in S_M, then A contains a nonzero idempotent, where $S_M = \mathrm{End}_R(M)$.

Proof: Since M is artinian, there exists a minimal submodule N = A'M with respect to the property $N' = A''M = A''^2M \neq 0$ for a right ideal $A'' \subset A$. Then A' is not nilpotent. Hence there exists an x in A' such that $xA'' \neq 0$. Again from the assumption, we can find a minimal submodule among submodules $x'M \neq 0$ $(x' \in A)$, say xM. Since $xA'A'M = xA'M \neq 0$, there exists a y in xA' such that $yA' \neq 0$. Then $yM \subset xA'M \subset xM$. Hence $yM = xM$ by the minimality of xM. Now, consider the diagram

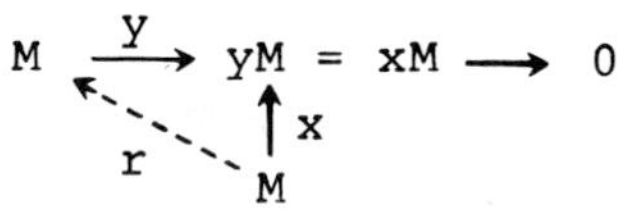

Since M is projective, we have an r in S_M such that $x = yr = xa$, where a is an element in A'. Hence $x = xa = xa^2 = \cdots$. Therefore, a is not nilpotent and $x(a - a^2) = 0$. Set $n = a^2 - a$. If $n = 0$, a is a nonzero idempotent. Suppose that $n \neq 0$. Put $A^* = \{z \in A' \mid xz = 0\}$; then $A' \supsetneq A^* \ni n$. We consider a chain $A^*M \supset A^{*2}M \supset \cdots \supset A^{*k}M \supset \cdots$. Since M is artinian, $A^{*t}M = A^{*t+1}M$ for some t. Since $A'M \supset A^*M$ and $A'M$ is minimal, $A'M = A^{*t}M$ or $A^{*t}M = 0$. On the other hand, $xA' \neq 0$ and $xA^* = 0$, and hence $A^{*t} = 0$, which implies that n is nilpotent. Next, put $a_1 = a + n - 2an$; then all a, n, and a_1 commute one another, since they are generated by a. Hence $-n + 2an$ is also nilpotent and a_1 is not nilpotent. Furthermore, $a_1^2 - a_1 = n^2(4n - 3)$. Repeating this argument, we get nonnilpotent elements $a_i \in A'$ such that $a_i = a_i^2 = n^{2i}z_i$; $z_i \in S_M$. Since n is nilpotent, we have a nonzero idempotent a_t in A'.

7.5.5 COROLLARY Let M be as in (7.5.4). Then S_M is a semiprimary ring.

Proof: Since M is artinian, M is a finite direct sum of indecomposable, projective modules M_i. First we assume that $M = M_1$. For any right ideal A in S_M, $A^nM = A^{n+1}M$ for some n. If $A^n \neq 0$, A contains a nonzero idempotent e by (7.5.4). Since M is indecomposable, $e = 1$. Therefore, S_M is a local ring with nilpotent radical. Next, we may assume that $M = \Sigma_{i=1}^n \Sigma_{j=1}^{s_i} \oplus M_{ij}$, where the M_{ij} are indecomposable and $M_{ij} \approx M_{ij'}$, $M_{ij} \not\approx M_{i'j'}$ if $i \neq i'$. Then $S_M = \{(s_{ij}) \mid s_{ij} \in S_{ij} = \mathrm{Hom}_R(M(j), M(i))\}$, where $M(i) = \Sigma_{\ell=1}^{s_i} \oplus M_{i1}$. Since M_{ij} is c.ind., from the above and (7.1.1), we have

$$J(S_M) = \begin{bmatrix} J(S_{11}) & S_{12} & \cdots & S_{1n} \\ S_{21} & J(S_{22}) & \cdots & S_{2n} \\ \cdots & \cdots & \cdots & \cdots \\ S_{n1} & \cdots & \cdots & J(S_{nn}) \end{bmatrix}$$

[cf. (3.7.4) and (5.5.9)]. Furthermore, all $J(S_{ii})$ are nilpotent and hence, $J(S_M)$ is nilpotent, and $S_M/J(S_M) \approx \Sigma_{i=1}^n \oplus S_{ii}/J(S_{ii})$. It is clear that $S_{ii}/J(S_{ii})$ is the ring of matrices over a division ring $S_{M_{i1}}/J(S_{M_{i1}})$.

7.5.6 THEOREM Let M_R be R-projective and artinian. Then M is a perfect and finitely generated, and S_M is a right artinian ring.

Proof: It is clear from the proof of (7.5.5) that $M = \Sigma_{i=1}^{n} \oplus M_i$, where the M_i are c.ind. Furthermore, since S_M is semiprimary by (7.5.5), $\{M_i\}$ is (locally) T-nilpotent with respect to $\underline{J}$. Therefore, M is perfect by (7.5.1) and (6.1.17) and each M_i is cyclic. Furthermore, (5.3.8) gives us a lattice monomorphism of the set of right ideals in S_M into the set of submodules of M. Hence S_M is right artinian.

7.6 PERFECT RINGS

If a ring R is a perfect module as a right R-module, R is called *right perfect* (Sec. 6.1). If eRe is a local ring for an idempotent e in R, e is called *local*, which is equivalent to the fact that eR is semiperfect and indecomposable, by (6.1.17) and (3.1.7). Thus we have from (7.5.1) the following.

7.6.1 THEOREM Let R be a ring. Then the following conditions are equivalent:

1. R is a right perfect ring.
2. $R = \Sigma_{i=1}^{n} \oplus e_iR$ with e_i local idempotents and J(R) is right T-nilpotent.

Next, we assume that R is right semiperfect as an R-module. Then $R = \Sigma_{i=1}^{n} \oplus e_iR$ with e_i local idempotents by (6.1.14), and $\mathrm{End}_R(eR) \approx \mathrm{End}_R(Re) \approx eRe$. Further, eR is indecomposable as a right R-module if and only if Re is indecomposable as a left R-module. Hence R is also a left R semiperfect module. We call R a *semiperfect ring*.

Summarizing the above, we get the following.

7.6.2 THEOREM Let R be a ring. Then we have the equivalent conditions:

1. R is a semiperfect ring.
2. $R = \Sigma_{i=1}^{n} \oplus e_iR = \Sigma_{i=1}^{n} \oplus Re_i$ with e_i local idempotents.
3. R/J(R) is semisimple and every idempotent in R/J(R) is lifted to R.

We note that every semiprimary ring is left and right perfect by (7.6.1). We shall study further properties of right perfect rings. Let $F = \Sigma \oplus u_iR$ be a free R-module and let $\{a_i\}$ be a set of elements in R. Let f_i be an element in $\mathrm{Hom}_R(u_iR, u_{i+1}R)$ such that $f_i(u_i) = a_i$. Put $G = \Sigma \oplus u_iR(f_i)$, where $u_iR(f_i) = \{u_ir - u_{i+1}a_ir \mid r \in R\}$. We denote this situation by $\{F, \{a_i\}, G\}$.

7.6.3 LEMMA Let $G_n = \Sigma_{i=1}^{n} \oplus u_i R(f_i)$. Then $F/G = \varinjlim F/G_n$ and hence F/G is flat.

Proof: We have the natural direct system $\{F/G_n, \pi_n^m : F/G_n \to F/G_m, n \le m\}$ and $\{G_n, \pi'^m_n : G_n \to G_m, n \le m\}$. From the exact sequence $0 \to G_n \to F \to F/G_n \to 0$, we have the exact sequence $0 \to G = \varinjlim G_n \to F \to \varinjlim F/G_n \to 0$ by (3.5.6). Since $F = G_n \oplus \Sigma_{k>n} \oplus u_k R$, F/G_n is flat (projective). Therefore, F/G is flat by (3.5.6) and (3.6.8).

7.6.4 LEMMA Let $\{F, \{a_i\}, G\}$ be as in (7.6.3). Put $A_k = \{r \in R \mid a_{n+k} \cdots a_k r = 0$ for some $n\}$. Then $A_k = r_R(z_k)$, where z_k is the residue class of u_k in F/G.

Proof: Since $z_{m+1}a_m = z_m$ for any m,

$$\text{(a)} \qquad z_{n+k}a_{n+k-1}\cdots a_n = z_n$$

Hence $A_k \subset r_R(z_k)$. Let r be in $r_R(z_k)$. Then $u_k r \in G$. Hence $u_k r = (u_1x_1 - u_2a_2x_1) + (u_2x_2 - u_3a_2x_2) + \cdots + (u_tx_t - u_{t+1}a_tx_t)$. Accordingly, $x_1 = \cdots = x_{k-1} = 0$, $x_k = r$, $x_ia_i = x_{i+1}$ for $i > k$ and $a_tx_t = 0$ for some $t \ge k$. Therefore, $0 = a_tx_t = a_ta_{t-1}\cdots a_kx_k = a_ta_{t-1}\cdots a_kr$.

7.6.5 LEMMA If G is a direct summand of F, $Ra_1 \supset Ra_2a_1 \supset \cdots \supset Ra_n\cdots a_1 \supset \cdots$ terminates.

Proof: We may assume that $F = G \oplus F/G$. Let $p : F \to F/G$ be the projection. Then since $p(u_n) = z_n$, $u_n = z_n + g_n$; $g_n \in G$. Being in $\mathrm{End}_R(F)$, p is expressed as a column-finite matrix (c_{ij}); i.e.,

$$z_k = p(u_k) = \Sigma_j u_jc_{jk}$$

(c_{ij}) is idempotent, for p is the projection. Let $I = \langle Rc_{i1}\rangle_i$. From (a) we have

$$\text{(b)} \qquad c_{jk} = c_{jn+1}a_n\cdots a_k \qquad \text{for any } j \ (n \ge k)$$

In particular, $c_{j1} = c_{jn+1}a_n\cdots a_1 \in Ra_n\cdots a_1$. Hence $I \subset \bigcap Ra_n\cdots a_1$. We shall show that $I \supset Ra_m\cdots a_1$ for a large m. Since (c_{ij}) is idempotent,

$$c_{j1} = \sum_{k=1}^{m+1} c_{jk}c_{k1} \qquad \text{for a large m}$$

$$= \sum_{k=1}^{m} c_{jm+1}a_m\cdots a_kc_{k1} + c_{jm+1}c_{m+1} \qquad \text{by (b)}$$

Put $\Sigma_{k=1}^{m} a_m \cdots a_k c_{k1} + c_{jm+1} = b \in I$. Then $c_{j1} = c_{jm+1}b$ for all j. Hence $z_1 = z_{m+1}b = z_{m+1}a_m \cdots a_1$. Therefore, $b - a_m \cdots a_1 \in A_{m+1}$. Accordingly, $a_p \cdots a_m \cdots a_1 = a_p \cdots a_{m+1} b \in I$ for some p by (7.6.4). Hence $I \supset Ra_p \cdots a_1$.

Now we are in a position to get the main theorem.

7.6.6 THEOREM Let R be a ring. Then the following conditions are equivalent:

1. R is a right perfect ring.
2. Every right flat R-module is projective.
3. R satisfies the descending chain condition on principal left ideals.
4. R has no infinite sets of orthogonal idempotents and every nonzero left R-module has a nonzero socle.

Proof: (1) ⇒ (2). Let M_R be a flat module and let $\nu : P \to M \to 0$ be a projective cover of M. Put $K = \ker \nu$. Then every f.g. submodule of K is a direct summand of P by (4.3.14). Hence, since K is small in P, K = 0.

(2) ⇒ (3). Let $Ra_1 \supset Ra_2a_1 \supset \cdots \supset Ra_n \cdots a_1 \supset \cdots$ be a descending chain of principal ideals. Take $\{F, \{a_i\}, G\}$. Since F/G is flat by (7.6.3), F/G is projective by (2). Hence G is a direct summand of P. Therefore, the above chain terminates by (7.6.5).

(3) ⇒ (4). It is clear that there are no infinite sets of orthogonal idempotents from (3). Let $_RM$ be any R-module. Take $m \neq 0$ in M. If Rm is not minimal, there exists a_1 in R such that $Rm \supsetneq Ra_1m$. Further, if Ra_1m is not minimal, we can take a_2 in R such that $Ra_1m \supsetneq Ra_2a_1$. Repeating this process, we obtain finally a minimal submodule in M from (3).

(4) ⇒ (1). First we shall show that J = J(R) is right T-nilpotent. We know that J has the lower Loewy series (Sec. 4.5) by (4) as a left R-module. Let $J = S^{\alpha}(J)$ and $n \in J$. We take the minimal ordinal γ such that $n \in S^{\gamma}(J)$. Then γ is not a limit ordinal by the definition of S^{γ}. We denote such a γ by h(n). Then since $JS^{\gamma}(J) \subset S^{\gamma-1}(J)$, $h(n'n) < h(n)$ for any $n' \in J$. Accordingly, J is right T-nilpotent, for $h(n_1) > h(n_2n_1) > \cdots > h(n_t \cdots n_1)$ are ordinals. Since J is T-nilpotent, J is nil. Therefore, every idempotent in $R/J = \bar{R}$ is lifted to R by (4.5.10). Furthermore, $S(\bar{R}) \neq 0$. Hence $\bar{R}$ contains a simple direct summand of $\overline{Re}$ by (4.5.2), say $\bar{R} = \overline{Re} \oplus \overline{R(1 - e)}$, where we may assume that $e = e^2$. $\overline{R(1 - e)}$ contains a simple direct summand, too, if $\overline{1 - e} \neq 0$. Repeating this argument, we know that $\bar{R}$ is artinian by (4) and the fact above. Since $\overline{eRe}$ is a division ring, e is local. Hence R is right perfect by (7.6.1).

7.6.7 PROPOSITION If a right ideal A is right T-nilpotent, then $M \neq MA$ for any module $M_R \neq 0$.

Proof: We assume that $M = MA$. Let m be in M. Then

$$m = \sum_{i_1=1}^{n_1} m_{i_1} a_{i_1} \qquad a_{i_1} \in A$$

$$m_{i_1} = \sum_{i_2'=1}^{n_2} m_{i_2' i_1} a_{i_2' i_1} \qquad a_{i_2' i_1} \in A$$

so

$$m = \sum_{i_1}^{n_1} \sum_{i_2'}^{n_2} m_{i_2' i_1} a_{i_2' i_1} a_{i_1}$$

Repeating this, we get a tree $\{a_{i_1}, a_{i_2' i_1} a_{i_1}, \ldots\}$. Since A is right T-nilpotent, $m = 0$ by (7.2.1).

7.6.8 COROLLARY If R is right perfect, $J(M) = MJ \neq M$ for every M_R; i.e., M contains a maximal submodule.

Proof: Since $MJ \subset J(M)$ by (3.3.11) and M/MJ is a semisimple R-module, $MJ = J(M)$.

We defined the lower Loewy series in Sec. 4.5. We now define the *upper Loewy series* for any module M_R.

Put $J^0(M) = M$, $J^\alpha(M) = J(J^{\alpha-1}(M))$ if α is not a limit ordinal and $J^\alpha(M) = \bigcap_{\alpha>\beta} J^\beta(M)$ if α is a limit ordinal. If R is perfect, $J^\gamma(M) = 0$ for some ordinal γ.

Let R be any ring. If every R-module $M_R \neq 0$ has a nonzero socle, we call R right *semiartinian*. Then if R is right perfect, R is left semiartinian by (7.6.6). We can define left T-nilpotent similarly to right T-nilpotent; i.e., for an infinite set $\{a_i\}$ of elements in R, $a_1 a_2 \cdots a_n = 0$ for some n.

7.6.9 THEOREM Let R be a ring. Then the following conditions are equivalent:

1. R is a right semiartinian ring.
2. R/J is a right semiartinian and J is left T-nilpotent, where $J = J(R)$.

Proof: (1) ⇒ (2). Since every R/J-module is an R-module, this implication is clear from the proof of (4) ⇒ (1) in (7.6.6).

(2) ⇒ (1). Let M_R be an R-module. If $MJ = 0$, $S(M) \neq 0$ by (2). We assume that $MJ \neq 0$. Then we use the same argument as in the proof (3) ⇒

(4) in (7.6.6), and we get a chain $mx_1R \supsetneqq mx_1x_2R \supsetneqq \cdots \supsetneqq mx_1x_2\cdots x_nR \supsetneqq \cdots$; $x_i \in J$. Hence, since $S(M) \supset S(MJ) \neq 0$ by (2), $S(M) \neq 0$.

Finally, we shall give a result for projective modules.

7.6.10 THEOREM Let R be a semiperfect ring. Then every projective R-module is a direct sum of c.ind. submodules isomorphic to e_iR, where the e_i are local.

Proof: Let P be R-projective. Then P is a direct summand of a free R-module. On the other hand, $R = \Sigma_{i=1}^n \oplus e_iR$; e_iR is c.ind. by (7.6.1). Hence $P = \Sigma_I \oplus P_\alpha$ with P_α countably generated by (7.3.1). Therefore, P_α is a direct sum of c.ind. modules by (7.3.4). The last assertion is clear, since R is semiperfect.

We give an example which shows that a right perfect ring is not left perfect in general. Let K be a field and $S = K_\infty$, the ring of column-finite matrices over K. We denote the matrix units by e_{ij}. Then

(c) $e_{12}e_{23}\cdots e_{nn+1} \neq 0$ for any n

However,

(d) $e_{i_ni_{n-1}}\cdots e_{i_3i_2}e_{i_2i_1} = 0$ for some n $(i_j < i_{j-1})$

Let R be the subring in S generated by $\{1, e_{ij}; i < j\}$. Then $J(R) = \Sigma_{i<j} e_{ij}K$ and J is right T-nilpotent from (d), but not left T-nilpotent by (c). Hence R is right perfect, but not left perfect.

Next, let $T = \Pi_1^\infty K$ and let $e_i = (0,\ldots,\overset{i}{\check{1}},0,\ldots)$. We have the subring in T generated by $\{1, e_i\}$. Then R is a regular ring and $R/\langle e_i \rangle_i \approx K$. Hence R is semiartinian, but $R = R/J(R)$ is not artinian.

7.7 GOLDIE DIMENSION

We shall apply the result in Sec. 6.4 to decomposition of an injective module E. Let $\underline{C}(E)$ be the category induced from E (Sec. 6.4) and $\bar{\underline{C}}(E) = \underline{C}(E)/\underline{J}$, where $\underline{J}$ is the Jacobson radical. Then $\bar{\underline{C}}(E)$ is a Grothendieck and spectral category by (6.4.1).

For any R-module M_R we put $\underline{Z}(M) = \{m \in M \mid r_R(m)$ is essential in R$\}$. Let m be in Z(M), let I be a nonzero right ideal, and $r \in R$. If $rI \neq 0$, $rI \cap r_R(m) \neq 0$, so $I \cap r_R(mr) \neq 0$. If $rI = 0$, $I \supset r_R(mr)$. Hence Z(M) is an R-submodule of M. We call Z(M) the *singular submodule* of M. If Z(M) = 0, we call M *nonsingular*, and if Z(M) = M, we call M *singular*.

If N_R is an essential submodule of M and $Z(N) = 0$, then $Z(M) = 0$.

7.7.1 LEMMA Let M be an injective module with $Z(M) = 0$; then $J(S_M) = 0$, where $S_M = \mathrm{End}_R(M)$.

Proof: Let $f \in J(S_M)$. Then ker f is essential in M, so $Z(M/\ker f) = M/\ker f$. On the other hand, $M/\ker f$ is isomorphic to a submodule of M. Hence $M = \ker f$.

7.7.2 PROPOSITION Let M be an injective R-module with $Z(M) = 0$. Then $\underline{C}(M)$ is a spectral and Grothendieck category with generator M. For any morphism f in $\underline{C}(M)$, ker f (im f) in $\underline{C}(M)$ is equal to ker f (im f) in $\underline{M}_R$.

Proof: From (7.7.1) we obtain $\underline{J} = 0$. Hence $\underline{C}(M)$ is a spectral and Grothendieck category. Furthermore, since M is a cogenerator in $\underline{C}(M)$, M is a generator. The remaining assertion is clear from Remark 2 in Sec. 6.4.

7.7.3 COROLLARY Let N be an R-module with $Z(N) = 0$ and let Q_1 and Q_2 be injective submodules in N. Then $Q_1 + Q_2$ and $Q_1 \cap Q_2$ are injective.

Proof: Let $E = E(N)$ and consider $\underline{C}(E)$. Then $Q_i \in \underline{C}(E)$ and $Q_1 + Q_2$ and $Q_1 \cap Q_2$ are an image and a kernel in $\underline{M}_R$ of some morphisms in $\underline{C}(E)$, respectively (cf. Sec. 2.1). Hence they are injective in $\underline{M}_R$ by (7.7.2).

7.7.4 LEMMA Let $\underline{B}$ be a full subadditive category in $\underline{M}_R$. Suppose that $\underline{B}$ contains a generator (cogenerator) in $\underline{M}_R$. Then every monomorphism (epimorphism) in $\underline{B}$ is a monomorphism (an epimorphism) in $\underline{M}_R$.

Proof: Let U be a generator in $\underline{M}_R$ contained in $\underline{B}$ and let $f : A \to B$ be a monomorphism in $\underline{B}$. Put $\ker f = C$ in $\underline{M}_R$. If $C \neq 0$, there exists $g \neq 0 \in [U,C]$ in $\underline{M}_R$ such that $ig \neq 0$, where $i : C \to A$ is the inclusion. However, $ig \in [U,A] \in \underline{B}$ and $fig = 0$, which is a contradiction.

7.7.5 PROPOSITION Let M be an R-injective module. We assume that M is a generator and cogenerator in $\underline{M}_R$ (e.g., R is a QF-ring). Then $\underline{C}(M)$ is an abelian category if and only if R is a semisimple artinian ring.

Proof: We assume that $\underline{C}(M)$ is abelian. We shall show for any morphism f in $\underline{C}(M)$ that $[\ker f \text{ in } \underline{C}(M)] = (\ker f \text{ in } \underline{M}_R)$. Let $f : N \xrightarrow{f'} \mathrm{im}\, f \xrightarrow{i} M'$ be a decomposition of f in $\underline{C}(M)$ and i is a monomorphism in $\underline{C}(M)$. Hence they are the same in $\underline{M}_R$ by (7.7.4). Hence $[\mathrm{im}\, f \text{ in } \underline{C}(M)] = (\mathrm{im}\, f \text{ in } \underline{M}_R)$. Set $K_1 = [\ker f \text{ in } \underline{C}(M)]$ and $K_2 = (\ker f \text{ in } \underline{M}_R)$. It is clear that $K_1 \subset K_2$ by (7.7.4). On the other hand, K_1 is R-injective and hence $N = K_1 \oplus N''$ in $\underline{M}_R$. Then $N'' \in \underline{C}(M)$, and N'' is isomorphic to im f via f in $\underline{M}_R$ from the above. Hence $K_1 = K_2$. Let A be any module; then there exists an R-exact sequence $0 \to A \to \Pi_{I_1} M \xrightarrow{f} \Pi_{I_2} M$. Since $\Pi M \in \underline{C}(M)$, $A = (\ker f \text{ in } \underline{M}_R) =$

[ker f in $\underline{C}(M)$]. Hence A is injective. Therefore, R is semisimple and artinian. The converse is clear.

7.7.6 LEMMA Let $\underline{B}$ be a full subadditive category in $\underline{M}_R$. We assume that every direct summand in $\underline{M}_R$ of an object in $\underline{B}$ belongs to $\underline{B}$. Then every finite coproduct in $\underline{B}/\underline{J}$ is lifted to $\underline{M}_R$, where $\underline{J}$ is the Jacobson radical.

Proof: Let B, B_1, and B_2 be in $\underline{B}$ and $\bar{B} = \bar{B}_1 \oplus \bar{B}_2$ in $\underline{B}/\underline{J}$. Then there exist morphisms $i_k : B_k \to B$ and $p_K : B \to B_k$ such that $1_{\bar{B}} = \bar{i}_1\bar{p}_1 + \bar{i}_2\bar{p}_2$ and $\bar{p}_k\bar{i}_k = 1_{\bar{B}_k}$. Since $\underline{J}$ is the radical, $p_k i_k$ is an isomorphism in $\underline{M}_R$. Hence $B = \text{im } i_1 \oplus \ker p_1$ in $\underline{M}_R$. By the assumption, im i_1 and ker p_1 are in $\underline{B}$ and it is clear that $\ker \bar{p}_1 = \bar{B}_2$ and $\bar{B} = \text{im } \bar{i}_1 \oplus \ker \bar{p}_1 = \bar{B}_1 \oplus \bar{B}_2$.

7.7.7 COROLLARY Let M be R-injective. Then an object $\bar{N}$ in $\underline{C}(M)/\underline{J}$ is minimal if and only if N is indecomposable.

Proof: This is clear from (6.4.1) and (7.7.6).

7.7.8 THEOREM Let R be a left perfect ring and let M be R-injective as a right R-module. Then $\underline{C}(M)/\underline{J}$ is a completely reducible and Grothendieck category.

Proof: Since R is left perfect, every right R-module contains minimal submodules by (7.6.6). Let N be in $\underline{C}(M)$ and let S(N) be the socle of N in $\underline{M}_R$; i.e., $S(N) = \Sigma \oplus I_\alpha$ and the I_α are minimal R-submodules. We know from the assumption that $\Sigma \oplus I_\alpha$ is essential in N. Hence $\bar{N} = \Sigma \oplus E(\bar{I}_\alpha)$ by Remark 1 in Sec. 6.4, and $\overline{E(I_\alpha)}$ is a minimal object in $\bar{\underline{C}}(M)$ by (7.7.7).

Let $\underline{E}$ be the full subadditive category of all injective modules in $\underline{M}_R$. By $\bar{\underline{E}}$ we shall denote $\underline{E}/\underline{J}$. We know from (7.7.8) that if R is a left perfect ring, then $\bar{\underline{E}}$ is completely reducible. We shall give a necessary and sufficient condition for $\bar{\underline{E}}$ to be completely reducible.

7.7.9 THEOREM Let $\bar{\underline{E}}$ be the category mentioned above. Then $\bar{\underline{E}}$ is completely reducible if and only if for any right ideal K, K always has a decomposition such that $K = K_1 \cap K_2$, K_1 is irreducible, and $R \supset K_2 \neq K$.

Proof: If $\bar{\underline{E}}$ is completely reducible, $E(R/K) = E_1 \oplus E_2$ by (7.7.6) and (7.7.7), where E_1 is indecomposable. Let $p_i : E \to E_i$ be the projection. Put $K_i/K = p_i^{-1}(E_i) \cap R/K$. Then $K = K_1 \cap K_2$ and K_1/K is isomorphic to a nonzero submodule in E_1 provided that $E_1 \neq 0$. Hence $E(K_1/K) = E_1$, so K_1 is irreducible by (4.7.16). Conversely, if $K = K_1 \cap K_2$ as in (7.7.9), then we have a natural exact sequence $0 \to R/K \xrightarrow{\phi} R/K_1 \oplus R/K_2$ and $\phi(K_2/K) \subset R/K_1$. Hence $E(R/K_1)$ [$\subset E(R/K)$] is indecomposable. Let E be injective.

Then $E \supset xR \approx R/r_R(x)$ for any $x \neq 0$ in E. Hence E contains an indecomposable direct summand from the above. Therefore, $\bar{E}$ contains a minimal object in $\underline{E}$; hence $\underline{\bar{E}}$ is completely reducible, since $\underline{\bar{E}}$ is a spectral, Grothendieck category.

Let N be an R-module. We consider the set $\underline{S}$ of submodules T of N such that $T = \Sigma_I \oplus K_\alpha$, where the K_α are uniform. Put $\dim N = \max_I (|I|)$ if this exists [we shall show in (7.7.11) that dim N exists for any N]. We call dim N the *Goldie dimension* of N.

7.7.10 THEOREM Let E be R-injective. Then dim E exists and we have a decomposition $E = E_1 \oplus E_2$ such that $\dim E = \dim E_1$, $\dim E_2 = 0$ and E_1 is a minimal injective submodule of E among injective submodules E' of E with decompositions as above. Furthermore, this decomposition is unique up to isomorphism.

Proof: Take the factor category $\underline{\bar{E}}$. Then $\dim E = 0$ if and only if the socle $S(\bar{E})$ of $\bar{E}$ in $\underline{\bar{E}}$ is zero. We assume that $S(\bar{E}) \neq 0$ and $S(\bar{E}) = \Sigma_I \oplus \bar{E}_\alpha = \overline{E(\Sigma_I \oplus E_\alpha)}$, where the E_α are indecomposable injectives. Then $E = E(\Sigma_I \oplus E_\alpha) \oplus E_2$ and $\dim E_2 = 0$. Let $N = \Sigma_J \oplus N_\alpha$ be a submodule in E, where the N_α are uniform. Then $E(N) = E(\Sigma_J \oplus E(N_\alpha))$ and $\overline{E(N_\alpha)}$ is minimal in $\underline{\bar{E}}$. Hence $\overline{E(N)} \subset S(\bar{E})$, so $|J| \leq |I|$. Therefore, $\dim E = |I|$. Let E' be an injective submodule of E such that $E = E' \oplus E_2'$, $\dim E' = \dim E$, and $\dim E_2' = 0$. Then $\bar{E}'$ contains $S(\bar{E}) = \Sigma_I \oplus \bar{E}_\alpha$. Hence E_1 is a minimal injective submodule among injectives with such decompositions. Let $E = E_1 \oplus E_2 = E_1' \oplus E_2'$ such that $\dim E_1 = \dim E_1'$, $\dim E_2 = \dim E_2' = 0$, and E_1, E_1' are minimal in such decompositions. Then $\bar{E}_1 = \bar{E}_1' = S(\bar{E})$ and hence $\bar{E}_2 \approx \bar{E}_2'$. Since $\underline{J}$ is the Jacobson radical, $E_i \approx E_i'$ for $i = 1, 2$.

7.7.11 THEOREM Let N be an R-module. Then dim N exists and N is an essential extension of a submodule $N_1 \oplus N_2$ such that $\dim N_1 = \dim N = |I|$, $\dim N_2 = 0$, and N_1 is an essential extension of $\Sigma_I \oplus T_\alpha$, where the T_α are uniform.

Proof: Put $E = E(N)$. Then $E = E_1 \oplus E_2$ as in (7.7.10). Put $N_i' = N \cap E_i$. Then N_i' is essential in E_i and $N_1' \oplus N_2'$ is essential in N' by (3.4.6). Hence $\dim E_2 = \dim N_2' = 0$ and $E_1 = E(N_1')$. Let $E_1 = E(\Sigma_I \oplus E_\alpha)$, where the E_α are indecomposable. Put $E_\alpha \cap N_1' = N$ and $N_1 = \Sigma_I \oplus N_\alpha$. Then the N_α are uniform, and N_1 is essential in N_1' by (3.4.6). Suppose that $N \supset T' = \Sigma_J \oplus T_\alpha$, where the T_α are uniform. Then $\overline{E(T')} = \overline{E(\Sigma_J \oplus E(T_\alpha))} = \Sigma_J \oplus \overline{E(T_\alpha)} \subset \bar{E}_1$. Hence $|J| \leq |I|$ and $\dim N = \dim E = |I|$.

7.8 RING-THEORETICAL PROOFS

The purpose of this section is to give ring-theoretical proofs of (7.2.6) and (7.3.14).

Let $\{M_\alpha\}_I$ be a set of c.ind. modules, and $M = \Sigma_I \oplus M_\alpha$. Put $S_M = \mathrm{End}_R(M)$. Let $\underline{A}$, $\underline{J}'$, and $\bar{\underline{A}}$ be as in Sec. 7.3.

7.8.1 LEMMA Let M, N be in $\underline{A}$, and $f : M \to N$. Then we have the following.

1. f is left (resp. right) regular (= nonzero division) mod $\underline{J}'$ if and only if there exists some $g : N \to M$ such that $gf = 1_M$ (resp. $fg = 1_N$) mod $\underline{J}'$.
2. f is an isomorphism mod $\underline{J}'$ if and only if f is left and right regular $\underline{J}'$.

Proof: 1. The "if" part is trivial. Conversely, we assume that f is left regular mod $\underline{J}'$. Since $S_M/S_M \cap \underline{J}'$ is a regular ring by (5.4.5), there exists some $g : N \to M$ such that $fgf = f$ (mod $\underline{J}'$). The left regularity of f mod $\underline{J}'$ implies that $gf = 1_M$ (mod $\underline{J}'$). The right regularity is similar.

2. This is clear.

7.8.2 LEMMA If $f: M \to N$ is left regular mod $\underline{J}'$ for M, N in $\underline{A}$ and $S_M \cap J'$ is equal to the Jacobson radical $J(S_M)$ of S_M, then f is an R-monomorphism and M is isomorphic to a direct summand of N as R-modules.

Proof: By (7.8.1), there exists some $g : N \to M$ such that $gf = 1_M$ mod $\underline{J}'$, since f is left regular mod $\underline{J}'$. Hence $1_M - gf \in S_M \cap \underline{J}' = J(S_M)$, so gf is an R-isomorphism by (3.3.14). Therefore, f is an R-monomorphism and M is isomorphic to a direct summand of N as R-modules by (1.7.5).

Let U and V be right R-modules and $f : U \to V$. We assume that $U = \Sigma_{\gamma \in K} \oplus U_\gamma$ is a direct sum of right R-submodules of U_γ. Then we consider the following condition.

f is an R-monomorphism and for any finite subset K' of K,

$f(\Sigma_{\gamma' \in K'} \oplus U_{\gamma'})$ is a direct summand of V. (7.8.1)

For example, let f, U, and V be as above. If f is an R-monomorphism and each U_γ is injective, then f satisfies (7.8.1) (with respect to the decomposition $U = \Sigma_K \oplus U_\gamma$).

From now on, monomorphisms satisfying (7.8.1) will be considered in $\underline{A}$. The following lemma is essential in this section.

7.8.3 LEMMA Let $M = \Sigma_K \oplus M_\alpha$ and N be in $\underline{A}$ and $f : M \to N$. Then f is left regular mod $\underline{J}'$ if and only if f is a monomorphism satisfying (7.8.1) (with respect to the decomposition $M = \Sigma_K \oplus M_\alpha$).

Proof: First, we assume that f is left regular mod $\underline{J}'$. Put $M_0 = \Sigma_{K'} \oplus M_{\alpha'}$, for any finite subset K' of K. Let i be the inclusion of M_0 to M. Then fi is left regular mod $\underline{J}'$ and $S_{M_0} \cap \underline{J}' = J(S_{M_0})$ by (7.1.1), because K' is a finite set. Hence fi is an R-monomorphism and $fi(M_0)$ is a direct summand of N by (7.8.2). Therefore, f is an R-monomorphism and $f(M_0)$ is a direct summand of N; i.e., f is a monomorphism satisfying (7.8.1). Conversely, let $g \in \mathrm{Hom}_R(T,M)$ for some module $T = \Sigma_L \oplus T_\gamma$ in $\underline{A}$ and assume that fg in $\underline{J}'$. Put $g_\gamma = gi_\gamma$, where the i_γ is the inclusion of T_γ to T, for all $\gamma \in L$. Then we can express g_γ as a column-summable matrix (vector) for all $\gamma \in L$ by (3.7.4). Hence g_γ is a column-summable matrix whose finite components are isomorphisms and the others are all nonisomorphisms. We can rearrange g_γ as follows: The first n components are isomorphisms. Put $M_0 = \Sigma_{i=1}^n \oplus M_i$. Let i be the inclusion of M_0 to M, and p the projection of M onto M_0. Then $fipg_\gamma = fg_\gamma = fgi_\gamma = 0$ (mod $\underline{J}'$). Since fi is left regular mod $\underline{J}'$, pg_γ is in $\underline{J}'$. Hence g_γ and thus g are in $\underline{J}'$, because $pg_\gamma + (1 - p)g_\gamma = g_\gamma$ (mod $\underline{J}'$). Therefore, f is left regular mod $\underline{J}'$.

We note that a monomorphism satisfying (7.8.1) does not depend on the decomposition of M, from (7.8.3).

7.8.4 COROLLARY

1. If $f : M \to N$ is left regular mod $\underline{J}'$, then f is an R-monomorphism.
2. For any f in $S_M \cap \underline{J}'$, $1_M - f$ is an R-monomorphism.

Proof: 1. This is clear by (7.8.3).

2. Since f is in $S_M \cap \underline{J}'$, $1_M - f$ is left regular mod $\underline{J}'$ and hence an R-monomorphism by (1).

7.8.5 COROLLARY Let M, N be in $\underline{A}$ and let $f : M \to N$ be an isomorphism mod $\underline{J}'$. Then f is an R-isomorphism provided that either $S_M \cap \underline{J}' = J(S_M)$ or $S_N \cap \underline{J}' = J(S_N)$. If, in particular, M is a finite direct sum of M_α in $\underline{M}_R$, then an isomorphism mod $\underline{J}'$ means an R-isomorphism.

Proof: Since f is an isomorphism mod $\underline{J}'$, there exists some $g : N \to M$ such that $gf = 1_M$ (mod $\underline{J}'$) and $fg = 1_N$ (mod $\underline{J}'$). Hence f and g are left regular mod $\underline{J}'$; that is, both are R-monomorphisms by (7.8.4). In case $S_N \cap \underline{J}' = J(S_N)$, $1_N - fg \in J(S_N)$. Hence fg is an R-isomorphism and so is f. On the other hand, if $S_M \cap \underline{J}' = J(S_M)$, then $1_M - gf \in J(S_M)$ and hence gf is an R-isomorphism. Therefore, f is an R-isomorphism. The latter assertion is clear by (7.1.1).

7.8.6 LEMMA Let M, N be in $\underline{A}$ and $f : M \to N$. Then there exists a locally direct summand N' of N in $\underline{A}$ via the inclusion $i : N' \to N$ and some $f' : M \to N'$ such that $f = if'$ (mod $\underline{J}'$), where i is left regular mod $\underline{J}'$ and f' is right regular mod $\underline{J}'$.

Proof: We begin with the case M = N and $f = f^2$ (mod $\underline{J}'$). There exist a projection e of $M = \Sigma_K \oplus M_\alpha$ onto $\Sigma_{K'} \oplus M_{\alpha'}$ for some subset of K' of K, and elements a, a' in S_M such that $f = aea'$ (mod $\underline{J}'$) and $aa' = a'a = 1$ (mod $\underline{J}'$), by (6.3.4), where $S_M = \mathrm{End}_R(M)$ (note that $S_M/S_M \cap \underline{J}'$ is isomorphic to a column-finite matrix over division rings). Put N' = aeM and N'' = eM, and consider the inclusions $i : N' \to M$ and $i' : N'' \to M$. They, by (7.8.3), N' is a locally direct summand of M and i is left regular mod $\underline{J}'$, since N' is isomorphic to N'' under ai' which is left regular mod $\underline{J}'$. Moreover, ea' is right regular mod $\underline{J}'$, and hence so is $f' = aea' : M \to N'$. Thus (7.8.6) holds. In the general case, for $f : M \to N$, there exist some homomorphisms $g : N \to N$ and $k : N \to M$ such that $g = g^2$ (mod $\underline{J}'$), $f = gf$ (mod $\underline{J}'$), and $g = fk$ (mod $\underline{J}'$), by (5.4.5). By the argument above, there exist a locally direct summand N' of N in $\underline{A}$, some $g' : N \to N'$, and the inclusion $i : N' \to N$, such that $g = ig'$ (mod $\underline{J}'$), and g' and i are right and left regular mod $\underline{J}'$, respectively. We can easily show that g'f is right regular mod $\underline{J}'$, since $g'i = 1_{N'}$ (mod $\underline{J}'$) and $f = ig'f$ (mod $\underline{J}'$).

7.8.7 LEMMA For M and N in $\underline{A}$, a homomorphism $f : M \to N$ is right regular mod $\underline{J}'$ if and only if there exist a locally direct summand M' of M in $\underline{A}$ and some $g : N \to M'$ such that $fig = 1_N$ (mod $\underline{J}'$) and g is an isomorphism mod $\underline{J}'$, where i is the inclusion of M' to M.

Proof: The "if" part is trivial. Conversely, suppose that f is right regular mod $\underline{J}'$. Then there exists some $g' : N \to M$ such that $fg' = 1_N$ (mod $\underline{J}'$) by (7.8.1). Since g' is left regular mod $\underline{J}'$, there exists a locally direct summand M' of M in $\underline{A}$ such that $g' = ig$ (mod $\underline{J}'$), where $g : N \to M'$ is right regular mod $\underline{J}'$ and the inclusion $i : M' \to M$ is left regular mod $\underline{J}'$ by (7.8.6). Therefore, $fig = 1_N$ (mod $\underline{J}'$) and g is an isomorphism mod $\underline{J}'$.

7.8.8 LEMMA Let M, N be in $\underline{A}$ and e an idempotent element in S_M, where N is contained in eM. Let the inclusion $i : N \to M$ be left regular mod $\underline{J}'$. Then there exists a locally direct summand N' of M in $\underline{A}$ satisfying the following conditions: $e = ip + i'p'$ (mod $\underline{J}'$), $pi = 1_N$ (mod $\underline{J}'$), $p'i' = 1_{N'}$ (mod $\underline{J}'$), and $pi' = p'i = 0$ (mod $\underline{J}'$), where i' is the inclusion of N' to M and p, p' are homomorphisms of M to N, N', respectively. Furthermore, the formal direct sum $N \oplus N'$ is R-isomorphic to a locally direct summand of eM.

Proof: For $i : N \to M$, there exists some $p_0 : M \to N$ such that $ip_0 i \equiv i \pmod{\underline{J}'}$. Since $ei = i$, $ip_0 ei \equiv i \pmod{\underline{J}'}$. Put $p = p_0 e$; i.e., $p : M \to N$ and $ipi \equiv i \pmod{\underline{J}'}$. Since i is left regular mod $\underline{J}'$, $pi \equiv 1_N \pmod{\underline{J}'}$. Now, we put $f = ip$ and $g = e - f$. Then $ef = fe = f$ and $eg = ge = g$. For $g : M \to M$ there exist a locally direct summand N' of M in $\underline{A}$ and some $p' : M \to N'$ such that $g \equiv i'p \pmod{\underline{J}'}$, the inclusion $i' : N' \to M$ is left regular mod $\underline{J}'$, and p' is right regular mod $\underline{J}'$ by (7.8.6). Therefore, $e = f + g \equiv ip + i'p' \pmod{\underline{J}'}$. Since $gf = fg = 0$, $ipi'p' \equiv i'p'ip \equiv 0 \pmod{\underline{J}'}$ implies that $pi' \equiv p'i \equiv 0 \pmod{\underline{J}'}$, because i and i' are left regular mod $\underline{J}'$ and p, p' are right regular mod $\underline{J}'$. Moreover, $g \equiv g^2 \pmod{\underline{J}'}$ implies that $p'i' \equiv 1_{N'} \pmod{\underline{J}'}$. Finally, we show that the formal direct sum $N \oplus N'$ is R-isomorphic to a locally direct summand of eM. Let $I_0 = (i, i') : N \oplus N' \to M$ and $t = \begin{bmatrix} t_1 \\ t_2 \end{bmatrix} : T \to N \oplus N'$ for any T in $\underline{A}$. Suppose that $I_0 t = it_1 + i't_2$ is in $\underline{J}'$. Then $pit_1 + pi't_2$ is in $\underline{J}'$. Since pi' is in $\underline{J}'$, t_1 is in $\underline{J}'$ and so is t_2. Hence t is in $\underline{J}'$. It follows that I_0 satisfies (7.8.1). Therefore, $N \oplus N'$ is a locally direct summand of M. On the other hand, $g = eg$ and $g \equiv i'p \pmod{\underline{J}'}$ imply that $ei' \equiv i' \pmod{\underline{J}'}$, so we may assume that $ei' = i'$ in the above. Since i' satisfies (7.8.1), im i' is contained in eM. Hence im I_0 is contained in eM, whence $N \oplus N'$ is R-isomorphic to a locally direct summand of eM.

7.8.9 COROLLARY Let N and M be in $\underline{A}$ and $N \subset M$. If the inclusion $i : N \to M$ is left regular mod $\underline{J}'$, then there exist a locally direct summand N' of M in $\underline{A}$, the inclusion $i' : N' \to M$, and some p, p' of M to N, N', respectively, such that $1_M - ip + i'p' \pmod{\underline{J}'}$, $pi \equiv 1_N \pmod{\underline{J}'}$, $p'i' \equiv 1_{N'} \pmod{\underline{J}'}$, and $p'i \equiv pi' \equiv 0 \pmod{\underline{J}'}$.

Proof: Put $e = 1_M$ in (7.8.8).

Let N be an R-module, and let $\{N_j = \Sigma_{i \in I_j} \oplus N_i^{[j]}\}_{j \in J}$ be the set of submodules of N in $\underline{A}$ which are locally direct summands of N. We define an order > in the set $\{N_j\}_J$ as follows: For each locally direct summand N_j of N, $N_j > N_k$ if and only if $\{N_i^{[j]}\}_{I_j} \supset \{N_i^{[k]}\}_{I_k}$, for any $j \neq k$ in J. Then there exists a maximal submodule of N among the set $\{N_j\}_J$ with respect to this order > by Zorn's lemma. We call this a *maximal locally direct summand* of N.

7.8.10 LEMMA Let N and M be in $\underline{A}$ and $N \subset M$. Then N is a maximal locally direct summand of M if and only if the inclusion i of N to M is an isomorphism mod $\underline{J}'$.

Proof: First, we assume that N is a maximal locally direct summand of M. If the inclusion $i : N \to M$ is not an isomorphism mod $\underline{J}'$, there exists a locally direct summand N' of M in $\underline{A}$ such that $1_M = ip + i'p'$ (mod $\underline{J}'$), where i' is the inclusion of N' to M, $p : M \to N$, and $p' : M \to N'$ by (7.8.8). Then $I_0 = (i,i') : N \oplus N' \to M$ satisfies (7.8.1). Hence the image of I_0 is equal to a locally direct summand $N \oplus \text{im } i'$ of M in $\underline{A}$ which contains N, contradicting the maximality of N. Hence $N' = 0$. Therefore, $1_M = ip$ (mod $\underline{J}'$), so i is an isomorphism mod $\underline{J}'$. Conversely, suppose that i is an isomorphism mod $\underline{J}'$. Then there exists some $p : M \to N$ such that $pi = 1_N$ (mod $\underline{J}'$) and $ip = 1_M$ (mod $\underline{J}'$), so N is a locally direct summand of M. If N is not maximal in M, there exists a locally direct summand N' of M in $\underline{A}$ such that $N \oplus N'$ is a locally direct summand of M in $\underline{A}$. Hence the inclusion $I_0 = (i,i') : N \oplus N' \to M$ is left regular mod $\underline{J}'$, where i' is the inclusion of N' to M. Therefore, there exists some $g : M \to N \oplus N'$ such that $gI_0 = 1_{N\oplus N'}$ (mod $\underline{J}'$) by (7.8.1). Let p_1 be the projection of $N \oplus N'$ onto N. Then $p_1 gi = 1_N = pi$ (mod $\underline{J}'$), so $p_1 g = p$ (mod $\underline{J}'$), which implies that $pi' = 0$ (mod $\underline{J}'$). Hence $N' = 0$, which is a contradiction. It follows that N is a maximal locally direct summand of M.

The submodule N in (7.8.10) was called a full submodule of M in Sec. 7.3. We note that $N \oplus N'$ in (7.8.10) is a full submodule of M.

7.8.11 COROLLARY Let M, N be in $\underline{A}$ and $f : M \to N$. Then there exist locally direct summands M' and N' of M and N in $\underline{A}$, respectively, such that the restriction $f|M'$ to N' is an R-isomorphism. In particular, f is an isomorphism mod $\underline{J}'$ if and only if M' and N' are full in M and N, respectively.

Proof: For $f : M \to N$, there exist a locally direct summand N'' of N in $\underline{A}$, the inclusion $i' : N'' \to N$, and some $f' : M \to N''$ such that $f = i'f'$ (mod $\underline{J}'$), where i' is left regular mod $\underline{J}'$ and f' is right regular mod $\underline{J}'$, by (7.8.6). Since f' is right regular mod $\underline{J}'$, there exist a locally direct summand M' of M in $\underline{A}$ and some $g : N'' \to M'$ such that $f'ig = 1_{N''}$ (mod $\underline{J}'$) and g is an isomorphism mod $\underline{J}'$, where i is the inclusion of M' to M, by (7.8.7). Since fig is left regular mod $\underline{J}'$ and g is an isomorphism mod $\underline{J}'$, fi is left regular mod $\underline{J}'$ and thus an R-monomorphism. Let N' be the image of fi in N. Then $fi : M' \to N'$ is an R-isomorphism, whence it follows that N' is a locally direct summand of N. Particularly, in case f is an isomorphism mod

$\underline{J}'$, i' and f' are isomorphisms mod $\underline{J}'$ by (7.8.1), and fig : N'' → N' is an isomorphism mod $\underline{J}'$, so that N' and M' are full in N and M, respectively, by (7.8.10). Conversely, if N' and M' are full in N and M, respectively, i and i' are isomorphisms mod $\underline{J}'$, and hence f is an isomorphism mod $\underline{J}'$ by (7.8.10).

7.8.12 COROLLARY If $S_M \cap \underline{J}' = J(S_M)$ for a module M, then M is the unique full submodule in M.

Proof: Let N be a full submodule of M. Then the inclusion i : N → M is an isomorphism mod $\underline{J}'$ by (7.8.10). Hence i is an R-isomorphism by (7.8.5), so N = M.

7.8.13 LEMMA Let e be an idempotent element in S_M for a module $M = \Sigma_K \oplus M_\alpha$ in $\underline{A}$. Then there exist a submodule N of eM in $\underline{A}$ and p : M → N such that e = ip (mod $\underline{J}'$) and pi = 1_N (mod $\underline{J}'$), where i : N → M is the inclusion.

Proof: Since eM is a direct summand of M, eM contains some M_α by (7.1.2). Hence there exists a maximal locally direct summand of eM in $\underline{A}$ by Zorn's lemma. Let N be the maximal summand and i the inclusion of N to M. Since i is left regular mod $\underline{J}'$ by (7.8.3), there exists a locally direct summand N' of M in $\underline{A}$ satisfying the following conditions: e = ip + i'p' (mod $\underline{J}'$), pi = $1_{N'}$ (mod $\underline{J}'$), and N ⊕ N' is R-isomorphic to a locally direct summand of eM, where p : M → N, p' : M → N', and i' is the inclusion of N' to M. Since N is maximal in eM, N' = 0 and hence e = ip (mod $\underline{J}'$).

7.8.14 THEOREM [See (7.3.14).] Let $P = \Sigma_L \oplus P_\alpha$ in $\underline{A}$ (each P_α is not necessarily in $\underline{A}$). Then there exists a submodule N_α of P_α in $\underline{A}$ such that $e_\alpha = i_\alpha p_\alpha$ (mod $\underline{J}'$), where $p_\alpha : P \to N_\alpha$, $i_\alpha : N_\alpha \to P$ is the inclusion, and $e_\alpha : P \to P_\alpha$ is the projection for each $\alpha \in L$. Moreover, $\Sigma_L \oplus N_\alpha$ is a maximal locally direct summand of P.

Proof: We can find a maximal locally direct summand N_α of $e_\alpha P = P_\alpha$ satisfying $e_\alpha = i_\alpha p_\alpha$ (mod $\underline{J}'$), where $p_\alpha : P \to N_\alpha$, $i_\alpha : N_\alpha \to P$ is the inclusion, and $e_\alpha : P \to P_\alpha$ is the projection for every $\alpha \in L$, by (7.8.13). Since a finite direct sum $\Sigma_{i=1}^n \oplus N_{\alpha_i}$ is a direct summand of P, $\Sigma_L \oplus N_\alpha$ is a locally direct summand of P. Hence the inclusion I : $\Sigma_L \oplus N_\alpha \to P$ is left regular mod $\underline{J}'$. In order to see that $\Sigma_L \oplus N_\alpha$ is full in P, we have only to prove that I is right regular mod $\underline{J}'$. Let t be a homomorphism of P to any module T in $\underline{A}$ and assume that tI is in $\underline{J}'$. If t is not in $\underline{J}'$, there exists some direct summand P_β in P such that the restriction $t|P_\beta$ is not in $\underline{J}'$. Since $1_P = \Sigma_L e_\alpha$, $\Sigma_K e_\alpha i = i$ and $e_\alpha i$ is a nonisomorphism for almost all $\alpha \in L$, where i is the inclusion of P_β to P. Hence, for some integer n, $\Sigma_{j=1}^n e_j i = i$

(mod $\underline{J}'$), so $ti = te_N i = tI_N p_N i = tIp_N i = 0$ (mod $\underline{J}'$), where $I_N : \Sigma_{i=1}^n \oplus N_{\alpha_j} \to P$, $p_N : P \to \Sigma_{j=1}^n \oplus N_{\alpha_j}$ and $e_N : P \to \Sigma_{i=1}^n \oplus P_{\alpha_j}$. Therefore, ti is in $\underline{J}'$, which is a contradiction. Thus t is in $\underline{J}'$, so I is right regular mod $\underline{J}'$.

7.8.15 COROLLARY Let M be in $\underline{A}$ and N a direct summand of M. If $S_M \cap \underline{J}' = J(S_M)$, then N is in A.

Proof: Since N is a direct summand of M, there exists a submodule N' of M such that $M = N \oplus N'$. Hence there exist full submodules N_0 and N_0' of N and N' in $\underline{A}$, respectively, such that $N_0 \oplus N_0'$ is full in M. Hence $N_0 \oplus N_0' = M$ by (7.8.12), which implies that N is isomorphic to a direct sum of completely indecomposable modules.

7.8.16 Theorem [See (7.2.6).] Let $M = \Sigma_K \oplus M_\alpha = \Sigma_J \oplus N_\beta$ be any two direct decompositions of a module M in $\underline{A}$ by completely indecomposable modules M_α and N_β, respectively, and assume that $S_M \cap \underline{J}' = J(S_M)$. Then for any subset K' of K, there exists a one-to-one mapping φ of K' into J such that $M = \Sigma_{\alpha' \in K'} \oplus N_{\varphi(\alpha')} \oplus \Sigma_{K-K'} \oplus M_{\alpha''}$ and $M_{\alpha'} \approx N_{\varphi(\alpha')}$ for $\alpha \in K'$.

Proof: For any subset K' of K, we put $M_0 = \Sigma_{K-K'} \oplus M_{\alpha''}$. Then, by Zorn's lemma, there exists a maximal member M^* in the set of locally direct summands of M, which are of the form $M_0 \oplus \Sigma_{K \in J_i} \oplus N_{\gamma_k}$, $J_i \subset J$. Since M is the unique full submodule of M by (7.8.12), M^* is a direct summand of M, say $M = M^* \oplus M'$ for some submodule M' of M. By (7.8.15), M' is in $\underline{A}$ if $M' \neq 0$. So by (7.1.2) there exists some N_β such that $M^* \oplus N_\beta$ is a direct summand of M. This contradiction shows that $M^* = M$. Since $\Sigma_{K'} \oplus M_{\alpha'} \approx M/M_0 \approx \Sigma_{J'} \oplus N_{\gamma'}$ for some subset J' of J by (7.1.2), we can find a one-to-one mapping φ of K' onto J' such that $M_{\alpha'} \approx N_{\varphi(\alpha')}$ for $\alpha' \in K'$.

8
Applications II: Exchange Property

In Chapter 7, we used some structures of the factor category $\underline{A}/\underline{J}'$ to prove the Krull-Remak-Schmidt-Azumaya theorem (7.1.2), and further, many results that are extensions of (7.1.2) were obtained. We already know that if $\underline{J}'$ is the Jacobson radical of $\underline{A}$, then (7.1.2) generalizes to the complete form (7.2.6).

In this chapter we extend these results to some direct summands of a module which is a direct sum of completely indecomposable modules, as a further application of the structure of $\underline{A}/\underline{J}'$. On the other hand, (7.1.2) is a special case of the exchange property for modules. Thus some modules with the exchange property will be dealt with in the last part of this chapter.

First we introduce the concept of the exchange property in a given module in Sec. 8.1. Next we study the property for a direct sum of completely indecomposable modules in Sec. 8.2, where we obtain one of the most general forms of the Krull-Remak-Schmidt-Azumaya theorem. Almost all of the known results concerning the Krull-Remak-Schmidt-Azumaya theorem are involved in Sec. 8.2. In Sec. 8.3 we investigate the exchange property for a direct summand of a direct sum of completely indecomposable modules and we give the complete result (8.3.6). In Sec. 8.4 we study the same problem in case of the $\aleph_0$-exchange property. In the above we are interested in the exchange property in a given module; however, we investigate the usual exchange property for modules in Sec. 8.5. We include a short section, Sec. 8.6, to illustrate the meaning of the exchange property. The final section, Sec. 8.7, is the ring-theoretical proof of (8.2.3).

8.1 EXCHANGE PROPERTIES IN MODULES

Let R be a ring and let M_R be an R-module. We define the exchange property in M. Let N_R be a direct summand of M. For any direct decomposition

$M = \Sigma_J \oplus T_\alpha$ of M with the cardinal $|J| \le a$, we assume that we have a direct decomposition $M = N \oplus (\Sigma_J \oplus T'_\alpha)$ of M, where T'_α is a submodule of T_α. In this case we say that N has the *a-exchange property in M*. If this holds for all finite a, we say that N has the *finite exchange property in M*, and if N has the *a-exchange property in M* for any cardinal a, we say that N has the *exchange property* in M. If, in particular, each T_α is a c.ind. module, we say that N has *the exchange property in M with respect to c.ind. modules*.

We can extend the definition above to any module K_R as follows. If, for module B_R containing K isomorphically as a direct summand, K has the a-exchange property in B, then we say that K has the a-*exchange property*. If K has the a-exchange property for any cardinal a, we say that K has the *exchange property*. It is clear that T'_α above is a direct summand of T_α by (1.7.4).

We start with the following property.

8.1.1 LEMMA Let T_R be an R-module and $T = N_1 \oplus N_2 = N'_1 \oplus N'_2$, where $\phi_i : N_i \approx N'_i$ for i = 1, 2. If N_1 has the a-exchange property in T, then N'_1 does.

Proof: We have an automorphism $\alpha = \phi_1 + \phi_2 : T \to T$. Hence (8.1.1) is clear from the definition.

8.1.2 LEMMA Let T be an R-module and N a direct summand of T. If N has the 2-exchange property in T, then N has the finite exchange property in T.

Proof: It is sufficient to show by induction that N has the 3-exchange property in T. Let $T = N_1 \oplus N = \Sigma_1^3 \oplus A_i$. Then

$$T = N \oplus A'_1 \oplus (A_2 \oplus A_3)'$$

where $A_1 = A'_1 \oplus A''_1$, $A_2 \oplus A_3 = (A_2 \oplus A_3)' \oplus (A_2 \oplus A_3)''$ and $(A_2 \oplus A_3)'' = (A_2 \oplus A_3) \cap (N \oplus A'_1)$. On the other hand, $N \approx A''_1 \oplus (A_2 \oplus A_3)''$ and $N_1 \approx A'_1 \oplus (A_2 \oplus A_3)'$. Hence $A''_1 \oplus (A_2 \oplus A_3)''$ has the 2-exchange property in T by (8.1.1). Accordingly, $T = (A_1 \oplus A_2) \oplus A_3 = A''_1 \oplus (A_2 \oplus A_3)'' \oplus (A_1 \oplus A_2)' \oplus A'_3$, where $(A_1 \oplus A_2)' \subset (A_1 \oplus A_2)$ and $A'_3 \subset A_3$. Hence since $(A_2 \oplus A_3)'' \oplus A'_3 \subset A_2 \oplus A_3$,

$$A_2 \oplus A_3 = (A_2 \oplus A_3)'' \oplus A'_3 \oplus D$$

where $D = (A_2 \oplus A_3) \cap (A_1 \oplus A_2)' \subset (A_2 \oplus A_3) \cap (A_1 \oplus A_2) = A_2$, namely, D is a direct summand of A_2. Put $N \oplus A'_1 = (A_2 \oplus A_3)'' \oplus K$. Then $T = N \oplus A'_1 \oplus (A_2 \oplus A_3)' = (A_2 \oplus A_3)'' \oplus K \oplus (A_2 \oplus A_3)' = (A_2 \oplus A_3) \oplus K = (A_2 \oplus A_3)'' \oplus A'_3 \oplus D \oplus K = N \oplus A'_1 \oplus D \oplus A'_3$.

8.1.3 LEMMA Let B_R be an indecomposable module. If B has the 2-exchange property, then B has the exchange property.

Proof: Let $A = B \oplus C = \Sigma_I \oplus D_\alpha$. For $x \neq 0$ in B we have $x = \Sigma_{i=1}^n x_{\alpha_i}$, where x_{α_i} is in D_{α_i}. Put $A = \Sigma_{i=1}^n \oplus D_{\alpha_i} \oplus \Sigma_{I-\{\alpha_i\}} \oplus D_\beta$. Since B has the finite exchange property by (8.1.2), $A = B \oplus \Sigma_{i=1}^n \oplus D'_{\alpha_i} \oplus (\Sigma_{I-\{\alpha_i\}} \oplus D_\beta)'$ and $B \approx \Sigma_i \oplus D''_{\alpha_i} \oplus (\Sigma_{I-\{\alpha_i\}} \oplus D_\beta)''$, where $D_\alpha = D'_\alpha \oplus D''_\alpha$, etc. On the other hand, since B is indecomposable, $\Sigma \oplus D''_{\alpha_i} = 0$ or $(\Sigma_{I-\{\alpha_i\}} \oplus D_\beta)'' = 0$. If $\Sigma \oplus D''_{\alpha_i} = 0$, $D_\alpha = D'_\alpha$. Hence since $B \cap \Sigma_i \oplus D_{\alpha_i} \neq 0$, $B \approx \Sigma_i \oplus D''_{\alpha_i}$, $(\Sigma_{I-\{\alpha_i\}} \oplus D_\beta)'' = 0$, so $B \approx D''_{\alpha_1}$ for some α_1, and $D''_{\alpha_j} = 0$ for $\alpha_j \neq \alpha_i$. Therefore, $A = B \oplus D'_{\alpha_1} \oplus \Sigma_{I-\{\alpha_1\}} \oplus D_\beta$.

8.1.4 THEOREM Let B_R be an indecomposable R-module. Then B has the exchange property if and only if B is completely indecomposable.

Proof: We assume that B is completely indecomposable and $A = B \oplus C = D \oplus E$. Let p and i be the projection and the injection of the direct sum, respectively. Then $1_A = i_D p_D + i_E p_E$. Hence $1_B = p_B i_B = p_B i_D p_D i_B + p_B i_E p_E i_B$. Put $\alpha = p_B i_D p_D i_B$ and $\beta = p_B i_E p_E i_B$. Since $S = \mathrm{End}_R(B)$ is local and $1 = \alpha + \beta$, we may assume that α is an isomorphism ($\alpha : B \to A \to D \to A \to B$). Hence $A = B \oplus \ker p_B i_B p_D = B \oplus C \cap D \oplus E$. Therefore, B has the exchange property by (8.1.3). Conversely, we assume that B has the exchange property. Let $1 = \alpha + \beta$ for some α, β in S. Put

$$\text{(a)}\quad A = B \oplus B \qquad A_1 = \{(b,b) \mid b \in B\} \qquad A_2 = \{(\alpha(b),-\beta(b)) \mid b \in B\}$$

Since $b = \alpha(b) + \beta(b)$, $A_i \approx B$ and $A_1 \cap A_2 = 0$. Further, $(b,0) = (\beta(b), \beta(b)) + (\alpha(b), -\beta(b))$. Hence $A = A_1 \oplus A_2$. Since A_2 has the 2-exchange property in A, $A = (B + 0) \oplus A_2$ or $A = (0 + B) \oplus A_2$ from (a) (note that B is indecomposable). We assume the former case. Then β is a monomorphism, since $(B + 0) \cap A_2 = 0$, and β is an epimorphism, since $A = (B + 0) \oplus A_2$. Hence β is an isomorphism. The latter case similarly implies that α is an isomorphism. Therefore, S is local.

8.1.5 COROLLARY Let B_R be completely indecomposable. Then if $A \oplus B \approx C \oplus B$ for modules A_R and C_R, $A \approx C$.

Proof: We may assume that $D = A \oplus B = C' \oplus B'$; $C \approx C'$ and $B \approx B'$. Then $D = B \oplus C'' \oplus B''$, $A \approx D/B \approx C'' \oplus B''$, and $B \approx D/(C'' \oplus B'') = (C' \oplus B')/(C'' \oplus B'') \approx C_0 \oplus B_0$, where $C' = C'' \oplus C_0$, $B' = B'' \oplus B_0$. Since B is indecomposable, either $B_0 = 0$ or $C_0 = 0$. If $B_0 = 0$, $B'' = B' \approx B \approx C_0$. Hence

$A \approx C'' \oplus B'' \approx C'' \oplus C_0 \approx C' \approx C$. If $C_0 = 0$, then $B' = B_0$, $C' = C''$, and $B'' = 0$. Hence $A \approx C'' \approx C' \approx C$.

8.2 EXCHANGE PROPERTIES IN $\underline{A}$

Let $\{M_\alpha\}_I$ be a set of c.ind. modules and $M = \Sigma_I \oplus M_\alpha$. Let $\underline{A}$ be the category induced from $\{M_\alpha\}_I$ and $\bar{\underline{A}} = \underline{A}/\underline{J}'$ (see Sec. 6.3). We shall study the exchange property in M for direct summands in $\underline{A}$.

First we shall summarize (7.2.6), (7.3.14), and (7.3.15) in terms of the exchange property.

8.2.1 THEOREM Let $\{M_\alpha\}_I$ and M be as above. Then the following conditions are equivalent:

1. $\{M_\alpha\}_I$ is locally semi-T-nilpotent.
2. $\underline{J}' = \underline{J}$
3. Every full submodule of M coincides with M.
4. Every direct summand of M is in $\underline{A}$ and has the exchange property in M.
5. $S/J(S)$ is a regular ring in the sense of von Neumann and idempotents can be lifted modulo $J(S)$, where $S = \mathrm{End}_R(M)$.

We shall generalize (8.2.1) for a direct summand of M.

8.2.2 PROPOSITION Let M be as in (8.2.1) and $M = N_1 \oplus N_2$. If either N_1 is a finitely generated R-module or its full submodule N' is a direct sum of c.ind. modules $\{M'_{\alpha'}\}_J$ such that $\{M'_{\alpha'}\}_J$ is a locally semi-T-nilpotent set, then $N_1 = N'$ and N_1 is in $\underline{A}$.

Proof: If N_1 is f.g., N_1 is contained in some $\Sigma_{i=1}^n \oplus M_{\alpha_i} < \oplus M$. Hence N_1 is a direct summand of $\Sigma_{i=1}^n \oplus M_{\alpha_i}$. Therefore, N_1 is in $\underline{A}$ by (7.1.2). If a full submodule N' of N is of the form in the assumption, then N' is a direct summand of N by (7.3.9) and hence N' = N, for N' is a full submodule of N.

8.2.3 THEOREM Let $\{M_\alpha\}_I$ be a set of c.ind. modules and $M = \Sigma_I \oplus M_\alpha = N_1 \oplus N_2$. Suppose that $N_1 = \Sigma_{I'} \oplus M'_{\alpha'}$, where the $M'_{\alpha'}$ are c.ind. If $\{M'_{\alpha'}\}_{I'}$ is a locally semi-T-nilpotent set, N_1 and N_2 have the exchange property in M.

Proof: First, we show that N_2 has the exchange property in M. Let $M = \Sigma_J \oplus Q_\alpha$ be any direct decomposition. By (7.3.14) each Q_α contains a full submodule $P_\alpha = \Sigma_{i\in T_\alpha} \oplus P_{\alpha_i}$. Since $\bar{\underline{A}}$ is a completely reducible and Grothendieck category by (6.3.8), we have

(a) $\bar{M} = \bar{N}_2 \oplus \sum_{\alpha\in J} \sum_{i'\in T'_\alpha} \oplus \bar{P}_{\alpha_{i'}}$ where $T'_\alpha \subset T_\alpha$

(see the notation in Sec. 7.3). It is clear that $\bar{N}_1 \approx \Sigma_J \Sigma_{T'_\alpha} \oplus P_{\alpha_i'}$. Hence $\{P_{\alpha_i'}\}_{J,T'_\alpha}$ is a locally semi-T-nilpotent set by the assumption. Let p_{N_1} be the projection of M to N_1 with $\ker p_{N_1} = N_2$. From (a) we know that $\bar{p}_{N_1} \mid \Sigma_J \Sigma_{T'_\alpha} \oplus \bar{P}_{\alpha_i'}$ is an isomorphism in $\bar{A}$ (consider a full submodule of N_2). Hence $p_{N_1} \mid \Sigma_I \Sigma_{T'_\alpha} \oplus P_{\alpha_i'}$ is an isomorphism in $\underline{M}_R$ by (8.2.1) and Remark 3 in Sec. 6.2. Therefore, $M = \Sigma_J \Sigma_{T'_\alpha} \oplus P_{\alpha_i'} \oplus \ker p_{N_1} = \Sigma_J \Sigma_{T'_\alpha} \oplus P_{\alpha_i'} \oplus N_2$. Since $\Sigma_{T'_\alpha} \oplus P_{\alpha_i'} \subset Q_\alpha$, N_2 has the exchange property in M. Next, we shall show that N_1 has the exchange property in M. By an argument similar to that above (a), we have a full submodule $P_\alpha = P'_\alpha \oplus P''_\alpha$ in Q_α such that:

(b) $\bar{M} = \bar{N}_1 \oplus \Sigma_J \oplus \bar{P}'_\alpha$

(c) $\bar{N}_1 \approx \Sigma_J \oplus \bar{P}''_\alpha$

Since $\bar{M} = \Sigma_J \oplus (\bar{P}'_\alpha \oplus \bar{P}''_\alpha)$, there exists a $p_\alpha \in [M, P''_\alpha]$ for each α such that $\bar{p}_\alpha$ is the projection of $\bar{M}$ onto $\bar{P}''_\alpha$ with $\ker \bar{p}_\alpha = \Sigma_J \oplus \bar{P}'_\alpha \oplus \Sigma_{J-\alpha} \oplus \bar{P}''_\alpha$. Then p_α splits by Remark 3 in Sec. 6.2, say

(d) $M = L_\alpha \oplus P''_\alpha$

It is clear that $\ker \bar{p}_\alpha = \bar{L}_\alpha$. From (d) we obtain

(e) $Q_\alpha = Q'_\alpha \oplus P''_\alpha$ where $Q'_\alpha = Q_\alpha \cap L_\alpha$

Then $\bar{Q}'_\alpha \subset \bar{Q}_\alpha \cap \bar{L}_\alpha = \bar{Q}_\alpha \cap \ker \bar{p}_\alpha = \bar{P}'_\alpha$ for $\bar{Q}_\alpha = \bar{P}'_\alpha \oplus \bar{P}''_\alpha$. On the other hand, since $\bar{Q}_\alpha = \bar{Q}'_\alpha \oplus \bar{P}''_\alpha = \bar{P}'_\alpha \oplus \bar{P}''_\alpha$ by (e),

(f) $\bar{Q}'_\alpha = \bar{P}'_\alpha$

Finally, let p be the projection of M onto $\Sigma_J \oplus P''_\alpha$ for the decomposition $M = \Sigma_J \oplus P''_\alpha \oplus \Sigma_J \oplus Q'_\alpha$ given by (e). Then $\ker \bar{p} = \Sigma_J \oplus \bar{Q}'_\alpha = \Sigma_J \oplus \bar{P}'_\alpha$ by (f), and hence $\ker \bar{p} \cap \bar{N}_1 = 0$ and $\bar{p}(\bar{N}_1) = \bar{p}(\bar{M}) = \Sigma_J \oplus \bar{P}''_\alpha$ by (b). Therefore, $p|N_1$ is an R-isomorphism, so $M = N_1 \oplus \ker p = N_1 \oplus \Sigma_J \oplus Q'_\alpha$.

8.2.4 LEMMA Let T_R be an R-module. Let N be a direct summand of T and N_1 a direct summand of N. If N_1 has the a-exchange property in T, then N_1 has the same in N.

Proof: Let $N = \Sigma_J \oplus T_\delta$ with $|J| \le a$. Then $T = N \oplus N' = \Sigma_J \oplus T_\delta \oplus N'$. Hence we have $T = N_1 \oplus \Sigma_J \oplus T'_\delta \oplus N''$, where $T'_\delta \subset T_\delta$ and $N'' \subset N'$ [cf. (8.1.2)]. Therefore, $N = N_1 \oplus \Sigma_J \oplus T'_\delta \oplus N \cap N''$ and $N \cap N'' \subset N \cap N' = 0$.

8.2.5 THEOREM Let M and $\{M_\alpha\}_I$ be as in (8.2.3) and $M = N_1 \oplus N_2$. Let f be the projection of M onto N_1. Then $fJ'f = fJf$ if and only if every direct summand of N_1 has the exchange property in M. In this case N_2 also has the exchange property in M, where $\underline{J}' = J' \cap S_M$, $J = J(S_M)$, and $S_M = \mathrm{End}_R(M)$.

Proof: Assume that $fJ'f = fJf$. Since $\bar{\underline{A}}$ is completely reducible, there exists a subset K of I such that im $\bar{f} \approx \Sigma_K \oplus \bar{M}_\alpha$ $(= \bar{M}_K)$. Let e be the projection of M to M_K. Then $fS_M/fJ' \approx eS_M/eJ'$ by (3.1.10). Hence there exist an a in eS_Mf and a b in fS_Me such that $ba \equiv f$ and $ab \equiv e$ (mod J') by (3.1.9). Put $f - ba = n \in J'$. Then $n = fnf \in fJ'f = fJf$, which is equal to the radical of $S_{N_1} = \mathrm{End}_R(N_1)$. Therefore, $ba : N_1 = fM \xrightarrow{a} eM \xrightarrow{b} N_1$ is an automorphism in S_{N_1}. Then $eM = a(fM) \oplus \ker b$ in $\underline{M}_R$. On the other hand, since $ab \equiv e$ (mod J'), $\bar{b} : \overline{eM} \to \bar{M}$ is a monomorphism (note that eM is in $\underline{A}$). Consider a full submodule of ker b. We know that $\overline{\ker b} = 0$ in $\bar{\underline{A}}$. Therefore, ker b = 0 by (7.3.11) and $eM \approx fM$ in $\underline{M}_R$. Since $fJ'f = fJf$, $\{M_\alpha\}_K$ is a locally semi-T-nilpotent set by (8.2.1). Hence every direct summand of $N_1 = fM$ $(\in \underline{A})$ has the exchange property in M by (8.2.3). Conversely, we assume that every direct summand N_1' of N_1 has the exchange property in M. Then $N = \Sigma_K \oplus T_\gamma$, where the T_γ are c.ind. and N_1' has the exchange property in N_1 by (8.2.4). Hence $\{T_\gamma\}_K$ is a semi-T-nilpotent set by (8.2.1). Therefore, $fJ'f = fJf$ by (3.1.8) and (8.2.1). The remaining assertions are clear from (8.2.3).

8.2.6 COROLLARY Let M and N_i be as in (8.2.3). We suppose that for every monomorphism g in S_{N_1}, im g is a direct summand of N_1; i.e., $gS_{N_1} = eS_{N_1}$ for some idempotent e. Then every direct summand of N_1 has the exchange property in M.

Proof: Let f be the projection of M to N_1 and a in $fJ'f$. Then $1 - a$ is a monomorphism by (7.3.12). Furthermore, $(1 - a) \mid N_2 = 1_{N_2}$ and im $(1 - a)$ = im $((1 - a) \mid N_1) \oplus N_2$. From the assumption, im $((1 - a) \mid N_1)$ is a direct summand of N_1, and hence im $(1 - a)$ is a direct summand of M. On the other hand, im $(1 - a)$ is a full submodule of M by (7.3.13). Therefore im $(1 - a) = M$, so im $((1 - a) \mid N_1) = N_1$. Hence a is quasi-regular in S_{N_1} and $fJ'f \subset fJf$. It is clear that $fJf \subset fJ'f$, since $\underline{J} \subset \underline{J}'$ by (7.1.3).

8.2.7 THEOREM Let $\{N_\alpha\}_J$ be a set of indecomposable injective modules and $N = \Sigma_J \oplus N_\alpha$. Suppose that $N = M_1 \oplus M_2$ and $Z(M_1) = 0$. Then M_i is a direct sum of c.ind. injective modules for i = 1, 2.

Proof: M_i contains a full submodule T_i by (7.3.14). Let $T_1 = \Sigma_I \oplus T_\alpha$, where the T_α are c.ind. Since $Z(M_1) = 0$, $Z(T_1) = 0$. Hence $\{T_\alpha\}_I$

is a T-nilpotent set by the proof of (7.7.1). Therefore, we have (8.2.7) by (7.3.9) and (8.2.3).

8.2.8 THEOREM Let $\{E_\alpha\}_I$ be a set of indecomposable injective modules and $E = \Sigma_I \oplus E_\alpha$. Then the following are equivalent:

1. $\{E_\alpha\}_I$ is a locally semi-T-nilpotent set with respect to $\underline{J}'$.
2. Every object in $\underline{C}'$ which is an extension of E contains E as a direct summand.
3. There is no proper and essential extension of E which is in $\underline{C}'$.
4. For each monomorphism g in $S_E = \mathrm{End}_R(E)$, im g is a direct summand of E, where $\underline{C}'$ is the category of all direct sums of c.ind. and injective modules.

Proof: (4) ⇒ (1). This is proved by (8.2.6).

(1) ⇒ (4). Let g be a monomorphism in S_E. Then im $g = \Sigma \oplus g(E_\alpha)$ and $E_\alpha \approx g(E_\alpha)$. Since $g(E_\alpha)$ are injective, im g is a locally direct summand of E. Hence im g is a direct summand of E by (1) and (7.3.14).

(1) ⇒ (2). This is clear from the proof of (1) ⇒ (4).

(2) ⇒ (3). This is also clear.

(3) ⇒ (1). Suppose that $\{E_\alpha\}_I$ is not a locally semi-T-nilpotent set. Then there exist a subset $\{E_{\alpha_i}\}_1^\infty$ of $\{E_\alpha\}_I$ and a set of nonisomorphisms $f_i : E_{\alpha_i} \to E_{\alpha_{i+1}}$ such that for some element x in E_{α_1}, $f_n f_{n-1} \cdots f_1(x) \neq 0$ for all n. We note that ker $f_i \neq 0$, since E_{α_i} are injective and indecomposable. Put $E_i' = \{x_i + f_i(x_i) \mid x_i \in E_{\alpha_i}\} \subset \Sigma_1^\infty \oplus E_{\alpha_i}$, a direct summand of E. Put $E = \Sigma_1^\infty \oplus E_{\alpha_i} \oplus E_0$. Then $E_{\alpha_i} \cap (\Sigma \oplus E_j') \supset \ker f_i \neq 0$ for each i. Hence $(\Sigma \oplus E_j') \oplus E_0$ is essential in E by (3.4.6). It is clear that $x \notin (\Sigma \oplus E_j') \oplus E_0$. Let E* be an injective hull of E. Since we have an isomorphism $t : (\Sigma \oplus E_j') \oplus E_0 \to E$, we can extend t to a monomorphism ϕ of E*. Therefore, $\phi((\Sigma \oplus E_j') \oplus E_0) = E \subsetneq \phi(E) = \Sigma \oplus \phi(E_\alpha) \in \underline{C}'$, which is a contradiction.

8.2.9 COROLLARY Let M be a module in $\underline{C}'$ and L a submodule of M. Suppose that L is a direct sum of injective modules and $Z(L) = 0$. Then L is a direct summand of M.

Proof: Since every injective module in L is in $\underline{C}'$ by (8.2.7), (8.2.9) is clear from (8.2.8).

Remark Let $\{E_i\}_1^3$ be a set of injective and indecomposable modules and $f_i : E_i \to E_{i+1}$ a nonisomorphism. Then ker $f_i \neq 0$ and im $f_1 \cap \ker f_2 \neq 0$ if $f_i \neq 0$, since E_2 is uniform. Hence $\ker f_1 \subsetneq \ker f_2 f_1$, provided that $f_1 \neq 0$. Therefore, $\{E_\alpha\}_I$ is a T-nilpotent set provided that E_α is noetherian. If E is quasi-injective and f in S_E is a monomorphism, then $f^{-1}|f(E)$ is extendible to an element g in S_E. Hence E = im f $\oplus$ ker g [cf. (8.2.6)].

8.3 DIRECT SUMMANDS WITH THE EXCHANGE PROPERTY

Let $\{M_\alpha\}_I$ be a set of completely indecomposable modules and $M = \Sigma_I \oplus M_\alpha$ as before. If a direct summand N of M is a direct sum of completely indecomposable modules $\{N_\gamma\}_J$ such that $\{N_\gamma\}_J$ is a locally semi-T-nilpotent set, then N and its direct complement have the exchange property by (8.2.3).

For the converse we consider the following example. Let Z be the ring of the integers and let p_1, p_2 be distinct primes. Put $N_i = \Sigma_{j=1}^{\infty} \oplus Z/p_i^j$ and $M = N_1 \oplus N_2$. Then N_i is the p_i-primary component of M. Hence N_i has the exchange property in M for i = 1, 2. However, $\{Z/p_i^j\}_j$ is not semi-T-nilpotent. This example gives us the information that the exchange property of N_i in M is related to the fact that $\mathrm{Hom}_Z(N_i, N_j) = 0$ for $i \neq j$.

We shall study this situation for any M. In this section we study principally the exchange property with respect to completely indecomposable modules. We denote this exchange property briefly by the exchange property with respect to ind. modules. We shall generalize the concept of the semi-T-nilpotent set.

Let $\{M_\alpha\}_I$ and $\{N_\beta\}_J$ be two sets of completely indecomposable modules. We take countable subsets $\{M_{\alpha_i}\}_1^\infty$ and $\{N_{\beta_j}\}_1^\infty$ of $\{M_\alpha\}_I$ and $\{N_\beta\}_J$, respectively. Take two sets $\{f_i\}$ and $\{g_i\}$ of nonisomorphisms, where $f_i : M_{\alpha_i} \to N_{\beta_i}$ and $g_i : N_{\beta_i} \to M_{\alpha_{i+1}}$. If for any element m in M_{α_1}, there exists an integer n, which depends on m, such that $g_n f_n \cdots g_1 f_1(m) = 0$, then we call $\{f_i, g_i\}$ (*locally* and) *relatively T-nilpotent*. If for all countable subsets $\{M_{\alpha_i}\}_1^\infty$ and $\{N_{\beta_j}\}_1^\infty$, every $\{f_i, g_i\}$ of nonisomorphisms is relatively T-nilpotent, then we call $\{M_\alpha\}_I$ and $\{N_\beta\}_J$ (locally and) *relatively semi-T-nilpotent*. We note that if $\{M_\alpha\}_I$ and $\{N_\beta\}_J$ are relatively semi-T-nilpotent, there exists an integer n', for any element x in N_{β_1}, such that $f_{n'+1} g_{n'} f_{n'} \cdots f_2 g_1(x) = 0$. If we have the same property in the above for $\alpha_i = \alpha_{i'}$ ($\beta_j = \beta_{j'}$) even if $i \neq i'$ ($j \neq j'$), we call $\{M_\alpha\}_I$ and $\{N_\beta\}_J$ *relatively T-nilpotent*. It is clear that if either $\{M_\alpha\}_I$ or $\{N_\beta\}_J$ is locally semi-T-nilpotent, then $\{M_\alpha\}_I$ and $\{N_\beta\}_J$ are relatively semi-T-nilpotent; however the converse is not true. If either I or J is finite, we assume, as a definition, that $\{M_\alpha\}_I$ and $\{N_\beta\}_J$ are relatively semi-T-nilpotent (in this case $\{M_\alpha\}_I$ or $\{N_\beta\}_J$ is locally semi-T-nilpotent).

We shall rearrange $\{M_\alpha\}_I$ and $\{N_\beta\}_J$ as follows:

$$\{M_\alpha\}_I = \{M_{kj}\}_{k\in K, j\in I_k} \qquad \{N_\beta\}_J = \{N_{kj}\}_{k\in K, j\in J_k}$$

where $M_{kj} \approx M_{kj'} \approx N_{kj} \approx N_{kj'}$ and $M_{kj} \not\approx M_{k'j'}$, $N_{kj} \not\approx N_{k'j'}$ provided that $k \neq k'$.

8.3.1 LEMMA Let $\{M_{kj}\}_{k\in K, j\in I_k}$ and $\{N_{kj}\}_{k\in K, j\in J_k}$ be as above and K' a subset of K. We assume that $\{M_{kj}\}$ and $\{N_{kj}\}$ are relatively semi-T-nilpotent. Then (1) if I_k and J_k are infinite for all $k \in K'$, then $\{M_{kj}\}_{K', I_k}$ is locally semi-T-nilpotent and (2) if either $|I_k| \leq |J_k|$ or $|I_k|$ is finite and $J_k \neq \emptyset$ for all $k \in K'$, then $\{M_{kj}\}_{K', I_k}$ is locally semi-T-nilpotent, where $|I|$ means the cardinal of a set I.

Proof: (1) and the first assertion of (2) are clear. We assume that I_k is finite and $J_k \neq \emptyset$. Let $\{M_i\}_1^\infty$ be a countable subset of $\{M_{kj}\}_{K', I_k}$ and $\{f_i : M_i \to M_{i+1}\}$ a set of nonisomorphisms. Assume that $M_1 = M_{k_1 j_1}$. Since I_{k_1} is finite, there exists an integer n_2 such that $M_{n'} \neq M_{k_1 j_1}$ for all $n' \geq n_2$. Next we assume that $M_{n_2} = M_{k_2 j_2}$ $(k_1 \neq k_2)$. Again there exists $n_3 \geq n_2$ such that $M_{n'} \neq M_{k_2 j_2}$ for all $n' \geq n_3$. Repeating these arguments, we obtain a subset $\{M_{n_i}\}$ of $\{M_i\}$ $(n_1 = 1)$ such that $M_{n_i} \not\approx M_{n_j}$ for all $i \neq j$ and a set $g_i = \{f_{n_{i+1}-1} f_{n_{i+1}-2} \cdots f_{n_i} : M_{n_i} \to M_{n_{i+1}}\}$. It is clear that no g_i is an isomorphism, by (6.3.1). Furthermore, $M_{n_i} \approx N_{k_j}$ for some k_j. Hence $\{g_i\}$ is locally T-nilpotent from the assumption. Therefore, $\{f_i\}$ is a locally T-nilpotent set.

8.3.2 LEMMA If $\{M_\alpha\}_I$ and $\{N_\beta\}_J$ are locally semi-T-nilpotent, then so is $\{M_\alpha, N_\beta\}_{I\cup J}$.

Proof: Let $\{T_i\}_1^\infty$ be a countable subset of $\{M_\alpha, N_\beta\}_{I\cup J}$ and $\{f_i : T_i \to T_{i+1}\}$ a set of nonisomorphisms. We assume that $\{T_i\}$ contains an infinite number of modules in $\{M_\alpha\}_I$, say $\{T_i\} \supset \{M_{\alpha_i}\}_1^\infty$. Then $\{g_i = f_{\alpha_{i+1}-1} f_{\alpha_{i+1}-2} \cdots f_{\alpha_i}\}$ is locally T-nilpotent and hence $\{f_i\}$ is. If $\{T_i\}$ contains only a finite number of modules in $\{M_\alpha\}_I$, then there exists an n such that $\{T_i\}_{i\geq n} \subset \{N_\beta\}_J$. Therefore, $\{f_i\}$ is also locally T-nilpotent.

8.3.3 LEMMA Let $\{M_\alpha\}_I$ and M be as in the beginning of this section. Let $M = T \oplus S$ and N a direct summand of M such that $N = \Sigma_J \oplus N_\beta$, where the N_β are completely indecomposable modules not isomorphic to any summand of S. If $\{N_\beta\}_J$ is locally semi-T-nilpotent, $M = N \oplus T' \oplus S$, where $T' \subset T$.

Proof: We have $M = N \oplus T' \oplus S'$, $T = T' \oplus T''$ and $S = S' \oplus S''$ by the assumption and (8.2.3). Since $T'' \oplus S'' \approx N$, $S'' = 0$.

8.3.4 LEMMA Let $\{M_\alpha\}_I$ and M be as in (8.3.3). We assume that $M = N_1 \oplus N_2$, $N_1 = \Sigma_{I'} \oplus M_{\alpha'}$, $N_2 = \Sigma_{I''} \oplus M_{\alpha''}$ and $M_{\alpha'} \not\approx M_{\alpha''}$ for any pair $(\alpha', \alpha'') \in I' \times I''$ $(I = I' \cup I'')$. Assume further that $M = T \oplus S$, $T = T_1 \oplus T_2$ and T (resp. S) contains a full submodule which is isomorphic to N_1 (resp. N_2). Then for any element t in T_1, there exists a direct summand $\Sigma_1^n \oplus T_i$ of T_1 such that $t \in \Sigma_1^n \oplus T_{i'}$, and $T_{i'} \approx M_{\alpha_i'}$ for some $\alpha_i' \in I'$.

Proof: Let t be in $\Sigma_1^{n_1} \oplus M_{\alpha_i'} \oplus \Sigma_1^{n_2} \oplus M_{\alpha_i''} = M'$. Then $M = M' \oplus T_1' \oplus T_2' \oplus S'$ by (8.2.3), where $T_i = T_i' \oplus T_i''$, $S = S' \oplus S''$, and $T_1'' = T_1 \cap (M' \oplus T_2' \oplus S') \ni t$. Since $T_1'' \oplus T_2'' \oplus S'' \approx M'$ and $T_1'' = \Sigma_1^n \oplus T_{i'}$, $T_{i'} \approx M_{\alpha_i'}$ by the assumption and (7.1.2).

8.3.5 LEMMA Let $M = N_1 \oplus N_2$ be as in (8.3.4). Then $M = T \oplus N_2 = N_1 \oplus S$ for any decomposition $M = T \oplus S$ as in (8.3.4) if and only if $\{M_{\alpha'}\}_{I'}$ and $\{M_{\alpha''}\}_{I''}$ are relatively semi-T-nilpotent. In this case $T \approx N_1$ and $S \approx N_2$.

Proof: We assume that $M = N_1 \oplus S = T \oplus N_2$ as in (8.3.5). We may assume that I' and I'' are infinite. Let $\{M_{2i-1}\}_1^\infty$ and $\{M_{2i}\}_1^\infty$ be countable subsets of $\{M_{\alpha'}\}_{I'}$ and $\{M_{\alpha''}\}_{I''}$, respectively, and let $\{f_n : M_n \to M_{n+1}\}$ be a set of nonisomorphisms. We shall make use of the same argument as in the proof (7.2.5). Put $M_n' = \{x + f_n(x) \mid x \in M_n\} \subset M_n \oplus M_{n+1} \subset M$. Then $M = \Sigma_1^\infty \oplus M_{2i-1}' \oplus N_1^* \oplus N_2 = N_1 \oplus \Sigma_1^\infty \oplus M_{2i}' \oplus N_2^*$, where $N_1^* = \Sigma_{I'-\aleph_0'} \oplus M_{\alpha'}$; $\aleph_0' = \{1, 3, \ldots, 2i-1, \ldots\}$ and $N_2^* = \Sigma_{I''-\aleph_0''} \oplus M_{\alpha''}$; $\aleph_0'' = \{2,4,\ldots,2i,\ldots\}$. It is clear that $\Sigma_1^\infty \oplus M_{2i-1} \overset{\varphi}{\approx} \Sigma_1^\infty \oplus M_{2i-1}'$. Define an automorphism ϕ of M by setting $\phi = \varphi + 1_{N_1^* \oplus N_2}$. Then $M = \phi^{-1}(M) = \phi^{-1}(N_1) \oplus \Sigma \oplus \phi^{-1}(M_{2i}')$. Hence we have $M = N_1 \oplus \Sigma \oplus \phi^{-1}(M_{2i}') \oplus \phi^{-1}(N_2^*)$ from the assumption. Therefore, $M = \phi(M) = \phi(N_1) \oplus \Sigma \oplus M_{2i}' \oplus N_2^* = \Sigma \oplus M_{2i}' \oplus \Sigma \oplus M_{2i-1}' \oplus N_1^* \oplus N_2^*$. Hence $\{f_i\}$ is locally T-nilpotent by (7.2.7), so $\{M_{\alpha'}\}_{I'}$ and $\{M_{\alpha''}\}_{I''}$ are relatively semi-T-nilpotent. We shall show the converse. Put $M^* = N_1 + S$. Let s be in $N_1 \cap S$. Then there exists a direct summand S_1 of S (and hence of M) such that $s \in S_1$ and $S_1 \approx \Sigma_1^\infty \oplus M_{\alpha''}$ by (8.3.4). On the other hand, $M = S_1 \oplus N_1 \oplus N_2'$ from (8.3.3), where $N_2' \subset N_2$, which is a contradiction. Hence $N_1 \cap S = 0$. If either I' or I'' is finite, $M = N_1 \oplus S' \oplus T'$ by (8.2.3), where $S' \subset S$ and $T' \subset T$. Since $N_2 \approx M/N_1 \approx S' \oplus T'$, $T' = 0$, so $M = N_1 \oplus S' \subset M^* \subset M$. We assume that I' and I'' are infinite and that $M \neq M^*$. Since $M^* \supset N_1$, there exists an α'' in I'' such that $M_{\alpha_1''} \not\subset M^*$. Let $m_{\alpha''} \in M_{\alpha_1''} - M^*$. From the decomposition

(a) $M = T \oplus S$

we can write $m''_{\alpha_1} = t + s$; $t \in T$, $s \in S$. Since s is in $S \subset M^*$, t is in $T - M^*$. We obtain, from (8.3.4), a direct summand $\Sigma_1^n \oplus T_i$ of T (and hence of M) such that $T_i \approx M_{\alpha'_i}$ and $t \in \Sigma \oplus T_i$. Let $p_i^{(1)}$ be the projection of M to T_i in this decomposition. Then there exists an i ($= \alpha'_2$) such that $p_{\alpha'_2}^{(1)}(m_{\alpha'_1}) = t_{\alpha'_2} \in T_{\alpha'_2} - M^*$. From (7.1.2) and the assumption, S contains a summand $S_{\alpha''(1)}$ isomorphic to $M_{\alpha''_1}$. We obtain again from (8.3.3)

(b) $M = N_1 \oplus \sum_{I''-\varepsilon_1} \oplus M_{\alpha''} \oplus S_{\alpha''(1)}$

where ε_1 is some index. Let $p_{\alpha''_i}^{(2)}$ be the projection of M to $M_{\alpha''_i}$ in (b). Since $N_1 \oplus S_{\alpha''(1)} \subset M^*$, there exists an $\alpha''_3 \in I'' - \varepsilon_1$ such that $p_{\alpha''_3}^{(2)}(t_{\alpha'_2}) = m_{\alpha''_3} \in M_{\alpha''_3} - M^*$. [It is possible that $\alpha''_1 = \alpha''_3$. In this case (b) implies that M contains at least two direct summands isomorphic to $M_{\alpha''_1}$.] We may assume that $T_{\alpha'_2} \approx M_{\alpha'(2)}$ for some $\alpha'(2) \in I'$. Then again by (8.3.3) and (a) we have

(c) $M = M_{\alpha'(2)} \oplus T' \oplus S$

where $T' \subset T$. Then $m_{\alpha''_3} = x + t$; $x \in M^*$, $t \in T' - M^*$. By the same argument after (a), we obtain α'_4 and a homomorphism $p_{\alpha'_4}^{(3)}$ of M to $T_{\alpha'_4}$ such that $p_{\alpha'_4}^{(3)}(m_{\alpha''_3}) = t_{\alpha'_4} \in T_{\alpha'_4} - M^*$ and $T_{\alpha'_4}$ is a direct summand of T'. [It is possible that $T_{\alpha'_2} \approx T_{\alpha'_4}$. In this case (c) implies that M contains at least two direct summands isomorphic to $M_{\alpha'(3)}$.] From the remark after (b) and (8.3.3), there exists an $\alpha''(3)$ such that

(d) $M = N_1 \oplus \sum_{I''-\{\varepsilon'_1,\varepsilon'_2\}} \oplus M_{\alpha''} \oplus S'_{\alpha''(1)} \oplus S_{\alpha''(3)}$

where $S \supset S_{\alpha''(3)} \approx M_{\alpha''_3}$ and $S'_{\alpha''(1)} \approx S_{\alpha''(1)}$. Hence we obtain an α''_5 in $I'' - \{\varepsilon'_1,\varepsilon'_2\}$ such that $p_{\alpha''_5}^{(4)}(t_{\alpha'_4}) = m_{\alpha''_5} \in M_{\alpha''_5} - M^*$. Similarly, we have

(e) $M = M'_{\alpha'(2)} \oplus M'_{\alpha'(4)} \oplus T'' \oplus S$

where $M'_{\alpha'(4)} \approx T_{\alpha'_4}$, $M'_{\alpha(2)} \approx M_{\alpha'(2)}$, and $T'' \subset T$. Repeating these arguments, we have the series of indecomposable modules $M_{\alpha''_1}$, $T_{\alpha'_2}$, $M_{\alpha''_3}$, $T_{\alpha'_4}$, ... and the series of homomorphisms $p_1 = p_{\alpha'_2}^{(1)} \mid M_{\alpha''_1}$, $p_2 = p_{\alpha''_3}^{(2)} \mid T_{\alpha'_2}$ [it is possible that $\alpha_i'' = \alpha_j''$ (resp. $T_{\alpha_k'} \approx T_{\alpha_\ell'}$) for $i \neq j$ (resp. $k \neq \ell$)]. Put $M^{(n)} = \Sigma_1^n \oplus M_{\alpha_{2i+1}''}$, $T^{(n)} = \Sigma_1^n \oplus T_{\alpha_{2i}'}$ (external direct sum). Then $M^{(n)}$ and $T^{(n)}$ are isomorphic to direct summands of M for all n from the decompositions (b), (c), (d), ... We try to find a contradiction to the assumption of relatively semi-T-nilpotency, and hence, after replacing $M_{\alpha_i''}$ (resp. $T_{\alpha_i'}$) by another isomorphic summand when $\alpha_i'' = \alpha_j''$ [resp. $\alpha'(h) = \alpha'(1)$] for $i \neq j$ (resp. $h \neq 1$), we may assume that $\alpha_i'' \neq \alpha_j''$ [resp. $\alpha'(k) \neq \alpha'(1)$]. It is clear that any p_i are nonisomorphisms and $p_n p_{n-1} \cdots p_1(m_{\alpha_1''}) \neq 0$ for all n. This is a contradiction to the relative semi-T-nilpotency. Therefore, $M = M^*$. Similarly, we have $M = T \oplus N_2$.

8.3.6 THEOREM Let $\{M_\alpha\}_I$ be a set of c.ind. modules and $M = \Sigma_I \oplus M_\alpha$. For a subset I' of I, we put $N_1 = \Sigma_{I'} \oplus M_\alpha$ and $N_2 = \Sigma_{I-I'} \oplus M_\alpha$ ($M = N_1 \oplus N_2$). Then the following statements are equivalent:

1. N_1 has the exchange property w.r.t. ind. modules.
2. N_2 has the exchange property w.r.t. ind. modules.
3. $\{M_{\alpha'}\}_{I'}$ and $\{M_{\alpha''}\}_{I-I'}$ are relatively semi-T-nilpotent.

Proof: Since condition (3) is symmetric, we may show that (1) is equivalent to (3). We know already, by the proofs of (7.2.7) and (8.2.3), that (1) implies (3). Now we assume (3). Let $\{M_\alpha\}_I = \{M_{kj}\}_{k\in K, j\in I_k}$ be as above (8.3.1), $N_1 = \Sigma_K \Sigma_{I_k} \oplus M_{kj}$ and $N_2 = \Sigma_K \Sigma_{J_k} \oplus N_{kj}$. Consider a partition of K as follows:

$$
\text{(f)}\quad
\begin{aligned}
K_1 &= \{k \in K \mid I_k \text{ and } J_k \text{ are infinite}\}\\
K_2 &= \{k \in K \mid I_k = \emptyset \text{ and } J_k \text{ is finite}\}\\
K_3 &= \{k \in K \mid I_k \text{ is finite and } J_k \text{ is infinite}\}\\
K_4 &= \{k \in K \mid I_k = \emptyset\}\\
K_5 &= \{k \in K \mid J_k = \emptyset\}
\end{aligned}
$$

We put $M(i) = \Sigma_{K_i} \Sigma_{I_k} \oplus M_{kj}$ and $N(i) = \Sigma_{K_i} \Sigma_{J_k} \oplus N_{kj}$. $\{M_{kj}\}_{K_1, I_k}$ and $\{N_{kj}\}_{K_1, J_k}$ are locally T-nilpotent and $\{M_{kj}\}_{K_3, I_k}$ and $\{N_{kj}\}_{K_2, J_k}$ are locally semi-T-nilpotent by the assumption and (8.3.1). Hence $\{M_{kj}\}_{K_1 \cup K_3, I_k}$ is locally semi-T-nilpotent by (8.3.2). Let $M = \Sigma_I \oplus L_\varepsilon$ be an direct decomposition with L_ε indecomposable. Then $M = M(1) \oplus M(3) \oplus \Sigma_{I'} \oplus L_{\varepsilon'}$ for some $I' \subset I$ by (8.2.3). Put $\bar{\bar{M}} = M/[M(1) \oplus M(3)]$; then

$$
\text{(g)}\quad \bar{\bar{M}} \approx M(2) \oplus M(5) \oplus \sum_{i=1}^{4} \oplus N(i) \approx \sum_{I'} \oplus \text{ I}
$$

Since $\{N_{kj}\}_{K_2,J_k}$ is locally semi-T-nilpotent, we obtain by (g) and (8.2.3), $\bar{\bar{M}} \approx M(2) \oplus M(5) \oplus \Sigma_{i'\neq 2} \oplus N(i) \oplus \Sigma_{I''} \oplus L_{\varepsilon''}$ for some $I'' \subset I'$. Hence $\bar{\bar{M}}/(\Sigma_{I''} \oplus L_{\varepsilon''}) \approx M(2) \oplus M(5) \oplus \Sigma_{i\neq 2} \oplus N(i) \approx \Sigma_{I'''} \oplus L_{\varepsilon'''}$, where $I''' = I' - I''$. We consider a partition of I''' as follows:

$$I_1''' = \{\varepsilon''' \in I''' \mid L_{\varepsilon'''} \approx M_{kj} \text{ for some } k \in K_2 \cup K_5\}$$

and $I_2''' = I''' - I_1'''$. Then $\Sigma_{I_1'''} \oplus L_{\varepsilon'''}$ is a full submodule of itself, which satisfies the assumption in (8.3.5). Hence $\bar{\bar{M}}/(\Sigma_{I''} \oplus L_{\varepsilon''}) \approx M(2) \oplus M(5) \oplus \Sigma_{I_2'''} \oplus L_{\varepsilon'''}$. Therefore, $M = N_1 \oplus \Sigma_{I''} \oplus L_{\varepsilon''} \oplus \Sigma_{I_2'''} \oplus L_{\varepsilon'''}$.

8.3.7 COROLLARY Let M and $M = N_1 \oplus N_2$ be as in (8.3.6). Then N_1 has the exchange property in M if and only if (3) in (8.3.6) is satisfied and any direct summand L of M has the direct decomposition $L = L_1 \oplus L_2$ such that a full submodule L_1' (resp. L_2') of L_1 (resp. L_2) is isomorphic to a direct summand of N_1 (resp. N_2).

Proof: We assume that N_1 has the exchange property in M. Let $M = L \oplus L'$. Then $M = N_1 \oplus L_1 \oplus L_1'$ and $L = L_1 \oplus L_2$, $L' = L_1' \oplus L_2'$. Since $L_1 \oplus L_1' \approx N_2$ and $L_2' \oplus L_2 \approx N_1$, L_1 (resp. L_2) contains a full submodule isomorphic to a direct summand of N_2 (resp. N_1). Conversely, we assume that the above fact and (3) in (8.3.6) hold. We use the same notation as in the proof of (8.3.6). Let $M = \Sigma_K \oplus L_\varepsilon$ be any direct decomposition of M. Then $\bar{\bar{M}} \approx M(2) \oplus M(5) \oplus \Sigma_{i=1}^{4} \oplus N(i) \approx \Sigma_K \oplus L_\varepsilon'$ by (g), where $L_\varepsilon = L_\varepsilon' \oplus L_\varepsilon''$. By (8.2.3) and (8.3.2), we have $\bar{\bar{M}} \approx M(2) \oplus M(5) \oplus \Sigma_{i=3,4} \oplus N(i) \oplus \Sigma_K \oplus L_\varepsilon'''$, where $L' = L_\varepsilon''' \oplus L_\varepsilon^{(4)}$. Hence

$$\text{(h)} \quad M = \bar{M}/\Sigma_K \oplus L_\varepsilon''' \approx M(2) \oplus M(5) \oplus \Sigma_{i=3,4} \oplus N(i) \approx \Sigma_K \oplus L_\varepsilon^{(4)}$$

Now we shall apply the assumption for $L_\varepsilon^{(4)}$. Put $L_\varepsilon^{(4)} = L$. Then $L = L_1 \oplus L_2$ and L_i contains a full submodule L_i' isomorphic to a direct summand of N_i. Hence $L' = M^*(2) \oplus M^*(5) \oplus N^*(3)$ and $L_2' = M^*(2)' \oplus N^*(3)' \oplus N^*(4)'$ by (h), where $M^*(\ )$ and $M^*(\ )'$ are isomorphic to direct summands of $M(\ )$, etc. Then $N^*(3)$ is isomorphic to a direct summand of N_1 and hence of $M(3)$. Therefore, $N^*(3)$ has the exchange property in M by (8.2.3) [note that the set of c.ind. modules T_α that appeared in $N^*(3) = \Sigma \oplus T_\alpha$ is locally semi-T-nilpotent]. Similarly, $M^*(2)'$ has the same property. Therefore, $N^*(3)$ and $M^*(2)'$ are direct summands of M (and hence of L) by (8.2.2). Hence $L_1 = N^*(3) \oplus L_1''$, $L_2 = M^*(2)' \oplus L_2''$ and L_1'' (resp. L_2'') contains a full submodule isomorphic to $M^*(2) \oplus M^*(5)$ [resp. $N^*(3)' \oplus N^*(4)'$]. Accordingly, from (h), $M(2) \oplus M(5) \oplus N(3) \oplus N(4) = \Sigma_K \oplus (L^{(4)}_{\varepsilon\ 1} \oplus L^{(4)}_{\varepsilon\ 2}) = \Sigma_K \oplus [L^{(4)}_{\varepsilon\ 1}{}'' \oplus N^*(3)_\varepsilon] \oplus \Sigma_K \oplus [L^{(4)}_{\varepsilon\ 2}{}'' \oplus M^*(2)_\varepsilon{}'] = \Sigma_K \oplus [M^*(2)_\varepsilon{}' \oplus L^{(4)}_{\varepsilon\ 1}{}''] \oplus \Sigma_K \oplus [N^*(3)_\varepsilon \oplus L^{(4)}_{\varepsilon\ 2}{}'']$. Therefore, we obtain from (8.3.5) that $\tilde{M} \approx M(2) \oplus M(5) \oplus \Sigma_K \oplus [N^*(3)_\varepsilon \oplus L^{(4)}_{\varepsilon\ 2}{}'']$. Thus $M = N_1 \oplus \Sigma_K[L'''_\varepsilon \oplus N^*(3) \oplus L^{(4)}_{\varepsilon\ 2}{}'']$ and $L'''_\varepsilon \oplus N^*(3)_\varepsilon \oplus L^{(4)}_{\varepsilon\ 2}{}''$ is a direct summand of L_ε.

8.3.8 COROLLARY Let M be as in (8.3.6). We assume that $M = \Sigma^n_{i=1} \oplus M^{(i)}$, where the $M^{(i)}$ are in $\underline{A}$. Then the following statements are equivalent:

1. $M^{(1)}$ has the exchange property in M w.r.t. ind. modules.
2. $M^{(1)}$ has the exchange property in $M^{(1)} \oplus M^{(i)}$ w.r.t. ind. modules for all i.

If $M^{(i)}$ and $M^{(j)}$ have the exchange property in M w.r.t. ind. modules, then $M^{(i)} \oplus M^{(j)}$ does.

Proof: This is clear by (8.3.6).

8.3.9 LEMMA Let P be an R-module. If P is itself T-nilpotent, P is a completely indecomposable module.

Proof: Put $S = \mathrm{End}_R(P)$. If e is in S and $e^2 = e$, then e = 1 or 0 by the assumption. Let x and y be two elements in S with x a nonunit. Then neither xy or yx is a unit in S from the above and (1.7.5). Furthermore, consider the sequence $\{x^n\}_n$ of nonunits in S. For any element p in P, there exists an integer n = n(p) such that $x^n(p) = 0$ from the assumption. Therefore, $X = 1 + \Sigma^\infty_{i=1} x^i$ is an element in S and $((1 - x)X)(p) = (1 - x)(1 + x + x^2 + \cdots + x^{n(p)-1})(p) = (1 - x^{n(p)})(p) = p$. Hence 1 - x is a unit in S. Let x and y be nonunits in S. We assume that x + y is a unit in S. Then we may assume that x + y = 1, which is a contradiction to the first observation. Therefore, S is a local ring.

8.3.10 THEOREM Let M be an R-module. We assume that $M = \Sigma_I \oplus P_\alpha = \Sigma_J \oplus Q_\beta$ are two relatively T-nilpotent decompositions of M. Then all P_α and Q_β are completely indecomposable modules, and hence these decompositions are unique up to isomorphism and every summand of M has the exchange property in M.

Proof: Put $I_1 = \{\alpha \in I \mid P_\alpha \approx Q_\beta$ for some $\beta \in J\}$ and $J_1 = \{\beta \in J \mid Q_\beta \approx P_\alpha$ for some $\alpha \in I\}$. We first show that $I_1 \neq \emptyset$, so $J_1 \neq \emptyset$. We assume the contrary. Let p_α and q_β be projections of M to P_α and Q_β, respectively. Let $x_1 \neq 0$ be in P_{α_1}. Then there exists a $\beta_2 \in J$ such that $q_{\beta_2}(x_1) = x_2 \neq 0$. Again there exists an α_3 in I such that $p_{\alpha_3}(x_2) = x_3 \neq 0$. Repeating this argument, we obtain a contradiction to the T-nilpotency, since $I_1 = J_1 = \emptyset$. Hence $I_1 \neq \emptyset$, so $J_1 \neq \emptyset$. Furthermore, $\{P_\alpha\}_{I_1}$, $\{Q_\beta\}_{J_1}$ are sets of completely indecomposable modules and locally T-nilpotent by (8.3.9) and the assumption. We put $M = \Sigma_{I_1} \oplus P_\alpha \oplus \Sigma_{I_1'} \oplus P_{\alpha'} = \Sigma_{J_1} \oplus Q_\beta \oplus \Sigma_{J_1'} \oplus Q_{\beta'}$, where $I_1' = I - I_1$ and $J_1' = J - J_1$. Let $\{P_{\alpha_i}\}_{i=1}^n$ be any finite subset of $\{P_{\alpha_i}\}_{I_1}$. Since $\Sigma_{i=1}^n \oplus P_{\alpha_i}$ has the exchange property by (8.1.4) and (8.4.5) below, we obtain

$$\text{(i)} \quad M = \sum_1^n \oplus P_{\alpha_i} \oplus \Sigma_{I_1'} \oplus P_{\alpha_i} = \Sigma_{J_1} \oplus Q_\beta \oplus \Sigma_{J_1'} \oplus Q_{\beta'}$$

where $Q_\beta = Q_\beta' \oplus Q_\beta''$. Then $\Sigma_{J_1} \oplus Q_\beta''$ is isomorphic to a direct summand of $\Sigma_1^n \oplus P_{\alpha_i}$. If $Q_{\beta'}'' \neq 0$, $Q_{\beta'}''$ contains a direct summand isomorphic to some P_{α_j} by (7.1.2), say $Q_{\beta'}'' = X \oplus Y$; $X \overset{\varphi}{\approx} P_{\alpha_j}$. Since $\beta' \in J_1'$, $Y \neq 0$. Let p be a projection of $Q_{\beta'}''$ to X and i the inclusion of X to $Q_{\beta'}''$; then p and i are not isomorphisms. However, $\varphi p i \varphi^{-1} = 1_{P_{\alpha_j}}$ and neither φp nor $i\varphi^{-1}$ is an isomorphism. This contradicts the relative T-nilpotency. Accordingly, $Q_{\beta'}'' = 0$, and $(\Sigma_1^n \oplus P_{\alpha_i}) \cap (\Sigma_{J_1'} \oplus Q_{\beta'}) = 0$. Therefore, $(\Sigma_{I_1} \oplus P_\alpha) \cap (\Sigma_{J_1'} \oplus Q_\beta) = 0$. Let $\bar{\bar{M}} = M/\Sigma_{J_1'} \oplus Q_{\beta'}$ and ψ the natural epimorphism of M to $\bar{M}$. Then $\bar{M} = \Sigma_{J_1} \oplus \psi(Q_\beta) \supset \Sigma_{I_1} \oplus \psi(P_\alpha)$ [note that $\psi(Q_\beta) \approx Q_\beta$ and $\psi(P_\alpha) \approx P_\alpha$]. On the other hand, $\Sigma_{I_1} \oplus \psi(P_\alpha)$ is a locally direct summand of M from (i), and $\{\psi(P_\alpha)\}_{I_1}$ is locally T-nilpotent. Hence $\Sigma_{I_1} \oplus \psi(P_\alpha)$ is a direct summand of M by (7.3.9), since $\{\psi(Q_\beta)\}_{J_1}$ is a set of c.ind. modules. Furthermore, $\Sigma_{I_1} \oplus \psi(P_\alpha)$ has the exchange property in $\bar{\bar{M}}$ by (8.2.3), so $\bar{\bar{M}} = \Sigma_{I_1} \oplus \psi(P_\alpha) \oplus \Sigma_{J_1''} \oplus \psi(Q_\beta)$, where $J_1'' \subset J_1$. Therefore, $M = \Sigma_{I_1} \oplus P_\alpha \oplus \Sigma_{J_1''} \oplus Q_{\beta'} \oplus \Sigma_{J_1'} \oplus Q_{\beta'} = \Sigma_{I_1} \oplus P_\alpha \oplus \Sigma_{I_1'} \oplus P_{\alpha'}$. Hence $\Sigma_{I_1''} \oplus Q_{\beta'} \oplus \Sigma_{J_1'} \oplus Q_{\beta'} \approx \Sigma_{I_1'} \oplus P_{\alpha'}$. However, $\{P_{\alpha'}\}_{I_1'}$ and $\{Q_{\beta'}\}_{J_1' \cup J_1''}$ are locally and relatively T-nilpotent, so some $P_{\alpha'}$ is isomorphic to some $Q_{\beta'}$ by the first observation, provided that $P_{\alpha_1'} \neq 0$. Therefore, $I_1' = J_1' = J_1'' = \emptyset$. The remaining assertions are clear by (8.2.3).

8.4 DIRECT SUMMANDS WITH THE $\aleph_0$-EXCHANGE PROPERTY

In the preceding section we studied the exchange property in a module M of a direct summand N of M which is a direct sum of completely indecomposable modules.

We investigate the $\aleph_0$-exchange property of N in this section. First we present some fundamental results on the exchange property for any module in $\underline{M}_R$. In the following, the capital roman letters denote right modules over a ring R.

8.4.1 LEMMA Let $A = \Sigma_I \oplus A_\alpha$ and $B_\alpha \subset A_\alpha$ for $\alpha \in I$. Then $A/\Sigma_I \oplus B_\alpha \approx \Sigma_I \oplus A_\alpha/B_\alpha$.

Proof: This is clear.

8.4.2 LEMMA Let $A = B \oplus C = \Sigma_I \oplus D_\alpha$. Let $D_\alpha' \subset D_\alpha$ and $A' = \Sigma_I \oplus D_\alpha'$. We assume that $A' \cap B = 0$ and $A/A' = \bar{B} \oplus \Sigma_I \oplus \bar{E}_\alpha$ $[\bar{B} = (B + A')/A', \bar{E}_\alpha \subset \bar{D}_\alpha = (D_\alpha + A')/A']$. Then there exists an R-module E_α' in D_α for all $\alpha \in I$ such that $A = B \oplus \Sigma_I \oplus E_\alpha'$ and $(E_\alpha' + D_\alpha')/D_\alpha' = \bar{E}_\alpha$.

Proof: It is clear that there exists E_α' in D_α such that $E_\alpha' \supset D_\alpha'$ and $E_\alpha'/D_\alpha' = \bar{E}$. Hence $A = B + \Sigma_I \oplus E_\alpha'$. Let b be in $B \cap \Sigma_I \oplus E_\alpha'$. Then $\bar{b} = 0$. Hence $b \in B \cap A' = 0$.

Similarly, we have the following.

8.4.3 LEMMA Let $A = B \oplus C \oplus E = \Sigma_I \oplus D_\alpha \oplus E$. We assume that $A/E = B \oplus \Sigma \oplus F_\alpha$, where $F_\alpha \subset D_\alpha$. Then $A = B \oplus \Sigma_I \oplus F_\alpha \oplus E$.

8.4.4 COROLLARY We assume that B has the a-exchange property. If $A = B \oplus C \oplus E = \Sigma_I \oplus D_\alpha \oplus E$ with $|I| \le a$, then $A = B \oplus \Sigma_I \oplus D_\alpha' \oplus E$ for some $D_\alpha' \subset D_\alpha$.

8.4.5 LEMMA Let a be a cardinal and n an integer < a. We assume that $B = \Sigma_{i=1}^n \oplus B_i$. Then B has the a-exchange property if and only if each B_i has.

Proof: We may show (8.4.5) in case n = 2. We assume that B_i has the a-exchange property. Let $A = B_1 \oplus B_2 \oplus C = \Sigma_I \oplus D_\alpha$ with $|I| \le a$. Then $A = B_1 \oplus \Sigma_I \oplus E_\alpha$; $E_\alpha \subset D_\alpha$. Hence $A = B_1 \oplus B_2 \oplus \Sigma_I \oplus F_\alpha$; $F_\alpha \subset E_\alpha \subset D_\alpha$ by (8.4.3), since B_2 has the a-exchange property. Conversely, we assume that $A = B_1 \oplus C = \Sigma_I \oplus D_\alpha$. Put $A' = B_1 \oplus B_2 \oplus C = \Sigma_I \oplus D_\alpha \oplus B_2$. Then $A' = B_1 \oplus B_2 \oplus \Sigma_I \oplus E_\alpha \oplus F$; $E_\alpha \subset D_\alpha$, $F \subset B_2$ [note that if a is finite, n may be large enough by (8.1.2)]. Since $F \subset B_2$, $F = 0$. Considering the natural epimorphism $A' \to A'/B_2$, we have $A = B_1 \oplus \Sigma_I \oplus E_\alpha$.

8.4.6 LEMMA We assume that B has the a-exchange property and a > 1. If B is generated by $\langle b_i \rangle_I$ with $|I| \le a$, then B has the exchange property.

Proof: Let $A = B \oplus C = \Sigma_K \oplus D_\alpha$. Then there exists an index set I with $|I| \le a$ such that $B \subset \Sigma_I \oplus D_\alpha$ and $\Sigma_I \oplus D_\alpha = B \oplus E$ by (1.7.4). Hence there exists $D'_\alpha \subset D_\alpha$ such that $B \oplus E = B \oplus \Sigma_I \oplus D'_\alpha$. Accordingly, $A = B \oplus \Sigma_I \oplus D'_\alpha \oplus \Sigma_{K-I} \oplus D_\gamma$.

8.4.7 LEMMA Let B be indecomposable. We assume that B is a direct summand of A and A is a direct sum of modules D_α with the 2-exchange property. Then B has the exchange property.

Proof: Let $A = \Sigma_I \oplus D_\alpha = B \oplus C$. Then there exists a finite subset J of I such that $(\Sigma_J \oplus D_\alpha) \cap B \neq 0$. Since $\Sigma_J \oplus D_\alpha$ has the finite exchange property by (8.1.2) and (8.4.5), $A = \Sigma_J \oplus D_\alpha \oplus B' \oplus C'$; $B' \subset B$ and $C' \subset C$. B being indecomposable, $B' = B$ or $B' = 0$. On the other hand, $B \cap \Sigma_J \oplus D_\alpha \neq 0$. Hence $B' = 0$. Therefore, $A/C' = \Sigma_J \oplus D_\alpha = B \oplus C''$; $C = C' \oplus C''$. Accordingly, B has the exchange property by (8.1.3) and (8.4.5).

We shall give a useful lemma.

8.4.8 LEMMA Let B be an R-module. We assume that $A = B \oplus C = \Sigma_I \oplus D_\alpha$; $|I| \leq a$ and D_α is isomorphic to a submodule of B for each α. If we have $A = B \oplus \Sigma_I \oplus D'_\alpha$; $D'_\alpha \subset D_\alpha$ for any decomposition above, then B has the a-exchange property.

Proof: Let $A = B \oplus C = \Sigma_I \oplus E_\alpha$. Put $C' = \Sigma_I \oplus (C \cap E_\alpha)$ and $E'_\alpha = E_\alpha \cap C$. Then $\bar{A} = A/C' = (B \oplus C)/C' = \Sigma_I \oplus E_\alpha/E'_\alpha$ by (8.4.1), and $\bar{E}_\alpha = E_\alpha/E'_\alpha$ is isomorphic to a submodule of B (take the projection on B). Hence, by the assumption, we have $\bar{A} = \bar{B} \oplus \Sigma \oplus E''_\alpha/E'_\alpha$, where $E'_\alpha \subset E''_\alpha \subset E_\alpha$. Therefore, $A = B \oplus \Sigma_I \oplus E''_\alpha$ by (8.4.2).

Next we study the a-exchange property in a given module.

8.4.9 LEMMA Let T be an R-module and $T = A_1 \oplus A_2 = M \oplus N$. We assume that $M = \Sigma_K \oplus M_\alpha$ and every M_α has the finite exchange property (in the usual sense) and $A_1 \cap M \neq 0$. Then there exists a finite subset $\{1,2,\dots,m\}$ in K such that $T = \Sigma_1^m \oplus M_1^* \oplus A_1^* \oplus A_2$, where $M_i^* \subset M_i$ ($M_j^* \neq 0$ for some j) and $A_1^* \subset A_1$.

Proof: There exists a finite subset $\{1,2,\dots,m\}$ in K such that $A_1 \cap (\Sigma_1^m \oplus M_i) \neq 0$. Put $M^* = \Sigma_1^m \oplus M_i$; then M^* has the finite exchange property by (8.4.5). Hence $T = M^* \oplus A'_1 \oplus A'_2$, where $A_i = A'_i \oplus A''_i$. Since $M^* \cap A_1 \neq 0$, $A'_1 \neq A_1$, so $A''_1 \neq 0$. Put $\bar{T} = T/(A'_1 \oplus A'_2) = \bar{A}''_1 \oplus \bar{A}''_2 = \bar{M}^*$. By (8.4.5), $\bar{A}''_2$ has the exchange property, and hence we obtain $\bar{A} = \bar{A}''_2 \oplus \Sigma_1^m \oplus \bar{M}'_1$. Therefore, $T = A'_1 \oplus \Sigma_1^m \oplus M'_1 \oplus A_2$ by (8.4.3).

By (8.4.3) we have the following.

8.4.10 LEMMA Let $T = B_1 \oplus B_2 \oplus B_3$ be an R-module. We assume that B_1 has the m-exchange property in T and B_2 has the m-exchange property in $B_2 \oplus B_3$. Then $B_1 \oplus B_2$ has the m-exchange property in T.

8.4.11 THEOREM Let $\{P_\alpha\}_I$ be an infinite set of R-modules which have the finite exchange property and $P = \Sigma_I \oplus P_\alpha$. Let I' be an infinite subset of I with infinite complement I - I'. We assume that $P_{I'} = \Sigma_{I'} \oplus P_\alpha$ has the 2-exchange property in P. We take any countable subsets $\{P_{2i-1}\}_1^\infty$ and $\{P_{2i}\}_1^\infty$ of $\{P_\alpha\}_{I'}$ and $\{P_\alpha\}_{I-I'}$, respectively, and any sets of homomorphisms $f_i : P_i \to P_{i+1}$ which satisfy one of the following two conditions for each i:

1. Let X be any direct summand of P_{2i-1}. Then $f_{2i-1}(X)$ is not a direct summand of P_{2i} provided that $f_{2i-1}(X) \neq 0$.
2. Let Y be any direct summand of P_{2i}. Then $f_{2i-1}^{-1}(Y)$ is not a direct summand of P_{2i-1} provided that $f_{2i-1}^{-1}(Y) \neq P_{2i-1}$.

Then there exists an integer n, depending on x in P, such that $f_n f_{n-1} \cdots f_1(x) = 0$.

Proof: We shall use the same notation as in the proof of (8.3.6), with M_α replaced by P_α. Put $P_i' = \{p_i + f_i(p_i) \mid p_i \in P_i\} \subset P_i \oplus P_{i+1}$. Then $P = \Sigma_1^\infty \oplus P_{2i-1}' \oplus P^{(1)} \oplus \Sigma_1^\infty \oplus P_{2i} \oplus P^{(2)} = \Sigma_1^\infty \oplus P_{2i-1} \oplus P^{(1)} \oplus \Sigma_1^\infty \oplus P_{2i}' \oplus P^{(2)}$. Since $\Sigma_1^\infty \oplus P_{2i-1}' \oplus P^{(1)}$ has the finite exchange property in P by the assumption and (8.1.2), we obtain from the decomposition above

$$\text{(a)}\quad P = \sum_1^\infty \oplus P_{2i-1}' \oplus P^{(1)} \oplus X \oplus Y \oplus Z$$

where $X \subset \Sigma_1^\infty \oplus P_{2i-1}$, $Y \subset \Sigma_1^\infty \oplus P_{2i}'$, and $Z \subset P^{(2)}$. We shall show X = 0. We assume on the contrary that $X \neq 0$. Then we have from (8.4.9),

$$\text{(b)}\quad P = \sum_1^t \oplus P_{2i-1}^* \oplus \sum_1^\infty \oplus P_{2i-1}' \oplus P^{(1)} \oplus X' \oplus Y \oplus Z$$

where $P_{2i-1}^* \subset P_{2i-1}$ ($P_{2j-1}^* \neq 0$ for some j) and $X' \subset X$. Consider the following modules and the decomposition of P:

$$\text{(c)}\quad P_1^* \oplus \sum_1^\infty \oplus P_{2i-1}' \oplus P^{(1)}$$

$$P = P_1 \oplus \left(\sum_{i\geq 2} \oplus P_{2i-1}\right) \oplus P^{(1)} \oplus P_2' \oplus \left(\sum_{i\geq 2} \oplus P_{2i}'\right) \oplus P^{(2)}$$

Since the former module has the finite exchange property in P by (8.4.10) and (8.4.5), we obtain from (c)

$$P = \left[P_1^* \oplus \sum_1^{\infty} \oplus P'_{2i-1} \oplus P^{(1)}\right] \oplus P_1^{**} \oplus A \oplus P_2'^{*} \oplus B \oplus C$$

where $P_1^* \subset P_1$, $A \subset \Sigma_{i\ 2} \oplus P_{2i-1}$, $P_2'^* \subset P_2'$, $B \subset \Sigma_{i\geq 2} \oplus P'_{2i}$, and $C \subset P^{(2)}$. Hence

(d) $P = (P_1^{\S} \oplus P_1') \oplus (P_2'^{*} \oplus D)$

where $P_1^{\S} = P_1^{**} \oplus P_1^*$ and $D = \Sigma_{i\geq 2} \oplus P'_{2i-1} \oplus P^{(1)} \oplus A \oplus B \oplus C \subset \Sigma_{i\geq 3} \oplus P_i$. Using only the fact that $D \subset \Sigma_{i\geq 3} \oplus P_i$ in (d), we shall show that $P_1^{\S} = 0$. Let x be in P_1 and $f_1(x) = 0$. Then $x \in P_1^{\S} \cap P_1' = 0$ by (d), so $f_1|P_1^{\S}$ is a monomorphism. Since P_2 is isomorphic to P_2', $P_2'^{*} = P_2^{*'}$, where $P_2^* \subset P_2$. Put $N = \{x \in P_2^* \mid f(x) \in D\}$. We shall show that $N \cap f_1(P_1^{\S}) = 0$. Take an element a in $N \cap f_1(P_1^{\S})$. Then $a = f_1(a_1)$ for some a_1 in $P_1^{\S}$. Hence $a_1 = a_1 + a - [a + f_2(a)] + f_2(a) \in P_1^{\S} \cap (P_1' \oplus P_2'^{*} \oplus D) = 0$ by (d). Therefore, $a = 0$. Let x be in P_2. Then $x = x_1 + x_2' + f_1(x_1') + x_2 + f_2(x_2) + d$ by (d), where $x_1 \in P_1^{\S}$, $x_1' \in P_1$, $x_2 \in P_2^*$, and $d \in D \subset \Sigma_{i\geq 3} \oplus P_i$. Hence $x_1 = -x_1'$, $x = x_2 + f_1(x_1')$, and $f_2(x_2) = -d$. Therefore,

(e) $P_2 = f_1(P_1^{\S}) \oplus N$

Furthermore,

(f) $P_1 = P_1^{\S} \oplus f_1^{-1}(N)$

since $f_1|P_1^{\S}$ is a monomorphism. Hence $P_1^{\S} = 0$ by (e), (f), and the assumptions. Therefore, $P_1^* = 0$. Next we consider, similarly to (c),

(c') $P_3^* \oplus \sum_1^{\infty} \oplus P'_{2i-1} \oplus P^{(1)}$

$$P = P_1 \oplus P_3 \oplus \left(\sum_{i\geq 3} \oplus P_{2i-1}\right) \oplus P^{(1)} \oplus P_2' \oplus P_4' \oplus \left(\sum_{i\geq 3} \oplus P'_{2i}\right) \oplus P^{(2)}$$

Then

$$P = \left[P_3^* \oplus \sum_1^{\infty} \oplus P'_{2i-1} \oplus P^{(1)}\right] \oplus P_1^{\triangle} \oplus P_3^{\triangle} \oplus A' \oplus P_2'^{*} \oplus P_4'^{*} \oplus B' \oplus C''$$

where $P_{2i}'^{*} \subset P'_{2i}$, $P^{\triangle}_{2i-1} \subset P_{2i-1}$, $A' \subset \Sigma_{i\geq 3} \oplus P_{2i-1}$, $B' \subset \Sigma_{i\geq 3} \oplus P'_{2i}$, and $C' \subset P^{(2)}$.

From the argument following (d), we know that $P_1^{\triangle} = 0$. Thus we have

$$(d')\quad P = (P_1' \oplus P_2'^{*}) \oplus [(P_3^{\S} \oplus P_3') \oplus (P_4'^{*} \oplus D')]$$

where $P_3^{\S} = P_3^{*} \oplus P_3^{\triangle}$ and $D' = \Sigma_{i\geq 3} \oplus P_{2i-1}' \oplus P^{(1)} = A' \oplus B' \oplus C' \subset \Sigma_{i\geq 5} \oplus P_i$. Applying the same argument for $P_3^{\S} \oplus P_3'$ and $P_4'^{*}$ in (d') as after (d), we obtain $P_3^{\S} = P_3^{*} = 0$. Continuing this argument, we have a contradiction of the assumption $X \neq 0$. Therefore, we have by (a),

$$(g)\quad P = \sum_1^{\infty} \oplus P_{2i-1}' \oplus P^{(1)} \oplus Y \oplus P^{(2)} \qquad Y \subset \sum_1^{\infty} \oplus P_{2i}'$$

Hence $\{f_i\}$ is locally T-nilpotent by (7.2.7).

Finally, we obtain the main theorem in this section.

8.4.12 THEOREM Let $\{M_\alpha\}_I$ be a set of completely indecomposable modules and $M = \Sigma_I \oplus M_\alpha$. We put $M_{I'} = \Sigma_{I'} \oplus M_{\alpha'}$ for some $I' \subset I$. Then the following statements are equivalent:

1. $M_{I'}$ has the 2-exchange property in M.
2. $M_{I'}$ has the finite exchange property in M.
3. $M_{I'}$ has the exchange property in M.
4. $M_{I-I''}$ has the exchange property in M.

Proof: (3) ⇒ (2) ⇒ (1). This is trivial. We assume (1). We shall show that $\{M_{\alpha'}\}_{I'}$ and $\{M_{\alpha''}\}_{I-I'}$ are relatively semi-T-nilpotent. We may assume that $I_1 = I'$ and $I_2 = I - I'$ are infinite. Let $\{M_{2i-1}\}_1^{\infty}$ and $\{M_{2i}\}_1^{\infty}$ be any countable subsets of $\{M_{\alpha'}\}_{I_1}$ and $\{M_{\alpha''}\}_{I_2}$, respectively, and $\{f_n : M_n \to M_{n+1}\}$ is a set of nonisomorphisms. We shall show that f_{2i-1} satisfies the assumptions in (8.4.11). Since M_i is c.ind., M_i has the (finite) exchange property by (8.1.4). If ker f_{2i-1} is a direct summand of M_{2i-1}, ker $f_{2i-1} = M_{2i-1}$ or ker $f_{2i-1} = 0$. The former case implies that $f_{2i-1} = 0$. We assume that ker $f_{2i-1} = 0$. If im f_{2i-1} is a direct summand of M_{2i}, then f_{2i-1} is an isomorphism. Hence im f_{2i-1} is not a direct summand of M_{2i}. Therefore, f_{2i-1} satisfies the assumption in (8.4.11). Hence $M_{I'}$ (resp. $M_{I-I'}$) satisfies the condition in (8.3.6). Therefore, $M_{I'}$ has the exchange property in M by (8.3.7).

8.4.13 COROLLARY Let $\{M_\alpha\}_I$ and M be as in (8.4.12). For any subset I' of I we put $M_{I'} = \Sigma_I \oplus M_{\alpha'}$. Then the following statements are equivalent:

1. $\{M_\alpha\}_I$ is locally semi-T-nilpotent.
2. $M_{I'}$ has the 2-exchange property in M for any $I' \subset I$.

3. $M_{I'}$ has the finite exchange property in M for any $I' \subset I$.
4. $M_{I'}$ has the exchange property in M for any $I' \subset I$.

Proof: This is clear by (8.2.1) and (8.4.12).

8.5 MODULES WITH THE EXCHANGE PROPERTY

In the preceding section we studied primarily modules with the exchange property in $\underline{A}$. In this section we study general modules with the exchange property over a ring R.

We defined a left or right suitable ring in Sec. 4.6. We shall give a connection between such rings and modules with the exchange property. Furthermore, we shall show that the definition of suitable ring is left- and right-sided symmetric.

8.5.1 THEOREM Let R be a ring. The following conditions are equivalent for an R-module M_R and $S = \mathrm{End}_R(M)$:

1. S is left suitable.
2. M has the finite exchange property.
3. S is right suitable.

Proof: (1) ⇒ (2). We show that M has the 2-exchange property, by (8.1.2). Assume that $X = M \oplus Y = N_1 \oplus N_2$. Let $\pi : X \to M$ and $\tau_i : X \to N_i$ be the projections following the above decompositions. Then $1 = \tau_1 + \tau_2$. Hence $\pi = \pi\tau_1\pi + \pi\tau_2\pi$. Since $S \approx \pi E\pi$ by (3.1.8), there exist orthogonal idempotents ν_i in $S \subset E$ such that $\pi = \nu_1 + \nu_2$; $\nu_i \in \pi E\pi\tau_i\pi$ by (4.6.6), where $E = \mathrm{End}_R(X)$. Let

$$\text{(a)} \quad \nu_i = \alpha_i\tau_i\pi \qquad \alpha_i \in \pi E\pi$$

We may assume that

$$\text{(b)} \quad \nu_i\alpha_i = \alpha_i \qquad \text{for } i = 1, 2$$

Put

$$\text{(c)} \quad \beta_i = \tau_i\alpha_i\tau_i$$

Then $\beta_i\beta_j = 0$ $(i \neq j)$ and $(\tau_i\alpha_i\tau_i)^2 = \tau_i\alpha_i\tau_i\alpha_i\tau_i = \tau_i(\alpha_i\tau_i\pi)\alpha_i\tau_i = \tau_i\nu_i\alpha_i\tau_i = \tau_i\alpha_i\tau_i$. Put $\beta = \beta_1 + \beta_2$. Since β is idempotent, $X = \mathrm{im}\,\beta \oplus \ker\beta$ by (3.1.6). On the other hand, $\beta|N_i = \beta_i$ for $i = 1, 2$. Hence $\ker\beta = \ker\beta_1 \oplus \ker\beta_2$ and $N_i = \mathrm{im}\,\beta_i \oplus \ker\beta_i$. Put $N_i' = \ker\beta_i$ and $N_i'' = \mathrm{im}\,\beta_i$. Then

(d) $X = N_1'' \oplus N_2'' \oplus N_1' \oplus N_2'$

We shall show $X = M \oplus \ker \beta = M \oplus N_1' \oplus N_2'$. Note that

(e) $\beta_i\pi = \tau_i\nu_i$

and so

(f) $\alpha_i\beta_i\pi = \alpha_i\tau_i\nu_i = (\alpha_i\tau_i\pi)\nu_i = \nu_i^2 = \nu_i$

Let x be in $M \cap (N_1' \oplus N_2')$. Then $x = x_1' + x_2'$; $x_i' \in N_i'$ and $\beta_i(x) = \beta_i(x_i) = 0$. Hence $x = \pi x = \Sigma_{i=1}^{2} \nu_i(x) = \Sigma\, \alpha_i\beta_i\pi(x) = 0$ by (f). Let x' be in X. Then $x' = x_1 + x_2 + y$, where x_i is in N_i'' and y in $N_1' \oplus N_2'$ by (d). Now we have

(g) $\beta_i\alpha_j\beta_j = \beta_i(\pi\alpha_j\beta_j) = (\beta_i\pi)(\nu_j\alpha_j)\beta_j = (\tau_i\nu_i)(\nu_j\alpha_j\beta_j)$
$= \delta_{ij}\tau_i\alpha_i\beta_i = \delta_{ij}\tau_i\alpha_i\tau_i\beta_i = \delta_{ij}\beta_i$

where δ_{ij} is the Kronecker delta. Hence since $x_i - \alpha_i x_i = \beta_i x_i - \alpha_i\beta_i x_i$, $x_i - \alpha_i x_i$ is in $\ker \beta$ by (g). Therefore, $x' = (x_1 - \alpha_1 x_1) + (x_2 - \alpha_2 x_2) + \alpha_1 x_1 + \alpha_2 x_2 + y \in M \oplus \ker \beta$ for $\alpha_i x_i \in M$.

(2) ⇒ (3). Let $1 = \alpha + \beta$ in S. We use the same method as in the proof of (8.1.4). Put $X = M \oplus M$, $D = \{(m,m) \mid m \in M\} \subset X$ and $M' = \{(\alpha(m), -\beta(m)) \mid m \in M\}$. Then $X = M' \oplus D$ and $M' \approx M$. Since $M \approx M'$, M' has the 2-exchange property and thus

(h) $X = M' \oplus M_1 \oplus M_2$ where $M_1 \subset M \oplus 0$ and $M_2 \subset 0 \oplus M$

Let $x = (m,m)$. Then from (h) we have

(i) $(m,m) = (\alpha(n), -\beta(n)) + (m_1,0) + (0,m_2)$

By (i) we can define

(k) $\alpha'(m) = m_2$ $\beta'(m) = m_1$ α', β' are in S

On the other hand, we have

(ℓ) $(\alpha\alpha'(m),\alpha\alpha'(m)) = (\alpha\alpha'(m), -\beta\alpha'(m)) + (0,0) + (0,\alpha'(m))$
$(\beta\beta'(m),\beta\beta'(m)) = (-\alpha\beta'(m), \beta\beta'(m)) + (\beta'(m),0) + (0,0)$

Hence $\alpha'\alpha\alpha' = \alpha'$ and $\beta'\beta\beta' = \beta'$ from (k) and (ℓ). Put $e_1 = \alpha\alpha'$ and $e_2 = \beta\beta'$. Then $e_i^2 = e_i$ and $e_1 \in \alpha S$, $e_2 \in \beta S$. Furthermore, from (i) we have $m = \alpha(n) + \beta'(m) = -\beta(n) + \alpha'(m)$. Hence $n = \alpha(n) + \beta(n) = \alpha'(m) - \beta'(m)$, so $m = \alpha(\alpha'(m) - \beta'(m)) + \beta'(m) = \alpha\alpha'(m) - (1 - \beta)\beta'(m) + \beta'(m) = (\alpha\alpha' + \beta\beta')(m)$. Thus $1 = \alpha\alpha' + \beta\beta'$ and S is right suitable.

(3) ⇒ (1). If R is a ring, $\mathrm{End}_R(R_R) \approx R \approx \mathrm{End}_R({}_RR)$. Hence ${}_SS$ has the finite exchange property by (1) ⇒ (2). Therefore, S is left suitable by (2) ⇒ (3), for $S \approx \mathrm{End}_S({}_SS)$.

By (8.5.1) we know that the concept of suitable ring is left- and right-sided symmetric.

If a ring R has the finite exchange property as a right R-module, then R has the exchange property by (8.4.6). In that case R also has this property as a left R-module. We call such a ring an *exchange ring*. However, an exchange ring is nothing but a suitable ring.

If R/J(R) is a regular ring in the sense of von Neumann and idempotents can be lifted modulo J(R), we say that R is *semiregular*.

8.5.2 PROPOSITION Every semiregular ring is suitable and hence is an exchange ring.

Proof: By (4.6.4) we may assume that R is regular. Let x be in R. Then $xyx = x$ for some $y \in R$. Put $e = xy \in xR$. Then $e^2 = e$ and $1 - e \in (1 - x)R$.

8.5.3 COROLLARY Let $\{M_\alpha\}_I$ be a set of c.ind. modules and $M = \Sigma_I \oplus M_\alpha$. If $\{M_\alpha\}_I$ is locally semi-T-nilpotent, M has the finite exchange property.

Proof: It is clear from (8.2.1), (8.5.1), and (8.5.2).

8.5.4 PROPOSITION Let M_R be an R-module. Then M has the finite exchange property if and only if, for any α in $S = \mathrm{End}_R(M)$, there exist γ and σ in S such that $\gamma\alpha\gamma = \gamma$, $(1 - \alpha\gamma)(1 - \alpha)\sigma = 1 - \alpha\gamma$.

Proof: We assume that such γ and σ exist for α in S. Put $e = \alpha\gamma \in \alpha S$. Then $e^2 = e$ and $(1 - e)(1 - \alpha)\sigma(1 - e) = 1 - e$ by direct computation. Put $f = 1 - (1 - \alpha)\sigma(1 - e)$. Then $f^2 = f = ef \in \alpha S$ from the above equality. Hence $1 - f \in (1 - \alpha)S$. Therefore, S is suitable. Conversely, let $e^2 = e \in \alpha S$ and $1 - e \in (1 - \alpha)S$ for α. Let $e = \alpha x$ and $1 - e = (1 - \alpha)\sigma$. Then we put $\gamma = x\alpha x$. These σ and γ satisfy the conditions in (8.5.4).

Next we consider quasi-injective modules.

8.5.5 THEOREM Every quasi-injective module has the exchange property.

Proof: Let M be quasi-injective and $A = M \oplus N = \Sigma_J \oplus B_\alpha$. In order to show (8.5.5), we may assume that B is isomorphic to a submodule of M by (8.4.8). In this case we note that any f in $\mathrm{Hom}_R(C,M)$ is extendible to an element in $\mathrm{Hom}_R(B_\alpha, M)$ for any $C_R \subset B_\alpha$. We consider the set of submodules $A'' = \Sigma_I \oplus B''_\alpha$; $B''_\alpha \subset B_\alpha$ and $A'' \cap M = 0$. We can find a maximal member among the above A'' by Zorn's lemma, say $A' = \Sigma_I \oplus B'_\alpha$. Put $\bar{A} = A/A'$. Then $\bar{A} = \Sigma \oplus \bar{B}_\alpha \supset \bar{M} \approx M$. Assume that $(\bar{B}_\alpha \cap \bar{M}) \cap \bar{C}_\alpha = \bar{M} \cap \bar{C}_\alpha = 0$ for some $B'_\alpha \subset C_\alpha \subset B_\alpha$. Then $\overline{(C_\alpha \oplus \Sigma_{I-\alpha} \oplus B'_\gamma) \cap M} \subset \bar{C}_\alpha \cap \bar{M} = 0$ which implies that $(C_\alpha \oplus \Sigma_{I-\alpha} \oplus B'_\gamma) \cap M \subset A'$, so $(C_\alpha \oplus \Sigma_{I-\alpha} \oplus B'_\gamma) \cap M = (C_\alpha \oplus \Sigma_{I-\alpha} \oplus B'_\gamma) \cap M \cap A' = 0$. Hence $C_\alpha = B'_\alpha$ by the maximality of A'. Therefore, $\bar{B}_\alpha \cap \bar{M} = \overline{B_\alpha \cap M}$ is essential in $\bar{B}_\alpha$ and $\bar{M}$ is essential in $\bar{A}$ by (3.4.6). Take the projections $p_\alpha : (M \cap B_\alpha) \oplus B'_\alpha \to M \cap B_\alpha$. Then p_α is extendible to $f_\alpha : B_\alpha \to M$ by the initial remark. Then $\ker f_\alpha \supset B'_\alpha$ and $f_\alpha|(M \cap B_\alpha)$ is a monomorphism. Hence $\ker f_\alpha = B'_\alpha$. Therefore, we may assume that $\bar{B}_\alpha \subset M$. Let $\pi_\alpha : \bar{A} \to \bar{B}_\alpha$ be the projection. Since M is quasi-injective and essential in $\bar{A}$, $\pi_\alpha(M) \subset M$ by (5.4.2). Hence $M = \Sigma_I \oplus M_\alpha$; $M_\alpha = \bar{B}_\alpha \cap M$. Take the inclusion $i_\alpha : M_\alpha \to M$. Then there exists $g_\alpha : \bar{B}_\alpha \to M$ with $g_\alpha|M_\alpha = i_\alpha$. Let $\rho_\alpha : M \to M_\alpha$ be the projection. Then $\rho_\alpha g_\alpha : \bar{B}_\alpha \to M_\alpha$ and $\rho_\alpha g_\alpha|M_\alpha = \rho_\alpha i_\alpha|M_\alpha = 1_{M_\alpha}$. Hence $\bar{B}_\alpha = M_\alpha \oplus \ker \rho_\alpha g_\alpha$, so $\bar{B}_\alpha = M_\alpha$. Therefore, $A = M \oplus \Sigma \oplus B'_\alpha$ by (8.4.2).

8.5.6 COROLLARY Let E be a quasi-injective module. If $\Sigma_{i=1}^\infty \oplus E_{2i}$ has the finite exchange property in $\Sigma_{i=1}^n \oplus E_i$, $E_i \approx E$, then the radical of $\mathrm{End}_R(E)$ is locally T-nilpotent on E.

Proof: This is clear from (5.4.2) and (8.4.11).

We obtain a similar proposition.

8.5.7 PROPOSITION Let P be a projective module with finite exchange property. If $\Sigma_{i=1}^\infty \oplus P_{2i}$ has the finite exchange property in $\Sigma_{i=1}^\infty \oplus P_i$, $P_i \approx P$, then the radical of $\mathrm{End}_R(P)$ is locally T-nilpotent on P.

Proof: This is clear from (6.1.2) and (8.4.11).

Finally, we study the exchange property for modules which are direct sums of c.ind. modules. Let $\{M_\alpha\}_I$ be a set of c.ind. modules and $M = \Sigma_I \oplus M_\alpha$. We denote the category induced from $\{M_\alpha\}_I$ by $\bar{A}$ (Sec. 6.3). We assume a strong condition on M.

For any submodule N of M, which is isomorphic to some M_β in $\{M_\alpha\}$, N is a direct summand of M. (8.5.1)

If each M_α is injective, then (8.5.1) is satisfied. We shall study M with (8.5.1) in the remaining part of this section.

8.5.8 LEMMA Let M and $\{M_\alpha\}_I$ be as after (8.5.7). We assume that $\{M_\alpha\}_I$ is locally semi-T-nilpotent and that M satisfies (<u>8.5.1</u>). Let A be a submodule of M. Then we have $A = A_1 \oplus A_2$, where A_1 is a direct summand of M (and hence A_1 is in $\underline{A}$) and A_2 does not contain any submodule isomorphic to some module in $\{M_\alpha\}_I$.

Proof: Let $\underline{S}$ be the set of submodules A' in $\underline{A}$ as follows: A' is in $\underline{A}$, say $A' = \Sigma_J \oplus A_\alpha$, where the A_α are isomorphic to some member in $\{M_\alpha\}_I$ and A' is a locally direct summand of M with respect to this decomposition. Then we can find a maximal element in $\underline{S}$ by Zorn's lemma, say A_1, with respect to the partial order defined by the members of direct components. Since $\{M_\alpha\}_I$ is locally semi-T-nilpotent, A_1 is a direct summand of M, $M = A_1 \oplus N_1$ by (7.3.9). Hence $A = A_1 \oplus (A \cap N_1)$ and $A \cap N_1$ does not contain any submodule in $\underline{A}$ from the assumption and the maximality of A_1.

8.5.9 LEMMA Let $\{M_\alpha\}_I$ and M be as in (8.5.8). We assume that M satisfies (<u>8.5.1</u>). Then M has the exchange property if and only if $\{M_\alpha\}_I$ is locally semi-T-nilpotent.

Proof: If M has the exchange property, then $\{M_\alpha\}_I$ is locally semi-T-nilpotent by (8.2.1). Conversely, we assume that $\{M_\alpha\}_I$ is locally semi-T-nilpotent. Let $A = M \oplus N = \Sigma_{\alpha\in J} \oplus A_\alpha$. We may assume by (8.4.8) that all A_α are isomorphic to submodules in M. Then from the assumption and (8.5.8), $A_\alpha = A_\alpha' \oplus A_\alpha''$, where $A_\alpha' \in \underline{A}$ and A_α'' does not contain any submodule isomorphic to some member in $\{M_\alpha\}_I$. Put $A' = \Sigma_J \oplus A_\alpha'$ and $A'' = \Sigma_J \oplus A_\alpha''$; then $A = A' \oplus A''$. Let $\varphi : A \to A/A''$ be the natural epimorphism. We shall show that M is a locally direct summand of A/A'' through φ. Let I' be a finite subset of I and $M' = \Sigma_{I'} \oplus M_\alpha$. Since M' has the exchange property by (8.1.4) and (8.4.5), $A = M' \oplus A_0' \oplus A_0''$, where $A_0' \subset A'$ and $A_0'' \subset A''$. Then $A'' = A_0'' \oplus K''$ and K'' is isomorphic to a direct summand of M'. If $K'' \neq 0$, K'' contains a completely indecomposable module K_1 (isomorphic to some member in $\{M_\alpha\}_I$) as a direct summand by (7.1.2). Since K_1 has the exchange property, we know from the argument above that some A_α'' contains a submodule isomorphic to K_1, which is a contradiction. Hence $A = M' \oplus A_0' \oplus A''$ and $\varphi(M) \approx M$ is a locally direct summand of A/A''. Since $A/A'' \approx A' \in \underline{A}$ and $\{M_\alpha\}_I$ is locally semi-T-nilpotent, $\varphi(M)$ is a direct summand of A/A'' by (7.3.9); i.e., $A/A'' = \varphi(M) \oplus \varphi(K)$ and $K \subset A'$. Furthermore, $\varphi(M)$ has the exchange property in $\underline{A}$ by (8.2.1), and hence $A/A'' = \varphi(M) \oplus \Sigma_J \oplus \varphi(A_\alpha''')$, where $A_\alpha''' \subset A_\alpha'$. Therefore, $A = M \oplus \Sigma_J \oplus (A_\alpha''' \oplus A_\alpha'')$ by (8.4.2).

Next we shall consider particular cases where M satisfies (<u>8.5.1</u>).

8.5.10 LEMMA Let $\{M_\alpha\}_I$, M be as in (8.5.8). Let N be a submodule of M isomorphic to some M_β in $\{M_\alpha\}_I$ and let $i : N \to M$ be the inclusion. Then N is a direct summand of M if and only if $p_\alpha i$ is an isomorphism for some α in I, where p_α is the projection of M to M_α.

Proof: This is clear.

8.5.11 PROPOSITION Let M_1 be a completely indecomposable module. Then $M = M_1^{(I)} = \Sigma_I \oplus M_\alpha$, $M_\alpha \approx M_1$, has the exchange property for any set I if and only if M_1 is itself locally T-nilpotent.

Proof: We shall show that M satisfies (8.5.1). We may assume that $N = M_1$ in the following. We put $f_\alpha = p_\alpha i$ and assume that f_α is not an isomorphism for all $\alpha \in I$. Let $m \neq 0$ be in N and $i(m) = \Sigma_{i=1}^n f_{\alpha_i}(m)$. Since i is a monomorphism, we may assume that $f_{\alpha_1}(m) = m_2 \neq 0$. Let $i(m_2) = \Sigma_{i=1}^{n'} f_{\alpha'_i}(m_2)$. Repeating this argument, we obtain the sequence $\{f_{\beta_i}\}$ such that $f_{\beta_n} f_{\beta_{n-1}} \cdots f_{\beta_1}(m) \neq 0$ for any n, which contradicts the T-nilpotency of $\{M_1\}$. Therefore, M satisfies (8.5.1) by (8.5.10). The converse is clear by (8.2.3).

8.5.12 THEOREM Let R be a ring. Then the following statements are equivalent:

1. R is a right perfect ring.
2. For every projective module P, we obtain that:
   (i) P has the exchange property.
   (ii) $\mathrm{End}_R(P)/J(\mathrm{End}_R(P))$ is a regular ring in the sense of von Neumann.
3. Put $P_0 = \Sigma_1^\infty \oplus R$. Then we obtain that:
   (i) P_0 has the exchange property.
   (ii) $\mathrm{End}_R(P_0)/J(\mathrm{End}_R(P_0))$ is a regular ring.
4. (i) P_0 has the exchange property.
   (ii) $R/J(R)$ is artinian.

Proof: (1) ⇒ (2). Let R be perfect and $R = \Sigma_{i=1}^n \oplus e_i R$, where $\{e_i\}$ is a complete set of mutually orthogonal primitive idempotents (Sec. 7.6). Let P be a projective R-module. Then $P \approx \Sigma_i \oplus (e_i R)^{(I_i)}$ and $\{e_i R\}$ is a T-nilpotent set of c.ind. modules by (7.5.1) and (7.6.10). Hence $e_i R^{(I_i)}$ has the exchange property by (8.5.11) and P does by (8.4.5). (ii) is obtained in (8.2.1).

(2) ⇒ (3). This is clear.

(3) ⇒ (4). Since P_0 has the exchange property, J(R) is right T-nilpotent by (8.5.7). Furthermore, $\mathrm{End}_R(P_0)$ is isomorphic to the ring of column-finite matrices over R with degree $\aleph_0$ by (3.7.4). Since J(R) is

right T-nilpotent, $J(\mathrm{End}_R(P_0))$ is isomorphic to the subring of column-finite matrices over $J(R)$ by (7.4.5). Hence $\mathrm{End}_R(P_0)/J(\mathrm{End}_R(P_0))$ is isomorphic to the ring of column-finite matrices over $R/J(R)$. Therefore, $R/J(R)$ is artinian by (6.5.4).

(4) ⇒ (1). This is clear by (7.6.1) and (8.5.7).

8.5.13 PROPOSITION Let R be a semiperfect ring and P a projective R-module. Then P is semiperfect if and only if P has the exchange property.

Proof: Since R is semiperfect, P is isomorphic to a module $\Sigma_{i=1}^{n} \oplus (e_iR)^{(J_i)}$ by (7.6.10). If P is semiperfect, $\{e_iR\}_{J_i}$ is semi-T-nilpotent by (7.5.1). Hence P has the exchange property by (8.4.5), (8.5.1), and (8.5.11). The converse is clear by (8.2.1).

Next, we add some remarks concerning (8.5.1).

8.5.14 LEMMA Let $\{M_\alpha\}_I$, M, and N be as in (8.5.1). If N is uniform, $p_\alpha i$ is a monomorphism for some α.

Proof: Let $n \neq 0$ be in N and $n = \Sigma_{i=1}^{n} m_{\alpha_i}$. Put $M_0 = \Sigma_{I-\{\alpha_i\}} \oplus M_\alpha$ and let $p_0 : M \to M_0$ be its projection. Then $0 = [\cap \ker(p_{\alpha_j} i)] \cap \ker(p_0 i)$. Since N is uniform and $\ker(p_0 i) \neq 0$, $\ker(p_{\alpha_j} i) = 0$ for some α_j.

8.5.15 PROPOSITION Let $\{M_\alpha\}_I$ and M be as in (8.5.1). We assume that all M_α are uniform and each M_α is not isomorphic to a proper submodule in M_β for all α, β (e.g., all M_α are injective). Then M has the exchange property if and only if $\{M_\alpha\}_I$ is a locally semi-T-nilpotent set.

Proof: This is clear from (8.5.9) and (8.5.14).

8.5.16 PROPOSITION We assume that all M_α are uniform and $\|M_\alpha\| \leq n$ for all α. Then $M = \Sigma_I \oplus M_\alpha$ has the exchange property.

Proof: Put $M(i) = \Sigma \oplus M_\gamma$, where $\|M_\gamma\| = i$. Then $M(i)$ satisfies (8.5.1) by (8.5.14). On the other hand, $\{M_\alpha\}_I$ is T-nilpotent by (4.5.21). Hence M has the exchange property by (8.5.9).

8.5.17 LEMMA Assume that a module N has the finite exchange property. Let $N \oplus N' = \Sigma_{i=1}^{\infty} \oplus A_i$ and $K_i = \Sigma_{j>i}^{\infty} \oplus A_j$. Then there exist direct summands A_i' and K_{i+1}' of A_i and K_{i+1}, respectively, such that $N \oplus N' = N \oplus \Sigma_{i=1}^{n} \oplus A_i' \oplus K_{n+1}'$ for any n.

Proof: We assume that we have found A_i' and K_{i+1}' $(i = 1, 2, \ldots, n)$ such that

$$\text{(m)} \quad N \oplus N' = N \oplus \sum_{i=1}^{m} \oplus A_i' \oplus K_{m+1}' \qquad \text{for all } m \leq n$$

We can set $K_{n+1} = K'_{n+1} \oplus K''_{n+1}$ with $K''_{n+1} \subset N \oplus \Sigma_{i=1}^{n} \oplus A'_i$. Then

$$(n)\quad N \oplus \sum_{i=1}^{n} \oplus A'_i = K''_{n+1} \oplus L$$

with some L. Let $A_i = A'_i \oplus A''_i$ for $i \le n$. Since $N \approx \Sigma_{i=1}^{n} \oplus A''_i \oplus K''_{n+1}$, K''_{n+1} has the finite exchange property by (8.4.5). Then, since $K_{n+1} = A_{n+1} \oplus K_{n+2}$, there exist direct summands A'_{n+1} and K'_{n+2} of A_{n+1} and K_{n+2}, respectively, such that

$$(o)\quad K_{n+1} = K''_{n+1} \oplus A'_{n+1} \oplus K'_{n+2}$$

Thus

$$N \oplus N' = N \oplus \sum_{i=1}^{n} \oplus A'_i \oplus K'_{n+1} = K''_{n+1} \oplus L \oplus K'_{n+1} = K_{n+1} \oplus L$$

$$= K''_{n+1} \oplus A'_{n+1} \oplus K'_{n+2} \oplus L = N \oplus \sum_{i=1}^{n+1} \oplus A'_i \oplus K'_{n+2}$$

We rearrange $\{M_\alpha\}_I$ as follows: $\{M_{\alpha\beta}\}_{\alpha\in K, \beta\in J_\alpha}$ such that $M_{\alpha\beta} \approx M_{\alpha\beta'}$ and $M_{\alpha\beta} \not\approx M_{\alpha'\beta'}$ if $\alpha \ne \alpha'$. Put $K^{(1)} = \{\alpha \in K \mid |J_\alpha| < \aleph_0\}$, $K^{(2)} = \{\alpha \in K \mid |j_\alpha| \ge \aleph_0\}$, and $M^{(i)} = \Sigma_{\alpha\in K^{(i)}, \beta\in J_\alpha} \oplus M_{\alpha\beta}$, where $|K|$ means the cardinal of K. Then $M = M^{(1)} \oplus M^{(2)}$.

8.5.18 THEOREM Let $M^{(1)}$ be as above. Then $M^{(1)}$ has the $\aleph_0$-exchange property if $\{M_\alpha\}_I$ is locally semi-T-nilpotent.

Proof: Let $M = \Sigma_{K^{(1)}} \Sigma_J \oplus M_{\alpha\beta}$. Then M has the finite exchange property by (8.5.3). Assume that $A = \Sigma_{i=1}^{\infty} \oplus A_i = M \oplus M'$ and put $K_i = \Sigma_{j=i}^{\infty} \oplus A_j$. By (8.5.17) there exist some A'_i and K'_{n+1} such that $A_i = A'_i \oplus A''_i$, $K_{i+1} = K'_{i+1} \oplus K''_{i+1}$ and $A = M \oplus \Sigma_{i=1}^{n} \oplus A'_i \oplus K'_{n+1}$ for any n. Since $(\Sigma_{i=1}^{n} \oplus A''_i) \oplus K''_{n+1} \approx M$, by (7.2.6) we have

$$(p)\quad A''_i = \sum_{K^{(1)}} \sum_{J'_\alpha(i)} \oplus M'_{\alpha\beta}(i)$$

where $J'_\alpha(i) \subset J$ and $M'_{\alpha\beta} \approx M_{\alpha\beta}$. We suppose that $M^* = M \oplus \Sigma_{i=1}^{\infty} \oplus A'_i \ne A$. Then noting that $A = \Sigma_{i=1}^{\infty} \oplus A'_i \oplus \Sigma_{i=1}^{\infty} \oplus A''_i$, we can find $M'_{\alpha_1\beta_1}(i_1)$ $(= B_1)$ not contained in M^*. Now let $a_1 \in B_1 - M^*$ and take the projection $p : A \to K'_{m+1}$ with respect to the decomposition $A = M \oplus \Sigma_{i=1}^{m} \oplus A'_i \oplus K'_{m+1}$, where $m = i_1$. Then $x_1 = p(a_1) \notin M^*$. For $x_1 \in K_{m+1} = \Sigma_{i=m+1} \oplus A_i \oplus \Sigma_{i=m+1} \oplus A''_i$, we can write as $x_1 = \Sigma x'_i + \Sigma x''_i$; $x'_i \in A'_i$ and $x''_i \in A''_i$. There exists some

$i_2 > i_1$ such that $x''_{i_2} \notin M^*$. From (p) we can find some $M'_{\alpha_2\beta_2}(i_2)$ $(= B_2)$ such that $q(x_1) \in B_2 - M^*$, where $q : A \to B_2$ is the projection with respect to the decomposition $A = \Sigma_{i=1}^{\infty} \oplus A'_1 \oplus \Sigma_{K(i)}\Sigma_{J'_\alpha(i)} \oplus M'_{\alpha\beta}(i)$. Putting $g_1 = gp|B_1 : B_1 \to B_2$, we have $a_2 = g_1(a_1) \in B_2 - M^*$. Repeating the same argument for a_2, we can find $g_k : B_k \to B_{k+1}$ $[B_k = M'_{\alpha_k\beta_k}(i_k),\ i_k < i_{k+1}]$ such that $g_k \cdots g_1(a_1) \notin M^*$. Since $|J'(i)| < \infty$, there exists a natural number k_1 such that $f_1 = g_{k_1} \cdots g_1 : B_1 \to B_{k_1+1}$ is not an isomorphism. Repeating the same procedure, we can find natural numbers $k_1 < k_2 < \cdots$ such that $f_n = g_{k_n} \cdots g_{k_{n-1}}$ is not an isomorphism for any n. Obviously, $\{f_i\}$ may be regarded as a sequence of nonisomorphisms for $\{M_\alpha\}_I$ and $f_n \cdots f_1(a_i) \neq 0$ for any n. This contradiction shows that $M \oplus \Sigma_{i=1}^{\infty} \oplus A'_i = A$.

8.5.19 COROLLARY If $|K^{(2)}| < \infty$, $M = M^{(1)} \oplus M^{(2)}$ has the $\aleph_0$-exchange property provided that $\{M_\alpha\}_I$ is locally semi-T-nilpotent.

Proof: Since $\{M_{\alpha\beta}\}$ is locally semi-T-nilpotent and $|K^{(2)}| < \infty$, $M^{(2)}$ has the $\aleph_0$-exchange property by (8.5.11) and (8.4.5). Hence M has the same property by (8.4.5).

8.5.20 THEOREM Let M and $\{M_\alpha\}_I$ be as above (8.5.1). We assume that each M_α is f.g. and $\{M_\alpha\}_I$ is locally semi-T-nilpotent, then M has the $\aleph_0$-exchange property.

Proof: Let $A = \Sigma_{i=1}^{\infty} \oplus A_i = M \oplus M'$. As in the proof of (8.5.18) we can find A'_i and K'_{n+1} such that $A_i = A'_i \oplus A''_i$, $\Sigma_{j=i+1}^{\infty} \oplus A_j = K_{i+1} = K'_{i+1} \oplus K''_{i+1}$ and

$$\text{(q)}\quad A = M \oplus \sum_{i=1}^{n} \oplus A''_i \oplus K'_{n+1} \qquad \text{for any } n$$

Let

$$\text{(r)}\quad p : A \to \sum_{j=1}^{n} \oplus A''_j$$

be the projection with respect to the decomposition $A = \Sigma_1^{\infty} \oplus A'_i \oplus \Sigma_1^{\infty} \oplus A''_i$. First we claim that if J is any finite subset of I, then $p(\Sigma_J \oplus M_\alpha)$ is a direct summand of $\Sigma_{i=1}^{\infty} \oplus A''_i$. Since $M^* = \Sigma_J \oplus M_\alpha$ is finitely generated, $M^* \subset \Sigma_{i=1}^{m} \oplus A'_i \oplus \Sigma_{i=1}^{m} \oplus A''_i$ for some m. Accordingly, if

$$\text{(s)}\quad q : A \to \sum_{i=1}^{m} \oplus A''_i \oplus K''_{i+1}$$

is the projection with respect to the decomposition $A = \Sigma_{i=1}^{m} \oplus A_i' \oplus K_{m+1}' \oplus \Sigma_{i=1}^{m} \oplus A_i'' \oplus K_{m+1}''$, then $p(M^*) = q(M^*)$. Since $q|M$ is a monomorphism by (q), $q(A) = q(M) = q(M^*) \oplus L = p(M^*) \oplus L$ with some L by (q), so $p(M^*)$ is a direct summand of $\Sigma_{i=1}^{m} \oplus A_i''$, and hence of $\Sigma_{i=1}^{\infty} \oplus A_i''$. Next, as in the proof of (8.5.18) we have by (p),

$$\text{(t)} \quad A_i'' = \sum_{I'} \oplus M_\alpha'$$

where $M_\alpha' \approx M_\alpha$. Since $\{M_\alpha\}_I$ is locally semi-T-nilpotent, $p(M)$ is a direct summand of $\Sigma_{i=1}^{\infty} \oplus A_i''$ by (7.3.9). On the other hand, $p(M)$ has the exchange property with respect to completely indecomposable modules by (8.2.3). Hence $\Sigma_{i=1}^{\infty} \oplus A_i'' = p(M) \oplus \Sigma_{i=1}^{\infty} \oplus A_i'''$ with some $A_i''' \subset A''_i$, so $A = \Sigma_{i=1}^{\infty} \oplus A_i' \oplus p(M) \oplus \Sigma_{i=1}^{\infty} \oplus A_i'''$. Noting that $p|M$ is a monomorphism, we obtain $A = M \oplus \Sigma_{i=1}^{\infty} \oplus (A_i' \oplus A_i''')$.

8.6 EQUIVALENCE REFINEMENTS

We have dealt with some modules with the exchange property. In this section we give the concept of the equivalence refinement, which is the origin of the idea of the exchange property.

Let M_R be a module over a ring R. We take two decompositions of M:

$$M = \Sigma_I \oplus A_\alpha = \Sigma_J \oplus B_\alpha \tag{8.6.1}$$

If there exist direct decompositions $A_\alpha = \Sigma_{I_\alpha} \oplus A_{\alpha\gamma}$, $B_\beta = \Sigma_{J_\beta} \oplus B_{\beta\delta}$ for all $\alpha \in I$ and all $\beta \in J$ such that there is a one-to-one mapping $\phi : \{A_{\alpha\gamma}\}_{I,I_\alpha} \to \{B_{\beta\delta}\}_{J,J_\beta}$ and $\phi(A_{\alpha\gamma}) \approx A_{\alpha\gamma}$, then we say that the two decompositions in (8.6.1) have the *equivalence refinement*.

8.6.1 PROPOSITION Let M_R be an R-module. Let $M = \Sigma_{i=1}^{n} \oplus B_i = \Sigma_J \oplus C_j$ be two decompositions of M. We assume that M has the a-exchange property and $|J| \leq a$. Then the above decompositions have the equivalence refinement.

Proof: Each B_i has the a-exchange property by (8.4.5). Hence we obtain

$$\text{(a)} \quad M = B_1 \oplus \Sigma_J \oplus C_{1j}' \qquad C_j = C_{1j} \oplus C_{1_j}' \text{ and } B_1 \approx \Sigma_J \oplus C_{1j}$$

Further we obtain from (a),

$$\text{(b)} \quad M = B_1 \oplus B_2 \oplus \Sigma_J \oplus C_{2j}' \qquad C_{ij} = C_{2j} \oplus C_{2j}' \text{ and } B_2 \approx \Sigma_J \oplus C_{2j}$$

Repeating the argument, we obtain

$$M = B_1 \oplus B_2 \oplus \cdots \oplus B_i \oplus \Sigma_J \oplus C'_{ij} \qquad C_{i-1j} = C_{ij} \oplus C'_{ij} \text{ and}$$
$$B_i \approx \Sigma_J \oplus C_{ij}$$

8.6.2 PROPOSITION Let $M = B_0 \oplus B_1 \oplus \cdots \oplus B_n \oplus \cdots = C_0 \oplus C_1 \oplus \cdots \oplus C_n \oplus \cdots$ be two direct decompositions with countably many direct summands. If all B_i and all C_j have the $\aleph_0$-exchange property, then the above decompositions have the equivalence refinement.

Proof: Since B_0 has the $\aleph_0$-exchange property,

(c) $M = B_0 \oplus C'_{00} \oplus C'_{01} \oplus \cdots \qquad C_j = C_{0j} \oplus C'_{0j}$ and $B_0 \approx C_{00} \oplus C_{01} \oplus \cdots,\ j \geq 0$

From (8.4.5) $B_0 \oplus C'_{00}$ has the $\aleph_0$-exchange property, so we obtain

(d) $M = B_0 \oplus C'_{00} \oplus B'_{10} \oplus B'_{20} \quad \cdots$

$B_i = B_{i0} \oplus B'_{i0}$ and $C'_{00} \approx B_{10} \oplus B_{20} \oplus \cdots,\ i \geq 1$

$B_0 \oplus C'_{00} \oplus B'_{10}$ has the $\aleph_0$-exchange property and hence we have by (c),

(e) $M = B_0 \oplus C'_{00} \oplus B'_{10} \oplus C'_{11} \oplus C'_{12} \oplus \cdots$

$C'_{0i} = C_{1i} \oplus C'_{1i}$ and $B'_{10} \approx C_{11} \oplus C_{12} \oplus \cdots,\ i > 1$

Continuing the above, we obtain

(f) $B_i = B_{i0} \oplus B_{i1} \oplus \cdots \oplus B_{ii-1} \oplus B'_{ii-1}$

(g) $C_j = C_{0j} \oplus C_{1j} \oplus \cdots \oplus C_{jj} \oplus C'_{jj}$

(h) $B'_{ii-1} \approx C_{ii} \oplus C_{ii+1} \oplus \cdots$

(i) $C'_{jj} \approx B_{j+1j} + B_{j+2j} \oplus \cdots$

We can inductively obtain the desired refinements of B_i and C_j from (c)-(i).

8.7 RING-THEORETICAL PROOF

We proved the main theorems for modules that are direct sums of completely indecomposable modules by making use of the structure of the factor category $\underline{A}/\underline{J}'$. However, we gave ring-theoretical proofs for them in Secs. 6.8 and 7.8. In this section we give a ring-theoretical proof of (8.2.3).

Let $\{M_\alpha\}_I$ be a set of completely indecomposable modules and $M = \Sigma_I \oplus M_\alpha$. Let $\underline{A}$ be the category-induced form $\{M_\alpha\}_I$. We shall use the same notation as in Sec. 7.8. Let $\underline{J}'$ be the ideal in $\underline{A}$ defined by (6.3.1) and $\bar{\underline{A}} = \underline{A}/\underline{J}'$.

8.7.1 LEMMA Let M be an object in $\underline{A}$ and N a direct summand of M. We assume that A_1, A_2 are in $\underline{A}$, $A_1 \oplus A_2$ is a submodule of N and e the projection of M onto N. Then there exist R-homomorphisms $p_j : M \to A_j$ which satisfy (1) $e \equiv i_1p_1 + i_2p_2$, (2) $p_ji_{j'} \equiv \delta_{jj'}1_{A_j}$, and (3) $p_je \equiv p_j$, $ei_j \equiv i_j$ (mod $\underline{J}'$), where $\delta_{jj'}$ is the Kronecker delta and $i_j = 1_{A_j}$ the inclusion.

Proof: By (7.8.13) there exists an R-homomorphism $p : M \to A_1 \oplus A_2$ such that $e \equiv ip$ and $pi \equiv 1_{A_1 \oplus A_2}$, where $i = i_{A_1 \oplus A_2}$. We may assume that $pe = p$. Let i_j', p_j' be the injections and projections of A_j for $A_1 \oplus A_2$, respectively. Then $1_{A_1 \oplus A_2} = i_1'p_1' + i_2'p_2'$. Put $p_j = p_j'p$; then $p_je = p_j$ and $i_1p_1 + i_2p_2 \equiv i(i_1'p_1' + i_2'p_2')p \equiv ip \equiv e$. $p_ji_j \equiv p_j'pi_j \equiv p_j'pii_j' \equiv p_j'i_j' \equiv 1_{A_j}$ (mod $\underline{J}'$). The remaining assertions are trivial.

8.7.2 LEMMA Let M and N be as in (8.7.1). We assume that $A_1 \oplus A_2$ and $A_1 \oplus A_2'$ are full submodules of N. Then A_2 is isomorphic to A_2' (mod $\underline{J}'$) via $p_2'i_2$, where the A_i and A_2' are in $\underline{A}$.

Proof: Let p_j and i_j be as in the proof of (8.7.1) and set $\delta = p_2'i_2'$. We shall show that δ is an isomorphism modulo $\underline{J}'$. Let $\alpha : B \to A_2$ and $B \in \underline{A}$. We assume that $\delta\alpha \equiv 0$. Then $i_2\alpha \equiv ei_2\alpha \equiv (i_1p_1' + i_2'p_2')i_2\alpha \equiv i_1p_1'i_2\alpha$, where $p_1' : M \to A_1 \subset A_1 \oplus A_2'$. Hence $\alpha \equiv p_2i_2\alpha \equiv p_2i_1p_1'i_2\alpha \equiv 0$. Next, let $\beta : A_2' \to B$ and $\beta\delta \equiv 0$. Then $\beta p_2' \equiv \beta p_2'e \equiv \beta p_2'(i_1p_1 + i_2p_2) \equiv \beta p_2'i_2p_2 \equiv 0$. Since p_2' is an epimorphism modulo $\underline{J}'$, $\beta \equiv 0$. Therefore, δ is an isomorphism modulo $\underline{J}'$ by (7.8.1).

8.7.3 LEMMA Let M be in $\underline{A}$ and $\{P_\alpha, Q_\beta\}_{I,J}$ a set of c.ind. modules. We assume that $\Sigma_I \oplus P_\alpha$ and $\Sigma_J \oplus Q_\beta$ are full submodules of M. Then, for any subset I' of I, there exists a subset J' of J such that $\Sigma_{I-I'} \oplus P_\alpha \oplus \Sigma_{J'} \oplus Q_\beta$ is also a full submodule of M.

Proof: Put $P = \Sigma_I \oplus P_\alpha$ and $Q = \Sigma_J \oplus Q_\beta$. We may assume $I' \neq I$. Let α be in $I - I'$ and put $P^* = \Sigma_{\delta\neq\alpha} \oplus P_\delta$. Then there exist p_α, p^* such that $1_M \equiv i^*p^* + i_\alpha p_\alpha$ by (8.7.1). Furthermore, there exists an isomorphism modulo $\underline{J}'$, $p_P : M \to P$, by (7.8.11). Let p be the projection of P onto P_α. Since $pp_Pi_Q \not\equiv 0$, there exists β in J such that $pp_Pi_\beta \not\equiv 0$, where $i_\beta = i_{Q_\beta}$ (cf. the definition of $\underline{J}'$). Consider the external direct sum $Q_\beta \oplus P^*$ and put $j = (i_\beta, i^*) : Q_\beta \oplus P^* \to M$. We shall show that j is a monomorphism modulo $\underline{J}'$. Let $\varphi = \begin{bmatrix} a \\ b \end{bmatrix} : L \to Q_\beta \oplus P^*$ and L in $\underline{A}$. Assume that $j\varphi \equiv 0$. Then $i_\beta a + i^*b \equiv 0 \equiv pp_P i_\beta a + pp_P i^*b \equiv pp_P i_\beta a + pp_P i_P i^{*\prime} b \equiv pp_P i_\beta a + p_i^{*\prime} b \equiv pp_P i_\beta a$, where $i^{*\prime} : P^* \to P$. Hence since $pp_P i_\beta$ is an isomorphism, $a \equiv 0$, so $i^*b \equiv 0$. Since i^* is a monomorphism modulo $\underline{J}'$, $b \equiv 0$. Therefore, j is a monomorphism modulo $\underline{J}'$. Thus im $j = Q_\beta \oplus P^*$ is the internal direct sum $Q_\beta \oplus P^*$ in M, and $Q_\beta \oplus \Sigma_{I'} \oplus P_\alpha$ ($\subset Q_\beta \oplus P^*$) is a locally direct summand in M by (7.8.3). Now we consider the set $\underline{B} = \{T_R = \Sigma_{I'} \oplus P \oplus \Sigma_K \oplus Q_\beta \mid K \subset J$ and T_K is a locally direct summand of M$\}$. Then $\underline{B}$ is not the empty set. We define a partial order $\geq$ in $\underline{B}$ by setting $T_K \geq T_{K'}$ if and only if $K \supset K'$. We can take a maximal element T in $\underline{B}$ by Zorn's lemma. If T is not a full submodule of M, we can take a full submodule $T \oplus L$ of M. If we use the first argument for a component of L, we can find Q_β such that $T \oplus Q_\beta$ is in $\underline{B}$, which contradicts the maximality of T. Therefore, T is a full submodule.

8.7.4 COROLLARY Let $M = N_1 \oplus N_2$, $N_i \in \underline{A}$, and let $\Sigma_I \oplus P_\alpha$ be as in (8.7.3). Then there exists a subset I' of I such that $N_1 \oplus \Sigma_{I'} \oplus P_\alpha$ is a full submodule of M.

Proof: Since the N_i are in $\underline{A}$, $N_1 \oplus N_2$ is a full submodule of M.

Proof of (8.2.3): First, we shall show that N_2 has the exchange property in M. Let $M = \Sigma_K \oplus Q_\gamma$ and $\Sigma_{J_\gamma} \oplus P_{\gamma\varepsilon}$ is a full submodule of Q_γ, where the $P_{\gamma\varepsilon}$ are c.ind. modules. Since $\Sigma_K \Sigma_{J_\gamma} \oplus P_{\gamma\varepsilon}$ is a full submodule in M by (7.8.14), there exists a subset J'_γ of J_γ for each $\gamma \in K$ such that $N_2 \oplus \Sigma_K \Sigma_{J'_\gamma} \oplus P_{\gamma\varepsilon}$ is full in M by (8.7.4). Let e_1 be the projection of $M = N_1 \oplus N_2$ onto N_1, $e_1 : M \xrightarrow{p_1} N_1 \xrightarrow{i_{N_1}} M$, and put $P' = \Sigma_K \Sigma_{J'_\gamma} \oplus P_{\gamma\varepsilon}$. Then $p_1 i_{P'}$ is an isomorphism modulo $\underline{J}'$ by (8.7.2). On the other hand, $\{M_\alpha\}_{I_1}$ is locally semi-T-nilpotent from the assumption. Hence $p_1 i_{P'}$ is an R-isomorphism, and $M = P' \oplus \ker p_1 = P' \oplus N_2$ (and $N_1 \approx P$). Finally, we shall show that N_1 has the exchange property in M. Similarly to the above, we obtain a subset J^*_γ of J_γ such that $N_1 \oplus \Sigma_K \Sigma_{J^*_\gamma} \oplus P_{\gamma\varepsilon}$ is a full submodule of M. We put $P^*_\gamma = \Sigma_{J^*_\gamma} \oplus P_{\gamma\varepsilon}$, $P''_\gamma = \Sigma_{J''_\gamma} \oplus P_{\gamma\varepsilon}$, and $P^* = \Sigma_K \oplus P^*_\gamma$, where $J''_\gamma = J_\gamma - J^*_\gamma$. Then $N_1 \approx \Sigma_K \oplus P''_\gamma$ by (8.7.2). Since $P''_\gamma \oplus P^*_\gamma \oplus \Sigma_{\alpha\neq\gamma} \Sigma_{J_\alpha} \oplus P_{\alpha\varepsilon}$ is full in M, there exist R-homomorphisms $p''_\gamma : M \to P''_\gamma$ and $p_0 : M \to P''_\gamma$ and $p_0 : M \to P^*_\gamma \oplus \Sigma_{\alpha\neq\gamma} \Sigma_{J_\alpha} \oplus P_{\alpha\varepsilon}$ such that

(a) $1_M \equiv i''_\gamma p''_\gamma + i_0 p_0 \pmod{\underline{J}'}$

where $i_0 : P^*_\gamma \oplus \Sigma_{\alpha\neq\gamma}\Sigma_{J_\alpha} \oplus P_{\alpha\varepsilon} \to M$. Furthermore, since $\{P_{\gamma\varepsilon}\}_{J''_\gamma}$ is locally semi-T-nilpotent and $p''_\gamma i''_\gamma \equiv 1_{P''_\gamma}$, $M = P''_\gamma \oplus \ker p''_\gamma$. Put $Q^*_\gamma = Q_\gamma \cap \ker p''_\gamma$; then $Q_\gamma = P''_\gamma \oplus Q^*_\gamma$ and

(b) $M = \Sigma_K \oplus P''_\gamma \oplus \Sigma_K \oplus Q^*_\gamma$

Let $Q^{*\prime}_\gamma$ be a full submodule of Q^*_γ via the inclusion $i^{*\prime}_\gamma$. Then $Q^{*\prime}_\gamma \oplus P''_\gamma$ and $P^*_\gamma \oplus P''_\gamma$ are full submodules of Q_γ. $i^{*\prime}_\gamma \equiv (i''_\gamma p''_\gamma + i_0 p_0) i^{*\prime}_\gamma \equiv i_0 p_0 i^{*\prime}_\gamma$ by (a), since $Q^{*\prime}_\gamma \subset Q^*_\gamma \subset \ker p''_\gamma$. Put $\delta_\gamma = p_0 i^{*\prime}_\gamma = \begin{bmatrix} a \\ b \end{bmatrix} : Q^*_\gamma \to P^*_\gamma \oplus \Sigma_{\alpha\neq\gamma}\Sigma_{J_\alpha} \oplus P_{\alpha\varepsilon}$. Then we have the commutative diagram modulo $\underline{J}'$,

(c) 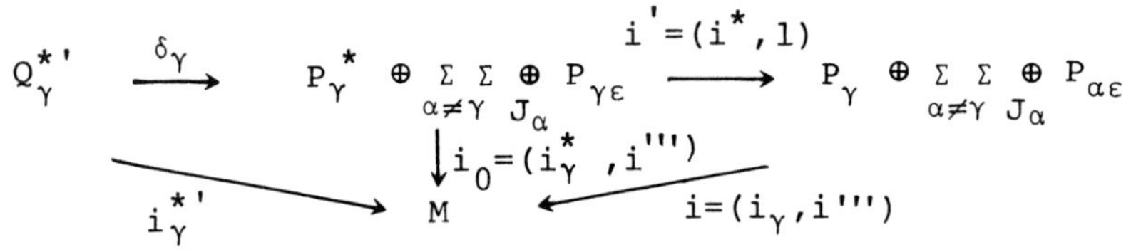

where i, i', etc., mean the inclusions. Furthermore, let e_1 and e_2 be the projections of $M = Q_\gamma \oplus (\Sigma_{\alpha\neq\gamma} \oplus Q_\gamma)$ onto each component. Then $0 \equiv e_2 i^{*\prime} \equiv e_2 i i' \delta_\gamma \equiv e_2 i^*_\gamma a + e_2 i''' b \equiv e_2 i''' b \equiv i''' b$, since $P^*_\gamma \oplus Q^{*\prime}_\gamma \subseteq Q_\gamma$ and $\Sigma_{\alpha\neq\gamma}\Sigma_{J_\alpha} \oplus P_{\alpha\varepsilon} \subset \Sigma_{\alpha\neq\gamma} \oplus Q_\alpha$. Hence $b \equiv 0$. Thus we obtain the commutative diagram modulo $\underline{J}'$,

(d) 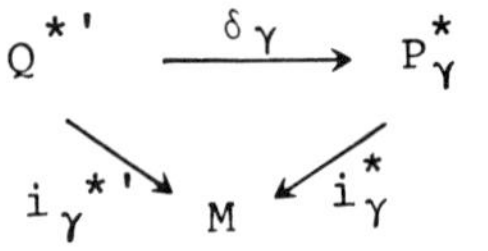

Now, since $P''_\gamma \oplus P^*_\gamma$ and $P''_\gamma \oplus Q^{*\prime}_\gamma$ are full in Q_γ, there exists, by (8.7.1) and (8.7.2), an R-homomorphism $p^*_\gamma : M \to P^*_\gamma$ such that $p^*_\gamma i^*_\gamma \equiv 1_{P^*_\gamma}$ and $p^*_\gamma i^{*\prime}_\gamma$ is an isomorphism modulo $\underline{J}'$. Hence $\delta_\gamma \equiv p^*_\gamma i^{*\prime}_\gamma$ is an isomorphism modulo $\underline{J}'$. Put $\delta = \Sigma_K \delta_\gamma : \Sigma_K \oplus Q^{*\prime}_\gamma \to \Sigma_K \oplus P^*_\gamma$. Since $\Sigma_K \oplus Q^{*\prime}_\gamma$ is a full submodule of $\Sigma_K \oplus Q^*_\gamma$, we have an R-homomorphism $p^{*\prime} : M \to \Sigma_K \oplus Q^{*\prime}_\gamma$ such that $1 - f \equiv i^{*\prime} p^{*\prime}$ by (7.8.13), where $i^{*\prime} : \Sigma \oplus Q^{*\prime}_\gamma \to M$ and $f : M \to \Sigma \oplus P''_\gamma$ is the projection in (b). On the other hand,

(e) $\Sigma \oplus P^*_\gamma \oplus N_1$

is a full submodule of M. Hence there exist R-homomorphisms $p^* : M \to \Sigma \oplus$

p_{γ}^{*} and p_{N_1} such that

$$\text{(f)} \quad 1_M \equiv i^*p^* + i_{N_1}p_{N_1} \pmod{\underline{J}'}$$

Since $\Sigma_K \oplus Q^{*\prime} \subset \Sigma \oplus Q_{\gamma}^{*} = \ker f$, $fi^{*\prime} = 0$. Using this result, the fact that δ is an isomorphism modulo $\underline{J}'$, and (d)-(f), we can prove similarly to (8.7.2) that fi_{N_1} is an isomorphism modulo $\underline{J}'$. Hence fi_{N_1} is an R-isomorphism and $M = N_1 \oplus \ker f = N_1 \oplus \Sigma_K \oplus Q_{\gamma}^{*}$.

9

Lifting Property and Extending Property

In Sec. 4.5 we studied some properties by which idempotents could be lifted modulo the Jacobson radical. Furthermore, these properties were extended to semiperfect modules in Secs. 6.1 and 7.5. In this chapter we generalize this concept for any module, as the lifting property modulo radical. This is a very elemental and fundamental property, which has not been studied for modules in general up to now. Nevertheless, we have used the property in Secs. 4.5, 6.1, and 7.5. We shall explore its fundamental nature and state some relationships between this property and generalized uniserial rings. As the concept dual to the lifting property modulo radical, we shall also investigate the extending property for modules.

Since this chapter is the starting point for the study of these properties, we are not able to consider them in a general form. Thus we shall deal with modules that are direct sums of cyclic hollow modules for the lifting property and are direct sums of uniform modules for the extending property. We hope that the reader succeeds in extending the results with the aid of this chapter. Finally, we note that the lifting property is a generalization of a property for projective modules and the extending property is that for injective modules.

As mentioned above, many new results are contained in this chapter. After defining the lifting property modulo radical in Sec. 9.1, we study the lifting property of simple modules modulo radical in Sec. 9.2, which is the starting point of this chapter. In Sec. 9.3 we extend this property to direct decompositions. We consider only the lifting property modulo radical for the direct sum of completely indecomposable modules. In Secs. 9.4 and 9.5 we investigate modules and rings with lifting property modulo radical, respectively. In particular, we show that these rings coincide with generalized uniserial rings. Sections 9.6-9.10 are the dual (extending property) to Secs. 9.1-9.5. When a ring is commutative we can study deeply the properties above. We give a characterization of a Dedekind domain in terms of the lifting (extending) property and give the complete classification of modules with the lifting or extending property over a Dedekind domain in Sec. 9.11.

9.1 DEFINITIONS CONCERNING THE LIFTING PROPERTY MODULO RADICAL

Let R be a ring with identity. We recall Theorems (4.5.10) and (4.5.11). We can consider R itself as a right R-module. Then $\mathrm{End}_R(R) = R$ by the multiplication of elements in R on the left. Let A be a two-sided ideal in R. Then A is a character submodule of R as a right R-module and $\mathrm{End}_R(R/A) = R/A$. Hence the two theorems above tell us that if $R/A = \Sigma_{i-1}^{n} \oplus M_i$, there exists a direct decomposition $R = \Sigma_{i=1}^{n} \oplus N_i$ such that $(N_i + A)/A = M_i$ by (3.1.6). We shall generalize this property to any right R-module M. However, we consider J(M) instead of the character submodule A of M, where J(M) is the Jacobson radical of M. Of course, we are interested in the case $M \neq J(M)$. We shall denote M/J(M) by $\bar{M}$. For a submodule N of M, we shall denote N + J(M)/J(M) by $\bar{N}$. [We note that if N is a direct summand of M, $\bar{N} \approx N/J(N)$.] If, for a direct decomposition $\bar{M} = \Sigma \oplus A_i$, there exists a direct decomposition $M = \Sigma \oplus M_i$ such that $\bar{M}_i - A_i$ for all i, then we say that the decomposition $\Sigma \oplus A_i$ of $\bar{M}$ is *lifted to a decomposition* $\Sigma \oplus M_i$ of M. If, for the above decomposition of $\bar{M}$ and a fixed A_1, there exists a direct summand N of M such that $\bar{N} = A_1$, then we say that A is *lifted to a direct summand* N of M. If every direct summand of $\bar{M}$ is lifted, we say that M has the *lifting property of direct summands modulo radical.*

We have already studied this property for semiperfect modules in Sec. 6.1. If P is a semiperfect module and $\bar{P} = \Sigma_I \oplus A_\alpha$ with A_α simple, there exists a direct decomposition $P = \Sigma_I \oplus P_\alpha$ such that $\bar{P}_\alpha = A_\alpha$ for all $\alpha \in I$ (6.1.1). In Sec. 6.1 we studied the following particular case.

1. $P_\alpha/J(P_\alpha)$ is semisimple and $J(P_\alpha)$ is small in P_α.
2. P_α is completely indecomposable.
3. A_α is simple.

Our aim in this chapter is to generalize this situation. In general, if every direct decomposition (not necessarily a finite direct sum) of $\bar{M}$ with a certain property is lifted to a direct decomposition of M, then we say that M has the *lifting property of direct decompositions* with the given property modulo radical. If every direct summand of $\bar{M}$ with a certain property is lifted to a direct summand of M, then we say that M has the *lifting property of direct summands* with the given property modulo radical.

Now we shall restrict ourselves to special cases of the property, as follows. If an R-module M has the unique maximal submodule J(M) as P_α in the above, then every proper submodule of M is small in M (see Sec. 3.4). If M has this property, we say that M is *hollow*. If further $M \neq J(M)$, then M is cyclic. If any simple direct summand of M is lifted to a cyclic and hollow direct summand N of M, then we say that M has the *lifting property of simple modules* modulo radical. Actually, we also need the assumption that N is completely indecomposable. From now on, if we say that M has the lifting property of simple modules modulo radical, then the lifting direct summand N is completely indecomposable unless otherwise stated. We are mainly interested in the lifting property modulo radical in this book, so we briefly call this property the *lifting property*.

The reason we assume the above comes from the following.

9.1.1 PROPOSITION Let M have the lifting property of simple modules. Then every direct summand of M has the same property.

Proof: Let $M = T_1 \oplus T_2$ and $A \subset \bar{T}_1 = T_1/J(T_1)$, where A is a simple summand of $\bar{T}_1$. Then there exists a completely indecomposable direct summand N_1 of M such that $\bar{N}_1 = A$. Since N_1 has the exchange property by (8.1.4), $M = N_1 \oplus T_1' \oplus T_2$ or $M = N_1 \oplus T_1 \oplus T_2'$, where $T_i' \subset T_i$. Since $\bar{N}_1 \subset \bar{T}_1$, we have the former case. Hence $M = N_1 \oplus T_1' \oplus T_2$ and $T_1 = T_1' \oplus T_1 \cap (N_1 \oplus T_2)$ by (1.7.4). Now $N_1 \approx T_1'' = T_1 \cap (N_1 \oplus T_2)$ and $T_1''/J(T_1'')$ is simple. Let $\bar{t}''$ be a generator of $\bar{T}_1''$ and $t'' = n + t_2$, where $n \in N_1$ and $t_2 \in T_2$. Since $\bar{n} \in A$ and $\bar{t}'' \in \bar{T}_1$, $\bar{t}_2 = 0$. Hence $A = \bar{n}R = \bar{t}''R = \bar{T}_1''$.

9.1.2 COROLLARY Let M be as in (9.1.1). We assume that $M/J(M) = \Sigma_{i=1}^n \oplus A_i$, where the A_i are simple. Then $T = \Sigma_{i=1}^n \oplus T_i \oplus S$ with $\bar{T}_i$ simple and $\bar{S} = 0$.

Proof: We can prove (9.1.2) by induction on n and (9.1.1).

9.1.3 PROPOSITION Let N be an R-module. We assume that N_1 and N_2 are completely indecomposable and direct summands of N. If $\bar{N}_1 = \bar{N}_2$ in $\bar{N}$, $N_1 \approx N_2$.

Proof: Let $N = N_1 \oplus N_1' = N_2 \oplus N_2'$. Since N_1 has the exchange property by (8.1.4), $N = N_1 \oplus N_2'$ as in the proof of (9.1.1). Hence $N_1 \approx N_2$.

9.2 LIFTING PROPERTY OF SIMPLE MODULES MODULO RADICAL

We shall now study the lifting property of simple modules (modulo radical) for a direct sum of cyclic hollow modules.

Let $\{M_\alpha\}_I$ be a set of submodules of M. If $M = \Sigma_I M_\alpha$ and $M \neq \Sigma_J M_\alpha$ for any proper subset J of I, we say that $\Sigma_I M_\alpha$ is an *irredundant sum.* If $\Sigma_K M_\gamma$ is a direct summand of M for every finite subset K of I, we say that $\Sigma_I M_\alpha$ is a locally direct summand of M (Sec. 7.3). We denote the natural epimorphism of M onto M/J(M) by φ. Let $M = \Sigma_I \oplus M_\alpha$. Then $\varphi_M|N = \varphi_N$ for every direct summand N of M and $\varphi|K = \varphi_K$ for a cyclic hollow submodule K with $K \not\subset J(M)$. If $\varphi(M_\alpha) \neq 0$ for all α and $\Sigma_I \varphi(M_\alpha) = \Sigma_I \oplus \varphi(M_\alpha)$, we say that $\Sigma_I M_\alpha$ is a *direct sum modulo* J(M).

In order to study further structures with the lifting property, we need some definitions. If an R-module M is noetherian, an epimorphism f of M onto itself is an isomorphism by (4.2.7) and if P is an R-projective and indecomposable module, then P has the same property, since f splits. Hence we consider the following condition for an R-module N.

Every epimorphism of N onto itself is an isomorphism. (9.2.1)

We note that if a hollow module N satisfies (9.2.1), N is completely indecomposable. Let $\{M_\alpha\}_I$ be a set of hollow modules satisfying (9.2.1). We define a partial order $\succcurlyeq$ for $\{M_\alpha\}_I$. If $M_\alpha \approx M_\beta$, we put $M_\alpha \equiv M_\beta$. If there exists an epimorphism f of M_α onto M_β, we put $M_\alpha \succcurlyeq M_\beta$. We know from (9.2.1) that $\equiv$ and $\succcurlyeq$ define a partial order for $\{M_\alpha\}_I$. Let $M_1 \succcurlyeq M_2$; then $\bar{M}_1 \approx \bar{M}_2$. If every element in $\mathrm{Hom}_R(\bar{M}_1,\bar{M}_2)$ is induced by some element in $\mathrm{Hom}_R(M_1,M_2)$, then we say that $\mathrm{Hom}_R(\bar{M}_1,\bar{M}_2)$ is induced from $\mathrm{Hom}_R(M_1,M_2)$.

First we give a proposition concerning (9.2.1).

9.2.1 PROPOSITION Let $\{M_i\}_1^n$ be a finite set of cyclic hollow modules satisfying (9.2.1). We assume that if $f : M_i \to M_j$ is an epimorphism, f is an isomorphism for any pair i and j. Then $M = \Sigma_{i=1}^n \oplus M_i$ satisfies (9.2.1).

Proof: We can express any element of $\mathrm{End}_R(M)$ by a matrix (f_{ij}), where $f_{ij} \in \mathrm{Hom}_R(M_j,M_i)$. Let $M = \Sigma_{i=1}^t \Sigma_{j=1}^{\rho(i)} \oplus M_{ij}$, where $M_{ij} \approx M_{i1}$ and $M_{i1} \not\approx M_{i'1}$ if $i \neq i'$. Then (f_{ij}) can be regarded as the block matrix $(f_{ij,kl})$. Let F be an epimorphism of M and $F = (f_{ij})$. We shall show that one f_{ij} among f_{ik}, k = 1, 2, ..., n, is an isomorphism. Since F is epimorphism, $M_i = \Sigma_j f_{ij}(M_j)$. However, M_i is hollow, so $M_i = f_{ij}(M_j)$ for some j. Hence f_{ij} is an isomorphism by the assumption. Since $M_{ij} \approx M_{i1}$, we may assume that $M_{ij} = M_{i1}$ for all j and the matrix units $e_{ij,ik}$ are elements in $End_R(M)$. Using these remarks and fundamental transformations of matrices, we know that there exist regular matrices P_1, Q_1 such that

$$P_1FQ_1 = \begin{bmatrix} 1 & 0 \\ 0 & F_2 \end{bmatrix} \qquad F_2 \in End_R\left(\sum_{i \geq 2} \oplus M_i\right)$$

Noting that F_2 is an epimorphism and repeating these arguments, we get regular matrices P and Q such that $PFQ = I_M$. Hence F is an isomorphism.

9.2.2 PROPOSITION Let M be an R-module and $\{M_\alpha\}_I$ a set of cyclic hollow submodules of M such that $M = \Sigma_I M_\alpha$. Then $M = \Sigma_I M_\alpha$ is an irredundant sum of M if $\varphi(M) = \Sigma_I \oplus \varphi(M_\alpha)$ and $\varphi(M_\alpha) \neq 0$ for all α. If J(M) is small in M, the converse is valid.

Proof: This is clear.

9.2.3 THEOREM Let $\{M_\alpha\}_I$ be a set of completely indecomposable and cyclic hollow modules and $M = \Sigma_I \oplus M_\alpha$. Then the following conditions are equivalent:

1. For any set $\{N_\gamma\}_J$ of indecomposable direct summands of M, if $\{\bar{N}_\gamma\}$ is independent in $\bar{M}$, $\{N_\gamma\}_J$ is independent (and $\Sigma_J \oplus N_\gamma$ is a locally direct summand of M).
2. If there exists an epimorphism f of M_α to M_β for any pair α and β in I, then f is an isomorphism.

Proof: (1) $\Rightarrow$ (2). We assume that there exists an epimorphism f of M_α onto M_β. We put $M_\alpha(f) = \{x + f(x) \mid x \in M_\alpha\} \subset M_\alpha \oplus M_\beta \subset M$. Then $M_\alpha \oplus M_\beta = M_\alpha(f) \oplus M_\beta$ and $M_\alpha(f) + M_\alpha$ is a direct sum modulo J(M), and hence $M_\alpha(f) + M_\alpha = M_\alpha(f) \oplus M_\alpha$ by (1). Therefore, $0 = M_\alpha(f) \cap M_\alpha = \ker f$.

(2) $\Rightarrow$ (1). We first note that if $M_\alpha \approx M_\beta$ for $\alpha \neq \beta$, M_α satisfies (9.2.1) by the assumption. Let $\Sigma_J N_\alpha$ be a direct sum modulo J(M) for indecomposable direct summands N_α of M. Let $K = \{1,2,\ldots,n\}$ be a finite subset of J and put $N(n) = \Sigma_{i=1}^n N_i$. We shall show, by induction on n, that N(n) is a direct summand of M and $N(n) = \Sigma_{i=1}^n \oplus N_i$. If n = 1, the assertion is clear by the assumption. We assume that $M = N(n - 1) \oplus M'$ and $N(n - 1) = \Sigma_{i=1}^{n-1} \oplus N_i$. Since N_n is a direct summand of M, N_n is isomorphic to some M_γ in $\{M_\alpha\}_I$ and $M' = \Sigma_{I'} \oplus M'_\beta$ by (7.1.2), where $I' = I - K$ and M'_β is isomorphic to some $M_{\rho(\beta)}$ in $\{M_\alpha\}_I$. Furthermore, since N_n has the exchange property in M by (7.1.2) either $M = N_n \oplus \Sigma_{j \neq k} \oplus N_j \oplus M'$ for some k or $M = N_n \oplus N(n - 1) \oplus \Sigma_{I'-\delta} \oplus M'_\beta$ for some δ. We have proved our assertion in the latter case. In the former case, $N_n \approx N_k \approx M_{\eta(k)}$. Let π_β be the projection for $M = N(n - 1) \oplus \Sigma_{I'} \oplus M'_\beta$ onto M'_β. Since

$\varphi(N_n) \not\subset \varphi(N(n-1))$, $\pi_\gamma|N_n$ is an epimorphism for some γ. If $\eta(k) \neq \rho(\gamma)$, $\pi_\gamma|N_n$ is an isomorphism by (2). If $\eta(k) = \rho(\gamma)$, M contains the direct summand $N_k \oplus M'_\gamma$ such that $N_n \approx N_k \approx M'_\gamma$. Hence N_n satisfies (9.2.1) by (2) and the initial remark. In either case $\pi_\gamma|N_n$ is an isomorphism. Accordingly, $M = N_n \oplus \ker \pi_\gamma = N_n \oplus N(n-1) \oplus \Sigma_{I'-\gamma} \oplus M'_\beta = N(n) \oplus \Sigma_{I'-\gamma} \oplus M'_\beta$.

9.2.4 THEOREM Let M and $\{M_\alpha\}_I$ be as in (9.2.3). Then the following conditions are equivalent:

1. For the set $\{N_\gamma\}_J$ in (9.2.3), $\Sigma_J \oplus N_\gamma$ is a direct summand of M.
2. $\{M_\alpha\}_I$ is a semi-T-nilpotent set and if there exists an epimorphism f of M_α to M_β for any pair α and β in I, then f is an isomorphism.

Proof: (1) ⇒ (2). Let $\{f_i : M_{\alpha_i} \to M_{\alpha_{i+1}}\}$ be a set of nonisomorphic homomorphisms. Then every f_i is a nonepimorphism by (9.2.3). Hence $f_i(M_{\alpha_i}) \subset J(M_{\alpha_{i+1}})$. We put $M_{\alpha_i}(f_i) = \{x + f_i(x) \mid x \in M_\alpha\} = M'_\alpha$ and $M' = M'_{\alpha_1} \oplus M'_{\alpha_2} \oplus \cdots \oplus M'_{\alpha_n} \oplus \cdots$. Then $\bar{M}'_\alpha = \bar{M}_\alpha$, so M'_α is a direct sum modulo J(M). Hence M' is a direct summand of M by (1). Therefore, $\{M_\alpha\}_I$ is a semi-T-nilpotent set by (7.2.7). The remaining assertions are clear by (9.2.3).

(2) ⇒ (1). This is clear from (7.3.9) and (9.2.3).

Remark If J(M) is small in M, $\{M_\alpha\}_I$ is a semi-T-nilpotent set by (7.4.1), provided that (2) in (9.2.3) is satisfied.

9.2.5 THEOREM Let M and $\{M_\alpha\}_I$ be as in (9.2.3). Then the following conditions are equivalent:

1. M has the lifting property of simple modules modulo J(M).
2. For any pair α and β in I such that $\bar{M}_\alpha \approx \bar{M}_\beta$ and any isomorphism f of $\bar{M}_\alpha$ onto $\bar{M}_\beta$, there exists an epimorphism g of either M_α onto M_β or M_β onto M_α such that $\bar{g} = f$ or $\bar{g} = f^{-1}$.

Proof: (1) ⇒ (2). Let M_1 and M_2 be two elements in $\{M_\alpha\}_I$. We assume that $\bar{M}_1 \approx \bar{M}_2$. Let f be an element in $\mathrm{Hom}_R(\bar{M}_1, \bar{M}_2)$ and put $K = \bar{M}_1(f)$. Since M has the lifting property, there exists a direct decomposition $M = M'_1 \oplus M'$ such that $\varphi(\bar{M}'_1) = K$. Let $\pi'_i : M \to M_i$ be the projection. Then $\pi'_i|M'_i$ is an epimorphism for i = 1, 2. M'_1 is isomorphic to some in $\{M_\alpha\}_I$ and $M = M'_1 \oplus (M_1 \oplus \overset{i}{\vee} \oplus M_n \oplus \cdots)$ by the exchange property of M'_1 by (7.1.2), where $\overset{i}{\vee}$ means that the ith component is omitted. If i were neither 1 nor 2, $\bar{M}'_1 \cap (\bar{M}_1 \oplus \bar{M}_2) = 0$. Hence i = 1 or 2. Thus we obtain $M = M'_1 \oplus M_2 \oplus M_3 \oplus \cdots$ or $M = M'_1 \oplus M_1 \oplus M_3 \oplus \cdots$. In the former case, let π_i be the projection of M onto M_i ($\pi_1 : M \to M_1'$) and $m_1 = \Sigma\pi_i(m_1)$ for $m_1 \in M_1$. Then $\bar{m}_1 = \Sigma\overline{\pi_i(m_1)} = \bar{m}_1' + [\overline{f(m_1')} + \overline{\pi_2(m_1)}] + \Sigma_{j\geq 3}\,\overline{\pi_i(m_1)}$; $m_1' \in M_1$. Hence $\bar{m}_1 = \bar{m}_1'$, $f(\bar{m}_1') = -\overline{\pi_2(m_1)}$ and $\Sigma_{i\geq 2}\overline{\pi_i(m_1)} = 0$. Accordingly, $-(\pi_2|M_1)$ is an element in $\mathrm{Hom}_R(M_2,M_1)$ and f is induced from $-(\Pi_2|M_1)$. In the latter case, $K = \{f^{-1}(y) + y \mid y \in \bar{M}_2\}$ and we know that $-(\pi_1|M_2)$ induces f^{-1} as above.

(2) ⇒ (1). Let A be a simple submodule of $\bar{M}$. Let n be the minimal integer among all m such that $A \subset \Sigma_{i=1}^m \oplus \bar{M}_{\alpha_i}$. Put $M_{\alpha_i} = M_i$ and let $\bar{\pi}_i$ be the projection for $\Sigma_{i=1}^m \oplus \bar{M}_i$ onto $\bar{M}_i$. Then $\bar{\pi}_i|A$ is an isomorphism and $A = \{\bar{m}_1 + f_2(\bar{m}_1) + \cdots + f_n(\bar{m}_1) \mid m_1 \in M_1,\ f_i = (\bar{\pi}_i|A)(\bar{\pi}_1|A)^{-1}\}$. We consider a set $\{(\bar{M}_i,\bar{M}_j),\ g_{ji} = f_jf_i^{-1}\}_{i\neq j}$. Then from (2) there exists either $h_{ji} \in \mathrm{Hom}_R(M_i,M_j)$ or $h_{ij} \in \mathrm{Hom}_R(M_j,M_i)$ such that $\bar{h}_{ji} = g_{ji}$ or $\bar{h}_{ij} = g_{ij}$. In the former case (resp. the latter case) we have $M_i \succcurlyeq M_j$ (resp. $M_i \preccurlyeq M_j$). We can easily see by induction and the fact $g_{ij}g_{jk} = g_{ik}$ that there exists a maximal member among the M_i with respect to the relation $\succcurlyeq$, say M_t. Then $A = \{g_{1t}(\bar{m}_t) + \cdots + g_{t-1t}(\bar{m}_t) + \bar{m}_t + \cdots + g_{nt}(\bar{m}_t) \mid m_t \in M_t\}$. We may assume that t = 1. Now from the construction above, there exist g_j in $\mathrm{Hom}_R(M_1,M_j)$ such that $\bar{g}_j = f_j$ for all j. Put $M_1' = \{m_1 + g_2(m_1) + \cdots + g_n(m_1) \mid m_1 \in M_1\} \subset M_1 \oplus M_2 \oplus \cdots \oplus M_n$. Then $M_1' \oplus M_2 \oplus \cdots \oplus M_n = M_1 \oplus \cdots \oplus M_n$ and $\bar{M}_1' = \{\bar{m}_1 + g_2(\bar{m}_1) + \cdots + g_n(\bar{m}_1)\} = A$.

9.2.6 COROLLARY Let $\{M_\alpha\}_I$ and M be as in (9.2.3). We assume that each M_α satisfies (9.2.1). Then the following conditions are equivalent:

1. M has the lifting property of simple modules modulo J(M).
2. In the subset $\{M_j\}$ of $\{M_\alpha\}_I$ such that $\bar{M}_j \approx \bar{M}_1$, the relation $\succcurlyeq$ is linear and $\mathrm{Hom}_R(\bar{M}_\alpha,\bar{M}_\beta)$ is induced from $\mathrm{Hom}_R(M_\alpha,M_\beta)$ for any pair $M_\alpha \succcurlyeq M_\beta$.

Proof: This is an immediate consequence of (9.2.5).

We shall restate the above fact by making use of matrices. Let $M = \Sigma_I \oplus M_\alpha$ be as in (9.2.3). We shall assume that I is countable for the sake of simplicity. Hence $M = \Sigma_{i=1}^\infty \oplus M_i$. Put $S = \mathrm{End}_R(M)$, $\bar{S} = \mathrm{End}_R(\bar{M})$, and $J_0(S) = \mathrm{Hom}_R(M,J(M))$. Then every element in S (or $\bar{S}$) is expressible as an infinite column-finite matrix with entries $f_{ij} \in \mathrm{Hom}_R(M_j,M_i)$ [or $\bar{f}_{ij} \in \mathrm{Hom}_R(\bar{M}_j,\bar{M}_i)$] (see Sec. 3.7). It is clear that $\bar{S} \supset S/J_0(S)$. We shall take a partition of I, $I = \cup_{i=1}^\infty I_i$, as follows: $\bar{M}_p \approx \bar{M}_q$ if and only if p and q are in the same I_j. Here we assume that all M_i satisfy (9.2.1) and M has the lifting property of simple modules. Then $\{M_p\}_{I_j}$ is a linearly ordered set with respect to $\succ$ by (9.2.6) for each j. Further, we obtain a partition of I_j, $I_j = \cup I_{jk}$, as follows: $M_p \approx M_q$ if and only if p and q are in the same I_{jk} and if $p \in I_{js}$, $q \in I_{jt}$, and $s < t$, then $M_p \not\geq M_q$.

Now, in general, $\bar{S} = \Pi_{i=1}^{\infty} \mathrm{End}_R(\Sigma_{I_i} \oplus \bar{M}_{ik})$ and $\mathrm{End}_R(\Sigma_{I_i} \oplus \bar{M}_{ik})$ is the ring of column-finite matrices over the division ring $\Delta_i = \mathrm{End}_R(\bar{M}_i)$. Following the partitions above, we shall consider the subring of $\mathrm{End}_R(\Sigma_{I_i} \oplus \bar{M}_{ij})$ which consists of all lower triangular matrices

$$
\begin{array}{r|c|c|c|c|c|}
 & & \overbrace{\quad}^{I_{it}} & & \overbrace{\quad}^{I_{ik}} & \\
\hline
 & \ddots & 0 & 0 & 0 & \cdots \\
\hline
I_{it}\left\{\right. & \Delta_i\Delta_i\cdots & \Delta_i(t)\ \cdots & 0 & 0 & \cdots \\
\hline
 & \Delta_i\cdots & \Delta_i\cdots & \ddots & \vdots & \cdots \\
\hline
I_{ik}\left\{\right. & \Delta_i\cdots & \Delta_i\cdots & & \Delta_i(k)\ \cdots & \\
\hline
 & & \Delta_i\cdots & & & \ddots \\
\hline
\end{array}
$$

where $\Delta_i(t) = \Delta_i$ if $|I_{it}| \geq 2$ and $\Delta_i(t) \approx \mathrm{End}_R(M_{it})/\mathrm{Hom}_R(M_{it}, J(M_{it}))$ if $|I_{it}| = 1$. We denote this ring by $T(\ldots, I_{it}, \ldots)$.

9.2.7 THEOREM Let $\{M_i\}_{i=1}^{\infty}$ and M be as in (9.2.3) and we assume that all M_i satisfy (9.2.1). Then M has the lifting property of simple modules modulo J(M) if and only if $S/J_0(S)$ is a subring $\Pi_i\ T(\ldots, I_{it}, \ldots)$ in $\bar{S}$ via the natural monomorphism $S/J_0(S) \to \bar{S}$.

Proof: This is clear.

We note that if $\mathrm{End}_R(\bar{M}_i)$ is lifted to $\mathrm{End}_R(M_i)$ for all i, then $\Delta_i(t) = \Delta_i$ in the above.

Finally, we obtain the following.

9.2.8 THEOREM Let $\{M_\alpha\}_I$ be a set of completely indecomposable and cyclic hollow modules. Put $M = \Sigma_I \oplus M_\alpha$ and assume that $\{M_\alpha\}_I$ is a semi-T-nilpotent set. Then the following conditions are equivalent:

1. M has the lifting property of simple modules modulo J(M).
2. M has the lifting property of direct summands modulo J(M).
3. Every direct summand of M has the above property.

Proof: (2) ⇒ (3) [cf. the proof of (9.1.1)]. Let $M = T_1 \oplus T_2$ and $\bar{T}_1 = A \oplus B$. We put $C = A \oplus \bar{T}_2$. Then from (2) we obtain direct summands K_i such that $M = K_1 \oplus K_2$ and $\bar{K}_1 = C$. Since $\{M_\alpha\}_I$ is semi-T-nilpotent, K_1 has the exchange property in M by (8.2.3). Hence $M = K_1 \oplus T_1' \oplus T_2'$; $T_i' \subset T_i$. Since $\bar{T}_2' \subset T_2 \subset C = \bar{K}_1$, $\bar{T}_2' = 0$, so $T_2' = 0$ by (7.1.2) and the assumption. Accordingly, $M = K_1 \oplus T_1'$ and $T_1 = T_1' \oplus (T_1 \cap K_1)$. On the other hand, since $\bar{M} = \bar{K}_1 \oplus \bar{T}_1' = C \oplus \bar{T}_1'$, $A \cap \bar{T}_1' = 0$, and $\overline{T_1 \cap K_1} \subset \bar{T}_1 \cap \bar{K}_1 = \bar{T}_1 \cap (A \oplus \bar{T}_2) = A$. Hence $A = \overline{T_1 \cap K_1} \oplus (A \cap \bar{T}_1') = \overline{T_1 \cap K_1}$ and $T_1 \cap K_1$ is a direct summand of T_1.

(3) ⇒ (1). This is clear.

(1) ⇒ (2). Since $\bar{M}$ is semisimple, we may assume that $\bar{M} = A \oplus B$ and $A = \Sigma_X \oplus A_\alpha$ with A_α simple. We assume that X is a well-ordered set. We show by transfinite induction that for each $\alpha > \beta \geq \kappa$ there exists an indecomposable direct summand N_κ of M such that $\Sigma_{\kappa\leq\beta} \oplus N_\kappa$ is a direct summand of M, $\overline{\Sigma_{\kappa\leq\beta} \oplus N_\kappa} = \Sigma_{\kappa\leq\beta} \oplus A_\kappa$ and $\{N_\delta\}_{\delta\leq\beta} \subset \{N_\gamma\}_{\gamma\leq\beta'}$ for $\beta \leq \beta'$. If $\alpha = 1$, we have $N_1 = M_1$ by (1). Since $\Sigma_{\beta<\alpha} \oplus N_\beta$ is a locally direct summand of M, by (7.3.9) it is a direct summand of M; $M = \Sigma_{\beta<\alpha} \oplus N_\beta \oplus N$ and $\overline{\Sigma_{\beta<\alpha} \oplus N_\beta} = \Sigma_{\beta<\alpha} \oplus A_\beta$. Put $A' = \Sigma_{\beta<\alpha} \oplus A_\beta \oplus A_\alpha$. Then $A' = \Sigma_{\beta<\alpha} \oplus A_\beta \oplus (A' \cap \bar{N})$ and $A' \cap \bar{N} \approx A_\alpha$ is simple. Hence there exists a direct summand N_α of N such that $\bar{N}_\alpha = A' \cap \bar{N}$ by (9.1.1). Therefore, $\Sigma_{\beta<\alpha} \oplus N_\beta \oplus N_\alpha$ is a direct summand of M and $\overline{\Sigma_{\beta<\alpha} \oplus N_\beta \oplus \bar{N}_\alpha} = \Sigma_{\beta<\alpha} \oplus A_\beta \oplus (A' \cap \bar{N}) = A'$.

9.3 LIFTING PROPERTY OF DIRECT DECOMPOSITIONS MODULO RADICAL

In the preceding section we studied the lifting property of simple modules modulo radical for a direct sum of completely indecomposable and cyclic hollow modules M_α. In the first part of this section, we assume only that $M_\alpha/J(M_\alpha)$ is simple (M_α is not necessarily either hollow or completely indecomposable). Again we denote $M/J(M)$ by $\bar{M}$. If $\bar{M} = \Sigma_{i=1}^n \oplus L_i$ is lifted to a direct decomposition of M, we say that M has the *lifting property of finite direct decompositions* or *of direct decompositions with finite direct summands modulo radical*. However, if we say that M has the lifting property of simple modules modulo radical, we always assume that the desired direct summand should be completely indecomposable and cyclic hollow.

First we consider a special case. Let $\{M_i\}_K$ be the subset of $\{M_\alpha\}_I$ such that $\bar{M}_i \not\approx \bar{M}_\alpha$ for any $\alpha \neq i$. Since $\bar{M}_i \not\approx \bar{M}_j$ for $i \neq j$ in K, $\Sigma_K \oplus M_i$ always has the lifting property, so M has the lifting property if

$\Sigma_{I-K} \oplus M_\beta$ does. Hence, when we study the lifting property, we may assume, without loss of generality, that

For each α in I, there exists $\rho(\alpha) \neq \alpha$ in I such that

$$\bar{M}_{\rho(\alpha)} \approx \bar{M}_\alpha. \tag{9.3.1}$$

9.3.1 THEOREM Let $\{M_\alpha\}_I$ be a set of R-modules with $M_\alpha/J(M_\alpha)$ simple (M_α is not necessarily either hollow or completely indecomposable). We put $M = \Sigma_I \oplus M_\alpha$. Then the following conditions are equivalent:

1. M has the lifting property of decompositions with finite direct summands (modulo radical).
2. M has the lifting property of decompositions with two direct summands (modulo radical).
3. $\mathrm{Hom}_R(\bar{M}_\alpha, \bar{M}_\beta)$ is lifted to $\mathrm{Hom}_R(M_\alpha, M_\beta)$ for $\alpha \neq \beta$ in I.
4. $\bar{S} = S/J_0(S)$ [under the assumption (9.3.1)], where $S = \mathrm{End}_R(M)$, $\bar{S} = \mathrm{End}_R(\bar{M})$, and $J_0(S) - \mathrm{Hom}_R(M, J(M))$.

Proof: (1) $\Rightarrow$ (2). This is clear.

(2) $\Rightarrow$ (3). Let f be in $\mathrm{Hom}_R(M_1, M_2)$ and $\bar{M}_1(f) = \{\bar{x} + f(\bar{x}) \mid x \in M_1\}$. Then $\bar{M} = \bar{M}_1(f) \oplus (\bar{M}_2 \oplus \bar{M}_3 \oplus \cdots)$. By (2) we obtain a decomposition $M = N_1 \oplus N_2$ such that $\bar{N}_1 = \bar{M}_1(f)$ and $\bar{N}_2 = (\bar{M}_2 \oplus \bar{M}_3 \oplus \cdots)$. Let $\pi_i : M \to N_i$ and $\rho_i : M \to M_i$ be the projections with respect to the decompositions $M = \Sigma_{i=1}^2 \oplus N_i$ and $M = \Sigma_I \oplus M_\alpha$, respectively. Put $g = \rho_2\pi_1|M_1 \in \mathrm{Hom}_R(M_1, M_2)$. Let m be in M_1. Then $m = \Sigma_{i=1}^2 \pi_i(m)$ and $\pi_i(m) = \Sigma_I \rho_\alpha(\pi_i(m))$. Hence since $\bar{m} = \Sigma \overline{\pi_i(m)}$ and $\overline{\pi_i(m)} = \bar{x} + f(\bar{x})$ for some $x \in M_1$, $\bar{m} = \bar{x}$. Accordingly, $\bar{g}(\bar{m}) = \overline{g(m)} = \overline{\rho_2\pi_1(m)} = \rho_2(\bar{x} + f(\bar{x})) = f(\bar{x}) = f(\bar{m})$. Hence f is lifted to g.

(3) $\Rightarrow$ (4). We assume (9.3.1). Let φ be an isomorphism of $\bar{M}_\alpha$ to $\bar{M}_{\rho(\alpha)}$ for $\alpha \neq \rho(\alpha)$. For any $g \in \mathrm{End}_R(\bar{M}_\alpha)$, $g = \varphi^{-1}(\varphi g)$. Then φ^{-1} and φg are lifted to $F \in \mathrm{Hom}_R(M_{\rho(\alpha)}, M_\alpha)$ and $H \in \mathrm{Hom}_R(M_\alpha, M_{\rho(\alpha)})$ by (3), respectively. Hence g is lifted to FH. Therefore, $\bar{S} = S/J_0(S)$ by (3).

(4) $\Rightarrow$ (1). First we shall show (4) $\Rightarrow$ (2). Let $\bar{M} = A \oplus B$ and $\bar{M} = A \oplus \Sigma_P \oplus \bar{M}_\delta$ (note that $\bar{M}_\alpha$ is simple). Now $M = \Sigma_{I-P} \oplus M_\gamma \oplus \Sigma_P \oplus M_\gamma$. Let π be the projection of $\bar{M}$ onto $\Sigma_P \oplus \bar{M}_\delta$ with respect to the decomposition $\bar{M} = A \oplus \Sigma_P \oplus \bar{M}_\delta$. Since $\bar{S} = S/J_0(S)$, we can choose, from the representation of column-finite matrices, a homomorphism $f : M \to \Sigma_P \oplus M_\delta$ such that $\bar{f} = -\pi$. We put $M_1(f) = \{x + f(x) \mid x \in \Sigma_{I-P} \oplus M\} \subset M$. Then $M = M_1(f) \oplus \Sigma_P \oplus M_\delta$ and $\overline{M_1(f)} = A$. Next we take the projection π' of $\bar{M}$ onto A with respect to the decomposition $\bar{M} = A \oplus B$. Since $A = \overline{M_1(f)}$, there exists $g : M \to M_1(f)$ such that $\bar{g} = -\pi'$ [note that $M_1(f) \approx \Sigma_{I-P} \oplus M_\gamma$ and

$M = M_1(f) \oplus \Sigma_P \oplus M_\delta]$. Put $M_2(g) = \{y + g(y) \mid y \in \Sigma_P \oplus M_\delta\}$. Then $M = M_1(f) \oplus M_2(g)$ and $\overline{M_2(g)} = B$. Finally, we shall show (4) ⇒ (1). Since $M_1(f) \approx \Sigma_{I-P} \oplus M_\gamma$ and $M_2(g) \approx \Sigma_P \oplus M_\delta$, we can prove the assertion by induction on the number of summands [note that (4) is satisfied for the direct summand $M_2(g)$].

We are mainly interested in the lifting property modulo radical in this book, so we call this property the lifting property.

We know, from the proof and the remark above (9.3.1), that (1)-(3) are equivalent without (9.3.1).

9.3.2 COROLLARY Let $\{M_\alpha\}_I$ and M be as in (9.3.1). We assume further that each M_α is completely indecomposable and $\{M_\alpha\}_I$ is semi-T-nilpotent. Then conditions 1-4 are equivalent to the fact:

5. M has the lifting property of direct decompositions.

Proof: (1) ⇒ (5). We shall use the same method as in the proof of (1) ⇒ (2) of (9.2.8). Let $\bar{M} = \Sigma_L \oplus B_\gamma$. We shall prove the assertion by transfinite induction on L. We assume that L is a well-ordered set and that for each $\rho \le \beta < \alpha$, there exist direct summands N_ρ such that $M = \Sigma_{\rho\le\beta} \oplus N_\rho \oplus M'$, $\bar{N}_\rho = B_\rho$ and $\{N_\gamma\}_\beta \subset \{N_\delta\}_{\beta'}$ for $\beta \le \beta' < \alpha$. We note that every direct summand of M is a direct sum of completely indecomposable modules by (8.2.1). Since $T = \Sigma_{\rho<\alpha} \oplus N_\rho$ is a locally direct summand, T is a direct summand of M by (7.3.9), say $M = T \oplus M'$. Now $\bar{M} = \Sigma_{\rho<\alpha} \oplus B_\rho \oplus \Sigma_{\alpha\le\gamma} \oplus B_\gamma$, $M = T \oplus M'$ and $\bar{T} = \Sigma_{\rho<\alpha} \oplus \bar{N}_\rho = \Sigma_{\rho<\alpha} \oplus B_\rho$. Then we know, from the proof (4) ⇒ (1) of (9.3.1), that there exists a decomposition $M = T \oplus M'(g)$ and $\overline{M'(g)} = \Sigma_{\alpha\le\gamma} \oplus B_\gamma$. $M'(g)$ also has the lifting property of decompositions with two direct summands by (9.3.1). Hence we obtain a direct summand N_α of M such that $\bar{N}_\alpha = B_\alpha$ and $M = \Sigma_{\beta\le\alpha} \oplus N_\beta \oplus M''$.

(5) ⇒ (1). This is clear.

If M_α is a cyclic hollow module with (9.2.1), then $\mathrm{Hom}_R(M_\alpha, J(M_\alpha)) = J(\mathrm{End}_R(M_\alpha))$. Further we assume that each M_α is c.ind. Then the condition 3 in (9.3.1) tells us the following fact: If $f : M_\alpha \to M_\beta$ is an epimorphism, then f is an isomorphism. We shall take the ideal $\underline{J}'$ in Sec. 6.3. If $f_{\alpha\beta} : M_\beta \to M_\alpha$ is not an isomorphism, $f_{\alpha\beta}(M_\beta) \subset J(M_\alpha)$, so $J'(S) \subset J_0(S)$. Hence $J'(S) = J(S)$ provided that $J_0(S) \subset J(S)$. It is clear that $J'(S) \supset J_0(S)$. Therefore, $J_0(S) \subset J(S)$ if and only if $\{M_\alpha\}_I$ is a semi-T-nilpotent set by (7.2.4). Furthermore, we know by (7.4.1) that if J(M) is small in M, $\{M_\alpha\}_I$ is semi-T-nilpotent. Thus we have the following.

9.3.3 COROLLARY Let $\{M_\alpha\}_I$ and M be as in (9.3.1). We assume that each M_α is a cyclic hollow and c.ind. module with (9.2.1). We assume one of the following:

(i) $J_0(S) \subset J(S)$.

(ii) $\{M_\alpha\}_I$ is a semi-T-nilpotent set.

(iii) J(M) is small in M (e.g., R is right perfect).

Then the conditions in (9.3.1) are equivalent to (5).

9.3.4 COROLLARY We further assume in (9.3.3) that $M_\alpha \approx M_1$ for all α in I. Then the conditions in (9.3.1) are equivalent to the fact:

6. M has the lifting property of simple modules.

9.4 MODULES WITH THE LIFTING PROPERTY MODULO RADICAL

In the preceding two sections, we studied principally the lifting property (modulo radical) for direct sums of cyclic hollow modules. Making use of these results, we shall study some modules that have the lifting property (modulo radical). If M = J(M), we have nothing to do, so we assume that $\bar{M} \neq 0$ and $\bar{M}$ is semisimple. (If we say that M has the lifting property of simple modules, we assume that every desired direct summand is c.ind. and cyclic hollow.)

9.4.1 THEOREM Let M be an R-module such that M/J(M) is semisimple. We assume that M has the lifting property of simple modules and satisfies one of the following conditions.

1. M has the lifting property of direct decompositions with two direct summands.
2. $\bar{S} = S/J_0(S)$, where $S = \mathrm{End}_R(M)$, $\bar{S} = \mathrm{End}_R(\bar{M})$, and $J_0(S) = \mathrm{Hom}_R(M,J(M))$.
3. For any cyclic hollow and c.ind. direct summand N of M and an element f in $\mathrm{Hom}_R(N,M)$, if $\bar{f} : \bar{N} \to \bar{M}$ is a monomorphism, f is a monomorphism.
4. For any two direct summands M_1 and M_2 as in (3) and $f \in \mathrm{Hom}_R(M_1,M_2)$, if $\bar{f}$ is an epimorphism, then f is an isomorphism.

Then M contains a submodule M' satisfying the following.

(i) $M' = \Sigma_I \oplus M_\alpha$, where the M_α are cyclic hollow and c.ind.

(ii) M' has the lifting property of simple modules and direct decompositions with finite direct summands.

(iii) $\Sigma_I \oplus M_\alpha$ is a locally direct summand of M.

(iv) M = M' + J(M).

Conversely, if M contains a submodule M' as above, then M has the lifting property of simple modules. Furthermore, if each M_α satisfies (9.2.1), then M satisfies (3) and (4).

Proof: We assume that M has the lifting property of simple modules. Let $\underline{N}$ be the set of submodules N of M such that $N = \Sigma_J \oplus N_\gamma$, where the N_γ are cyclic hollow and c.ind. and N is a locally direct summand of M (with respect to $\Sigma_J \oplus N_\gamma$). Since M has the lifting property of simple modules, $\underline{N}$ is not empty. We can define a partial order in $\underline{N}$ by the members of direct summands of N. Then we can find a maximal element, say $M' = \Sigma_I \oplus N_\alpha$, in $\underline{N}$ by Zorn's lemma. Since M' is a locally direct summand of M, $\bar{M}' \subset \bar{M}$.

Case 1. We put $\bar{M} = \bar{M}' \oplus K$ and show that $K = 0$. If $K \neq 0$, we have a simple submodule A such that $\bar{M} = (\bar{M}' \oplus K') \oplus A$. Since M has the lifting property of decompositions with two direct summands, $M = L \oplus N$ and $\bar{L} = \bar{M}' \oplus K'$, $\bar{N} = A$. Let J be any finite subset of I. Then $M = \Sigma_J \oplus N_\gamma \oplus P$. Since $\Sigma_J \oplus N_\gamma$ has the exchange property by (8.1.4) and (8.4.5), $M = \Sigma_J \oplus N_\gamma \oplus L' \oplus N'$, where $L = L' \oplus L''$ and $N = N' \oplus N''$. Since $\bar{N} = \bar{N}' \oplus \bar{N}'' = A$, $\bar{N}' = 0$ or $\bar{N}'' = 0$. If $\bar{N}' = 0$, $\bar{M} = \Sigma_J \oplus \bar{N}_\gamma \oplus \bar{L}' \subset \bar{L}$. Hence $\bar{N}' \neq 0$, so $\bar{N}'' = 0$. On the other hand, N'' is isomorphic to a direct summand of $\Sigma_J \oplus N_\gamma$. Hence $N'' = 0$, so $N = N'$. Therefore, $\Sigma_J \oplus N_\gamma \oplus N$ is a direct summand of M, which contradicts the maximality of M' in $\underline{N}$. Therefore, $M = M' + J(M)$. M' has the lifting property of simple modules by (9.2.5) and (9.1.1). Next we shall show that M' has the lifting property of decompositions with finite direct summands. Let N_1 and N_2 be in $\{N_\alpha\}_I$; then $N_1 \oplus N_2$ is a direct summand of M by (iii), say $M = N_1 \oplus N_2 \oplus M^*$. We can apply the argument in the proof (2) ⇒ (3) of (9.3.1) to the decomposition $M = N_1 \oplus N_2 \oplus M^*$, instead of $M = \Sigma_I \oplus M_\alpha$. Hence $\mathrm{Hom}_R(\bar{N}_1, \bar{N}_2)$ is lifted to $\mathrm{Hom}_R(N_1, N_2)$. Therefore, M' has the lifting property of decompositions with finite direct summand by (9.3.1) and has the lifting property of simple modules by (9.1.1) and (9.2.5).

Case 2. We also show that $K = 0$. Put $N_0 = \Sigma_J \oplus N_i$ for a finite subset J of I. Then $M = N_0 \oplus P$ and $\bar{M} = \bar{N}_0 \oplus \bar{P} = \bar{N}_0 \oplus \Sigma_{I-J} \oplus \bar{N}_\delta \oplus K$. Let π' be the projection of $\bar{M}$ onto $\bar{N}_0$ with respect to the latter decomposition and $\pi = \pi'|\bar{P}$. Since $\bar{S} = S/J_0(S)$ and

$$S = \begin{bmatrix} \mathrm{Hom}_R(N_0, N_0) & \mathrm{Hom}_R(P, N_0) \\ \mathrm{Hom}_R(N_0, P) & \mathrm{Hom}_R(P, P) \end{bmatrix}$$

there exists f in $\mathrm{Hom}_R(P, N_0)$ with $\bar{f} = -\pi$. Then $M = P(f) \oplus N_0$ and $\overline{P(f)} = \Sigma_{I-J} \oplus \bar{N}_\delta \oplus K$ [cf. the proof of (9.3.1)]. We assume that $K \neq 0$ and K_α is a simple component of K. Since M has the lifting property of simple modules, there exists a direct summand M_α of M such that $\bar{M}_\alpha = K_\alpha$. M_α has the exchange property, so $M = M_\alpha \oplus P(f)' \oplus N_0$, and $P(f)' \subset P(f)$, since $\bar{M}_\alpha \subset \overline{P(f)}$. Hence $M_\alpha \oplus N_0$ is a direct summand of M for any finite subset J of I. Hence $M_\alpha \subset M'$ is a locally direct summand of M, which is a contradiction. Therefore, K = 0 and $\bar{M} = \bar{M}'$. It is clear that every direct summand of M satisfies condition 2. Hence M' has the lifting property of simple modules and direct decompositions with finite direct summands by (9.3.1).

Case 3 or 4. Put $\bar{M} = \Sigma_I \oplus A_\alpha$, where the A_α are simple. Let M_α be a cyclic hollow and c.ind. direct summand of M with $\bar{M}_\alpha = A_\alpha$. Put $M' = \Sigma_I M_\alpha$. Then $\bar{M}' = \bar{M}$. We shall show that $\Sigma_I M_\alpha$ is a direct sum [cf. the proof of (9.2.3)]. Let $\Sigma_{i=1}^n M_i = M(n)$ be any finite sum in $\Sigma_I M_\alpha$. We show that $M(n) = \Sigma_{i=1}^n \oplus M_i$ and M(n) is a direct summand of M by induction of n. If n = 1, the assertion is clear. We assume that $M = M(n-1) \oplus T$. Let π be the projection of M onto T and $f = \pi|M_n$. Since $f(M_n) \not\subset J(T)$, $\overline{f(M_n)}$ is a simple component of $\bar{T}$ and T has the lifting property by (9.1.1), so there exists a cyclic hollow and c.ind. direct summand T_1 of T such that $\bar{T}_1 = \overline{f(M_n)}$. Let $T = T_1 \oplus T_2$ and π_1 be the projection of T onto T_1. Put $g = \pi_1 f$. Then $\overline{g(M_n)} = \bar{T}_1$. Hence $\bar{g}$ is a monomorphism, so g is a monomorphism by (3) or (4). Accordingly, $M = M_n \oplus \ker \pi_1\pi = M_n \oplus T_2 \oplus M(n-1) = M(n) \oplus T_2$ and $M(n) = \Sigma_{i=1}^n \oplus M_i$. Cases 3 and 4 imply that each M_α satisfies (9.2.1). Hence M' has the lifting property of direct decompositions with finite direct summands by (9.3.1) and (9.2.6).

Conversely, we assume that M has the submodule M'. Since $M' = \Sigma_I \oplus M_\alpha$ is a locally direct summand of M, $\bar{M} = \Sigma_I \oplus \bar{M}_\alpha = \bar{M}'$. Let A be a simple submodule of $\bar{M}$. Then there exists a finite subset J of I such that $A \subset \Sigma_J \oplus \bar{M}_\gamma$. Since M' has the lifting property of simple modules, $\Sigma_J \oplus M_\gamma$ contains a direct summand N with $\bar{N} = A$ by (9.1.1), so N is also a direct summand of M. Finally, we assume that each M_α satisfies (9.2.1). Let N_α be any direct summand of M as in (3). Since $\bar{M} = \bar{M}'$, there exists a direct summand M_α of M' such that $\bar{M}_\alpha = \bar{N}_\alpha$. Hence $M_\alpha \approx N_\alpha$ by (9.1.3). Accordingly, (4) is satisfied by (9.3.1). Let f be as in (3). Since $\bar{M} = \bar{M}'$ and M' has the lifting property of simple modules, there exists a direct summand T_α with $\bar{T}_\alpha = \overline{f(N)}$. T_α is cyclic, so T_α is contained in a direct summand $\Sigma_J \oplus M_\alpha$ of M by (iii). Hence T_α is a direct summand of M. Let π be the projection of M onto T_α. Then $\pi f : N_\alpha \to T_\alpha$ is an isomorphism by the above and (9.3.1). Hence f is a monomorphism.

9.4.2 COROLLARY Let R be a right artinian ring and M an R-module. Then M has the lifting property of direct decompositions if and only if M has the lifting property of direct decompositions with two direct summands.

Proof: If R is right artinian, every R-module N with N/J(N) simple is a cyclic hollow and c.ind. module. Hence if M has the lifting property of direct decompositions with two direct summands, then M has the lifting property of simple modules, so $M = \Sigma_I \oplus M_\alpha$ by (9.4.1), where the M_α are cyclic hollow modules. Hence M has the lifting property of direct decompositions by (9.3.3).

Remarks 1. We note that if J(M) is small in M, M = M' in (9.4.1). If, further, each M_α satisfies (9.2.1), then all conditions (1)-(4) in (9.4.1) are equivalent when M has the lifting property of simple modules. 2. We assume that $M = M' \oplus K$ with $J(K) = K$. Then M satisfies (1) and (2) if M' satisfies (i) and (ii) in (9.4.1). We do not know whether this fact is true or not without the assumption.

9.4.3 PROPOSITION Let M be an R-module. We assume that M is countably generated, M/J(M) is semisimple, and J(M) is small in M. If M has the lifting property of simple modules, then $M = \Sigma_{i=1}^{\infty} \oplus M_i$ with M_i indecomposable.

Proof: Let $\{m_1, m_2, \ldots, m_n, \ldots\}$ be a set of generators. Since J(M) is small, we may assume that $m_i \notin J(M)$ for all i. Further, we may assume that $\overline{m_i R}$ is simple. We assume that there exists a set of indecomposable direct summands M_i such that $M(n) = \Sigma_{i=1}^{t_n} \oplus M_i$ is a direct summand of M and $\Sigma_{i=1}^{n} \overline{m_k R} \subset \overline{M(n)}$. Let $M = M(n) \oplus T$ and $m_{n+1} = x + t$; $x \in M(n)$, $t \in T$. If $t \notin J(T)$, there exists an indecomposable direct summand $M_{t_{n+1}}$ of T such that $\bar{M}_{t_{n+1}} = \overline{tR}$ by (9.1.1). Hence there exists a direct summand $M(n+1)$ of M such that $\overline{M(n+1)} \supset \Sigma_{i=1}^{n+1} \overline{m_i R}$. Accordingly, $\cup_n M(n) = \Sigma_1^{\infty} \oplus M_{t_i} = M$, since J(M) is small in M.

We come back to the lifting property of simple modules.

9.4.4 LEMMA Let M be an R-module and let $\Sigma_{I_1} \oplus N_{\alpha_1}$, $\Sigma_{I_2} \oplus N_{\alpha_2}$, ... be submodules in M. We assume that the N_{α_i} are c.ind. and $\Sigma_{I_i} \oplus N_{\alpha_i}$ is a locally direct summand of M for all i. If $N_{\alpha_i} \not\approx N_{\alpha_j}$ for any $\alpha_i \in I_i$ and any $\alpha_j \in I_j$, then $\Sigma_i \Sigma_{I_i} \oplus N_{\alpha_i}$ is a locally direct summand of M.

Proof: Let J_i be any subset of I_i. Then $M = \Sigma_{J_i} \oplus N_{\alpha_i'} \oplus M(i)$ for i = 1, 2, ... Since $\Sigma_{J_1} \oplus N_{\alpha_1'}$ has the exchange property, $M = \Sigma_{J_1} \oplus N_{\alpha_1'} \oplus \Sigma_{J_2} \oplus N_{\alpha_2'} \oplus M^*(2)$, since $N_{\alpha_1'} \not\approx N_{\alpha_2'}$, where $M^*(2) \subset M(2)$. Repeating this argument, we know that $\Sigma_i \Sigma_{I_i} \oplus N_{\alpha_i}$ is a locally direct summand of M.

9.4.5 COROLLARY Let M be an R-module such that $M/J(M) = \Sigma_I \oplus A_\alpha$ with A_α simple and $A_\alpha \not\approx A_\beta$ for $\alpha \neq \beta \in I$. Then M has the lifting property of simple modules if and only if M contains a submodule M' which has the lifting property of simple modules and satisfies (i), (iii), and (iv) in (9.4.1).

Proof: If M has the lifting property of simple modules, we obtain a cyclic hollow and c.ind. module M_α with $\bar{M}_\alpha = A_\alpha$. Then $M_\alpha \not\approx M_\beta$ if $\alpha \neq \beta$. Hence $\Sigma_I \oplus M_\alpha = M'$ satisfies (i), (iii), and (iv) by (9.4.1). The remaining assertion is clear.

9.4.6 THEOREM Let M be an R-module with M/J(M) semisimple. We assume that every cyclic and hollow direct summand of M is noetherian (e.g., R is right noetherian). Then M has the lifting property of simple modules if and only if M contains a submodule M' such that (1) $M' = \Sigma_I \oplus M_\alpha$, where the M_α are cyclic hollow and c.ind., (2) M' has the lifting property of simple modules, (3) $\Sigma_I \oplus M_\alpha$ is a locally direct summand of M, and (4) $M = M' + J(M)$.

Proof: The "if" part is clear by (9.4.1). We assume that M has the lifting property of simple modules. Let M_1 and M_2 be c.ind. and cyclic hollow direct summands of M, say $M = M_1 \oplus N_1 = M_2 \oplus N_2$. Then $M_1 \approx M_2$ or $M = M_1 \oplus M_2 \oplus N_2'$ by (9.4.4). In the latter case, since $M_1 \oplus M_2$ has the lifting property of simple modules by (9.1.1), $M_1 \preccurlyeq M_2$ or $M_1 \succcurlyeq M_2$ if $\bar{M}_1 \approx \bar{M}_2$ by (9.2.5). Therefore, we may assume that in any cases $M_1 \succcurlyeq M_2$ or $M_1 \preccurlyeq M_2$, whenever $\bar{M}_1 \approx \bar{M}_2$. We note that every cyclic and hollow direct summand satisfies (9.2.1) by the assumption. Now let $\{M_\gamma\}_L$ be a representative set of cyclic hollow and completely indecomposable direct summands of M with $\bar{M}_\gamma \approx \bar{M}_{\gamma_0}$ for all $\gamma \in L$ (γ_0 is fixed). Then $\{M_\gamma\}_L$ is a linearly ordered set with respect to $\preccurlyeq$. Since M_γ is noetherian, we have the minimal member M_1 and a finite chain $M_\gamma \succ M_{\gamma-1} \succ \cdots \succ M_1$. Hence $\{M_\gamma\}_L$ is a well-ordered set, so we may assume that $\{M_\gamma\}_L = \{M_n\}$ with $M_n \succ M_{n-1}$.

Step 1. We define the set $\underline{N}_1$ of submodules of M as follows: $\underline{N}_1 = \{T = \Sigma_{\alpha \in I_1'} \oplus M_{1\alpha} \mid T$ is a locally direct summand of M and $M_{1\alpha} \approx M_1$ for all $\alpha \in I_1'\}$. Let $T_1 = \Sigma_{I_1} \oplus M_{1\alpha}$ be a maximal member in $\underline{N}_1$. Then $\bar{M} = \bar{T}_1 \oplus \bar{K}_1$ and $\bar{T} = \Sigma_{I_1} \oplus \bar{M}_{1\alpha}$. Let A_1 be a simple submodule of $\bar{M}$ not contained in $\bar{T}_1$. Since M has the lifting property of simple modules, there exists a completely indecomposable and cyclic hollow direct summand N_1 of M such that $\bar{N}_1 = A_1$. We shall show that $N_1 \not\approx M_1$. Let J_1 be any finite subset of I_1. Then $M = \Sigma_{J_1} \oplus M_{1\alpha} \oplus M'$. Let $\pi_{M'} : M \to M'$ be the projection. Since

$\bar{N}_1 \not\subset \overline{\Sigma_{J_1} \oplus M_{1\alpha}}$, $\bar{N}_1 \approx \overline{\pi_{M'}(N_1)}$, so M' contains a cyclic hollow direct summand N_1' with $\bar{N}_1' = \overline{\pi_{M'}(N_1)}$ by (9.1.1), say $M' = N_1' \oplus M''$. Then $M = \Sigma_{J_1} \oplus M_{1\alpha} \oplus M'' \oplus N_1'$. Let $\pi : M \to N_1'$ and $\pi' : \bar{M}' \to \bar{N}_1'$ be the projections, respectively. Then $\overline{\pi(N_1)} = \overline{(\pi'\pi_{M'})(N_1)} = \pi'(\overline{\pi_{M'}(N_1)}) = \pi'(\bar{N}_1') = \bar{N}_1'$. Hence $\pi|N_1 : N_1 \to N_1'$ is an epimorphism. If $N_1 \approx M_1$, $N_1' \approx M_1$, since M_1 is minimal, so $\pi|N_1$ is an isomorphism. Therefore, $M = N_1 \oplus \ker \pi = N_1 \oplus \Sigma_{J_1} \oplus M_{1\alpha} \oplus M''$. Since J_1 is any finite subset of I_1, $N_1 \oplus T_1$ is a locally direct summand of M and $N_1 \approx M_1$, which is a contradiction to the maximality of T_1 in $\underline{N}_1$. Hence $N \not\approx M_1$. Now $\bar{M} = \bar{T}_1 \oplus K_1$ and $M_2 \oplus T_1$ is a locally direct summand of M by (9.4.4).

Step 2. Let $\underline{N}_2 = \{T = \Sigma_{I_1} \oplus M_{1\alpha} \oplus \Sigma_{I_2'} \oplus M_{2\beta} \mid T$ is a locally direct summand of M and $M_{2\beta} \approx M_2\}$. Let T_2 be a maximal member in $\underline{N}_2$. Then $M = \bar{T}_1 \oplus \Sigma_{I_2} \oplus \bar{M}_{2\beta} \oplus K_2$. Let A_2 be a simple submodule not contained in $\bar{T}_2$ and N_2 a cyclic hollow direct summand of M with $\bar{N}_2 = A_2$. Since $A_2 \not\subset \bar{T}_1$, $N_2 \not\approx M_1$ by the first step. We assume that $N_2 \approx M_2$. Let J_2 be any finite subset of I_2 and $M = \Sigma_{J_2} \oplus M_{2\beta} \oplus M^*$. Let $\pi_{M^*} : M \to M^*$ be the projection. Since $\bar{N}_2 = A_2 \not\subset \bar{T}_2$ and $\overline{\Sigma_{I_1} \oplus M_{1\alpha}} \oplus \overline{\Sigma_{J_2} \oplus M_{2\beta}} \subset \bar{T}_2$, $\overline{\pi_{M^*}(N_2)} \not\subset \bar{T}_2$. Let N_2' be a cyclic hollow direct summand of M^* with $\bar{N}_2' = \overline{\pi_{M^*}(N_2)}$. Since $\overline{\pi_{M^*}(N_2)} \not\subset \bar{T}_1$, $N_2' \not\approx M_1$ from the first step. We obtain the epimorphism $\pi|N_2 : N_2 \to N_2'$, similarly to step 1. Since $N_2' \not\approx M_1$, $N_2 \approx N_2'$. We have a contradiction, similarly to step 1. Hence $N_2 \not\approx M_i$ $(i = 1, 2)$.

Step 3. We can define $T(i) = T_1 \oplus \cdots \oplus T_i$ inductively, since $\{M_\gamma\}_L$ is a well-ordered set [take $T(\alpha) = \Sigma_{\beta<\alpha} T(\beta)$ for a limit ordinal α]. Thus we obtain a locally direct summand $T(\gamma_0) = \Sigma_{I_1} \oplus M_{1\alpha_1} \oplus \Sigma_{I_2} \oplus M_{2\alpha_2} \oplus \cdots \oplus \Sigma_{I_n} \oplus M_{n\alpha_n} \oplus \cdots$, which has the lifting property of simple modules and each $M_{i\alpha_i}$ is a cyclic hollow and c.ind. direct summand of M.

Step 4. Let $\{S_\varepsilon\}_P$ be a representative set of simple modules in $\bar{M}$. For each S_ε, we can obtain $T(\gamma_\varepsilon)$ by step 3, where $S_\varepsilon \approx \bar{M}_{\gamma_\varepsilon}$. Then $\Sigma_P T(\gamma_\varepsilon)$ is a locally direct summand of M by (9.4.4) and $\bar{M} = \Sigma_P \oplus \overline{T(\gamma_\varepsilon)}$ from the steps above. Thus $M' = \Sigma_P \oplus T(\gamma_\varepsilon)$ is the desired submodule of M by (9.2.5). The converse is clear.

9.4.7 COROLLARY Let R be a right artinian ring and M an R-module. Then M has the lifting property of simple modules if and only if $M = \Sigma_I \oplus M_\alpha$, where the M_α are indecomposable hollow, and $\{M_\alpha\}_I$ satisfies the conditions in (9.2.5).

Proof: If R is right artinian, J(M) is small in M and M/J(M) is semisimple.

9.5 RINGS WITH THE LIFTING PROPERTY

We have studied the lifting property of R-modules modulo radical. In this section we study rings R which have the lifting property modulo radical as R-modules. Further, a characterization of generalized uniserial rings, related to the lifting property, is given.

We denote the Jacobson radical, injective envelope, and the socle of an R-module M by J(M), E(M), and S(M), respectively. $M^{(J)}$ means the direct sum of $|J|$ copies of M, and $\|M\|$ and $|J|$ mean the composition length of M and the cardinal of J, respectively.

We put $J^n(M) = J(J^{n-1}(M))$ inductively. If α is a limit ordinal, we put $J^\alpha(M) = \cap_{\beta<\alpha} J^\beta(M)$. If $0 = J^\gamma(M)$ for some γ, we say that M has the *upper Loewy series* (see Sec. 7.6).

For any primitive idempotent e in R, if the set of R-submodules of eR is linear with respect to the inclusion, namely, eR is serial and has the upper Loewy series, we say that R is a right *generalized uniserial ring* (we usually consider a semiperfect ring).

9.5.1 THEOREM Let R be a ring. Then the following conditions are equivalent:

1. Every projective module has the lifting property of simple modules modulo radical.
2. R has the above property as a right R-module.
3. R contains a right ideal A satisfying the following conditions:
   (i) $A = \Sigma_I \oplus e_\alpha R$ is a locally direct summand of R and the $e_\alpha R$ are hollow (local).
   (ii) $S(R/J(R)) = [A + J(R)]/J(R)$.

Proof: (1) ⇒ (2). This is clear.

(2) ⇒ (3). We may assume that $S(\bar{R}) \neq 0$. We note that every indecomposable and cyclic hollow projective module M_α satisfies (9.2.1) and if $\bar{M}_\alpha \approx \bar{M}_\beta$, $M_\alpha \approx M_\beta$. Hence, noting the remark above, we can show, by the same argument given in the proof of (9.4.6), that R contains a right ideal A as in (3).

(3) ⇒ (1). Let P be a projective module. Then P is a direct summand of $R^{(J)}$ for some J. It is clear that $A^{(J)} = \Sigma_I \oplus e_\alpha R^{(J)}$ is a locally direct summand of $R^{(J)}$ and $S(\bar{R}^{(J)}) = \bar{A}^{(J)}$. Since $\mathrm{End}_R(e_\alpha R) \to \mathrm{End}_R(\overline{e_\alpha R}) \to 0$ is exact, A has the lifting property [cf. the proof of (9.4.1)]. Therefore, since J(P) = PJ(R), P has the lifting property of simple modules by (9.1.1).

9.5.2 COROLLARY Let R be a ring such that R/J(R) is artinian. Then R has the lifting property of simple modules as a right or left R-module if and only if R is semiperfect.

Proof: Since R/J(R) is artinian, R is equal to A in (9.5.1) if and only if R has the lifting property of simple modules as a right R-module. The concept of semiperfect ring is left- and right-sided symmetric.

9.5.3 THEOREM Let R be a semiperfect ring. Then the following conditions are equivalent:

1. For any primitive idempotent e, $eR \supset eJ \supset eJ^2 \supset \cdots \supset eJ^n \supset \cdots \supset e(\cap_n J^n)$ is a unique composition series of right ideals eA such that $\|eR/eA\| < \infty$ in eR.
2. Every direct sum of cyclic hollow modules with finite length has the lifting property of simple modules.

Proof: Since R is semiperfect, we have a complete set of mutually orthogonal primitive idempotents $\{e_i\}$ with $1 = \Sigma e_i$.

(1) ⇒ (2). Let M be a cyclic hollow module with finite length and let $M \approx e_iR/e_iA$ for some e_i in $\{e_j\}$ and a right ideal A. Then $e_iA = e_iJ^n$ for some integer n. Hence e_iA is a character submodule of e_iR. Accordingly, we have (2) by (9.2.5).

(2) ⇒ (1). We shall show that e_iJ^n/e_iJ^{n+1} is simple or zero by induction on n. We put $e = e_i$. Assume that $eJ^t \supsetneq eJ^{t+1}$. We take two right ideals eA_i (i = 1, 2) such that $eJ^t \supset eA_i \supset eJ^{t+1}$ and eJ^t/eA_i is simple. Since $\|eR/eA_1\| = \|eR/eA_2\|$, $eR/eA_1 \approx eR/eA_2$ by (9.2.5). Let f be any element in $\mathrm{End}_R(eR/eJ)$. Then f is given by the left-sided multiplication by a regular element x in eRe. f is lifted to an element in $\mathrm{Hom}_R(eR/eA_1, eR/eA_2)$ by (9.2.5). This homomorphism is given by the left-sided multiplication by x + j, where j is in eJe. Hence $(x + j)eA_1 = eA_2$. Since $jeA_1 \subset eJeJ^t \subset eA_2$, $xeA_1 \subset eA_2$, so $xeA_1 = eA_2$ for $\|eR/eA_1\| = \|eR/eA_2\|$. Now e is regular in eRe. Hence $eA_1 = eeA_1 = eA_2$. Accordingly, eJ^t/eJ^{t+1} is simple, since eJ^t/eJ^{t+1} is semisimple. Therefore, if $\|eR/eA\| < \infty$, $eA = eJ^k$ for some k. Let $\|eR/eA\| = \infty$. Since eJ^n is a unique maximal submodule

of eJ^{n+1}, $eA \subset eJ^n$ for all n. Therefore, $eA \subset \bigcap_n eJ^n = e(\bigcap_n J^n)$. Hence every right ideal in eR is either contained in $e(\bigcap_n J^n)$ or equal to eJ^n.

9.6 EXTENDING PROPERTY OF SIMPLE MODULES

We shall study the concept dual to the lifting property of R-modules modulo radical. The module dual to the hollow module is the uniform module and the module dual to the cyclic hollow module is a uniform module with nonzero socle. If a submodule N of an R-module M is essential in M, we shall indicate it by $M \supset_e N$.

Let K be a submodule of M. If there exists a direct summand N of M with $K \subset_e N$, we say that K is *extendible to a direct summand* of M. If every submodule with a certain property is extendible to a direct summand of M, we say that M has the *extending property of submodules* with the given property. Let $\{K_\gamma\}_J$ be a set of independent submodules of M. If there exists a direct decomposition $M = \Sigma_J \oplus N_\gamma \oplus N$ such that $K_\gamma \subset_e N_\gamma$ for each $\gamma \in J$, then we say that $\Sigma_J \oplus K_\gamma$ is *extendible to a direct decomposition* of M. If the sum of every set of independent submodules with a certain property is extendible to a direct summand of M. we say that M has the *extending property of the sum of independent set of submodules* with the given property.

Next we assume that $S(M) \neq 0$. If every simple submodule (or uniform submodule) K is extendible to a (uniform) direct summand of M which is completely indecomposable, then we say that M has the *extending property of simple modules* (*uniform modules*). Finally, we assume that $M = \Sigma_I \oplus M_\alpha$. If a submodule N of M is contained in a finite direct sum $\Sigma_{i=1}^n \oplus M_{\alpha_i}$, we say that N is *finitely contained* (with respect to the decomposition $M = \Sigma_I \oplus M_\alpha$). We shall denote this briefly by f.c. submodule.

We shall introduce the condition dual to (9.2.1).

Every monomorphism of an R-module M into itself is an isomorphism. (9.6.1)

If a uniform R-module M satisfies (9.6.1), M is a completely indecomposable module. In particular, indecomposable quasi-injective R-modules are completely indecomposable modules with (9.6.1) [see (9.6.12) below]. An artinian R-module clearly satisfies (9.6.1).

For a set $\{M_\alpha\}_I$ of R-modules with (9.6.1), we define a partial order $\lesssim^*$ as follows. If $M_\alpha \approx M_\beta$, we put $M_\alpha \equiv M_\beta$. If there exists a monomorphism of M_α to M_β, we define $M_\alpha \lesssim^* M_\beta$.

We start with the dual to (9.1.1).

9.6.1 PROPOSITION Let N be an R-module with extending property of uniform modules. Then every direct summand of N has the same property.

Proof: Let $N = N_1 \oplus N_2$ and A a uniform submodule in N_1. Then $N = K_1 \oplus K_2$ with $K_1 \supset_e A$. Since K_1 has the exchange property by (8.1.4), $N = K_1 \oplus N_1' \oplus N_2$ and $N_1 = N_1' \oplus [N_1 \cap (K_1 \oplus N_2)]$. Let $x \neq 0$ be in $N_1 \cap (K_1 \oplus N_2)$ and $x = k_1 + n_2$; $k_1 \in K_1$, $n_2 \in N_2$. Since $k_1 \neq 0$, there exists an r in R such that $k_1 r \neq 0 \in A$. Hence $xr - k_1 r = n_2 r \in N_1 \cap N_2 = 0$. Therefore, $A \subset_e N_1 \cap (K_1 \oplus N_2)$.

Remark In the proof above we know that A is essentially extendible to a direct summand of N_1 without the assumption "uniform for A," if K_1 has the exchange property.

9.6.2 COROLLARY Let N be as in (9.6.1). If the Goldie dimension of N is finite, $N = \Sigma_{i=1}^{n} \oplus N_i$ and the N_i are uniform and completely indecomposable modules.

Proof: This is clear.

9.6.3 COROLLARY Let N be an R-module with $S(N) \neq 0$. We assume that $S(N) = \Sigma_{i=1}^{n} \oplus A_i$, where the A_i are simple and that every simple submodule in S(N) is essentially extendible to a completely indecomposable direct summand of N. Then $N = \Sigma_{i=1}^{n} \oplus N_i \oplus K$, where the N_i are uniform, $\Sigma_{i=1}^{n} \oplus S(N_i) = S(N)$, and $S(K) = 0$.

Proof: This is clear.

9.6.4 PROPOSITION Let $\{E_\alpha\}_I$ be a set of injective and indecomposable R-modules and $E = \Sigma_I \oplus E_\alpha$. Then we have the following.

1. Each E_α is a uniform module with (9.6.1).
2. E has the extending property of f.c. uniform submodules.
3. Let $\{N_\beta\}_J$ be an independent set of f.c. uniform submodules. Then there exists a set $\{F_\beta\}_J$ of direct summands of E such that $N_\beta \subset_e F_\beta$ and $\Sigma_J \oplus F_\beta$ is a locally direct summand of E.

Proof: 1. This is clear by (4.4.10).

2. Let K be a f.c. uniform submodule of E, say $K \subset \Sigma_{i=1}^{n} \oplus E_i$. Since $\Sigma_{i=1}^{n} \oplus E_i = E'$ is injective, E(K) in E' is a direct summand of E', and hence of E. It is clear that E(K) is completely indecomposable, for K is uniform.

3. Let $\{F_\beta\}_J$ be as in (3). We shall show that $\Sigma_J F_\beta$ is a direct sum. We assume that $\{F_1, F_2, \ldots, F_n\}$ is independent. If $F_{n+1} \cap (\Sigma_{i=1}^{n} \oplus F_i) \neq 0$,

$N_{n+1} \cap (\Sigma_{i=1}^n \oplus F_i) = N_{n+1} \cap F_{n+1} \cap (\Sigma_{i=1}^n \oplus F_i) \neq 0$. Hence $N_{n+1} \cap (\Sigma_{i=1}^n \oplus N_i) \neq 0$ by (3.4.6), which is a contradiction. Therefore, we know by induction that $\{F_\beta\}$ is independent. Since F_β is injective, $\Sigma_J \oplus F_\beta$ is a locally direct summand of E.

From now on we assume that $M = \Sigma_I \oplus M_\alpha$ and each M_α is uniform and completely indecomposable, unless otherwise stated.

9.6.5 THEOREM Let M and $\{M_\alpha\}_I$ be as above. Then the following conditions are equivalent:

1. M has the extending property of f.c. uniform modules.
2. For any pair α and β in I, any homomorphism f of a submodule A_α of M_α to M_β is extendible to an element in $\mathrm{Hom}_R(M_\alpha, M_\beta)$ or $f^{-1}|f(A_\alpha)$ is extendible to an element in $\mathrm{Hom}_R(M_\beta, M_\alpha)$ provided that ker f = 0.

Proof: (1) ⇒ (2). Let M_1 and M_2 be two members of $\{M_\alpha\}_J$ and f : $A_1 \to M_2$. Put $A_1(f) = \{x + f(x) \mid x \in A_1\} \subset M_1 \oplus M_2$. Then $A_1(f)$ is a f.c. uniform submodule of M. Hence there exists a direct summand M' of M such that $M' \supset_e A_1(f)$. Since M' is indecomposable by (4.4.10), M' has the exchange property by (7.1.2) and (8.1.4). Hence we have $M = M' \oplus \Sigma_{I-\gamma} \oplus M_\delta$. If $\gamma \neq 1, 2$, $M = M' \oplus M_1 \oplus M_2$ and $M' \subset M_1 \oplus M_2$, which is a contradiction. Therefore, either $M = M' \oplus \Sigma_{I-\{1\}} \oplus M_\delta$ or $M = M' \oplus \Sigma_{I-\{2\}} \oplus M_\delta$. Assume that $0 \neq \ker f \subset M_1 \cap M'$. Then we have the former decomposition. Let π : $M \to M_2$ be the projection on this decomposition. Then $\pi|M_1$ is an extension of f. We assume that ker f = 0. If we have the first decomposition, we are done as above. Hence we assume the second decomposition. Put $A_2 = f(A_1)$. Then $A_1(f) = A_2(f^{-1})$. Hence f^{-1} is extendible to an element in $\mathrm{Hom}_R(M_2, M_1)$.

(2) ⇒ (1). Let N be a f.c. uniform submodule in M, say $N \subseteq \Sigma_{i-1}^n \oplus M_i$. Let $\pi_\alpha : M \to M_\alpha$ be the projection. Since $\cap \ker (\pi_i|N) = 0$ and N is uniform, there exists an i such that $\ker (\pi_i|N) = 0$. Hence by (2) and the proof of (9.2.5), there exists a j in $\{1,2,\ldots,n\}$, say 1, such that $N = \{x + f_2(x) + \cdots + f_n(x) \mid x \in \pi_1(N)\}$ and $f_i : M_1 \to M_i$ is an extension of $\pi_i (\pi_1|N)^{-1}$. Put $M' = \{y + f_2(y) + \cdots + f_n(y) \mid y \in M_1\}$. Then $\Sigma_{i=1}^n \oplus M_i = M' \oplus \Sigma_{i=2}^n \oplus M_i$ and $N \subset_e M'$.

9.6.6 THEOREM Let M and $\{M_\alpha\}_I$ be as in (9.6.5). We assume that each M satisfies (9.6.1) and $S(M_\alpha) \neq 0$. Then the following conditions are equivalent:

1. M has the extending property of simple modules.

2. The relation $\lesssim^*$ is linear for the subset $\{M_\beta \in \{M_\alpha\}_I \mid S(M_\beta) = S(M_{\beta_0})\}_{I_0}$, and $\mathrm{Hom}_R(S(M_i),S(M_j))$ is extendible to $\mathrm{Hom}_R(M_i,M_j)$ for any pair i and j in I_0 with $M_i \lesssim^* M_j$, where β_0 is a fixed index in I and runs through I.

Proof: (1) ⇒ (2). In the proof of (9.6.5), if A_1 is simple, so is $A_1(f)$. Hence we can use the same argument and the assumption (9.6.1).

(2) ⇒ (1). Every simple submodule is f.c.

9.6.7 PROPOSITION Let M be an R-module. We assume that $M = M_1 \oplus M_2 \oplus M_3$, N is a submodule of M_1, and $f \in \mathrm{Hom}_R(N,M_2)$. If there exists a direct summand T of M such that $M = T \oplus M_2 \oplus M_3$ and $T \supset N(f)$, then f is extendible to an element in $\mathrm{Hom}_R(M_1,M_2)$. Conversely, we assume that $M = T \oplus T'$. Let A be a submodule of T and g an element in $\mathrm{Hom}_R(A,T')$. If g is extendible to an element f in $\mathrm{Hom}_R(T,T')$, we have a decomposition $M = T'' \oplus T'$ such that $T'' \supset A(g)$.

Proof: Let π_T and π_{M_2} be the projections of M with respect to the decomposition $M = T \oplus M_2 \oplus M_3$ and $M = M_1 \oplus M_2 \oplus M_3$, respectively. Then $-\pi_{M_2}(\pi_T|M_1) \in \mathrm{Hom}_R(M_1,M_2)$ is an extension of f. Put $T'' = T(f)$ for the second assertion.

Since (9.6.1) is dual to (9.1.1), we obtain the following, similarly to (9.2.8).

9.6.8 THEOREM Let $\{M_\alpha\}_I$ and M be as in (9.6.6). We assume that $\{M_\alpha\}_I$ is locally semi-T-nilpotent. Then the following conditions are equivalent:

1. M has the extending property of simple modules.
2. M has the extending property of submodules in S(M).
3. Every direct summand of M has the above property.

9.6.9 THEOREM Let $M = \Sigma_I \oplus M_\alpha$ be as in (9.6.6). Then the following conditions are equivalent:

1. For any independent family $\{N_\beta\}_J$ of uniform direct summands of M, $N = \Sigma_J \oplus N_\beta$ is a locally direct summand of M.
2. Every monomorphism of M_α to M_β is an isomorphism for any pair α and β in I.

Proof: (1) ⇒ (2). We assume that $f : M_\alpha \to M_\beta$ is a monomorphism, where $\alpha,\beta \in I$ and $\alpha \neq \beta$. Then $M_\alpha(f) \cap M_\beta = 0$ and $M_\alpha(f) \oplus M_\beta = M_\alpha \oplus M_\beta$, from which we see that $M_\alpha(f) \approx M_\alpha$ and $M_\alpha(f)$ is a direct summand of $M_\alpha \oplus M_\beta$. Furthermore, $M_\alpha(f) \cap M_\alpha = 0$. Hence, by (1), $M_\alpha(f) \oplus M_\alpha = M_\alpha \oplus M_\beta$, and hence f is an epimorphism.

(2) ⇒ (1). We note that if $M_\alpha \approx M_\beta$ for $\alpha \neq \beta$, then M_α satisfies (9.6.1). To show (1), we may show the following. If $\{N_1,N_2,\ldots,N_n\}$ is an

independent set of uniform direct summands of M, then $N_1 \oplus \cdots \oplus N_n$ is also a direct summand of M. If $n = 1$, this is clear. Assume that $n > 1$ and $N_1 \oplus \cdots \oplus N_{n-1}$ is a direct summand of M. Since each N_i is a direct summand of M and is uniform, it is isomorphic to some member in $\{M_\alpha\}_I$ by (7.1.2). Since $N_1 \oplus \cdots \oplus N_{n-1}$ has the exchange property, we have

$$M = N_1 \oplus \cdots \oplus N_{n-1} \oplus \Sigma_J \oplus M_\gamma \qquad \text{for some } J \subset I$$

Since N_n has the exchange property, we have either $M = N_1 \oplus \cdots \oplus N_{k-1} \oplus N_{k+1} \oplus \cdots \oplus N_{n-1} \oplus N_n \oplus \Sigma_J \oplus M_\gamma$ for some k or $M = \Sigma_{i=1}^{n} \oplus N_i \oplus \Sigma_{J-\{\sigma\}} \oplus M_\gamma$ for some $\sigma \in J$. We are done in the latter case. In the former case $N_k \approx N_n$. Let $\pi_\gamma : N_1 \oplus \cdots \oplus N_{n-1} \oplus \Sigma_J \oplus M_\gamma \to M_\gamma$ be the projection. Noting that $N_n \not\subset N_1 \oplus \cdots \oplus N_{n-1}$, we can easily see that there exists $0 \neq x \in N_n$ and some ρ in $I - J$ for which $\pi_\rho|xR$ is a monomorphism. Since $xR \subset_e N_n$, it follows that $\pi_\rho|N_n$ is a monomorphism. Since $N_n \approx N_k \approx M_\alpha$ for some α in $I - J$, $\pi_\rho|N_n$ is an isomorphism by (2). Thus $M = N_n \oplus \ker \pi_\rho = N_1 \oplus \cdots \oplus N_n \oplus \Sigma_{J-\{\rho\}} \oplus M_\beta$.

9.6.10 COROLLARY Assume that each M_α is uniform. Then the following conditions are equivalent:

1. If N is as in (9.6.9), then N is a direct summand of M.
2. $\{M_\alpha\}_I$ is a locally semi-T-nilpotent set and (2) of (9.6.9) holds.

Proof: Use the same argument as in (9.3.2).

We obtain another theorem which is similar to (9.6.5).

9.6.11 THEOREM Let $\{M_\alpha\}_I$ be a set of uniform and completely indecomposable modules and $M = \Sigma_I \oplus M_\alpha$. We assume that for any pair α, β in I, every monomorphism of M_α to M_β is an isomorphism. Then the following conditions are equivalent:

1. M has the extending property of uniform modules.
2. For any $\alpha \in I$ and any submodule A_α of M_α, $\mathrm{Hom}_R(A_\alpha, \Sigma_{I-\{\alpha\}} \oplus M_\beta)$ is extendible to $\mathrm{Hom}_R(M_\alpha, \Sigma_{I-\{\alpha\}} \oplus M_\beta)$.

Proof: We denote the projection of M onto M_α by π_α for $\alpha \in I$.

(1) ⇒ (2). Let f be an element in $\mathrm{Hom}_R(A_\alpha, \Sigma_{I-\{\alpha\}} \oplus M_\beta)$. First we assume that $\ker f \neq 0$. Since $A(f)$ is uniform, M has a direct summand M' such that $M' \supset_e A(f)$ and M' is completely indecomposable. Since $\ker f \neq 0$, we have a decomposition $M = M' \oplus \Sigma_{I-\{\delta\}} \oplus M_\beta$ as the proof of (9.6.5). Hence f is extendible to $\mathrm{Hom}_R(M_\alpha, \Sigma_{I-\{\alpha\}} \oplus M_\beta)$ by (9.6.7). Next, we assume that $\ker f = 0$. For $x \neq 0$ in A_α, there exists a finite subset

$K = \{1,2,\ldots,n\} \subset I - \{\alpha\}$ such that $\pi_\beta f(x) = 0$ if $\beta \notin K$ and $\pi_{\beta'} f(x) \neq 0$ if $\beta' \in K$. Put $\pi_{I-K} = \Sigma_{I-K}\, \pi_\beta$. Then ker $\pi_{I-K} f$ contains x. Now $f = \pi_K f + \pi_{I-K} f$ and A_α is uniform. Hence we may assume ker $\pi_i f = 0$ for all $i \in K$. Then there exists a $g_i : M_\alpha \to M_i$ such that $g_i | A_\alpha = \pi_i f$ by (9.6.5) and the assumption. Since ker $\pi_{I-K} f \neq 0$, we obtain, from the first observation, $g : M_\alpha \to \Sigma_{I-K-\{\alpha\}} \oplus M_\beta$ such that $g|A_\alpha = \pi_{I-K} f$. Hence $\Sigma\, g_i + g$ is an extension of f.

(2) ⇒ (1). Let N be a nonzero uniform submodule of M. Then there exists a finite subset $K' = \{1,2,\ldots,n'\}$ such that ker $\pi_i | N \neq 0$ for $i \notin K'$ and ker $(\pi_j | N) = 0$ for $j \in K'$. Hence $N = \{x_1 + f_2(x_1) + \cdots + f_{n'}(x_1) + f(x_1) \mid x_1 \in \pi_1(N),\ f_i = (\pi_i|N)(\pi_1|N)^{-1}$, and $f = \pi_{I-K'}(\pi_1|N)^{-1}\}$. Since $F = f_2 + \cdots + f_{n'} + f$ is an element in $\mathrm{Hom}_R(\pi_1(N), \Sigma_{I-\{1\}} \oplus M_\beta)$, $M_1(g)$ is a direct summand of M by (9.6.7) and $M_1(g) \supset_e N$, where g is an extension of f in $\mathrm{Hom}_R(M_1, \Sigma_{I-\{1\}} \oplus M_\beta)$.

9.6.12 COROLLARY Let T be a completely indecomposable and uniform module. Then T is quasi-injective if and only if $T \oplus T$ has the extending property of uniform modules and T satisfies (9.6.1).

Proof: Let T be a uniform and quasi-injective module. If $f : T \to T$ is a monomorphism, f is extendible to a g in $\mathrm{End}_R(E(T))$. Hence g is an isomorphism, since E(T) is indecomposable and g is a monomorphism. Hence $g^{-1}(T) \subset T$, so $T = g(T) = f(T)$ by (5.4.2). Therefore, T satisfies (9.6.1). The converse is clear.

9.6.13 THEOREM We further assume in (9.6.11) that $\{M_\alpha\}_I$ is locally semi-T-nilpotent. Then the following conditions are equivalent:

1. M has the extending property of submodules.
2. For any subset J of I and any submodule A of $\Sigma_J \oplus M_\beta$, $\mathrm{Hom}_R(A, \Sigma_{I-J} \oplus M_\alpha)$ is extendible to $\mathrm{Hom}_R(\Sigma_J \oplus M_\beta, \Sigma_{I-J} \oplus M_\alpha)$.

Proof: (1) ⇒ (2). Let J be a subset and A a submodule of $\Sigma_J \oplus M_\beta$. Let f be an element in $\mathrm{Hom}_R(A, \Sigma \oplus M_\alpha)$. Put $P = \Sigma_J \oplus M_\beta$ and $Q = \Sigma_{I-J} \oplus M_\alpha$. To show that $\mathrm{Hom}_R(A,Q)$ is extendible to $\mathrm{Hom}_R(P,Q)$, we may assume that $A \subset_e P$ (3.4.7). Now, by (1) there exists a direct summand B of M such that $A(f) \subset_e B$. Then $B \cap Q = 0$ and $B \oplus Q \subset_e M$. Since B is a direct summand of M, B is also a direct sum of completely indecomposable modules by (7.3.15). Hence we see that $M = B \oplus Q$ by (9.6.10). Hence f is extendible to an element in $\mathrm{Hom}_R(P,Q)$ by (9.6.7).

(2) ⇒ (1). Let $A \neq 0$ be a submodule of M. If A is not essential in M, there exists M_{α_1} with $A \cap M_{\alpha_1} = 0$ by (3.4.6). By Zorn's lemma we can

find $J \subset I$ such that $A \cap \Sigma_{I-J} \oplus M_\alpha = 0$ and $A \cap \Sigma_K \oplus M_\gamma \neq 0$ for any $K \supsetneqq I - J$. Again we put $P = \Sigma_I \oplus M_\beta$ and $Q = \Sigma_{I-J} \oplus M_\alpha$. Let π and π' be the projections: $M = \Sigma_I \oplus M_\alpha \to P$ and $M = \Sigma_I \oplus M_\alpha \to Q$, respectively. Then we see $\pi(A) \subset_e P$, since $\pi(A) \oplus Q = A \oplus Q \subset_e M$. Since $A \cap Q = 0$, the mapping $f : \pi(A) \to Q$ given by $f(\pi(a)) = \pi'(a)$ is a homomorphism. By (2), f is extendible to a homomorphism f' of P to Q. Then $A \subset P(f')$ is a direct summand of M by (9.6.7). Moreover, $\pi(A) \subset_e P$ shows $A \subset_e P(f')$.

9.7 EXTENDING PROPERTY OF DECOMPOSITIONS

In this section we study modules that are direct sums of uniform modules.

9.7.1 THEOREM Let $\{M_\alpha\}_I$ be a set of uniform modules (not necessarily completely indecomposable) and $M = \Sigma_I \oplus M_\alpha$. Then the following conditions are equivalent:

1. M has the extending property of direct sums of two independent submodules.
2. M has the extending property of direct sums of finite independent submodules.
3. Let N_1 and N_2 be any two independent submodules of M. Then the projection $N_1 \oplus N_2$ to N_1 is extendible to an element in $\mathrm{End}_R(M)$.
4. Let N_i be as in (3). Then any element in $\mathrm{Hom}_R(N_1, N_2)$ is extendible to an element in $\mathrm{End}_R(M)$.

In this case for every direct summand K of M, there exists a subset J of I such that $M = K \oplus \Sigma_J \oplus M_\gamma$. If the N_i in (3) are direct summands of M, so is $N_1 \oplus N_2$. Further, if f is a monomorphism of N_1 to N_2, then im f is a direct summand of N_2.

Proof: Let N be a submodule of M. Then we can find, by Zorn's lemma, a subset J of I such that $\{M_\gamma, N\}_J$ is independent and $M \supset_e N \oplus \Sigma_J \oplus M_\gamma$.

(1) ⇒ (3). Let N_i be as in (3). Then by (1) we have a decomposition $M = T_1 \oplus T_2 \oplus T_3$ with $T_i \supset_e N_i$ $(i = 1, 2)$. The projection of M onto T_1 is a desired extension.

(3) ⇒ (2). First we assume that S_1 and S_2 are independent and $M \supset_e S_1 \oplus S_2$. Then there exists a subset J of I such that $M \supset_e S_1 \oplus \Sigma_J \oplus M_\gamma$ $(= L)$. Let $f : L \to \Sigma_J \oplus M_\gamma$ be the projection. Then there exists an element g in $\mathrm{End}_R(M)$ with $g|L = -f$. Let π_J be the projection of M onto $\Sigma_J \oplus M_\gamma$ with respect to $M = \Sigma_J \oplus M_\gamma \oplus \Sigma_{I-J} \oplus M_\varepsilon$. Put $F = \pi_J g$ and $M_{I'}(F) = \{x + F(x) \mid x \in \Sigma_{I'} \oplus M_\varepsilon\}$, where $I' = I - J$. Then $M = M(F) \oplus \Sigma_J \oplus M_\gamma$. Let s be an element in S_1. Then $s = \pi_J(s) + \pi_{I'}(s)$ and $0 = F(s) = F\pi_J(s) +$

$F\pi_{I'}(s) = -\pi_J(s) + F\pi_{I'}(s)$. Hence S_1 is essential in $M_{I'}(F)$ for $S_1 \oplus \Sigma_J \oplus M_\gamma \subset_e M = M_{I'}(F) \oplus \Sigma_J \oplus M_\gamma$. Since $S_1 \subset_e M_{I'}(F)$ and $S_1 \cap S_2 = 0$, $S_2 \cap M_{I'}(F) = 0$, and $S_2 \oplus M_{I'}(F) \subset_e M$. Further, $M = M_{I'}(F) \oplus \Sigma_J \oplus M_\gamma$ and $M_{I'}(F) \approx \Sigma_{I-J} \oplus M_\varepsilon$. Therefore, we can obtain similarly to the above that $M = M_{I'}(F) \oplus M_J(F')$ and $M_J(F') \subset_e S_2$ [see the proof of (9.3.1)]. We note that $M_{I'}(F) \approx \Sigma_{I-J} \oplus M_\varepsilon$ and condition 3 is valid for a direct summand. Thus we can prove (1) and (2) by the first observation and induction on the number of independent submodules.

(2) ⇒ (1). This is clear.

(1) ⇒ (4). Let f be an element in $\mathrm{Hom}_R(N_1,N_2)$. Then $N_1 \oplus N_2 = N_1(f) \oplus N_2$. There exists a decomposition $M = T_1 \oplus T_2 \oplus T_3$ such that $N_1(f) \subset_e T_1$ and $N_2 \subset_e T_2$. Then $-\pi_2|N_1$ is a desired extension of f, where $\pi_2 : M \to T_2$ is the projection.

(4) ⇒ (1). We shall quote the same argument as (3) ⇒ (2). We use the same notation. Let $M \supset_e S_1 \oplus \Sigma_J \oplus M_\gamma$. Since $S_1 \cap \Sigma_J \oplus M_\gamma = 0$, $\pi_{I-J}|S_1$ is an isomorphism. Put $S_1' = \mathrm{im}\,(\pi_{I-J}|S_1)$. Then $S_1 = \{a + f(a) \mid a \in S_1'$, $f = \pi_J(\pi_{I-J}|S_1)^{-1} : S_1' \to \Sigma_J \oplus M_\gamma\}$. Let g in $\mathrm{End}_R(M)$ be an extension of f. Put $G = \pi_J g \pi_{I-J}$ and $M_{I'}(G) = \{b + G(b) \mid b \in \Sigma_{I-J} \oplus M_\varepsilon\}$. Then $M = M_{I'}(G) \oplus \Sigma_J \oplus M_\gamma$ and $M_{I'}(G) \supset_e S_1$. Similarly, we obtain $M = M_{I'}(G) \oplus M_J(G')$ and $M_J(G') \supset_e S_2$. For the remaining assertions, we assume that S_1 is a direct summand of M. Then $S_1 = M_{I'}(G)$, so $M = S_1 \oplus \Sigma_J \oplus M_\gamma$. Let f in $\mathrm{Hom}_R(N_1,N_2)$ be a monomorphism. Then $N_1(f) \cap N_1 = 0$, so $N_1 \oplus N(f)$ is a direct summand of $N_1 \oplus N_2$. Let π be the projection of $N_1 \oplus N_2$ onto N_2. Them im f = $\pi(N_1(f))$ is a direct summand of N_2.

9.7.2 THEOREM Let $\{M_\alpha\}_I$ and M be as in (9.7.1). Then the following conditions are equivalent:

1. M has the extending property of finite direct sums of f.c. uniform modules.
2. $\mathrm{Hom}_R(A_\alpha,M_\beta)$ is extendible to $\mathrm{Hom}_R(M_\alpha,M_\beta)$ for any $\alpha \neq \beta$ in I and $A_\alpha \subset M_\alpha$.

Proof: (1) ⇒ (2). Let f be in $\mathrm{Hom}_R(A_\alpha,M_\beta)$ and put $A(f) = \{a + f(a) \mid a \in A_\alpha\}$. We consider the direct sum $A(f) \oplus M_\beta$. Then there exists a decomposition $M = S_\alpha \oplus S_\beta \oplus S$ such that $S_\alpha \supset_e A(f)$ and $S_\beta \supset_e M$. Since M_β is a direct summand of M, $S_\beta = M_\beta$. Let $\pi : T \to S_\beta = M_\beta$ be the projection for the decomposition $T = S_\alpha \oplus M_\beta \oplus S$. Then $-\pi|M_\alpha$ is an extension of f [cf. (9.6.7)].

(2) ⇒ (1). Let $N = \Sigma_{i=1}^{n} \oplus M_i$ be a submodule of M with N_i f.c. uniform. Then we may assume that $N \subset \Sigma_{i=1}^{m} \oplus N_i \subset M$. Hence we assume that $M = \Sigma_{i=1}^{m} \oplus M_i$. First we assume that there exists a set of uniform direct summands T_i of M for $i < k$ such that $T_i \supset_e N_i$, $T_i \approx M_{\rho(i)}$ and $M = \Sigma_{i=1}^{k} \oplus T_i \oplus \Sigma_{j=k+1}^{m} \oplus M_{\rho(j)}$, where ρ is a permutation of $\{1,2,\ldots,m\}$. Let π_p be the projection of M onto T_p $(p \geq k)$ or onto $M_{\rho(p)}$ $(p \geq k+1)$ for the decomposition above. Since $\bigcap_p \ker(\pi_p|N_{k+1}) = 0$, $\pi_q|N_{k+1}$ is a monomorphism for some q. Assume that $L = \bigcap_{s\geq k+1} \ker(\pi_s|N_{k+1}) \neq 0$. Then $L \subset \Sigma_{i=1}^{k} \oplus T_i$ and $\Sigma_{i=1}^{k} \oplus N_i \subset_e \Sigma_{i=1}^{k} \oplus T_i$, which is a contradiction. Hence we may assume that $q = \rho(k+1)$. Then if we put $N'_{k+1} = \operatorname{im}(\pi_{k+1}|N_{k+1}) \subset M_{\rho(k+1)}$, $N_{k+1} = \{f_1(a) + f_2(a) + \cdots + f_k(a) + a + f_{k+2}(a) + \cdots + f_m(a) \mid a \in N'_{k+1}\}$, where $f_s = \pi_s(\pi_{k+1}|N_{k+1})^{-1}$ [cf. the proof of (9.6.11)]. Since $T_i \approx M_{\rho(i)}$, there exists a set of homomorphisms $\{g_j \in \operatorname{Hom}_R(M_{\rho(k+1)},K_j) : K_j = T_j$ or $K_j = M_{\rho(j)}\}$ such that $g_j|N'_{k+1} = f_j$. Put $M_{\rho(k+1)}(g) = \{g_1(b) + \cdots + g_k(b) + b + g_{k+2}(b) + \cdots + g_n(b) \mid b \in M_{k+1}\}$. Then $M = \Sigma_{i=1}^{k} \oplus T_i \oplus M_{\rho(k+1)}(g) \oplus \Sigma_{p\geq k+2} \oplus M_{\rho(p)}$. It is clear that $N_{k+1} \subset_e M_{\rho(k+1)}(g) \approx M_{\rho(k+1)}$. Therefore, we can prove (9.7.2) by induction.

9.7.3 COROLLARY We assume that each M_α is uniform and completely indecomposable and further $\{M_\alpha\}_I$ is a locally semi-T-nilpotent. We put $M = \Sigma_I \oplus M_\alpha$. Then the following conditions are equivalent:

1. M has the extending property of direct sums of f.g. uniform modules.
2. $\operatorname{Hom}_R(A_\alpha,M_\beta)$ is extendible to $\operatorname{Hom}_R(M_\alpha,M_\beta)$ for $\alpha \neq \beta$ in I and any f.g. submodule A_α of M_α.

Proof: (1) ⇒ (2). We can use the same argument as the proof of (1) ⇒ (2) in (9.7.2).

(2) ⇒ (1). Let $\{A_\alpha\}_I$ be a set of independent and f.g. uniform submodules of M with $M \supset_e \Sigma_I \oplus A_\alpha$. We may assume that I is a well-ordered set and we shall use the same argument as in the proof of (9.2.8). We assume for each $\kappa \leq \beta < \alpha$ that there exist direct summands T_κ such that $T_\kappa \supset_e A_\kappa$ and $\Sigma_{\kappa\leq\beta} \oplus T_\kappa$ is a locally direct summand of M and $\{T_\kappa\}_{\kappa\leq\beta} \subset \{T_{\kappa'}\}_{\kappa'\leq\beta'}$ if $\beta \leq \beta'$. Then $M = \Sigma_{\beta<\alpha} \oplus T_\beta \oplus T$ and $T_\beta \supset_e A_\beta$ by (7.3.9), since $\{M_\alpha\}_I$ is semi-T-nilpotent. We may assume that $T = \Sigma_P \oplus N_\delta$, where each N_δ is isomorphic to some in $\{M_\alpha\}_I$ by (7.3.15). Let $\pi : M \to T$ and $\pi_\delta : T \to N_\delta$ be the projections of M and T, respectively. Since $(\Sigma_{\beta<\alpha} \oplus A_\beta) \cap A_\alpha = 0$ and $\Sigma_{\beta<\alpha} \oplus T_\beta \supset_e \Sigma_{\beta<\alpha} \oplus A_\beta$, $\pi|A_\alpha$ is a monomorphism. A_α being f.g. uniform, $\pi_\delta\pi|A_\alpha$ is a monomorphism for some δ. Making use of the method in the proof of (9.7.2), we can obtain $M = \Sigma_{\beta<\alpha} \oplus T_\beta \oplus \Sigma_{\delta'\neq\delta} \oplus$

$T_{\delta'} \oplus T_\delta(f)$ and $T_\delta(f) \supset_e A_\alpha$. Hence we have proved (2) ⇒ (1) by transfinite induction.

9.7.4 COROLLARY Let $\{M_\alpha\}_I$ be a set of uniform modules with nonzero socles and $M = \Sigma_I \oplus M_\alpha$. Then the following conditions are equivalent.

1. M has the extending property of finite direct sums of simple modules.
2. $\mathrm{Hom}_R(S(M_\alpha), S(M_\beta))$ is extendible to $\mathrm{Hom}_R(M_\alpha, M_\beta)$ for any $\alpha \neq \beta$ in I.

Proof: This is clear from the proof of (9.7.2).

Remark Let R be a self-injective local, but not perfect, ring (see Example in Sec. 10.3 [191]). Then $R^{(I)}$ has the extending property of finite direct sums of f.c. uniform modules; however, $R^{(I)}$ does not have the extending property of infinite direct sums of uniform modules if I is infinite [see (9.7.5) below].

Let T_R and N_R be R-modules. We take a set of R-submodules N_n of N. Let $\{f_n \in \mathrm{Hom}_R(T, N_n)\}_n$ be a set of homomorphisms. If, for $t \in T$, $f_n(t) = 0$ for almost all n, $\{f_n\}$ is called summable.

9.7.5 THEOREM Let $\{M_\alpha\}_I$ be a set of completely indecomposable and uniform modules and $M = \Sigma_I \oplus M_\alpha$. Then the following conditions are equivalent:

1. M has the extending property of direct sums of independent uniform submodules.
2. $\{M_\alpha\}_I$ is locally semi-T-nilpotent, and for any set of summable homomorphisms $\{f_\beta \in \mathrm{Hom}_R(A_\alpha, M_\beta)_{\beta\neq\alpha} : \alpha, \beta \in I\}$, there exists a set of summable homomorphisms $\{F_\beta \in \mathrm{Hom}_R(M_\alpha, M_\beta) \mid F_\beta$ is an extension of $f_\beta\}$, where the A_α are submodules of the M_α.

Proof: (1) ⇒ (2). Let $\{M_i\}_{i=1}^\infty$ be a countable subset of $\{M_\alpha\}_I$ and $\{f_i : M_i \to M_{i+1}\}$ a family of nonisomorphisms. Put $M_i' = M_i(f_i)$. Then $\{M_i'\}$ is an independent set of uniform submodules. Hence by (1) there exists a decomposition $M = \Sigma_{i=1}^\infty \oplus N_i \oplus N$ and $N_i \supset_e M_i'$ for all i. Since M_i' is a direct summand of M, $N_i = M_i'$. Therefore, $\Sigma_{i=1}^\infty \oplus M_i'$ is a direct summand of $\Sigma_{i=1}^\infty \oplus M_i$. Accordingly $\{M_\alpha\}_I$ is locally semi-T-nilpotent by (7.2.7). Let $\underline{F}$ be any set of summable homomorphisms f_β in $\mathrm{Hom}_R(A_\alpha, M_\beta)$; $\beta \neq \alpha$ and $A_\alpha \subset M_\alpha$. Since $\underline{F}$ is summable, $A_\alpha(\underline{F}) = \{a + \Sigma_{\underline{F}} f_\beta(a) \mid a \in A\}$ is an R-submodule of M and $M \supset_e A(\underline{F}) \oplus \Sigma_{\beta\neq\alpha} \oplus M_\beta$. Then we have a direct decomposition $M = M_1' \oplus \Sigma_{\beta\neq\alpha} \oplus M_\beta$ by (1). Let $\pi_\beta : M \to M_\beta$ be the projection. Then $\{F_\beta = -\pi_\beta|M_\alpha\}$ is the desired set.

(2) ⇒ (1). Let $M = \Sigma_I \oplus M_\alpha$ be any decomposition as in (9.7.5) and B a uniform submodule of M. Let $\pi_\alpha : M \to M_\alpha$ be the projection for each $\alpha \in I$.

Then $\cap_\alpha \ker \pi_\alpha = 0$. Let $b \neq 0$ be an element in B and $b = \Sigma_{i=1}^n \pi_{\alpha_i}(b)$. Then $b \in \cap_{\beta \in I - \{\alpha_i\}} \ker \pi_\beta$. Hence there exists π_{α_i} such that $\pi_{\alpha_i}|B$ is a monomorphism. Now by (2), the proof of (9.7.2) and (9.7.3), and the above observation, we can obtain (1) by making use of transfinite induction.

We add here a result similar to (9.7.5).

9.7.6 THEOREM We further assume in (9.7.5) that each M_α satisfies (9.6.1). Then the following conditions are equivalent:

1. M has the extending property of direct sums of f.c. uniform modules.
2. (i) $\{M_\alpha\}_I$ is a locally semi-T-nilpotent set.
   (ii) Every monomorphism of M_α to M_β is an isomorphism for any pair α and β in I.
   (iii) For any pair α and β in I and any submodule A_α of M_α, $\mathrm{Hom}_R(A_\alpha, M_\beta)$ is extendible to $\mathrm{Hom}_R(M_\alpha, M_\beta)$.
3. (i) $\{M_\alpha\}_I$ is a locally semi-T-nilpotent set.
   (ii) Every monomorphism of M_α to M_β is an isomorphism for any pair α and β in I.
   (iii) M has the extending property of f.c. uniform modules.

Proof: (1) ⇒ (2). We have (i) and (ii) in (2) from the proof of (9.7.5) and (9.6.9), and (iii) is clear by (9.6.5). The converse is also clear by (9.6.5), (9.6.9), and (7.3.15). Similarly, we can show the remaining assertions.

9.8 MODULES WITH THE EXTENDING PROPERTY

We shall present results dual to those in Sec. 9.3.

9.8.1 THEOREM Let M be an R-module. We assume that:

a. S(M) is essential in M.
b. $\mathrm{End}_R(S(M))$ is extendible to $\mathrm{End}_R(M)$.

Then the following conditions are equivalent:

1. M has the extending property of simple modules.
2. M contains a submodule M' as follows:
   (i) $M' = \Sigma_I \oplus M_\alpha$, where the M_α are uniform and completely indecomposable and S(M') = S(M).
   (ii) M' is a locally direct summand of M and has the entending property of finite direct sums of simple modules.

In this case M has the extending property of finite direct sums of simple modules.

Proof: (2) ⇒ (1). Let S be a finite direct sum of simple submodules in M. Then $S \subset \Sigma_{i=1}^{n} \oplus S(M_{\alpha_i})$ by (i). Since $\Sigma_{i=1}^{n} \oplus M_{\alpha_i}$ is a direct summand of M by (ii) and has the extending property of finite direct sums of simple submodules by (9.7.4), M has the same property

(1) ⇒ (2). We can use the argument dual to the proof (2) ⇒ (1) of (9.4.1). Let $\underline{N}$ be the set of submodules N' in M such that $N' = \Sigma_J \oplus N_\gamma$, where the N_γ are uniform and completely indecomposable and N' is a locally direct summand of M. Let N be a maximal element in $\underline{N}$. We shall show $N \subset_e M$. There exists a submodule A of S(M) such that $M \supset_e N \oplus A$ by (a). Let K be any finite subset of J and put $N_0 = \Sigma_K \oplus N_\gamma$. Then $M = N_0 \oplus P$ and $S(M) = S(N_0) \oplus \Sigma_{J-K} \oplus S(N_\gamma) \oplus A$. Let π' be the projection of S(M) onto $S(N_0)$ and $\pi = \pi'|S(P)$. Then we obtain an f in $\mathrm{Hom}_R(P,N_0)$ such that $f|S(P) = -\pi$ by (b). Hence $M = P(f) \oplus N_0$ and $S(P(f)) = \Sigma_{J-K} \oplus S(N_\delta) \oplus A$. If $A \neq 0$, there exists a direct summand T of P(f) with S(T) = A by (9.6.1). Hence $\Sigma N_\gamma \oplus T$ is a locally direct summand of M, which contradicts the maximality of N. Therefore, A = 0. Let N_1 and N_2 be in $\{N_\gamma\}_J$. Then $M = N_1 \oplus N_2 \oplus M_0$ and we know from (b) that $\mathrm{Hom}_R(S(N_1),S(N_2))$ is extendible to $\mathrm{Hom}_R(N_1,N_2)$. Hence N has the extending property of finite direct sums of simple submodules by (9.7.4).

9.8.2 THEOREM Let M be an R-module. We assume that $M \supset_e S(M)$ and every uniform direct summand of M is artinian. Then M has the extending property of simple modules if and only if M contains a submodule M' satisfying the following:

1. $M \supset_e M'$, so $S(M) = S(M')$.
2. $M' = \Sigma_I \oplus M_\alpha$ with M_α uniform.
3. $\Sigma_I \oplus M_\alpha$ is a locally direct summand of M.
4. M' has the extending property of simple modules.

Proof: We note that every artinian module satisfies (9.6.1), and (9.8.2) is dual to (9.4.6). Therefore, we can prove (9.8.2) by making use of the argument dual to the proof of (9.4.6).

9.8.3 PROPOSITION Let R be a right noetherian ring and M an R-module. We assume that every uniform direct summand of M is artinian and $M \supset_e S(M)$. If M has the extending property of simple modules, then M is a direct sum of uniform submodules.

Proof: From the assumption and (9.8.2), M contains a submodule M' such that $M' = \Sigma_I \oplus M_\alpha$ is a locally direct summand of M and $S(M) = S(M')$, where the M_α are uniform. We shall show $M = M'$. If $M \neq M'$, there exist an x in M - M' and a nonzero right ideal A such that $xA \subset M'$ for $M \supset_e S(M) = S(M')$. Since R is right noetherian, we can find a maximal member B among A above; i.e., $m \in M - M'$ and $mB \subset M'$. Now, mB is finitely generated, so $mB \subset \Sigma_J \oplus M_{\alpha_i}$ $(= M_J)$ for some finite subset J of I. M' being a locally direct summand of M, $M = M_J \oplus M^*$. Let $m = m_1 + m_2$; $m_1 \in M_J$, $m_2 \in M^*$. Then $m_2B \subset mB + m_1B \subset M_J$ and $m_2B \subset M_J \cap M^* = 0$. Since $m_2 \notin M'$ and B is a maximal member, $m_2R \approx R/B$. Let $S(m_2R) \approx C/B$. Then $m_2C = S(m_2R) \subset S(M) = S(M') \subset M'$, which is a contradiction. Therefore, M = M'.

9.8.4 COROLLARY Let R be a right artinian ring such that every indecomposable R-injective is artinian. Then M has the extending property of simple modules if and only if M is equal to M' in (9.8.2).

Proof: This is clear by (9.8.3).

9.9 QUASI-INJECTIVE MODULES AND THE EXTENDING PROPERTY

We shall consider special modules with extending property.

9.9.1 PROPOSITION If M is a quasi-injective R-module, then M has the extending property of submodules. If we assume further that M is a direct sum of completely indecomposable modules, say $M = \Sigma_I \oplus M_\alpha$, then M has the extending property of direct sums of submodules.

Proof: Let N be a submodule of M. Then $E(M) = E(N) \oplus K$ for some K. Since M is quasi-injective, $M = [E(N) \cap M] \oplus (K \cap M)$ and then $N \subset_e E(N) \cap M$. Next, since M is quasi-injective, M has the exchange property by (8.5.5). Hence we see by (7.2.6) that $\{M_\alpha\}_I$ is locally semi-T-nilpotent. For any pair α and β in I, we show that every monomorphism $f : M_\alpha \to M_\beta$ is an isomorphism. Let ρ be an isomorphism of $E(M_\alpha)$ to $E(M_\beta)$ which is an extension of f. Since $M_\alpha \oplus M_\beta$ is quasi-injective, $\rho^{-1}(M_\beta) \subset M_\alpha$ by (5.4.2). Hence $M_\beta \subset \rho(M_\alpha) = f(M_\alpha)$, and hence f is an isomorphism. Now, let $\{N_\beta\}_J$ be an independent set of submodules of M. Then there exists a direct summand N'_β of M such that $N_\beta \subset_e N'_\beta$ for all β in J. Since $\{M_\alpha\}_I$ is locally semi-T-nilpotent and N'_β is a direct summand of M, each N' is a direct sum of completely indecomposable modules. Consequently, we see by (9.6.10) that $\Sigma_J \oplus N_\beta$ is a direct summand of M; whence $\Sigma_J \oplus N_\beta$ is extendible to a decomposition of M.

9.9.2 THEOREM Let $\{M_\alpha\}_I$ be a set of completely indecomposable R-modules and put $M = \Sigma_I \oplus M_\alpha$. Then the following conditions are equivalent:

1. M is quasi-injective.

2. $M \oplus M$ has the extending property of direct sums of submodules.

Proof: (1) ⇒ (2). Since M is quasi-injective, $M \oplus M$ is also quasi-injective by (5.4.2). Hence (2) holds by (9.9.1).

(2) ⇒ (1). This is easily seen by (9.6.13) and (9.7.5).

9.9.3 COROLLARY Let T be a completely indecomposable R-module and put $M = T^{(I)}$. Then M is quasi-injective if and only if M has the extending property of direct sums of submodules.

Proof: If the cardinal $|I| = \infty$, then $M \oplus M \approx M$; whence the assertion follows from (9.9.2). If $|I| < \infty$, then M is quasi-injective if and only if T is quasi-injective. Hence in this case, the assertion also follows from (9.9.2).

9.9.4 PROPOSITION Let $\{M_\alpha\}_I$ be a set of uniform and completely indecomposable R-modules satisfying (9.6.1) and $|I| \neq 1$. Put $M = \Sigma_I \oplus M_\alpha$. We assume that $E(M_\alpha) \approx E(M_\beta)$ for all α and β in I and M has the extending property of direct sums of f.c. uniform modules. Then if either $|I| < \infty$ or R is right noetherian, M is quasi-injective.

Proof: Let ρ be any isomorphism of $E(M_1)$ to $E(M_2)$ for 1 and 2 in I. Then there exists an isomorphism $f : M_1 \to M_2$ such that $f = \rho$ on $\rho^{-1}(M_2) \cap M_1$ by (9.7.6). Let h be an element in $\mathrm{Hom}_R(A, M_1)$ for $A \subset M_1$. Then fh is extendible to a homomorphism $t : M_1 \to M_2$ by (9.7.6). Hence $f^{-1}t$ is an extension of h, so M_1 is quasi-injective. Further, since we may assume that $f\rho^{-1} \in \mathrm{End}_R(E(M_2))$, $M_2 = f(\rho^{-1}(\rho(M_1))) = \rho(M_1)$ by (5.4.2). If $|I| < \infty$ or R is noetherian, $E(M) = \Sigma_I \oplus E(M_\alpha)$. Hence M is quasi-injective by (5.4.2) and (4.7.18).

9.9.5 COROLLARY Let $\{M_\alpha\}_I$ and M be as in (9.9.4). We assume further that each M_α is nonsingular and M has the extending property of f.c. uniform modules. Then if either $|I| < \infty$ or R is right noetherian, $\Sigma_{I'} \oplus M_\alpha$ is quasi-injective, where $I' = \{\alpha \in I \mid$ there exists a β ($\neq \alpha$) in I such that $E(M_\alpha) \approx E(M_\beta)\}$.

Proof: Since M_α is nonsingular, $\mathrm{Hom}_R(E(M_\alpha), E(M_\beta)) = 0$ if $E(M_\alpha) \not\approx E(M_\beta)$. Hence we have (9.9.5) by (9.9.4).

9.10 RINGS WITH THE EXTENDING PROPERTY

This section is the dual to Sec. 9.5.

9.10.1 THEOREM Let R be a ring. We assume that every uniform direct summand M with $S(M) \neq 0$ of R is artinian. Then the following conditions are equivalent:

1. Every projective module has the extending property of simple modules.
2. $R \oplus R$ has the above property as a right R-module.
3. R contains a right ideal A satisfying the following.
   (i) $A = \Sigma_{\alpha\in I}\Sigma_{\beta\in I(\alpha)} \oplus e_{\alpha\beta}R$ and $\Sigma\Sigma \oplus e_{\alpha\beta}R$ is a locally direct summand of R, where the $e_{\alpha\beta}R$ are uniform modules with $S(e_{\alpha\beta}R) \neq 0$
   (ii) $S(R) = S(A)$.
   (iii) $S(e_{\alpha 1}R) \approx S(e_{\alpha\beta}R)$ for any β and $S(e_{\alpha 1}R) \not\approx S(e_{\alpha' 1}R)$ if $\alpha \neq \alpha'$.
   (iv) $I(\alpha)$ is a linearly ordered set and $\mathrm{Hom}_R(S(e_{\alpha\beta}R),S(e_{\alpha\beta'}R))$ is extendible to $\mathrm{Hom}_R(e_{\alpha\beta}R,e_{\alpha\beta'}R))$ for $\beta < \beta'$ (and $e_{\alpha\beta}R$ is isomorphic to a submodule of $e_{\alpha\beta'}R$).

9.10.2 PROPOSITION Let R be a right artinian ring. Then R has the extending property of simple modules as a right R-module if and only if $R = \Sigma_{i=1}^{n} \oplus e_iR$ satisfies the following.

1. Each e_iR is uniform.
2. For any element f in $\mathrm{Hom}_R(S(e_iR),S(e_jR))$, either f is extendible to an element in $\mathrm{Hom}_R(e_iR,e_jR)$ of f^{-1} is extendible to an element in $\mathrm{Hom}_R(e_jR,e_iR)$ for $i \neq j$.

Proof: This is clear by (9.6.2) and (9.6.6).

9.10.3 PROPOSITION Let R be a self-injective ring as a right R-module. Then we have the following.

1. Every projective module has the extending property of simple modules.
2. R contains a right ideal A satisfying the following.
   (i) $A = \Sigma_I \oplus e_\alpha R$ and $\Sigma_I \oplus e_\alpha R$ is a locally direct summand of R, where the $e_\alpha R$ are indecomposable and injective.
   (ii) $S(R) = S(A)$.

Proof: Put $S(R) = \Sigma_I \oplus S_\alpha$, where the S_α are simple. Then $A = \Sigma_I \oplus E(S_\alpha)$ is a desired right ideal. Hence every direct sum $R^{(J)}$ has the extending property of simple modules. Therefore, every projective module has the same property by (9.6.1).

We shall define right generalized co-uniserial rings. If every indecomposable and injective module with nonzero socle is serial and has the lower Loewy series, we say that R is a right *generalized co-uniserial ring*.

9.10.4 THEOREM Let R be any ring. Then the following conditions are equivalent:

1. Every direct sum of uniform modules of finite length has the extending property of simple modules.

2. For every indecomposable injective module E, $0 \subset S_1(E) \subset S_2(E) \subset \cdots \subset S_n(E) \subset \cdots$ is the unique composition series of submodules of finite length of E.

Proof: (1) ⇒ (2). Put $T = \mathrm{End}_R(E)$ and $J' = J(T)$. Then $\mathrm{End}_R(S_1(E)) = T/J'$ if $S_1(E) \neq 0$. We assume that S_i/S_{i-1} is simple for all $i \leq n$. If $\|B\| \leq n$ for a submodule B of E, $B \subset S_n(E)$. Let A_i $[\supset S_n(E)]$ be a submodule of E such that $A_i/S_n(E)$ is simple for $i = 1, 2$. Then taking the identity in T, we obtain some f in J' such that $(1 + f)(A_i) \subset A_{i'}$, by (9.6.6), where $(i,i') = (1,2)$. Since $\|f(A_i)\| < \|A_i\|$, $f(A_i) \subset S_n(E)$. Hence $A_i \subset A_{i'}$, so $A_i = A_{i'}$. Therefore, $S_{n+1}(E)/S_n(E)$ is simple or zero.

(2) ⇒ (1). Let M be a uniform module of finite length. Then E(M) has a nonzero socle. Hence we have (1) by (9.6.6).

9.10.5 COROLLARY Let R be a ring such that every indecomposable injective module E with $S(E) \neq 0$ is equal to $\bigcup_i S_i(E)$ (e.g., R is a semiprimary or commutative noetherian ring). Then the following conditions are equivalent:

1. Every direct sum of uniform modules with nonzero socles has the extending property of simple modules.
2. R is a right generalized co-uniserial ring.

Proof: If R is a right generalized co-uniserial ring, then for every uniform module M with $S(M) \neq 0$, either $M = E(M)$ or $M = S_i(E)$, so M is completely indecomposable.

Remarks 1. If R is semiprimary, EJ^i/EJ^{i+1} is semisimple and $EJ^n \subset S_1(E)$ if $EJ^{n+1} = 0$.

2. If R is a commutative noetherian ring, then $E = \bigcup S_i(E)$ by (4.7.20).

Let R be an algebra over a field K of finite dimension (more generally, R is an artinian algebra with duality).

9.10.6 PROPOSITION Let R be an algebra over a field of finite dimension. Then the following conditions are equivalent:

1. Every direct sum of hollow R-modules has the lifting property of simple modules modulo radical.
2. Every direct sum of uniform left R-modules has the extending property of simple modules.
3. R is a right generalized uniserial ring.
4. R is a left generalized co-uniserial ring.

Proof: Since R has duality (see Sec. 5.6), (9.10.6) is clear by (9.5.3) and (9.10.4).

If R is a right noetherian and generalized co-uniserial ring, then every injective module is a direct sum of injective hulls of simple modules. Hence R is right artinian by (4.7.19).

9.10.7 THEOREM Let R be a right artinian and generalized co-uniserial ring. Then an R-module M has the extending property of simple modules if and only if M is a direct sum of uniform submodules.

Proof: Every uniform module is artinian if R is co-uniserial, and $M \supset_e S(M)$ if R is right artinian. Hence we have (9.10.7) by (9.8.4) and (9.10.5).

9.10.8 THEOREM Let R be a right and left artinian ring. We assume that R is self-dual (e.g., an algebra over a field of finite dimension). Then the following conditions are equivalent:

1. R is a left and right generalized uniserial ring.
2. Every right R-module has the lifting property of simple modules modulo radical.
3. Every right R-module has the extending property of simple modules.

Proof: (1) ⇒ (2) and (3). This is clear by (9.10.4) and (9.4.7).

(2) ⇒ (1). R is right generalized uniserial by (9.5.3). Let e be a primitive idempotent in R. Since Re is hollow, the dual $(Re/Ae)^*$ of Re/Ae is uniform for any left ideal Ae of Re. Further, $(Re/Ae)^*$ is hollow by (2). Hence Re/Ae is uniform. Therefore, Re is serial.

(3) ⇒ (1). R is right co-uniserial by (9.10.5). Since eR/eA is indecomposable for any right ideal eA, eR/eA is uniform by (3).

9.11 LIFTING PROPERTY AND EXTENDING PROPERTY OVER COMMUTATIVE RINGS

We assume that R is a commutative ring. Then every cyclic hollow module M is isomorphic to R/A, where the ideal A is contained in a unique maximal ideal P = P(M). Hence R/A is a local ring. Conversely, if R is a local ring, then R/A is a cyclic hollow module for any ideal A. Let $M \approx R/A$ and P = P(M). Since R/A is a local ring, R/A is an R_P-module and every R-submodule of R/A is an R_P-module. Hence $R/A = R/A \otimes R_P = R_P/A_P$.

9.11.1 LEMMA Let R be a commutative ring and P a prime ideal. Let A be an ideal in R_P. Then every R-submodule of R_P/A is an R_P-module if and only if $R_P = A + \tilde{R}$ as R-modules, where $\tilde{R} = \rho(R)$ and ρ is the natural homomorphism of R into R_P. If there exists such an ideal $A \neq R_P$, then P is maximal.

Proof: We assume that $R_P = A + \tilde{R}$. Let B be an R-submodule intermediate between R_P and A, and let $x = a + \tilde{r}$ in R_P; $a \in A$, $r \in R$. Then $Bx \subset Ba + Br \subset B$. Hence B is an R_P-module. Conversely, we assume that every R-submodule of R_P/A is an R_P-module. Then $(A + \tilde{R})/A$ is an R_P-module. Hence $A + \tilde{R} = R_P$. Since $A \subset P_P$, $R_P = P_P + \tilde{R}$, which implies that $R_P \approx R_P/P_P$, and hence P is maximal.

9.11.2 PROPOSITION Let R be as in (9.11.1). Let P be a maximal ideal and A an ideal contained in P. Then the following conditions are equivalent:

1. P is a unique maximal ideal containing $B = \rho^{-1}(A_P)$.
2. R_P/A_P is a cyclic hollow R-module.
3. $R_P = A_P + \tilde{R}$ as R-modules.

Proof: (1) ⇒ (2). Since $B_P = A_P$, we obtain (2) by the remark above (9.11.1).

(2) ⇒ (3) and (1). If R_P/A_P is a cyclic hollow R-module, $R_P/A_P \approx R/C$ for some ideal C with $Q = P(R/C)$. Hence we have an R-isomorphism f of R_Q/C_Q $(= \bar{R}_Q)$ to R_P/A_P $(= \bar{R}_P)$. If $P \neq Q$, $\bar{1}_Q P = \bar{1}_Q P R_Q = \bar{R}_Q$. Hence $f(\bar{1}_Q)P = \bar{R}_P$, which is a contradiction. Accordingly, $Q = P$ and $R_P/A_P \approx R_P/C_P$ as R_P-modules. Therefore, $A_P = C_P$ and $A_P + \tilde{R} = R_P$ by (9.11.1), and P is a unique maximal ideal containing $\rho^{-1}(A_P) = \rho^{-1}(C_P)$.

(3) ⇒ (2). If $R_P = A_P + R$, R_P/A_P is a cyclic hollow module by (9.11.1).

9.11.3 THEOREM Let R be a commutative ring. Then the following conditions are equivalent:

1. Every direct sum of cyclic hollow modules has the lifting property of simple modules modulo radical.
2. For each maximal ideal P, the set of ideals A such that P is a unique maximal ideal containing A is a linearly ordered set with respect to inclusion.

Proof: (2) ⇒ (1). Let $M_i \approx R/A_i$ be a cyclic hollow module for i = 1, 2. If $\bar{M}_1 \approx \bar{M}_2$, $P(M_1) = P(M_2)$ as in the proof of (9.11.2). Hence $A_1 \subset A_2$ or $A_2 \subset A_1$. Therefore, we have (1) by (9.2.5).

(1) ⇒ (2). Let A_1 and A_2 be ideals in R as in (2). Since R/A_i is a cyclic R-hollow module and $\overline{R/A_1} \approx \overline{R/A_2}$, we obtain an element x in R - P such that $xA_1 \subset A_2$ or $xA_2 \subset A_1$ by (9.2.5). Let Q be a maximal ideal ($\neq$ P). Then $A_{iQ} = R_Q$ for i = 1, 2. Hence $A_1 \subset A_2$ or $A_2 \subset A_1$. Therefore, the set of these ideals A_i is a linearly ordered set.

9.11.4 THEOREM Let R be a commutative local ring. Then (2) in (9.11.3) is equivalent to the fact:

2'. R is a generalized uniserial ring (namely, a valuation ring).

9.11.5 PROPOSITION Let R be a commutative perfect ring and M an R-module. Then M has the lifting property of direct decompositions modulo radical if and only if M is quasi-projective.

Proof: If M has the above property, $M = \Sigma_I \oplus M_\alpha$ with M_α cyclic hollow. Since R is commutative and perfect, $M_\alpha \approx eR/eA_\alpha$ is quasi-projective by (5.4.10), where e is a primitive idempotent and A_α is an ideal. Furthermore, $eR/eA_\alpha \approx eR/eA_\beta$ for $\alpha,\beta \in I$, so $eA_\alpha = eA_\beta$. Hence M is quasi-projective by (5.4.10). The converse is clear by (5.4.11).

9.11.6 THEOREM Let R be a commutative ring. Then the following conditions are equivalent:

1. Every direct sum of uniform modules with finite length has the extending property of simple modules.
2. For each maximal ideal P in R, the set of submodules with finite length of E(E/P) is serial.

Proof: Each uniform module with finite length is isomorphic to a submodule of E(R/P) for some maximal ideal P in R. Hence (9.11.6) is clear by (9.10.4).

We shall assume that R is a commutative noetherian ring. Then every indecomposable and injective R-module E is of the form E(R/P) for some prime ideal P and $E = \cup_i A_i$, where $A_i = \{x \in E \mid r_R(x) = P^i\}$ (see Sec. 4.7). If $S(E) \neq 0$, P is a maximal ideal in R. Then E(R/P) is artinian, and every R-submodule is quasi-injective by (5.5.36).

9.11.7 THEOREM Let R be a commutative noetherian ring. Let M be an R-module with $M \supset_e S(M)$. Then M has the extending property of simple modules if and only if $M \approx \Sigma_P \Sigma_{I(P)} \oplus M_{P\alpha}$, where:

1. The $M_{P\alpha}$ are submodules of E(R/P) for each P.
2. $\{M_{P\alpha}\}_\alpha$ is linearly ordered with respect to inclusion for each P, and P runs through all maximal ideals in R.

Proof: If $M \approx \Sigma_P \Sigma_{I(P)} \oplus M_{P\alpha}$, M has the extending property of simple modules by (1), (2), and (9.6.6), since $M_{P\alpha}$ is quasi-injective. Conversely, we assume that M has the extending property of simple modules. Let N be a uniform submodule of M. Since $M \supset_e S(M)$, $S(N) \neq 0$. Hence N is a submodule of some E(R/P) and N is artinian. Therefore, M is isomorphic to $M' = \Sigma_{P,\alpha} \oplus M_{P\alpha}$ satisfying (1) and (2) by (9.8.2) and (9.8.3).

Let R be, in particular, a Dedekind domain and M a f.g. R-module. Then $M = M(t) \oplus M_0$, where M(t) is the torsion part of M and $S(M_0) = 0$ by (4.7.36). Hence M has the extending property of simple modules if and only if M(t) does. We note that $M \supset_e S(M)$ if and only if M is torsion.

9.11.8 COROLLARY Let R be a Dedekind domain and M a torsion R-module. Then M has the extending property of simple modules if and only if M is a direct sum of injective modules and cyclic P-groups R/P^n, where P runs through all nonzero primes in R.

Proof: This is clear.

9.11.9 PROPOSITION Let R be a commutative and noetherian ring and let M be an R-module with $S(M) \subset_e M$. Then M has the extending property of direct decompositions of S(M) if and only if M is a quasi-injective module. In this case, M has the extending property of direct sums of uniform modules.

Proof: If M has the extending property above, $M = \Sigma_{P,\alpha} \oplus M_{P\alpha}$ as in (9.11.7) by the definition. Furthermore, $M_{P\alpha} \approx M_{P\beta}$ by (9.7.6). Since R is noetherian, $\Sigma_\alpha \oplus M_{P\alpha}$ is quasi-injective by (5.4.2). Hence M is also quasi-injective. The converse is clear by (9.9.1).

9.11.10 PROPOSITION Let R be a commutative and noetherian ring. Then R has the extending property of simple modules as an R-module if and only if R is a direct sum of Frobenius rings and a ring with zero socle. In this case, every projective module has the extending property of simple modules.

Proof: We assume that R has the extending property of simple modules. We may assume by (9.6.1) that R is directly indecomposable. If $S(R) \neq 0$, R is uniform, and R is artinian by (5.5.36). Hence R is a Frobenius ring. The remaining assertions are clear by (9.10.1).

9.11.11 THEOREM Let R be a commutative and noetherian ring. Then the following conditions are equivalent:

1. Every direct sum of cyclic hollow modules has the lifting property of simple modules modulo radical.
2. Every direct sum of uniform modules with nonzero socle has the extending property of simple modules.
3. R_P is a (generalized uni-) serial ring for every maximal ideal P.
4. R is a direct sum of artinian uniserial rings and Dedekind domains.

Proof: (1) ⇒ (3). Let P be a maximal ideal. Then all P^n are contained in a unique maximal ideal P. Hence P^n/P^{n+1} is simple by (9.11.3). Therefore, $R_P \supset P_P \supset P_P^2 \supset \cdots$ is a unique series of ideals, since $\cap_n P_P^n = 0$.

(3) ⇒ (1). This is clear by (9.2.5), (9.11.3), and taking localizations with respect to maximal ideals.

(2) ⇒ (3). Let P be a maximal ideal in R and put E = E(R/P). Then $A_i = S_i(E)$. P_P^n/P_P^{n+1} is simple or equal to zero by (4.7.23) and (9.10.4).

(3) ⇒ (2). If R_P is a serial ring for a maximal ideal P, A_i/A_{i-1} is simple by (4.7.23). Hence $E = \cup_i A_i$ is serial. Hence by the definition of A_n we have (2) by (9.10.5).

(4) ⇒ (3). This is clear.

(3) ⇒ (4). We note that every factor ring of R satisfies (3). Since R is noetherian, we may assume that R is an irreducible ring as an R-module. We know Krull dim R ≤ 1 by (3). Let $0 = A_1 \cap \cdots \cap A_k \cap B_1 \cap \cdots \cap B_m \cap C_1 \cap \cdots \cap C_n$ be an irredundant representation by primary ideals A_i, B_j, and C_q such that the associated prime ideal $P(A_i)$ of A_i is maximal and minimal, $P(B_j)$ is maximal but not minimal, and $P(C_q)$ is a minimal but not maximal ideal in R [we may assume that all P( ) are distinct] [see (4.7.14)]. Let P be a maximal ideal in R. We note that P does not contain two distinct minimal ideals by (3). Since $P(A_i)$ is maximal and minimal, A_i is relatively prime to the other ideals A_j $(j \neq i)$, B_j, and C_q. Hence $A_i = 0$ or $0 = B_1 \cap \cdots \cap B_m \cap C_1 \cap \cdots \cap C_n$ by (3.2.5), since R is irreducible. We assume the latter. Since $P(B_i)$ contains a unique minimal ideal $P(C_{i'})$ from the above, we have $0 = (B_1 \cap C_{1'}) \cap (B_2 \cap C_{2'}) \cap \cdots \cap (B_m \cap C_{m'}) \cap C_{m+1'} \cap \cdots \cap C_{n'}$, where $\{1',2',\ldots,n'\} \equiv \{1,2,\ldots,n\}$. We know again, from the above observation, that $0 = B_1 \cap C_{i'}$ or $C_{m+i'} = 0$. In the former case, $P(C_{i'}) = D$ is the unique minimal ideal in R. Let P be any maximal ideal in R. Since $D \subset P$ and R_P is serial, $D_P = 0$, since $\cap P_P^n = 0$. Hence D = 0, which is a contradiction. Thus we have shown that R is a primary ring with a unique minimal prime ideal Q. If R is artinian, R is local, and so $R = R_P$ is a serial ring by (3). If R is not artinian, R_P is a Dedekind domain for any maximal ideal P. Hence Q = 0. Accordingly, R is a Dedekind domain.

Next we shall determine the types of R-modules that have the extending property of direct sums of uniform modules over a Dedekind domain.

9.11.12 THEOREM Let R be a Dedekind domain and M an R-module. Then M has the extending property of direct sums of uniform modules if and only if either

1. M is quasi-injective or
2. $M = K \oplus E$, where E is torsion and injective and $K \subsetneq Q^* = E(R)$.

Proof: If M has the extending property of direct sums of uniform modules, then $M = \Sigma_I \oplus M_\alpha$ with M_α uniform, since R is noetherian. We shall complete the proof by making use of elementary properties of abelian groups as follows.

9.11.13 LEMMA Let K be an R-submodule of Q* and $\{f_i : K \to E(R/P_i)^{(I_i)}\}$ a set of homomorphisms. We assume that $P_i \neq P_j$ if $i \neq j$, and for some $a \neq 0$ in K, $f_i(a) = 0$ for almost all i. Then $\{f_i\}$ is summable, where the P_i are nonzero prime ideals.

Proof: We may assume that a = 1 and $K \supset R$. Then $f_i \in \mathrm{Hom}_R(K/R, E(R/P_i)^{(I_i)})$ for almost all i. Hence, since $K/R = \Sigma \oplus K/R[[S]]$, where $K/R[[S]]$ is the S-primary component of K/R, $\{f_i\}$ is summable $(E(R/P)[[S]] = 0$ if $P \neq S)$.

9.11.14 LEMMA Let K be as in (9.11.12) and L an R-submodule of E(R/P). If $\mathrm{Hom}_R(A,L)$ is extendible to $\mathrm{Hom}_R(K,L)$ for any submodule A in K, then $L = E(E/P)$.

Proof: We may assume that $K \supset R$. Assume that $L \neq E(R/P)$. Then $L = \bar{p}^{-n}R$, where $E(R/P) = \cup_m \bar{p}^{-m}R$ and $\bar{p}^{-m} \in Q^*/R$. Put $A = p^nR \supset B = p^{2n}R$. Then the natural epimorphism $f : A \to A/B \approx L$ is in $\mathrm{Hom}_R(A,L)$. Let $g \in \mathrm{Hom}_R(R,L)$ be an extension of f. Then $R = \ker g + A$. Hence $R_P = (\ker g)_P + A_P = (\ker g)_P$. Since $(\ker g) \cap A = B$, $A_P = B_P$, which is a contradiction. Hence $L = E(R/P)$.

The following lemma is similar to (9.6.7).

9.11.15 LEMMA Let $T = \Sigma_I \oplus T_\alpha$ be any direct decomposition of an R-module, where the T_α are uniform. If T has the extending property of direct sums of f.c. modules, then any element in $\mathrm{Hom}_R(A_\alpha, T_\beta)$ is extendible to an element in $\mathrm{Hom}_R(T_\alpha, T_\beta)$ for any α and β in I and any submodule A_α in T_α.

Proof: Let f be in $\mathrm{Hom}_R(A_1, T_2)$. Put $A(f) = \{a + f(a) \mid a \in A\}$ and consider an essential submodule $A(f) \oplus T_2 \oplus \Sigma_{I-\{1,2\}} \oplus T_\alpha$ of T. Then there exists a direct decomposition $T = \Sigma_I \oplus S_\alpha$ such that $S_\alpha \supset_e T_\alpha$ for $\alpha \neq 1$ and $S_1 \supset_e A(f)$. Since T_α is a direct summand of T, $S_\alpha = T_\alpha$ for $\alpha \neq 1$. Thus $T = S_1 \oplus \Sigma_{I-\{1\}} \oplus T_\alpha$, so $-\pi|T_1$ is an extension of f, where $\pi : T \to T_2$ is the projection for the decomposition [cf. (9.6.7)].

Proof of (9.11.12): We assume that M has the extending property of direct sums of uniform modules. Then $M \approx \Sigma_{I_1} \oplus M_\alpha \oplus \Sigma_{I_2} \oplus M_\beta$, where the M_α (resp. the M_β) are torsion-free (resp. torsion) uniform submodules of M. We may assume that $M_\alpha \subset Q^*$ for all α in I_1. First we assume that $|I_1| \geq 2$.

Take $x \neq 0$ in $M_{\alpha_1} \cap M_{\alpha_2}$. For any $m \neq 0 \in R$, a homomorphism $f : xmR \to xR$ given by setting $f(xmr) = xr$ is extendible to a $g : M_{\alpha_1} \to M_{\alpha_2}$ by (9.11.15). Hence $xR(1/m) \subset M_{\alpha_1}(1/m) = g(M_{\alpha_1}) \subset M_{\alpha_2}$. Therefore, $M_{\alpha_2} = Q^*$, so each $M_\alpha = Q^*$ for α in I_1. Furthermore, we know that $M_\beta = E(R/P)$ for any β in I_2 by (9.11.14). Hence M is injective. If $|I_1| = 1$ and $M_{\alpha_1} = Q^*$, M is also injective as above. Next assume that $|I_1| = 1$ and $M_{\alpha_1} \subsetneqq Q^*$. Then M is of the form (2). Finally, assume that M is torsion. Each M_β is a completely indecomposable module with (9.6.1). Hence M is quasi-injective by (9.9.4), the fact that $\mathrm{Hom}_R(E(R/P),E(R/S)) = 0$ provided that $P \neq S$, and (5.4.2), where P and S are prime ideals. Conversely, if M is quasi-injective, M has the extending property of direct sums of submodules by (9.9.1). Let $M = K \oplus \Sigma_P \oplus E(R/P)^{(I_P)}$ as in (2). We assume that $N = N_1 \oplus \Sigma_{I_2} \oplus N_\beta$ is an essential submodule of M, where N_1 (resp. N_β) is torsion free (resp. torsion) and uniform. Let π and π_P be the projections of M onto K and $E(R/P)^{(I_P)}$, respectively. Then $\pi|N_1$ is an isomorphism. Put $f_P = \pi_P(\pi|N_1)^{-1} : \pi(N_1) \to E(R/P)^{(I_P)}$. Then $\{f_P\}$ is summable. Since $E(R/P)^{(I_P)}$ is injective, we obtain an extension $g_P \in \mathrm{Hom}_R(K,E(R/P)^{(I_P)})$ of f_P. Then $\{g_P\}$ is also summable by (9.11.13). Put $K' = \{x + \Sigma_P\, g_P(x) \mid x \in K\} \subset M$. Then $M = K' \oplus \Sigma_P \oplus E(R/P)^{(I_P)}$, $K' \supset N_1$, and $\Sigma_P \oplus E(R/P)^{(I_P)} \supset_e \Sigma_{I_2} \oplus N_\beta$. Hence $N_1 \oplus \Sigma_{I_2} \oplus N_\beta$ is extendible to a direct decomposition of M by (9.9.1).

We have determined the types of R-modules that have the extending property of direct sums of uniform modules in (9.11.12). We shall determine the types of R-modules that have the extending property of uniform modules.

We put $Q^* = E(R)$. Let M be a torsion-free and uniform module. Then we may assume that $M \supset R$. $Q^*/R \supset M/R = \Sigma_P \oplus \bar{P}^{-n(P)}R$, where n(P) is finite or infinite. Put $\underline{P} = \{P_i \mid n_i(P_i) < \infty\}$ for M and denote M by $F(\underline{P})$. Then M is completely indecomposable if and only if $\underline{P}$ is either a singleton or empty (i.e., $M = Q^*$). We note that $F(\underline{P})_P \neq Q^*$ for $P \in \underline{P}$ and $F(\underline{P})_S = Q^*$ for $S \notin \underline{P}$. An R-module N is called P^∞-*divisible* if $P^nN = N$ for all n, and we denote the unique maximal P^∞-divisible submodule of N by N[P].

In the preceding sections we assumed that a direct summand of M which is an extension of a uniform submodule is completely indecomposable. In the following, we drop this assumption. We consider only the extending property of uniform modules, so we call it simply the extending property.

9.11.16 THEOREM Let R be a Dedekind domain and M an R-module. Then M has the extending property of uniform modules if and only if M satisfies one of the following.

1. M is torsion and $M = \Sigma_P \oplus [M_1(P)^{(J_1)} \oplus M_2(P)^{(J_2)}]$ and $|\|M_1(P)\| - \|M_2(P)\|| \leq 1$, where the $M_i(P)$ are completely indecomposable [$\|M_1(P)\| = \infty$ means that $M_1(P)$ is injective] and $M_i(P) \subset E(R/P)$.
2. $E \oplus M$, where E is injective, and M is torsion free and uniform or zero.

Proof: We shall prove (9.11.16) by making use of several lemmas below.

9.11.17 LEMMA Let M be an R-module with the extending property (resp. of cyclic and uniform modules) and $M = T_1 \oplus T_2$. Then:

1. If M is torsion free, T_1 has the extending property (resp. of cyclic and uniform modules).
2. If T_2 is the torsion submodule M(t) of M, T_1 and T_2 have the same property.

Proof: 1. Let N be a uniform submodule (resp. cyclic and uniform submodule) of T_1. Then there exists a direct decomposition $M = L_1 \oplus L_2$ and $L_1 \supset_e N$. Let $\pi_i : M \to T_i$ be the projection. Then since $\ker \pi_2 \supset N$, $\pi_2(L_1) = 0$. Hence $L_1 \subset T_1$ and $T_1 = L_1 \oplus T_1 \cap L_2$.

2. Let N be a uniform submodule (resp. cyclic and uniform submodule) of M. Then $M = L_1 \oplus L_2$ and $L_1 \supset_e N$. If $N \subset T_2$, $L_1 \subset T_2$. Hence T_2 has the extending property. We assume that $N \subset T_1$. Then L_1 is torsion free. Hence $T_2 \subset L_2$ and $L_2 = T_2 \oplus T_1 \cap L_2$. Now, $M = T_1 \oplus T_2 = L_1 \oplus T_1 \cap L_2 \oplus T_2$. Therefore, $T_1 = \pi_1(L_1) \oplus \pi_1(T_1 \cap L_2)$ and $\pi_1(L_1) \supset_e \pi_1(N) = N$.

Since R is a Dedekind domain, every R-module M contains a unique maximal injective submodule E, say $M = E \oplus K$ and K is reduced by (4.7.31).

9.11.18 LEMMA Let $M = E \oplus K$ be as above. If M has the extending property, then K does.

Proof: Let N be a uniform submodule of K. Then $M = L_1 \oplus L_2$ and $L_1 \supset_e N$. Since L_1 is indecomposable, L_1 is either injective or reduced. If L_1 is injective, L_1 has the exchange property by (8.5.5), so there exists a direct summand K_1 of K with $K_1 \supset_e N$ by the remark below (9.6.1). We assume that L_1 is reduced. Let E' be the unique maximal injective submodule of L_2. Then $E \supset E'$ and $M = L_1 \oplus L_2' \oplus E'$, $L_2' \subset L_2$. Accordingly, $E = E' \oplus E \cap (L_1 \oplus L_2')$ and the injective module $E \cap (L_1 \oplus L_2')$ has the exchange property. However, L_1 and L_2' are reduced. Hence $E \cap (L_1 \oplus L_2') = 0$, so

$E = E'$. Therefore, $M = L_1 \oplus L_2' \oplus E = K \oplus E$. Since $N \subset K \cap L_1$, $K = \pi(L_1) \oplus \pi(L_2')$ and $\pi(L_1) \supset_e N$, where $\pi : M \to K$.

9.11.19 LEMMA Let $M = M_1 \oplus M_2$ be torsion free. We assume that the M_i are completely indecomposable uniform modules. Then M has the extending property of cyclic uniform modules if and only if either M_1 or M_2 is injective. In this case M has the extending property.

Proof: The "if" part is clear by (9.6.5). We assume that $M_i = F(\underline{P}_i)$, where $\underline{P}_i = \{P_i\}$. Then considering multiplication by x^{-m} ($x \in P - P^2$) and using the proof (1) ⇒ (2) of (9.6.5), we have $x^{-m}M_1 \subset M_2$ or $x^{-m}M_2 \subset M_1$. Hence either M_1 or M_2 is injective.

9.11.20 LEMMA Let M be torsion free and reduced. If M has the extending property of cyclic uniform modules, then M is uniform.

Proof: Since every direct summand of M has the extending property of cyclic uniform modules by (9.11.17), M has a direct summand $M_1 \oplus M_2$, where the M_i are uniform if M is not uniform. Let $M_i = F(\underline{P}_i)$, $i = 1, 2$. Since $M_1 \oplus M_2$ has the extending property of cyclic uniform modules, we may assume that $M = M_1 \oplus M_2$. If $\underline{P}_1 \cap \underline{P}_2 \neq \emptyset$ (say $P \in \underline{P}_1 \cap \underline{P}_2$), $M_{1P} \neq Q^*$ and $M_{2P} \neq Q^*$. However, M_P has the extending property of cyclic uniform modules, which is a contradiction by (9.11.19). Hence there exists $P \in \underline{P}_1 - \underline{P}_2$. Put $N = \{x + x' \mid x \in R \subset M_1,\ x = x' \in R \subset M_2\}$. Then we obtain that $M = L_1 \oplus L_2$ and $L_1 \supset_e N$. Now, $M[P] = M_2$, so $M_1 \approx M/M[P] \approx L_1/L_1[P] \oplus L_2/L_2[P]$. Since M_1 is uniform, $L_1 = L_1[P]$ or $L_2 = L_2[P]$. We assume that $L_1 = L_1[P]$. Then $L_2/L_2[P]$ is torsion free. Since L_2 is uniform, $L_2[P] = 0$. Hence $M_2 = M[P] = L_1[P] \subset L_1$. However, $0 = N \cap M_2 = N \cap L_1 = N$, which is a contradiction. If $L_2 = L_2[P]$, $M_2 = L_2$ as above. Therefore, the identity map $R \to R$ is extendible to $\mathrm{Hom}_R(M_1,M_2)$ by (9.6.7). We may assume from the first observation that there exists $S \in \underline{P}_2 - \underline{P}_1$. Then $g(M_{1S}) = g(Q^*) \subset M_{2S} \subsetneq Q^*$, which is a contradiction. Hence M is uniform.

Next, we shall study torsion modules. If M_α is torsion and uniform, $M_\alpha \subset E(R/P)$ for some P. We denote such M_α by $M_\alpha(P)$.

9.11.21 LEMMA Let M be torsion. We assume that $M = \Sigma_I \oplus M_\alpha$ with M_α uniform. Then M has the extending property if and only if $M = \Sigma_P \oplus (M_{\alpha_1}(P)^{(B_1)} \oplus M_{\alpha_2}(P)^{(B_2)})$ and $\big|\,\|M_{\alpha_1}(P)\| - \|M_{\alpha_2}(P)\|\,\big| \leq 1$ for each P.

Proof: Since E(R/P) is serial and M_α is completely indecomposable, we have (9.11.21) by (9.6.5).

9.11.22 LEMMA Let M be torsion. If M has the extending property, then M is of the form in (9.11.21), provided that M is reduced.

Proof: Since M is torsion, every indecomposable uniform submodule is completely indecomposable. M is a direct sum of P-primary components M[[P]] and it is clear that M[[P]] has the extending property. Therefore, we may assume that M is P-primary and reduced. Let x be in M. Put $r_R(x) = \{r \in R \mid xr = 0\} = P^n$ and put $n = n(x)$. We first show that $\{n(x)\}_{x \in M}$ is bounded. Let $N = xR$ be a uniform submodule of M. Then $M = L_1 \oplus L_2$ and $L_1 \supset_e N$. Since M is reduced, $L_1 = yR$. If $\{n(x)\}_{x \in M}$ is not bounded, there exists z in L_2 such that $n(z) \geq n(y) + 1$. Since L_2 has the extending property by the remark following (9.6.1), $L_2 = L_2' \oplus L_3$ and $L_2' \supset_e zR$. $L_1 \oplus L_2'$ has the extending property, too, which is a contradiction by (9.11.21). Next, we shall show that $M = \Sigma_I \oplus M_\alpha$ with M_α uniform. We denote the least upper bound of $\{n(x)\}_{x \in M}$ by m. Put $\underline{A} = \{\Sigma \oplus A_\alpha \subset M \mid A_\alpha = Ry \text{ and } n(y_\alpha) = m\}$. We can find a maximal member in $\underline{A}$ with respect to the number of direct components by Zorn's lemma, say $A = \Sigma_I \oplus A_\alpha$. Then we can find a submodule B of M such that $M \supset_e A \oplus B$ and $B = \Sigma \oplus B_\beta$ with B_β uniform. Then $E(M) = \Sigma \oplus E(A_\alpha) \oplus \Sigma \oplus E(B_\beta) \supset_e M \supset_e A \oplus B$. Let $x \in M$ and $x = \Sigma x_\alpha + \Sigma x_\beta$; $x_\alpha \in E(A_\alpha)$ and $x_\beta \in E(B_\beta)$. Since $n(x) \leq m$, $n(x_\gamma) \leq m$ and $A_\alpha = \{z \in E(A_\alpha) \mid n(z) \leq m\}$ [$E(A_\alpha)$ is serial]. Hence Σx_α is in A. Therefore, $M = A \oplus M \cap [\Sigma \oplus E(B_\beta)]$ and $n(y) < m$ for any $y \in M \cap [\Sigma \oplus E(B_\beta)]$ by the maximality of A. Use induction on m. Hence M is of the form in (9.11.21).

9.11.23 LEMMA Let M be an R-module. If M has the extending property of cyclic uniform modules, then M/M(t) does.

Proof: Put $\bar{M} = M/M(t)$. Let $\overline{aR}$ be a uniform submodule of $\bar{M}$. Since $a \notin M(t)$, aR is a torsion-free and uniform submodule of M. Hence we have a direct decomposition $M = L_1 \oplus L_2$ and $L_1 \supset_e aR$. Then $L_2 \supset M(t)$, so $\bar{M} = \bar{L}_1 \oplus L_2/M(t)$ and $L_1 \approx \bar{L}_1 \supset_e \overline{aR}$.

9.11.24 LEMMA Let $M = M(t) \oplus M_\beta$ have the extending property and let M_β be torsion free and uniform. Then M(t) is injective.

Proof: Let N be a uniform submodule of M_β and $f \in \mathrm{Hom}_R(N, M(t))$. Then $M = L_1 \oplus L_2$ and $L_1 \supset_e N(f)$. Since L_1 is torsion free, $L_2 = M(t)$. Then f is extendible to an element in $\mathrm{Hom}_R(M_\beta, M(t))$ by (9.6.7). Hence M(t) is injective by (9.11.14), (9.11.17), and (9.11.22).

9.11.25 LEMMA Let $M = E \oplus T$, where E is injective and torsion free, and let $T = E' \oplus T'$, where E' is torsion and injective and T' is torsion free. If T has the extending property, then M does.

Proof: Let N be a uniform submodule of M. If N is torsion, $N \subset T$. Hence N is essentially extendible to a direct summand of M. Let N be torsion free. If $N \subset E \oplus E'$, N is essentially extendible to a direct summand by (9.6.4). We assume that $N \not\subset E \oplus E'$. Let $\pi : M \to T'$. Then $\pi|N$ is an epimorphism and $N = \{x + f(x) \mid x \in \pi(N)\}$, where $f \in \mathrm{Hom}_R(\pi(N), E \oplus E')$. Since T' has the extending property by (9.11.17), $T' = D_1 \oplus D_2$ and $D_1 \supset_e \pi(N)$. $E \oplus E'$ being injective, f is extendible to $g \in \mathrm{Hom}_R(D_1, E \oplus E')$. Since $N = \pi(N)(g)$, $M = D_1(g) \oplus D_2 \oplus E \oplus E'$ and $D_1(g) \supset N$ by (9.6.7).

Proof of (9.11.16): We assume that M has the extending property. Let $M = E \oplus K$ with E injective and K reduced. Then K has the extending property by (9.11.18). Assume that K is torsion. Then K is of the form (1) by (9.11.21). In this case every indecomposable module is completely indecomposable. Hence every direct summand of M has the extending property by the remark below (9.6.1). We assume that $E \neq 0$. If E is not torsion, K is injective by (9.11.24). If E is torsion, K = 0 by (9.11.21). In either case, K = 0 if $E \neq 0$, so M is injective and is of the form (2). Next, we assume that K has a torsion-free uniform submodule N. Then $K = L_1 \oplus L_2$ and $L_1 \supset_e N$. If L_2 is not torsion, L_2 contains a torsion-free and uniform submodule N'. Let $K = L_2' \oplus L_3$ and $L_2' \supset_e N'$. Then since $L_2' \cap L_2 \supset N'$, $L_2' \subset L_2$ [see the proof of (9.11.17)]. Hence $K = L_1 \oplus L_2' \oplus L_2''$, where L_1 and L_2' are uniform and torsion free. Since $K(t) \subset L_2''$, $L_1 \oplus L_2'$ is isomorphic to a direct summand of K/K(t). K/K(t) has the extending property of cyclic uniform modules by (9.11.23) and $L_1 \oplus L_2'$ does by (9.11.17), which is a contradiction by (9.11.20). Hence $L_2 = K(t)$, and K(t) is injective by (9.11.24). Thus M is of the form (2). Conversely, if M is of the form (1), M has the extending property by (9.11.21). Let M be of the form (2) and $E = E(t) \oplus E'$. Then $E(t) \oplus K$ has the extending property by (9.11.12). Therefore, $M = E' \oplus E(t) \oplus K$ has the extending property by (9.11.25).

10

Lifting Property for Injectives and Extending Property for Projectives

In Chap. 9, we studied the lifting property and the extending property of modules. We also noted that the former property is a generalization of projectives and the latter is a generalization of injectives.

In this chapter we reciprocally study rings whose projectives have the extending property and whose injectives have the lifting property. Of course, these rings give us a new class of rings. They are closely related to the QF-2, QF-2*, and QF-rings of Sec. 5.8.

In Sec. 10.1 we define a new module, a small module, and define a functor Z* in $\underline{M}_R$ related to small modules. Defining a condition (10.2.1) on a nonsmall module in Sec. 10.2, we show that a perfect ring satisfying the condition (10.2.1) is artinian, and we give a structure theorem for these rings in Sec. 10.3. In Sec. 10.4 we consider weaker conditions (10.4.1) and (10.4.2) than (10.2.1) and show that every injective module has the lifting property of simple modules if and only if (10.4.2) is satisfied, provided that R is right artinian. Concerning the lifting property, we define a new property of modules—max. quasi-projectivity—in Sec. 10.5 and show that every injective module has the lifting property of direct decompositions modulo radical if and only if R is right QF-2* and a minimal injective cogenerator is max. quasi-projective. Sections 10.6-10.9 are dual to Secs. 10.1-10.5, and finally, in Sec. 10.10 we study conditions under which the rings in Secs. 10.5 and 10.6 are QF.

10.1 FUNCTOR Z*

We know from (4.7.37) that if the Krull dimension of a commutative ring R is zero, then R is never small in any ring extension as an R-module. We shall consider an analogous property for R-modules. First, we take any ring R, not necessarily commutative.

10.1.1 PROPOSITION Let M be an R-module. Then the following conditions are equivalent:

1. M is not small in any extension module M' of M.
2. M is not small in an injective envelope E(M) of M.
3. There exists an injective module E containing M such that M is not small in E.

Proof: (1) ⇒ (2) ⇔ (3). This is clear.

(2) ⇒ (1). We assume that $M' \supset M$. Then $E(M') = E(M) \oplus E_1$. Hence M is not small in E(M'). Therefore, M is not small in M' by (3.4.1).

If M satisfies one of the three equivalent conditions in (10.1.1), we say that M is *nonsmall* and otherwise we say that M is *small*.

10.1.2 LEMMA Let $0 \to M \to Q$ and $Q' \to M \to 0$ be exact. If M is nonsmall, then Q and Q' are also nonsmall.

Proof: This is clear from the definition for the first assertion. Take a commutative diagram

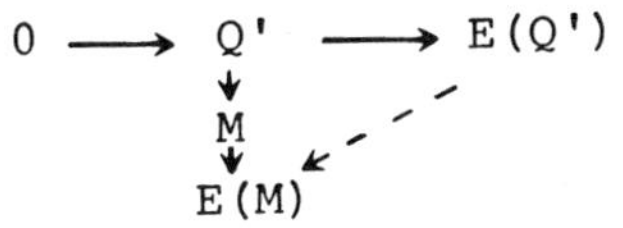

Since M is not small in E(M), Q' is nonsmall by (3.4.1).

We shall define a subfunctor in the category of all right R-modules $\underline{M}_R$ [cf. the functor Z(·)]. Let M be an R-module. We put

$$Z^*(M) = \{m \in M \mid mR \text{ is small}\} \qquad \underline{(10.1.1)}$$

Since J(M), the Jacobson radical of M, is the union of all small submodules in M by (3.4.3), $Z^*(E) = J(E)$ for any injective E, and $Z^*(M) = M \cap J(E(M)) = M \cap J(E')$ for an injective $E' \supset M$. It is clear by (10.1.2) that $Z^*(\cdot)$ is a subfunctor of the identity functor. If $M \neq Z^*(M)$, M is nonsmall; however, the converse is not true. If R is a right perfect ring, J(M) is a unique maximal small submodule in M by (7.6.8), so $M \neq Z^*(M)$ if and only if M is nonsmall. We have $Z^*(M) \supset J(M)$ and, in general, $Z^*(M) \neq J(M)$. If R is right artinian, $\ell_R(J(R)) = S(R_R)$, the socle of R, is a unique minimal essential right ideal, where $\ell_R(J(R)) = \{x \in R \mid xJ(R) = 0\}$. Hence $Z(M) = \ell_M(\ell_R(J(R))$ and $Z^*(M) = \ell_M(r_R(J))$ by (5.5.7). We can define functor Z_n^* as follows: $Z_n^*(M)/Z_{n-1}^*(M) = Z^*(M/Z_{n-1}^*(M))$. It is well known that $Z_2 = Z_3 = \cdots$ for a singular submodule Z(M). We do not know whether

$Z_2^* = Z_3^* = \cdots$ or not. However, we have the following.

10.1.3 PROPOSITION We assume that $R/J(R)$ is right artinian. Then $Z_2^* = Z_3^*$ and $M/Z_2^*(M)$ is semisimple and injective for every R-module M.

Proof: Since $Z^*(E) = J(E) = EJ(R)$ for an injective E, $E/Z^*(E)$ is semisimple. Let $E/Z^*(E) = \Sigma_I \oplus S_\alpha \oplus \Sigma_J \oplus S_\beta$, where the S_α are injective and minimal, and the S_β are small and minimal. Hence $Z_2^*(E)/Z^*(E) \approx \Sigma_J \oplus S_\beta$ and $Z_2^*(E) = Z_3^*(E)$. Put $\bar{E} = E/Z_2^*(E)$. Since $Z^*(\bar{E}) = 0$, $J(E(\bar{E})) \cap \bar{E} = 0$. Hence $J(E(\bar{E})) = 0$ and $E(\bar{E})$ is also semisimple. Therefore, $\bar{E} = E(\bar{E})$. Let $E \supset M$. Then $E/Z^*(E)$ contains $M/Z^*(M)$ isomorphically. Hence $Z_2^*(M)/Z^*(M) = [Z_2^*(E)/Z^*(E)] \cap M/Z^*(M)$. Therefore, $M/Z_2^*(M)$ is isomorphic to a submodule of $E/Z_2^*(E)$, which is semisimple and injective. Thus $Z_2^*(M) = Z_3^*(M)$.

From now on in this section we assume that R is (left and) right perfect unless otherwise stated. Then there exists a complete set $\{g_i\}$ of mutually orthogonal primitive idempotents such that $1 = \Sigma\, g_i$. Let $E = E(R)$ and x in $E - J(E)$. Then we obtain an epimorphism $f : R \to xR \subset E$. Since xR is nonsmall by (10.1.1), R is nonsmall by (10.1.2). Thus we shall divide $\{g_i\}$ into two parts $\{g_i\} = \{e_i\}_{i=1}^n \cup \{f_j\}_{j=1}^m$, where the e_iR are nonsmall and the f_jR are small. We know $n \geq 1$ from the above. We call an idempotent g *nonsmall* (resp. *small*) if gR is nonsmall (resp. small). When we denote the primitive idempotents by e and f, we mean that e is nonsmall and f is small, respectively.

10.1.4 LEMMA Let R be right perfect. Then every injective module is a homomorphic image of the form $\Sigma \oplus e_kR$, where the e_k are nonsmall.

Proof: Let E be injective and let $\varphi : \Sigma_I \oplus g_iR \to E$ be a projective cover of E. Assume that $\{g_i\}_{I'}$ are small for $I' \subset I$. Then $\varphi(\Sigma_{I'} \oplus g_iR)$ is a small submodule in E by (10.1.2). Since $\Sigma_I \oplus g_iR$ is a projective cover, $I' = \emptyset$.

10.1.5 LEMMA Let R be as in (10.1.4). If M is not small in $\Sigma \oplus g_iR/g_iA_i$, then there exists π_i such that $\pi_i(M) = g_iR/g_iA_i$, where the A_i are right ideals and the π_i are the projections on g_iR/g_iA_i.

Proof: Since $M \not\subset \Sigma \oplus g_jJ(R)/g_jA_j$, $\pi_i(M) \not\subset g_iJ(R)/g_iA_i$ for some i. Hence $\pi_i(M) = g_iR/g_iA_i$, since g_iR is hollow.

We call R a right QF-3 ring if E(R) is projective as a right R-module (Sec. 5.8).

10.1.6 THEOREM Let R be right perfect. Then R is right QF-3 if and only if each e_iR is injective, where the e_i are nonsmall primitive idempotents.

Proof: We assume that R is right QF-3. Then $E = E(R) = \Sigma \oplus e_kR$ by (7.6.10). Let e be a nonsmall primitive idempotent. Then eR is epimorphic to some e_kR by (10.1.5). Hence $eR = e_kR$ is injective. Conversely, let f be a small idempotent. Then we have an exact sequence $\Sigma \oplus e_kR \to E(fR) \to 0$ by (10.1.4). Accordingly, we have a diagram

$$\begin{array}{ccccc} 0 & \longleftarrow & E(fR) & \longleftarrow & \Sigma \oplus e_kR \\ & & i\uparrow & \overset{h}{\nearrow} & \\ & & fR & & \end{array}$$

where i is the inclusion. Since fR is projective and i is a monomorphism, we obtain a monomorphism h of fR to $\Sigma \oplus e_kR$. Therefore, E(R) is projective, since the e_kR are injective and fR is cyclic.

10.1.7 LEMMA Let R be right artinian and M an R-module. Then M is small if and only if $Mr_R(J) = 0$.

Proof: This is clear from (5.5.7).

10.1.8 COROLLARY Let R be a right artinian and QF-3 ring. Then R is QF-ring if and only if $Z^*(R)[= \ell_R(r_R(J(R)))] = J(R)$.

Proof: $Z^*(R) = \ell_R(r_R(J(R)))$ by (10.1.7). Hence every idempotent is nonsmall provided that $Z^*(R) = J(R)$. Therefore, R is QF by (10.1.6). The converse is clear by (5.7.3).

10.2 CONDITION (*)

It is clear by (10.1.1) and (10.1.2) that every module containing an injective submodule is nonsmall. We shall study the converse. For example, if R is a QF-ring, every nonsmall module contains an injective module [see (10.3.2) below]. In this section we investigate some rings with this property. Hence we consider two conditions.

Every nonsmall module contains a non-zero injective module. (*) (10.2.1)

Every indecomposable injective module E is hollow; i.e., every proper submodule is small in E (right QF-2*). (10.2.2)

If R is right perfect, (*) is equivalent to "every finitely generated nonsmall module contains a nonzero injective module," since $M \neq Z^*(M)$ if M is nonsmall.

We shall make use of the notation in Sec. 10.1.

10.2.1 LEMMA We assume that R satisfies (*). Then every injective module contains a cyclic injective module, and R contains a nonzero injective right ideal.

Proof: Let E be injective. We consider an exact sequence $\Sigma_{x \in E} \oplus E_x \xrightarrow{\varphi} E \to 0$, where $E_x = E$ and $\varphi | E_x = 1_E$. Then $\varphi(\Sigma_{x \in E} \oplus xR) = E$, so $\Sigma \oplus xR$ is nonsmall by (10.1.2). Hence $\Sigma \oplus xR$ contains an injective module F. Therefore, some xR ($\approx$ R) contains an injective submodule isomorphic to a direct summand of F by (8.5.5). If we replace $\Sigma \oplus E_x$ by a free R-module, we obtain the last assertion.

10.2.2 PROPOSITION Let R be a right noetherian ring satisfying (*). Then R is right artinian.

Proof: Let E = E(R). Then E is finite direct sum of indecomposable injective modules E_i. Then E_i is cyclic by (10.2.1), so E is noetherian. Hence R is right artinian by (5.8.2).

10.2.3 PROPOSITION We assume that R contains no infinite set of mutually orthogonal idempotents modulo Z*(R) and R satisfies (*). Then R is a right QF-3 of finite Goldie dimension.

Proof: Since R contains a nonzero injective right ideal by (10.2.1), we may assume that $R = \Sigma \oplus e_iR \oplus hR$, where the e_iR are indecomposable and injective and hR is small. Let $E_1 = E(hR)$ and let $\Sigma_k \oplus R \xrightarrow{\varphi} E_1 \to 0$ be exact. If $\varphi(\Sigma \oplus hR)$ is not small in E_1, hR contains an injective module by (*) and (10.1.2). Hence $\varphi(\Sigma_k \oplus (1 - h)R) = E_1$. Thus we have the diagram

$$\begin{array}{ccccc} 0 & \longleftarrow & E_1 & \longleftarrow & \Sigma \oplus (1-h)R \\ & & i \uparrow & \nearrow_{f} & \\ & & hR & & \end{array}$$

where i is the inclusion, since hR is projective. Therefore, f being a monomorphism, $E(R) \approx \Sigma \oplus e_{ij}R$, where $e_{ij}R \approx e_iR$, and R is of finite Goldie dimension.

10.2.4 THEOREM Let R be perfect. Then (*) holds if and only if there exists an integer n_i for each nonsmall primitive idempotent e_i such that $e_iR/e_i\ell_R(J^t)$ is injective for $0 \le t \le n_i$ and $e_iR/e_i\ell_R(J^{n_i+1})$ is small. In this case $e_iR/e_i\ell_R(J^t) \not\approx e_iR/e_i\ell_R(J^{t'})$ for $t < t' \le n_i$ and every submodule of e_iR either contains $e_i\ell_R(J^{n_i+1})$ or is equal to some $e_i\ell_R(J^t)$, $t < n_i + 1$, where $J = J(R)$.

Proof: We assume (*). Then e_iR is injective and indecomposable from (*). We assume that $e_i = e$ and eR/eB is nonsmall for some right ideal B. Then eR/eB is injective, since eR/eB is indecomposable. Since eR is injective and R is perfect, eR contains a unique minimal submodule $e\ell_R(J)$ and $eB \supset e\ell_R(J)$. We have a natural epimorphism $eR/e\ell_R(J) \to eR/eB$, and eR/eB is injective. Hence $eR/e\ell_R(J)$ is injective from (*) and (10.1.2). Therefore, $eR/e\ell_R(J)$ contains a unique minimal submodule $e\ell_R(J^2)/e\ell_R(J)$, which is contained in $eB/e\ell_R(J)$. Repeating this argument, we obtain $eR \supset eB \supset e\ell_R(J^k) \supset \cdots \supset e\ell_R(J) \supset 0$. First we shall note that $eR/e\ell_R(J^t) \not\approx eR/e\ell_R(J^{t'})$ for $k \geq t > t'$, since $\ell_R(J^t)$ is a two-sided ideal. Now R/J is artinian and the representative set of minimal modules is finite. Therefore, the length of the above series is finite. Accordingly, $eB = e\ell_R(J^s)$ for some s and $eR/e\ell_R(J^n)$ is small for some n. Conversely, we assume that the n_i in (10.2.4) exists. Let M be a nonsmall module and $E = E(M)$. Then $M \not\subset EJ$. Let m be in $M - EJ$; then mR is nonsmall by (10.1.1). Let $1 = \Sigma e_i + \Sigma f_j$. Now $me_iR \approx e_iR/e_iB$ for some right ideal B. Since $e_iR/e_i\ell_R(J^t)$ is injective for $t \leq n_i$, $e_i\ell_R(J^{t+1})/e_i\ell_R(J^t)$ is a unique minimal submodule of $e_iR/e_i\ell_R(J^t)$. Hence either $e_iB = e_i\ell_R(J^s)$ for some s or $e_iB \supset e_i\ell_R(J^{n_i+1})$. In the latter case, we have an epimorphism $e_iR/e_i\ell_R(J^{n_i+1}) \to e_i/e_iB$, which is a contradiction by (10.1.2). Hence me_iR is injective. The last assertion is clear from the above.

Finally, we shall show that a left and right perfect ring satisfying (*) is right artinian.

10.2.5 THEOREM Let R be a perfect and right QF-2* ring. We assume that every indecomposable injective module satisfies (9.2.1). Further, if every injective module E has the lifting property of simple modules modulo radical, then R is right artinian.

Proof: Let E be an indecomposable injective module, say $E = E(S)$; S is simple. We shall show that any direct sum of copies of E is injective. Put $T = \Sigma_I \oplus E_\alpha$; $E_\alpha = E$ and $Q = E(T)$. It is clear that $\Sigma \oplus \bar{E}_\alpha$ is a direct summand of $\bar{Q} = Q/QJ$, since R is perfect. Now we can express $\bar{Q} = \Sigma_I \oplus \bar{E}_\alpha \oplus \Sigma_J \oplus A_\beta$, where the A_β are simple. Then $Q = \Sigma_I \oplus E_\alpha + \Sigma_L \oplus E(S_\beta)$ by the assumption, and $\overline{E(S_\beta)} = A_\beta$. We shall show $L = \emptyset$. Otherwise, there would exist an $A_\beta \neq 0$. $\Sigma_I \oplus S$ is the socle of T and hence of Q. Hence $S_\beta \approx S$ and $E(S_\beta) \approx E$. Let $S_\beta \subset \Sigma_K \oplus S$ for some finite subset K of I and $E_K = \Sigma_K \oplus E_\gamma$. Then $Q = E_K \oplus F$. We denote the projection of Q onto F by π.

$\ker \pi|E(S_\beta) \supset S_\beta$ and $\pi(E(S_\beta)) \not\subset FJ$, since $Q \neq \Sigma_I \oplus E_\alpha + \Sigma_{\gamma\neq\beta} E(S_\gamma)$. It is clear $\pi(E(S_\beta)J) \subset FJ$. Hence $F/FJ = C/CJ \oplus B_2 \cdots$, where $C = \pi(E(S_\beta)) + FJ$ and the B_i are simple. Then we obtain an indecomposable direct summand E'_δ of F such that $\bar{E}'_\delta = C/CJ$. It is clear, as above, that $E'_{\delta_1} \approx E$. Put $F = E'_{\delta_1} \oplus L$ and π_1 the projection of F onto E'_{δ_1}. Since $C = E'_{\delta_1} + FJ$, $\pi_1\pi$: $E(S_\beta) \to E'_{\delta_1}$ is an epimorphism. Hence $\pi_1\pi|E(S_\beta)$ is an isomorphism by the assumption on E. However, $\ker \pi_1\pi|E(S_\beta) \supset S_\beta$, which is a contradiction. Thus T is injective. Let V be any injective module and $\Sigma_{i=1}^t \Sigma_{\alpha_i} \oplus S_{i\alpha_i}$ be the socle of V, where $S_{i\alpha} \approx S_{i\beta}$ and $S_{i\alpha} \not\approx S_{j\beta}$ if $i \neq j$. Then $V = E(\Sigma_{i=1}^t \Sigma_{\alpha_i} \oplus S_{i\alpha_i}) = \Sigma_i \oplus E(\Sigma_{\alpha_i} \oplus S_{i\alpha_i}) = \Sigma_{i=1}^t \oplus \Sigma_{\alpha_i} \oplus E(S_{i\alpha_i})$ by the above. Hence R is right artinian by (4.7.19).

We shall now show in the proof (1) $\Rightarrow$ (2) in (10.4.3) below that every injective module has the lifting property of simple modules modulo radical provided that R satisfies (*). Thus we have the following.

10.2.6 THEOREM Let R be a (left and right) perfect ring. If R satisfies (*), R is right artinian.

Proof: Let E be an injective module and $\bar{A}$ a simple submodule of E/EJ, where A ($\supset$ EJ) is a submodule of E. Since $A \neq EJ$, A contains an injective submodule E'_1 by (*). Hence $E = E_1 \oplus E'$ and $\bar{E}_1 = \bar{A}$. If E is indecomposable, E is hollow and hence E satisfies (9.2.1).

10.3 RIGHT ARTINIAN RINGS WITH (*)

Let R be a right artinian ring. Then we have the standard decomposition of R:

$$R = \sum_{i=1}^{p} \sum_{j=1}^{\rho(i)} \oplus g_{ij}R$$

where $\{g_{ij}\}$ is a set of mutually orthogonal primitive idempotents such that $g_{ij}R \approx g_{ij'}R$ and $g_{ij}R \not\approx g_{kj'}R$ if $i \neq k$. As in Sec. 10.1, we denote nonsmall idempotents by e_{ij} and put $E_i = \Sigma_{j=1}^{\rho(i)} e_{ij}$.

Now it is clear that R satisfies (*) if and only if every module M is a direct sum of an injective submodule and a small submodules. We can restate (10.2.4) as follows.

10.3.1 THEOREM Let R be right artinian. Then the following conditions are equivalent:

1. R satisfies (10.2.1).

2. There exists an $n_i > 0$ for each nonsmall idempotent $e_i = e_{i1}$ such that $e_iR/e_iJ^{k_i}$ is injective for all $n_i \le k_i$, and $e_iR/e_iJ^{n_i-1}$ is small.
3. $R/r_R(J)J^k$ is a direct sum of an injective module and a small projective module for all $k > 0$ as R-modules.
4. R/A is a direct sum of an injective module and a small module for every right ideal A contained in $r_R(J)$.

In this case $A = \Sigma \oplus A_i$ and the A_i are right ideals in $E_ir_R(J)$ where $J = J(R)$ and $r_R(J) = \{x \in R \mid Jx = 0\}$.

Proof: Let $R = \Sigma_{i=1}^n \oplus e_iR \oplus \Sigma_{i=1}^m \oplus f_iR$ as in Sec. 10.1 and $D = r_R(J)$. Since f_jR are small, $f_jD = 0$ by (10.1.7). Hence $D = \Sigma_{i=1}^n \oplus e_iD$ and $DJ^k = \Sigma \oplus e_iDJ^k$ for any integer.

(1) ⇒ (2). We assume that $e_i = e$, eR is injective, and $eJ^{q-1} \ne 0$, $eJ^q = 0$. Then eJ^{q-1} is the unique minimal submodule in eR. Hence $eJ^{q-1} = e\ell_R(J)$. Similarly, we obtain $eJ^{q-t} = e\ell_R(J^t)$ if eR/eJ^{q-t+1} is injective. If eR/eJ^s is small and eR/eJ^{s+1} is injective, $eD \subset eJ^s$, and hence $eD = eJ^s$, since eJ^s/eJ^{s+1} is unique minimal. Hence we have proved (1) ⇔ (2) by (10.2.4).

(1) ⇒ (3). $D = \Sigma_{i=1}^n \oplus e_iD = \Sigma_{i=1}^n \oplus e_iJ^{n_i}$ for some n_i by the proof of (1) ⇔ (2). Hence $R/DJ^k = \Sigma_{j=1}^m \oplus f_jR \oplus \Sigma_{i=1}^n \oplus e_iR/e_iJ^{n_i+k}$ and the $e_iR/e_iJ^{n_i+k}$ are injective for $k > 0$ by (10.2.4).

(3) ⇒ (1). We always have $DJ^k = \Sigma_{i=1}^n \oplus e_iDJ^k$. Hence $R/DJ^k = \Sigma_{j=1}^m \oplus f_jR \oplus \Sigma_{i=1}^n \oplus e_iR/e_iDJ^k$. Therefore, the e_iR/e_iDJ^k are injective for any $k > 0$ by (7.2.1), since e_iR is nonsmall. If $e_iDJ^t = 0$ and $e_iDJ^{t-1} \ne 0$, then e_iR is injective, e_iDJ^{t-1} is the unique minimal submodule in e_iR, and $e_iDJ^{t-1} = e_i\ell_R(J)$. Repeating this argument as in the proof of (10.2.4), there exist an integer n_i and a unique series of submodules $e_i\ell_R(J^t)$ of e_iR such that $e_iR/e_i\ell_R(J^t)$ is injective for $t < n_i$ and $e_iD = e_i\ell_R(J^{n_i})$. Therefore, R satisfies (*) by (10.2.4).

(1) ⇒ (4). This is clear by the fact mentioned before (10.3.1).

(4) ⇒ (1). Since $DJ^k \subsetneq D$ for $k > 0$, e_iR/e_iDJ^i is not small by (10.1.7). Hence we can use the same method as in (3) ⇒ (1). Finally, we assume (*) and $e_i = e_{i1}$. Let $e_iD \supset e_iA_i \supsetneq e_iB_i$ be right ideals. Then $e_iB_i = e_iJ^{t_i}$ for some t_i and e_iR/e_iB_i is injective by (10.2.4). If $e_iA_i/e_iB_i \overset{\psi}{\approx} e_jA_j/e_jB_j$, we can extend ψ to an isomorphism of e_iR/e_iB_i to e_jR/e_jB_j, since e_pR/e_pB_p is indecomposable and injective for $p = i, j$. Hence $i = j$. Since $e_{ij}D \approx e_{ij'}D$, the set of nonisomorphic simple factor modules of composition series of E_iD coincides with that of $e_{i1}D$. Therefore, $A = \Sigma_{i=1}^n \oplus (E_iD \cap A)$.

Example There exists a commutative ring R with R/J artinian, which satisfies the condition in (10.2.4). Let Z be the ring of integers and $Z_{p^\infty} = E(Z/p)$, where p is prime. Let $R = Z_{(p)} \oplus Z_{p^\infty}$, where $Z_{(p)} = \mathrm{End}_Z(Z_{p^\infty})$. Then R has a ring structure as follows.

$$(f + x)(g + y) = fg + f(y) + g(x) \quad \text{where } f,g \in Z_{(p)} \text{ and } x,y \in Z_{p^\infty}$$

(see [191]). $J(R) = pZ_{(p)} \oplus Z_{p^\infty}$ and $r_R(J) = \{(0,1/p)\}$. Furthermore, $0 \to \{(0,1/p)\} \to R \to pZ_{(p)} \oplus Z_{p^\infty}$ $[\approx J(R)] \to 0$ is exact. We can show directly that R is self-injective and $R/r_R(J)$ is a small module. Hence R does not satisfy (*) by (10.2.1), since $Q_{(p)}$ does not contain any cyclic injective module, where $Q_{(p)}$ is the quotient field of $Z_{(p)}$.

10.3.2 PROPOSITION Let R be right and left artinian. Then the following conditions are equivalent:

1. R is a QF-ring.
2. (*) holds and $e_iRf_j = 0$ for every nonsmall e_i and every small f_j.
3. (*) holds and $\ell_R(J) \subset r_R(J)$.
4. e_iR is injective, and $e_iR/e_i\ell_R(J)$ is small whenever $e_i\ell_R(J) \neq 0$ for every nonsmall e_i.

Proof: (1) ⇒ (2). Let R be a QF-ring. Then $\ell_R(J) = r_R(R)$ by (5.7.3), so $\ell_R(J)J^k = 0$. Hence (*) holds by (10.3.1). Since $f_j = 0$, $e_iRf_j = 0$.

(2) ⇒ (1). We assume (*); then R is QF-3 and f_jR is monomorphic to some submodule of $\Sigma \oplus e_{ij}R$ by (10.1.6) and its proof, where $e_{ij}R \approx e_iR$. Hence $e_iRf_j = 0$ implies that $f_j = 0$.

(3) ⇒ (1). Let g be a primitive idempotent. $RS(gR) \subset r_R(J)$ by the assumption, where S(·) is the socle of (·). Hence $gr_R(J) \supset gRS(gR) = gRgS(gR) \neq 0$. Therefore, $g \notin \ell_Rr_R(J) = Z^*(R)$, so g is nonsmall. Accordingly, R is a QF-ring by (10.2.4) and (10.1.6).

(1) ⇒ (4). If $e_iR/e_i\ell_R(J)$ is nonsmall, $e_iR/e_i\ell_R(J)$ is injective by (2). Hence, since R is a QF-ring, $e_iR/e_i\ell_R(J)$ is projective by (5.7.7). Therefore, $e_i\ell_R(J) = 0$.

(4) ⇒ (1). Let $\{e_iR\}_1^t$ be a complete set of nonisomorphic right ideals in $\{e_iR\}_1^n$. Since e_iR is injective, $S(e_iR) = e_i\ell_R(J)$ is the unique minimal submodule. Let E be an indecomposable injective module. Then we obtain a projective cover of E: $\Sigma_{i=1}^t \oplus e_iR^{(n_i)} \to E$, by (10.1.4). Hence $E \approx e_iR$ for some i by (4). Therefore, $\{e_iR/e_iJ\}_{i=1}^t$ is the complete set of nonisomorphic right minimal modules. Let g be a primitive idempotent. Then gR/gJ

is isomorphic to a member of $\{e_iR/e_iJ\}_{i=1}^{t}$. Hence $gR \approx e_iR$ is injective.

(1) ⇒ (3). This is clear.

10.4 LIFTING PROPERTY OF SIMPLE MODULES FOR INJECTIVES

We show in this section that the condition (*) guarantees the lifting property of simple modules modulo radical for injectives.

Now we shall study injective modules over a right QF-2* ring (Sec. 5.8). We define two conditions weaker than (*).

Every nonsmall module which is a homomorphic image of an injective module contains a nonzero injective module. (10.4.1)

For every nonsmall module M, there exists an indecomposable direct summand E_1 of E(M) such that $\overline{E(M)} \supset \bar{M} \supset \bar{E}_1$, where $\bar{E} = E/J(E)$. (10.4.2)

We know from (10.2.2) that every indecomposable and injective module has the form eR/eA, where e is a nonsmall primitive idempotent and eA is a right ideal of R.

10.4.1 LEMMA Let R be a right QF-2* and artinian ring. Then for every nonsmall primitive idempotent e, there exists a right ideal A such that eR/eA is injective.

Proof: Let $E(eR) = \Sigma_{i=1}^{k} \oplus e_i'R/e_i'A_i$, where e_i' is nonsmall and the $e_i'A_i$ are right ideals. Since eR is nonsmall and $e_i'R/e_i'A_i$ is hollow, $\pi_i(eR) = e_i'R/e_i'A$ for some i by (10.1.5), where $\pi_i : E(eR) \to e_i'R/e_i'A_i$ is the projection. Hence $eR \approx e_i'R$ by (10.1.5).

10.4.2 THEOREM Let R be a right QF-2* and artinian ring. Then the following conditions are equivalent:

1. R satisfies (10.4.1).
2. Every irredundant sum of direct summand E_α in an injective module E such that $\overline{\Sigma E_\alpha} = \Sigma \oplus \bar{E}_\alpha$ ($\bar{E}_\alpha \neq 0$) is injective.
3. For each nonsmall idempotent e, there exist right ideals $eA_1 \subset eA_2 \subset \cdots \subset eA_{t-1} \subset eA_t$ that satisfy the following:
   (i) eA_t/eA_1 is a uniserial module such that eA_i/eA_{i-1} is the socle of eR/eA_{i-1}.
   (ii) eR/eB is small for all right ideals $eB \supsetneq eA_t$.
   (iii) The set of these eR/eA_i is the representative set of indecomposable injectives, where $\bar{E} = E/J(E)$.

(It is possible that eR contains more than one such increasing series.)

Proof: (1) ⇒ (2). Let E be injective and ΣE_α a sum in E with E_α injective and indecomposable as in (2). Since R is artinian (hence noetherian) we may show that $\Sigma_K E_\alpha$ is injective for every finite subset K of I. Let $\{\alpha_1, \alpha_2, \ldots, \alpha_p\}$ be a finite subset of I and $E(p) = \Sigma_{i=1}^p E_{\alpha_i}$. We shall show the above fact by induction on p. We assume that $E(p - 1)$ is injective and $E = E(p - 1) \oplus K'$. Let $\pi : E \to K'$ be the projection. Then $\pi(E_{\alpha_p})$ is not a small submodule in K', since $\overline{\Sigma E_\alpha} = \Sigma \oplus \bar{E}_\alpha$. Hence $\pi(E_{\alpha_p})$ is injective by (10.4.1) for E_{α_p}, is hollow. Therefore, $E(p) = E(p - 1) \oplus \pi(E_{\alpha_p})$ is injective.

(2) ⇒ (3). We shall show that if eR/eA is injective and eR/eB is nonsmall for eB ⊃ eA, then eR/eB is injective, where $e = e_i$. Put E = E(eR/eB) and F = eR/eA ⊕ E. Let f : eR/eA → eR/eB be the natural epimorphism and $G = \{x + f(x) \mid x \in eR/eA\} \subset F$. Then G ≈ eR/eA and $\varphi_F(eR/eA + G) = \varphi_F(eR/eA) \oplus \varphi_F(G)$, since eR/eB is nonsmall, and hence $\varphi_F(x) \neq 0$ for some x, where φ_F is the natural epimorphism of F to F/J(F). Hence eR/eA + G = eR/eA ⊕ f(eR/eA) is injective. Therefore, eR/eB = f(eR/eA) is injective. By (10.4.1) we have the injective module eR/eC for each e. We may assume that eA_1 is minimal among eC. Let eA_2/eA_1 be the socle of eR/eA_1. If eR/eA_2 is nonsmall, eR/eA_2 is injective by the above. Repeating this argument, we get a series of right ideals $eA_1 \subset eA_2 \subset \cdots \subset eA_t$ such that eR/eA_j is injective and eR/eB is small for all $eB \supsetneq eA_t$ by (10.1.2). Hence we have proved (2) ⇒ (3) by (10.2.2).

(3) ⇒ (1). Let M be a nonsmall module which is a homomorphic image of injective E. Let $E = \Sigma \oplus E(S_\alpha)$. Then there exists an $E(S_\alpha)$ ($\approx eR/eA_i$) whose image is a nonsmall submodule M_0 of M. Hence $M_0 \approx eR/eC$, $eC \supset eA_t$. On the other hand, either $eC = eA_j \supset eA_i$ for some j or $eC \supsetneq eA_t$ from (3),(i). Hence $eC = eA_j$ by (3),(i).

10.4.3 THEOREM Let R be a right (QF-2*) artinian ring. Then the following conditions are equivalent:

1. R satisfies (10.4.2).
2. Every injective module has the lifting property of simple modules modulo radical.
3. (i) For each nonsmall idempotent e, there exists a chain of right ideals A_i such that eR/eA_i is injective, $eA_1 \subset eA_2 \subset \cdots \subset eA_t$, and each element in $\mathrm{End}_R(eR/eJ)$ is induced from some element in $\mathrm{Hom}_R(eR/eA_i, eR/eA_j)$ for $i \leq j$.

(ii) The set of $\{e_iR/e_iA_{ij}\}_{i=1,j=1}^{n\ \ t_i}$ is the representative set of indecomposable injectives, i.e., $\Sigma \oplus e_iR/e_iA_{ij}$ is the minimal injective cogenerator.

Proof: (1) ⇒ (2). Let E be injective and $E/EJ = \Sigma \oplus A_\alpha$; the A_α are simple. We take an element x in E such that xR is nonsmall by the definition. Hence there exists an indecomposable direct summand E_α of E such that $A_\alpha = \overline{xR} \supset \bar{E}_\alpha$ by (10.4.2). Therefore, $\bar{E}_\alpha = A_\alpha$.

(2) ⇒ (3). We know from (10.2.2) and (10.4.1) that the representative set of injectives is $\{e_iR/e_iB_{ij}\}_{i=1,j=1}^{n\ \ n(i)}$. We put $E(i) = \Sigma_{j=1}^{n(i)} \oplus e_iR/e_iB_{ij}$, $e_i = e$, and $B_{ij} = B_j$. Then there exists either an epimorphism $f_{\ell m} : eR/eB_m \to eR/eB_\ell$ or $f_m : eR/eB_\ell \to eR/eB_m$ for any pair ℓ and m by (2) and (9.2.5). We denote this by $eR/eB_m \succcurlyeq$ (resp. $\preccurlyeq$) eR/eB_ℓ. Then the relation $\succcurlyeq$ is linear. We take the maximal member among eR/eB_j, say eB_1 $(= eA_1)$ with respect to the relation $\succcurlyeq$. Let eR/eB_2 be the second element. Since there exists an epimorphism f_{21}, there exists a right ideal A_2 such that $eA_2 \supset eA_1$ and $eR/eA_2 \approx eR/eB_2$. Repeating this argument, we have a chain of right ideals A_i satisfying (i) by (9.2.5).

(3) ⇒ (2). This is clear by (3) and (9.2.5).

(2) ⇒ (1). This is clear from the definition.

10.5 MAX. QUASI-PROJECTIVE MODULES

We shall continue to study the lifting property of direct decompositions modulo radical for injectives. We shall define a new class of modules. For any maximal submodule N of M, we consider a diagram

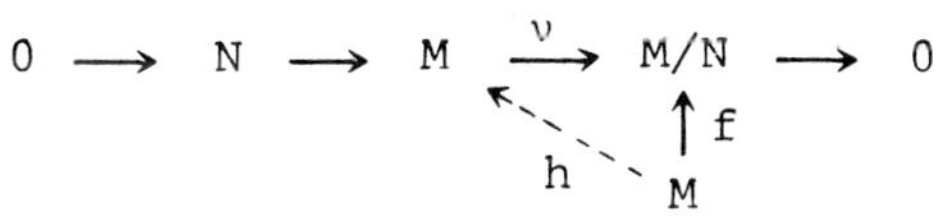

where ν is the natural epimorphism. If there exists h in $\mathrm{End}_R(M)$ for any f in $\mathrm{Hom}_R(M,M/N)$ which makes the diagram commute, we say that M is a *max. quasi-projective module.* It is clear that every quasi-projective module is max. quasi-projective, and the converse is not true in general. For example, let R be a local algebra over a field K such that $R/J(R) \approx K$ and let A be a right ideal of R not two-sided. Then R/A is max. quasi-projective, but not quasi-projective.

Let N and N' be two maximal submodules of M. Then the definition above is equivalent to the following.

$$\mathrm{Hom}_R(M,M) \supset \mathrm{Hom}_R(M,M)^* \to \mathrm{Hom}_R(M/N,M/N')$$

is an epimorphism via the natural epimorphisms $\nu : M \to M/N$ and $\nu' : M \to M/N'$, where $\mathrm{Hom}_R(M,M)^* = \{f \in \mathrm{Hom}_R(M,M) \mid f(N) \subset N'\}$. Put $S = \mathrm{End}_R(M)$, $\bar{S} = \mathrm{End}_R(\bar{M})$, and $J_0(S) = \mathrm{Hom}_R(M,J(M))$, where $\bar{M} = M/J(M)$. Then we have the natural monomorphism $\theta : S/J_0(S) \to \bar{S}$.

10.5.1 THEOREM Let R be right artinian and let M be an R-module. We assume that $M = \Sigma_I \oplus M_\alpha$, where the M_α are completely indecomposable [i.e., $\mathrm{End}_R(M_\alpha)$) is local]. Then the following conditions are equivalent:

1. θ is an epimorphism.
2. M is max. quasi-projective and each M_α is hollow.
3. M has the lifting property of direct decompositions of $\bar{M}$ [under the assumption (9.3.1)].

Proof: (1) ⇒ (2). It is clear that θ induces an epimorphism θ_α : $\mathrm{End}_R(M_\alpha) \to \mathrm{End}_R(\bar{M}_\alpha)$. Since $\mathrm{End}_R(M_\alpha)$ is local, so is $\mathrm{End}_R(\bar{M}_\alpha)$. Furthermore, $\bar{M}_\alpha$ is semisimple, so $\bar{M}_\alpha$ is simple. Therefore, M_α is hollow. Let N_1 and N_2 be two maximal submodules of M. Then $N_i \supset J(M)$. Hence M/N_i ($\approx \bar{M}/\bar{N}_i$) is a direct summand of $\bar{M}$. Accordingly, M is max. quasi-projective.

(2) ⇒ (3). Since each M_α is hollow, $J(M_\alpha) \oplus \Sigma_{\beta\neq\alpha} \oplus M_\beta$ is a maximal submodule of M. Hence $\mathrm{Hom}_R(M_\alpha,M_\beta) \to \mathrm{Hom}_R(\bar{M}_\alpha,\bar{M}_\beta)$ is an epimorphism for α and β in I. We assume that $\bar{M}_\alpha \approx \bar{M}_\beta$. Then M_α and M_β being hollow, there exist epimorphisms $f : M_\alpha \to M_\beta$, $g : M_\beta \to M_\alpha$ by the fact above. Since R is artinian, $M_\alpha \approx M_\beta$. Hence $\{M_\alpha\}_I$ is (semi-) T-nilpotent. Therefore, M has the lifting property of direct decompositions of M by (9.3.2).

(3) ⇒ (1). This is clear from (9.3.2).

10.5.2 THEOREM Let R be a right artinian ring. Then the following conditions are equivalent:

1. Every injective E has the lifting property of direct decompositions (modulo radical).
2. An injective cogenerator is max. quasi-projective and a direct sum of hollow submodules [i.e., right QF-2* (Sec. 5.8)].

Proof: Every injective is a direct sum of completely indecomposable modules by (4.7.18). Hence (10.5.2) is clear by (10.5.1).

10.5.3 PROPOSITION Let $\{M_i\}$ be a set of indecomposable modules of finite length.

1. We assume that $M = M_1 \oplus M_2$ is max. quasi-projective and $M_1 \not\approx M_2$; then every simple submodule in $\bar{M}_1$ is not isomorphic to any simple submodule in $\bar{M}_2$.

2. We assume that M_1 is max. quasi-projective and $M_1/N_1 \approx M_1/N_2$ for maximal submodules N_i of M_1; then there exists an automorphism f of M_1 with $f(N_1) = N_2$.

Proof: 1. We assume that there exists a maximal submodule N_i of M_i such that $M_1/N_1 \approx M_2/N_2$. Then there exists an f_i in $\mathrm{Hom}_R(M_i, M_{i'})$ ($i \neq i'$) which induces the above isomorphism and satisfies $f_i(N_i) \subset N_{i'}$, by the assumption. Put $h = f_2 f_1 \in \mathrm{End}_R(M_1)$. Then $h(N_1) \subset N_1$ and $M_1 = h(M_1) + N_1$. If h is not an isomorphism, h is nilpotent, for M_1 has finite composition length (4.5.19). Hence $N_1 = h^n(M_1) + N_1 = N$ for some n, which is a contradiction. Therefore, h is an isomorphism and f_i is also an isomorphism, which is a contradiction.

2. If we apply the above argument for $M_1 = M_2$, we have (2).

10.5.4 LEMMA Let M_1 be an indecomposable module of finite length. Put $M = M_1^{(I)}$, a direct sum of $|I|$ copies of M_1. We assume that M_1 is max. quasi-projective and N a maximal submodule of M. Then there exists a direct decomposition $M = \Sigma_I \oplus M'_\alpha$ such that $N = N_1 \oplus \Sigma_{I-\{1\}} \oplus M'_\alpha$, where $M'_\alpha \approx M_1$ for all $\alpha \in I$ and N_1 is a maximal submodule of M'_1.

Proof: We shall show that N contains a nonzero direct summand of M if $|I| \geq 2$. Since N is maximal, $N \supset J(M)$. We denote $M/J(M)$ by $\bar{M}$, and $M = \Sigma_I \oplus M_\alpha$. Put $M_\alpha \cap N = N_\alpha$. If $M_\alpha \subset N$ for some $\alpha \in I$, we are done. Hence we may assume that $M = N + M_\alpha$ for any $\alpha \in I$. Then $M/N = (N + M_\alpha)/N \approx M_\alpha/N_\alpha$, so N_α is a maximal submodule of M_α. Now $\bar{M}_\alpha = \bar{N}_\alpha \oplus \bar{A}_\alpha$, where $J(M_\alpha) \subset A_\alpha \subset M_\alpha$ and $\bar{A}_\alpha$ is a simple submodule. Since $N \supset \Sigma \oplus N_\alpha \supset J(M)$, $\bar{N} = \Sigma_I \oplus \bar{N}_\alpha \oplus \bar{N}_0$ and $N_0 \neq 0$, for N is maximal and $|I| \geq 2$. Let $\bar{N}_0^*$ be a simple submodule of N_0; $N_0^* \supset J(M)$. We take the direct decomposition

$$\bar{M} = (\Sigma_I \oplus \bar{N}_\alpha) \oplus (\Sigma_I \oplus \bar{A}_\alpha)$$

Let p and p_α be the projections of $\bar{M}$ onto $\Sigma_I \oplus \bar{N}_\alpha$ and $\bar{A}_\alpha$, respectively. Since $\bar{N}_0^*$ is simple, there exists a finite subset I' of I such that $p(\bar{N}_0^*) = \bar{A}_\alpha$ for $\alpha \in I'$ and $p_\beta(\bar{N}_0^*) = 0$ for $\beta \in I - I'$. We may assume that $I' = \{1,2,\ldots,n\}$. Hence there exists a set of isomorphisms $g_i : \bar{A}_1 \to \bar{A}_i$ ($i \geq 2$) such that $N_0^* = \{p(n) + a_1 + g_2(a_1) + \cdots + g_n(a_1) \mid n \in \bar{N}_0^*, a_1 = p_1(n) \in \bar{A}_1\}$. On the other hand, there exists $f_i : M_1 \to M_i$ such that $f_i(N_1) = N_i$ and f_i induces g_i on $\bar{A}_1$ by (10.5.3). We put $M_1(f) = \{m_1 + f_2(m_1) + \cdots + f_n(m_1) \mid m_1 \in M_1\}$. Then $M = M_1(f) \oplus \Sigma_{I-\{1\}} \oplus M_\alpha$. Let $\bar{m}_1 = n_1 + \bar{a}_1$; $\bar{n}_1 \in \bar{N}_1$, $\bar{a}_1 \in \bar{A}_1$. Then $\overline{m_1 + f_2(m_1) + \cdots + f_n(m_1)} = \bar{n}_1 + f_2(\bar{n}_1) + \cdots + f_n(\bar{n}_1) + \bar{a}_1 + f_2(\bar{a}_1) + \cdots + f_n(\bar{a}_1) \in \Sigma_I \oplus \bar{N}_\alpha \oplus \bar{N}_0^* \subset \bar{N}$. Hence $M_1(f) \subset N$.

Let $\underline{F}$ be the set of all direct sums of indecomposable modules isomorphic to M_1 that are contained in N and are locally direct summands of M. Then $\underline{F} \neq \emptyset$ by the fact above, and we can find a maximal element with respect to the members of direct components in $\underline{F}$ by Zorn's lemma, say $\Sigma_J \oplus M'_\alpha$ ($\subset N$). Since M_1 has finite length, $\{M_1\}$ is a (semi-) T-nilpotent set by (4.5.21). Hence $\Sigma_J \oplus M'_\alpha$ is a direct summand of M by (7.3.9), say $M = \Sigma_J \oplus M'_\alpha \oplus M^*$. Hence $N = \Sigma \oplus M'_\alpha \oplus (N \cap M^*)$ and $N \cap M^*$ is a maximal submodule of M^*. M^* is also a direct sum of $\{M''_\beta\}$, where the M''_β are isomorphic to M_1 by (7.1.2). Therefore, $|I - J| = 1$ by the maximality of $\Sigma_J \oplus M'_\alpha$.

We assume that an R-module M is a direct sum of indecomposable modules M_α of finite length. Then we can rearrange them as follows.

$$M \approx \sum_{\alpha \in I} \oplus M_\alpha^{(J_\alpha)} \qquad M_\alpha \not\approx M_\beta \text{ if } \alpha \neq \beta$$

10.5.5 THEOREM Let M be as above. Then M is max. quasi-projective if and only if M_α is max. quasi-projective for all α and each simple submodule in $\bar{M}_\alpha$ is not isomorphic to any simple submodules in $\bar{M}_\beta$ if $\alpha \neq \beta$, and hence $|I|$ is finite.

Proof: We assume that M is max. quasi-projective. Then any direct summand of M is also max. quasi-projective by the definition. Hence the M_α satisfy the condition in (10.5.5) by (10.5.3). Conversely, we assume the property above. Since R is artinian, I is finite. Let N be a maximal submodule in M. By the assumption $\bar{M}_\alpha^{(J_\alpha)}$ is a direct sum of the homogeneous components in $\bar{M}$. Hence $\bar{N} = \bar{N}_1 \oplus \Sigma_{\alpha\neq 1} \oplus \bar{M}_\alpha^{(J_\alpha)}$, where N_1 is a maximal submodule of $M_1^{(J_1)}$, and so $M/N \approx M_1^{(J_1)}/N_1$. We take another maximal submodule N' of M such that $M/N \approx M/N'$. Then we obtain similarly $\bar{N} = \bar{N}'_1 \oplus \Sigma_{\alpha\neq 1} \oplus \bar{M}_\alpha^{(J_\alpha)}$, where N'_1 is a maximal submodule of $M_1^{(J_1)}$. Now we have two decompositions of $M_1^{(J_1)}$ by (10.5.4):

$$M_1^{(J_1)} = M_1'^{(J_1-1)} \oplus M'_1 \supset N = M_1'^{(J_1-1)} \oplus N'_1$$

$$M_1^{(J_1)} = M_1''^{(J_1-1)} \oplus M''_1 \supset N' = M_1''^{(J_1-1)} \oplus N''_1$$

where $N'_1 \subset M'_1$ and $N''_1 \subset M''_1$ and $M_1 \approx M'_1 \approx M''_1$. Hence we obtain an automorphism f of $M_1^{(J_1)}$ by (10.5.3), which induces a given isomorphism of

M_1'/N_1' to M_1''/N_1''. Hence M is max. quasi-projective.

10.5.6 COROLLARY Let M be as in (10.5.5) and let N be a submodule of M containing J(M). We assume that M is max. quasi-projective and each M_α is cyclic hollow. Then there exists a direct decomposition of M such that $M = \Sigma_I \oplus M_\alpha' \supset N = \Sigma_{I_1} \oplus M_\beta' \oplus \Sigma_{I_2} \oplus J(M_\gamma')$; $I = I_1 \cup I_2$ and the M_α' are indecomposable. Let N' be another submodule of M containing J(M). If $M/N \approx M/N'$, there exists an automorphism f of M which induces the above isomorphism [f(N) = N'].

Proof: We take the same argument as in the proof of (10.5.4). Since M is hollow, $\bar{A}_\alpha$ is either simple or zero. Hence if $N \neq J(M)$, N contains a nonzero direct summand of M from the proof of (10.5.4). We can use the same argument for the remainder.

10.5.7 COROLLARY Let $E = \Sigma_{i=1}^n \oplus E_i$ be a minimal injective cogenerator with E_i indecomposable. We assume that E is finitely generated and max. quasi-projective. Then (1) all simple submodules in $\bar{E}_i$ are isomorphic to one another and the simple module in $\bar{E}_i$ is not isomorphic to any simple submodules in $\bar{E}_j$ for $i \neq j$. (2) If $\bar{E}_i$ is simple for all i, then every primitive idempotent e in R is nonsmall, R contains all simple submodules up to isomorphism, and $r_R(J) \subset \ell_R(J)$, where $J = J(R)$.

Proof: 1. Since E is a minimal cogenerator, $E_i \not\approx E_j$ for $i \neq j$ and $\Sigma_{i=1}^n \oplus \bar{E}_i$ contains all simple R-modules up to isomorphism by (10.5.3). Hence we obtain (1) from (10.5.3).

2. If $\bar{E}_i$ is simple for all i, R is right QF-2*. Hence we obtain the nonisomorphic representative set of indecomposable and injective modules $\{e_1R/e_1A_1, e_2R/e_2A_2, \ldots, e_nR/e_nA_n\}$ from (9.3.1), (10.4.3), and (10.5.2), where $\{e_i\}$ is the set of mutually orthogonal and nonisomorphic primitive idempotents and $\{A_i\}$ is a set of right ideals. We assume that e_i is small. Then $\tilde{E} = E(e_iR) \supset J(E) \supset e_iR \supset e_iA_i$. Since e_iR/e_iA_i is injective, $\tilde{E}/e_iA_i$ contains a direct summand e_iR/e_iA_i contained in $J(\tilde{E}/e_iA_i)$, which is a contradiction. Let $E(R) \approx \Sigma_{j\in K} \oplus (e_jR/e_jA_j)^{m_j}$, where $(e_jR/e_jA_j)^{m_j}$ is a direct sum of m_j copies of e_jR/e_jA_j. Since e_i is nonsmall for all i, K contains all i. Hence E(R) is a cogenerator. We may assume that $R \subset \Sigma_{i=1}^k \oplus e_iR/e_iA_i$ and $r_R(J) \subset \Sigma_{i=1}^k \oplus \bar{a}_ir_R(J)$: $a_i \in e_iR$, where $\bar{a}_i = a_i + e_iA_i$ and e_i may be equal to e_j for some i and j. We assume that $\bar{a}_1r_R(J) \not\subset$ the socle $S(e_1R/e_1A_1)$ of e_1R/e_1A_1. Since e_1R/e_1A_1 is uniform, $\bar{a}_1r_R(J) \supset S_1 = S(e_1R/e_1A_1)$. E(R) being an injective cogenerator, there exists an f in

$Hom_R(e_1R/e_1A_1, e_iR/e_iA_i)$ for some i such that $f(S_1) = 0$ and $f(\bar{a}_1r_R(J)) \neq 0$. f is given by left-sided multiplication by an element b in e_iRe_1. Since $f(S_1) = 0$, f is not an isomorphism. Hence $b \in e_iJe_1$ by the construction of $\{e_iR/e_iA_i\}$. $f(\bar{a}_1r_R(J)) = b\bar{a}_1r_R(J) \subset \bar{e}_iJr_R(J) = 0$, which is a contradiction. Therefore, $\bar{a}_1r_R(J) \subset S_1$ and $\bar{a}_1r_R(J)J \subset S_1J = 0$. Thus $r_R(J) \subset \ell_R(J)$.

10.6 CONDITION (**)

In this section we consider the concept dual to nonsmall modules. The following propositions and lemmas are obtained directly from the definition and we shall omit some proofs.

10.6.1 PROPOSITION Let M be an R-module. Then the following conditions are equivalent:

1. For any module T and any epimorphism $f : T \to M$, ker f is not essential in T.
2. There exist a projective module P and an epimorphism $f : P \to M$ such that ker f is not essential in P.

If M satisfies one of the conditions above, we call M a *noncosmall module* and otherwise we call M a *cosmall module*.

10.6.2 LEMMA Let $0 \to M \to Q$ and $Q' \to M \to 0$ be exact. If M is noncosmall, then Q and Q' are also noncosmall.

10.6.3 PROPOSITION M is noncosmall if and only if $M \neq Z(M)$.

Proof: Let $F = \Sigma_{m \in M} \oplus u_mR \xrightarrow{f} M \to 0$ be the mapping given in (1.7.2) and put K = ker f. Then K is essential in F if and only if $K \cap u_mR = r_R(m)$ is essential in $R = u_mR$ for all $m \in M$ by (3.4.6).

We know from (10.6.3) that a cosmall module is nothing but a singular module. Every projective module is a noncosmall module by (10.6.1), so every module containing a projective submodule (as a direct summand) is a noncosmall module. We shall consider the converse.

Every noncosmall module contains a direct summand which is projective. (**) (10.6.1)

Every indecomposable projective module is uniform [right QF-2 (Sec. 5.8)]. (10.6.2)

10.6.4 LEMMA We assume (**). Then every indecomposable semiperfect module is uniform.

Proof: Let P be as in (10.6.4). Then J(P) is the unique maximal submodule of P by (6.1.17). Hence P/K is indecomposable for any submodule K of P. We assume that $K_1 \cap K_2 = 0$. Then P/K_1 is noncosmall by (10.6.1) if $K_2 \neq 0$. Hence P/K_1 is projective from the above and (**). Therefore, $K_1 = 0$.

10.6.5 PROPOSITION If R satisfies (**), R contains a projective and injective right ideal.

Proof: Let E = E(R). Then E ≠ Z(E), so E contains a projective direct summand P. Since P is a summand of a free module and is injective, R contains a direct summand isomorphic to a direct summand of P by (8.5.5).

10.6.6 PROPOSITION Let R be a QF-2 semiperfect ring. Then M is a noncosmall module if and only if N contains a projective module.

Proof: Let M be noncosmall. Then M contains a cyclic noncosmall submodule mR by (10.6.3). Let $0 \leftarrow mR \xleftarrow{f} \Sigma \oplus e_iR$ be a projective cover of mR. Since ker f is not essential in $\Sigma \oplus e_iR$, ker f ∩ e_jR is not essential in e_jR for some j. Hence ker f ∩ $e_jR = 0$ and M contains a submodule isomorphic to e_jR.

10.6.7 PROPOSITION We assume (**). Then for every uniform projective P, either Z(P) = 0 or Z(P/Z(P)) = P/Z(P). Every submodule not contained in Z(P) is projective.

Proof: Let $P \supset T$ and $Z(P) \not\supset T$. Then T is noncosmall by (10.6.3) and indecomposable. Hence T is projective. Let K_i (i = 1, 2) be submodules property containing Z(P). Then K_i is projective. Put $K = K_1 \cap K_2$ and consider an epimorphism $\varphi : K_1 \oplus K_2 \to K_1 + K_2 \to 0$, where $\varphi = 1_{K_1} - 1_{K_2}$. Then ker $\varphi \approx K$. Since $K_1 + K_2$ is projective, so is K. Hence K ≠ Z(P). Therefore, Z(P) is irreducible, so P/Z(P) is indecomposable. Hence P/Z(P) = Z(P/Z(P)) if Z(P) ≠ 0.

10.6.8 THEOREM Let R be semiperfect. Then (**) holds if and only if there exist sets of primitive idempotents $\{e_i\}$ and integers $\{n_i\}$ such that

1. The e_iR are injective.
2. $e_kJ^{t_i}$ is projective for $t \leq n_i$ and $e_iJ^{n_i+1}$ is singular.
3. Every indecomposable projective is isomorphic to some $e_iJ^{t_i}$; i.e., $\Sigma_i\Sigma_{t_i} \oplus e_iJ^{t_i}$ is a projective generator.

In this case every submodule e_iB of e_iR is either contained in $e_iJ^{n_i+1}$ or equal to some e_iJ^t, $t \leq n_i + 1$, where J = J(R).

Proof: We assume that (**) holds. Then we can take a complete set of primitive idempotents e_i such that e_iR is injective by (10.6.5). Let $e = e_i$ and eK be a proper and projective submodule of eR. Then $eR \supset eJ \supset eK$. Since eR is uniform and $eK \not\subset Z(eR)$, eJ is projective by (10.6.7). Now $eJ \approx fR$ by (7.1.2), so eJ^2 is the unique maximal submodule of eJ. Therefore, we have inductively a unique chain $eR \supset eJ \supset \cdots \supset eJ^t \supset eK$ with eJ^i projective. If $eJ^i \approx eJ^j$, this isomorphism is extendible to an automorphism of eR. Hence $i = j$. Thus we can find some m such that eJ^m is projective and eJ^{m+1} is cosmall, i.e., singular by (10.6.3). If f_jR is not injective, f_jR is contained in some e_iR. To see this, let $1 = \Sigma e_i + \Sigma f_j$, where $\{e_i, f_j\}$ is a complete set of mutually orthogonal primitive idempotents such that the e_iR are injective. Since f_jR is uniform by (10.6.4), $E_j = E(f_jR)$ is indecomposable, and $Z(E_j) \neq E_j$. Hence E_j is projective, so $E_j \approx e_iR$ for some i. Hence $f_jR \approx e_iJ^{t_i}$ for some t_i. Conversely, let M be a noncosmall. Then there exist $m \in M$ and a primitive idempotent g such that mgR is cosmall by (10.6.3). Since gR is uniform from (1) and (3), $mgR \approx gR$ and $gR \approx e_iJ^t$ by (3). Thus we have a diagram

$$\begin{array}{ccccc} 0 & \longrightarrow & mgR & \longrightarrow & M \\ & & \approx\downarrow & \swarrow h & \\ & & e_iJ^t & & \\ & & \downarrow & & \\ & & e_iR & & \end{array}$$

Then we can find a homomorphism h of M to e_iR, since e_iR is injective. Since $\operatorname{im} h \supset e_iJ^t$ and $e_iR \supset e_iJ \supset e_iJ^2 \supset \cdots$ is a unique chain by the first half, $\operatorname{im} h = e_iJ^{t_i}$ is projective. Hence M contains a projective module isomorphic to $e_iJ^{t_i}$ as a direct summand. The last assertion is clear from the proof above.

10.7 EXTENDING PROPERTY OF SIMPLE MODULES FOR PROJECTIVES

In this section we are interested only in right QF-2 rings, so from now on, we always assume that R satisfies (10.6.2) unless otherwise stated. Furthermore, we assume that R is semiperfect and we shall denote the Jacobson radical of R and primitive idempotents by J and e, respectively. Let P be projective. Then $P = \Sigma \oplus P_\alpha$; the P_α are indecomposable. Hence the socle S(P) of P is essential in P by (10.6.2).

10.7.1 LEMMA Let R be a right QF-2 and semiperfect ring and e a primitive idempotent. Let $eR \supset eJ^n \underset{\neq}{\supseteq} eJ^{n+k}$ be projective. Then $eJ^n \not\approx eJ^{n+k}$ if J is nil or eR is injective.

Proof: Since eJ^n is projective and $S(eJ^n)$ is simple, $eJ^n \approx fR$ for some idempotent f. If $eJ^n \approx eJ^{n+k}$, $fR \approx fJ^k$. This isomorphism is induced by an element in fJf. If J is nil, we have a contradiction. If eR is injective, the isomorphism $eJ^n \approx eJ^{n+k}$ is extendible to an automorphism of eR. Hence $eJ^n = eJ^{n+k}$, which is a contradiction.

We shall define two conditions dual to (10.4.1) and (10.4.2).

Every noncosmall module which is contained in a projective module contains a nonzero projective summand. (10.7.1)

For every finitely generated projective P with essential socle S(P), P/T contains a nonzero projective summand for any submodule $T \underset{\neq}{\subseteq} S(P)$. (10.7.2)

10.7.2 THEOREM Let R be a semiperfect and right QF-2 ring with nil Jacobson radical. Then the following conditions are equivalent:

1. R satisfies (10.7.1).
2. Let $\{P_\alpha\}_I$ be a set of direct summands of a projective P such that $P = P_\alpha \oplus P'$ and $S(P')$ is simple for all α. If $\cap_I S(P_\alpha)$ is irredundant, $\cap_K P_\alpha$ is a direct summand of P for any finite subset K of I.
3. (i) For some primitive idempotent e, there exists a positive integer t(e) such that $eR/eJ^{t(e)}$ is a serial module, eB $[= eJ^s, s \le t(e)]$ is projective for any $eR \supset eB \supset eJ^{t(e)}$, and $Z(eC) = eC$ and $eC \underset{\neq}{\subseteq} eJ^{t(e)}$ for every nonprojective right ideal eC in eR.
   (ii) $\{eJ^s\}_{e,s=0}^{t(e)}$ is the representative set of indecomposable projectives.

Proof: (1) ⇒ (2). Let $K = \{1,2,\ldots,n\}$ be a finite subset of I and put $P(n) = \cap_{i=1}^n P_i$. We shall show that P(n) is a direct summand of P by induction on n. If n = 1, this is clear by the assumption. Put $P = P_n \oplus P'_n$ with P'_n indecomposable and $\pi_n : P \to P'_n$ the projection. We note that $S(\cap P_\alpha) = \cap S(P_\alpha)$. Since $S(P(n-1)) = \cap_{i=1}^{n-1} S(P_i) \not\subseteq S(P_n)$, $\pi_n(S(P(n-1)) \neq 0$. Hence $\pi_n(P(n-1))$ is a noncosmall module in P'_n. Then there exists an indecomposable summand P_0 of $\pi_n(P(n-1))$ by (1). Since $S(P'_n)$ is simple, $\pi_n(P(n-1)) = P_0$. Therefore, $P(n-1) = P'_0 \oplus \ker \pi_n \mid P(n-1) = P'_0 \oplus P(n)$, where $P'_0 \approx P_0$. Since $P = P(n-1) \oplus P'$, P(n) is a direct summand of P.

(2) ⇒ (3). Let e be a primitive idempotent. We assume that eA is projective and eB (⊂ eA) is noncosmall for right ideals eA and eB. Then

there exists a projective module P such that $0 \leftarrow eB \xleftarrow{f} P \leftarrow K \leftarrow 0$ is exact and $S(P) \not\subset K$ by the definition [cf. (10.6.2)]. If S(P) is simple, K = 0 and eB is projective. We assume that $P = P_1 \oplus \Sigma_I \oplus P_\alpha$ such that the P_α are indecomposable and $S(P_1)$ is a simple module not contained in K. We put $Q = P \oplus eA$ (external direct sum) and $P' = \{x + f(x) \mid x \in P\} \subset Q$. Then $S(P') = [S(P) \cap K] \oplus S((1 + f)(P_1))$ and $S(P) = S(P_1) \oplus [S(P) \cap K]$. Since $S(P) \cap S(P')$ is irredundant, $P \cap P' = K$ is a direct summand of Q and hence of P. Accordingly, eB is projective. Now if eJ is noncosmall, eJ is projective from the above. Hence eJ contains a unique maximal submodule eJ^2, since eJ is indecomposable by (10.6.2). Repeating this argument, we obtain a unique chain $eR \supset eJ \supset eJ^2 \supset \cdots \supset eJ^t$ of projectives and eB is cosmall for any $eB \subsetneq eJ^t$ by (10.7.1). Hence eB = Z(eB) by (10.6.3). The remaining assertion is clear from the construction of eJ^i.

(3) ⇒ (1). Let P be a projective module which contains a noncosmall module M. Then $P = \Sigma \oplus e_iJ^{t_{ij}}$. Let $\pi_{ij} : P \to e_iJ^{t_{ij}}$ be the projection. Since $M \neq Z(M)$, $\pi_{k_j}(M) \not\subset Z(e_kJ^{t_{kj}}) \subsetneq e_kR$ for some k, j. Hence $\pi_{k_j}(M)$ is projective, so $M = \ker \pi_{k_j} | M \oplus M'$; $M' \approx \pi_{k_j}(M)$.

10.7.3 THEOREM Let R be a right QF-2 and semiperfect ring. Then the following conditions are equivalent:

1. R satisfies (10.7.2).
2. Every projective module has the extending property of simple modules.
3. (i) For some primitive idempotent e, there exists a chain of projective right ideals eA_i such that $eR = eA_1 \supset eA_2 \supset \cdots \supset eA_t$, and $\mathrm{Hom}_R(S(eA_i), S(eA_j))$ is extendible to $\mathrm{Hom}_R(eA_i, eA_j)$ for any pair $i \geq j$.

   (ii) $\{eA_i\}_{e,i}$ is the representative set of indecomposable projectives such that $S(eR) \not\approx S(e'R)$ if $e \neq e'$; i.e., $\Sigma_k\Sigma_{i(k)} \oplus e_kA_{ki(k)}$ is a projective generator.

Proof: (1) ⇒ (2). Let P be a projective module and $P = \Sigma \oplus P_\alpha$, where the P_α are uniform. Let S be a simple submodule of S(P). Then there exists a finite subset $K = \{1,2,\ldots,n\}$ of I such that $S \subset S(\Sigma_K \oplus P_i)$. If n = 1, the assertion is clear. Hence we assume that $S \subsetneq S(\Sigma_K \oplus P_i)$ and put $P(n) = \Sigma_{i=1}^n \oplus P_i$. Then $P^{(n)}/S = P_0 \oplus Q$ and P_0 is projective by (1). Considering an epimorphism $p^{(n)} \to P/S \to P_0$, we obtain $p^{(n)} = P_0' \oplus L$, where $P_0' \approx P_0$ and $L \supset S$. Since by (7.1.2) $L = \Sigma_{i=1}^{n-1} \oplus P_i'$, the P_i' are projective and uniform, and we can use induction.

(2) ⇒ (3). Let eR and fR be uniform projectives with isomorphic socle. Then there exists a monomorphism $f : eR \to fR$ (or $fR \to eR$) by (9.6.6); i.e., $eR \prec^* fR$ or $fR \prec^* eR$. Let eR be a maximal projective among uniform projectives P with isomorphic socle with respect to the relation $\prec^*$. Then these P are isomorphic to right ideals eA in eR. Since the relation $\prec^*$ is linear on {eA}, by repeatedly taking projectives we get a chain of projective right ideals $eR = eA_1 \supset eA_2 \supset \cdots \supset eA_t$. The second condition is clear by (9.6.6).

(3) ⇒ (2). This is clear from (9.6.6).

(2) ⇒ (1). Let $P = P_1 \oplus P_2 \oplus \cdots \oplus P_n$ be projective; the P_i are uniform. Let $T \subsetneq S(P)$ and $T = S_1 \oplus S_2 \oplus \cdots \oplus S_t$, where the S_i are simple. Then there exists a direct summand P_1' of P such that $S(P_1') = S_1$. Let $P = P_1' \oplus K_1$. Then $T = S_1 \oplus \pi_1(T)$, where $\pi_1 : P \to K_1$. Hence $S(K_1) \supsetneq \pi_1(T)$ and $P/T \approx P_1'/S_1 \oplus K_1/\pi_1(T)$. Repeating the same argument on $K_1/\pi_1(T)$, finally we obtain $P/T \approx P_1'/S \oplus \cdots \oplus P_i'/S_i' \oplus K_i$ and K_i is projective, since $\pi_j(T) = 0$ for some $j \leq n$.

10.8 MINI-INJECTIVE MODULES

First we generalize the concept of injective module. Let M be an R-module and let I be a right ideal in R. We take a nonzero R homomorphism f of I to M. Put M_1 = im f and consider a diagram

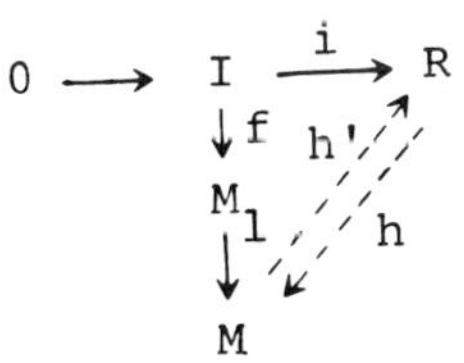

We shall introduce two conditions.

There exists $h \in \mathrm{Hom}_R(R,M)$ such that $hi = f$. (10.8.1)

There exists either $h \in \mathrm{Hom}_R(R,M)$ of $h' \in \mathrm{Hom}_R(M,R)$ such that $hi = f$ or $if^{-1} = h'|M_1$ provided that f is a monomorphism. (10.8.2)

If M satisfies (10.8.1) [resp. (10.8.2)] for every minimal right ideal I of R and any f in $\mathrm{Hom}_R(I,M)$, we say that M is right (resp. *weakly*) *mini-injective*. Similarly, if M satisfies (10.8.1) [resp. (10.8.2)] for every uniform right ideal I of R and any f in $\mathrm{Hom}_R(I,M)$, then we say that M is right (resp. *weakly*) *uni-injective*.

It is clear that every injective is uni-injective and every uni-injective is mini-injective. The converse is not true in general. Every semisimple module is weakly mini-injective, but not mini-injective. If R is a local and right QF-2 ring, then every right ideal is uniform. Hence every uni-injective is injective in this case.

10.8.1 PROPOSITION Let M be an R-module and $M = M_1 \oplus M_2$. Then we have the following.

1. M is mini-injective (resp. uni-injective) if and only if each M_i is.
2. If M is weakly mini-injective (resp. weakly uni-injective), then each M_i is.

Proof: This is clear.

10.8.2 THEOREM Let R be a right artinian ring and M an R-module. Then M is mini-injective (resp. uni-injective) if and only if for any minimal (resp. uniform) right ideal I in e_iR and any f in $\mathrm{Hom}_R(I,M)$, f is extendible to an element in $\mathrm{Hom}_R(e_iR,M)$, where e_i runs through all primitive idempotents.

Proof: The "if" part: First we take a minimal right ideal I of $R = \Sigma_{i=1}^{m} \oplus e_iR$. Let f be in $\mathrm{Hom}_R(I,M)$ and $\pi_i : R \to e_iR$ the projection. We may assume that $I_i = \pi_i(I) \neq 0$ for $i \leq$ some t and $I_j = 0$ for $j > t$. Since $\pi_1|I$ is a monomorphism, we put $f_1 = f(\pi_1|I)^{-1}$. Then there exists F_1 in $\mathrm{Hom}_R(e_1R,M)$ such that $F_1|I_1 = f_1$ by the assumption. Put $F_j = 0$ [$\in \mathrm{Hom}_R(e_jR,M)$] for $j \neq 1$ and $F = \Sigma F_i$. Let x be in I and $x = \Sigma_{i=1}^{t} \pi_i(x)$. Then $F(x) = \Sigma F_i\pi_i(x) = f(\pi_1|I)^{-1}\pi_1(x) = f(x)$. If I is uniform, $\bigcap_i \ker(\pi_i|I) = 0$ implies that some $\pi_i|I$ is a monomorphism. Hence we can use the same argument in this case, too.

10.9 EXTENDING PROPERTY OF DIRECT DECOMPOSITIONS OF THE SOCLE FOR PROJECTIVES

Let R be a right artinian ring. We assume in this section that every idempotent is primitive and we denote it by e. We put $R/J = \bar{R}$, and $\bar{e}$ means the residue class of e in $\bar{R}$, where $J = J(R)$.

First we shall study the extending property for R-projectives. Let $R = \Sigma_{i=1}^{n}\Sigma_{j=1}^{\rho(i)} \oplus e_{ij}R$ be as in Sec. 10.3.

10.9.1 THEOREM Let R be right artinian. Then we have the following.

1. Every projective has the extending property of simple modules if and only if R is right QF-2 and R is weakly mini-injective as a right R-module.

2. Every projective has the extending property of direct decompositions of the socle if and only if R is right QF-2 and mini-injective as a right R-module.

Proof: 1. We assume that every projective has the extending property of simple modules. Then R is right QF-2. Let $R = \Sigma_{i=j}^{m} \oplus e_iR$ with e_i primitive and let $\pi_i : R \to e_iR$ be the projection. We take two minimal right ideals K_1 and K_2 and assume that $f : K_1 \to K_2$ is an isomorphism. We assume that $K_i \subset \Sigma_{j=1}^{t_i} \oplus I_{ij(i)}$ $(i = 1, 2)$, where $I_{ij(i)} = \pi_{j(i)}(K_i) \neq 0$. Since $I_{ij(i)} \approx K_1$ for all i, j, by (9.6.6) we can find a minimal ideal among $e_{j(i)}R$ with respect to the relation $\prec^*$ in Sec. 9.6, say $e_{j(i)}R = e_1R$ and $i = 1$. Take $p_k = \pi_k f \pi_1^{-1} : I_{11} \to e_kR$. If $k \notin \{2(1), 2(2), \ldots, 2(t_2)\}$, $p_k = 0$. Hence since e_1R is minimal, there exists $F_k \in \mathrm{Hom}_R(e_1R, e_kR)$ such that $F_k|I_{11} = p_k$ by (9.6.6). Put $h = (\Sigma_{k=1}^{m} F_k)\pi_1 \in \mathrm{Hom}_R(R,R)$. Then $h|K_1 = [(\Sigma_k F_k)\pi_1]|K_1 = (\Sigma_k \pi_k f)|K_1 = f$. If $i = 2$ for the minimal ideal above, then we take f^{-1} in the above. Then we can find $h' \in \mathrm{Hom}_R(R,R)$ such that $h'|K_2 = f^{-1}$. The converse is clear by (9.6.6).

2. We can show this similarly by making use of (9.7.4) and the proof of (9.7.3), instead of (9.6.6).

For the uni-injective case, we have the following.

10.9.2 THEOREM Let R be right artinian. Then we have the following.

1. Every projective has the extending property of uniform submodules if and only if R is right QF-2 and weakly uni-injective as a right R-module.
2. Every projective has the extending property of direct sums of uniform submodules if and only if R is right QF-2 and uni-injective as a right R-module [see (10.10.1) below].

Proof: First we note that every uniform submodule in a projective module P is finitely generated. Let $P = \Sigma_I \oplus P_\alpha$, $P_\alpha \approx e_{i(\alpha)}R$, and U a uniform submodule. Let $x \neq 0$ be in U. Then $x = \Sigma_{i=1}^{n} p_{\alpha_i}$; $p_{\alpha_i} \in P_{\alpha_i}$. Hence $U \cap \Sigma_{i=1}^{n} \oplus P_{\alpha_i} \neq 0$, so $U \cap \Sigma_{I-\{\alpha_i\}} \oplus P_\beta = 0$. Accordingly, U is isomorphic to a submodule of $\Sigma_{i=1}^{n} \oplus P_{\alpha_i}$. Furthermore, $U \approx \pi_i(U)$ for some i, where $\pi_i : P \to P_{\alpha_i}$ is the projection. Therefore, we can apply the same argument given in the proof of (10.9.1) by making use of (9.6.7) and (9.7.3).

Next we shall study self-mini-injective rings.

10.9.3 THEOREM Let R be right artinian and mini-injective as a right R-module. Then we have the following.

1. If $e_1R \not\approx e_2R$, each minimal submodule in e_1R is not isomorphic to any minimal submodule in e_2R.
2. $S(e_1R) = e_1J^k$ and all minimal submodules in e_1R are isomorphic to one another.
3. $r_R(J) \supset \ell_R(J)$ and $J = Z(R)$.

Here $J = J(R)$, the e_i is a primitive idempotent, $r_R(J) = \{x \in R \mid Jx = 0\}$, $\ell_R(J) = \{x \in R \mid xJ = 0\}$, $Z(R)$ is the right singular ideal of R, and $S(\cdot)$ is the socle of $(\cdot)$.

Proof: Let $e_1R \not\approx e_2R$ and let I_i be a minimal right ideal in e_iR for $i = 1, 2$. If $I_1 \approx I_2$, there exist a y in $e_2Re_1 = e_2Je_1$ and a z in e_1Je_2 such that $I_2 = yI_1$, $I_1 = zI_2$ by the assumption. Hence $I_1 = zyI_1$ and $zy \in J$, which is a contradiction. Therefore, $\{I_i\}_{i=1}^n$ is the representative set of minimal R-modules. Let S be a minimal right ideal in e_1R. Then S must be isomorphic to I_1 from the first observation. Let $e_1J^k \neq 0$ and $e_1J^{k+1} = 0$. We take a minimal right ideal K in e_1J^k. Since $K \approx S$, there exists an x in e_1Re_1 such that $S = xK \subset e_1J^k$. Hence $S(e_1R) = e_1J^k$. We have obtained (1) and (2).

3. We take I_1 in $S(e_1R)$. Let $I_1 = xR$ and $x \in e_1R$. Now $Jx \subset \Sigma_{i=1}^m e_iJx = \Sigma_{i=1}^m e_iJe_1x$. If $e_iR \not\approx e_1R$, $e_jJe_1xR = 0$ by (1). If $e_jR \approx e_1R$, we take a z in e_1Re_j which induces an isomorphism of e_jR to e_1R. Then $ze_jJe_1R \subset e_1Je_1xR = 0$ by (2). Hence $e_jJe_1R = 0$. Therefore, $Jx = 0$ and $\ell_R(J) = S(R_R) \subset r_R(J)$. Furthermore, $Z(R) = \{x \in R \mid x\ell_R(J) = 0\} \supset J$, so $Z(R) = J$, since every ideal properly containing J contains a projective submodule.

10.9.4 PROPOSITION Let R be a right artinian ring. Then R is mini-injective as a right R-module if and only if R is weakly mini-injective as a right R-module and $\ell_R(J) \subset r_R(J)$.

Proof: The "if" part: We assume that $I_1 \approx I_2$ for minimal right ideals I_i in e_iR. Then there exists an element x in either e_1Re_2 or e_2Re_1 which induces an isomorphism between I_1 and I_2. Hence $x \notin J$ by the assumption. Therefore, x induces an isomorphism between e_1R and e_2R. Accordingly, R is mini-injective for $\text{Hom}_R(e_iR, e_jR) = e_jRe_i$. The converse is clear from (10.9.3).

Similarly we obtain the following.

10.9.5 PROPOSITION Let R be right artinian. Then R is uni-injective as a right R-module if and only if R is weakly uni-injective as a right R-module and $\ell_R(J) \subset r_R(J)$.

Proof: Since a uni-injective is mini-injective, the "only if" part is clear from (10.9.3). Let U_i be a uniform submodule of e_iR and $f : U_1 \to U_2$ a homomorphism. If $\ker f \neq 0$, f is extendible to an element in $\mathrm{Hom}_R(e_1R, e_2R)$ by the assumption. We assume that $\ker f = 0$. We know from (10.9.4) that R is mini-injective as a right R-module. Hence $e_1R \approx e_2R$ by (10.9.3). Therefore, f and f^{-1} are extendible to elements in $\mathrm{Hom}_R(e_1R, e_2R)$ and $\mathrm{Hom}_R(e_2R, e_1R)$, respectively. Thus R is uni-injective by (10.8.2).

10.9.6 THEOREM Let R be a right artinian and basic ring. Then R is a right QF-2 and self-mini-injective ring if and only if $\ell_R(J) = Ru = uR$.

Proof: Let $R = \Sigma_{i=1}^n \oplus e_iR$, where the e_i are primitive idempotents. We assume that R is a right QF-2 and self-mini-injective ring. Then $e_i\ell_R(J) = u_iR$ and $u_i \in e_iRe_{i'}$, where $\bar{R} = R/J$. We know by (10.9.3) that $\ell_R(J) \subset r_R(J)$. Hence $Ru_i = \bar{R}u_i = \Sigma_j \overline{e_jRe_j}u_i = \overline{e_iRe_i}u_i$ by the assumption. Therefore, since u_iR is the unique minimal right ideal, $Ru_i \subset u_iR$. $\overline{e_{i'}Re_{i'}}$, being a division ring, a mapping $u_i \to u_ir$ $(r \in \overline{e_{i'}Re_{i'}})$ is extendible to an element in $\mathrm{Hom}_R(u_iR, u_iR)$. Then since R is right self-mini-injective, there exists an element x in e_iRe_i with $xu_i = u_ir$. Therefore, $Ru_i = u_iR$. Put $u = \Sigma_i u_i$. Then $e_iu = ue_{i'} = u_i$. Hence $uR = \Sigma_{i=1}^n \oplus u_iR = \Sigma_i \oplus Ru_i = Ru$. Conversely, we assume that $\ell_R(J) = uR = Ru$. Then $\ell_R(J)$ is a homomorphic image of R/J as a right R-module. Hence $uR \approx R/J$ by considering the composition length. Therefore, R is right QF-2, $\ell_R(J) = \Sigma_{i=1}^n \oplus e_i\ell_R(J)$ and $e_i\ell_R(J)$ is the unique minimal right ideal. Furthermore, $e_i\ell_R(J) \not\approx e_j\ell_R(J)$ if $i \neq j$. Hence $e_i\ell_R(J)$ is a two-sided ideal, so $\ell_R(J) = uR = Ru$ implies that $Re_iu = e_iuR = e_i\ell_R(J)$. Therefore, R is right self-mini-injective, since $\mathrm{End}_R(e_iuR) = \overline{e_iRe_i}$.

10.9.7 THEOREM Let R be a right artinian ring. Then the following conditions are equivalent:

1. R is self-mini-injective as a right R-module.
2. R satisfies the following:
   (i) If $e_1R \not\approx e_2R$, no minimal right ideal in e_1R is isomorphic to one in e_2R.
   (ii) There exists a minimal right ideal I contained in e_1J^k $(e_1J^{k+1} = 0)$ such that $\mathrm{End}_R(I) = \{a \in \overline{e_1Re_1} \mid aI \subset I\}$, and $S(e_1R) = e_1Re_1I$ for each e_1, where the e_i are primitive idempotents.

Proof: (1) ⇒ (2). This is clear by (10.9.3).

(2) ⇒ (1). The second assertion of (ii) implies that each minimal right ideal I' in e_1R is isomorphic to I contained in e_iJ^k. We assume that $I \approx \overline{e_2R}$, I = xR, and I' = x'R. Then we may assume that $xe_2 = x$ and $x'e_2 = x'$. We obtain by (ii) that $x' = x'e_2 = \Sigma\, y_i x r_i$, where $y_i \in e_1Re_1$ and $r_i \in Re_2$. Now $xr_ie_2 = xe_2r_ie_2 = \overline{xe_2r_ie_2}$. Since a mapping $xz \to \overline{xe_2r_ie_2}z$ is an R-endomorphism of I, there exists an $\bar{a}_i$ in $\overline{e_1Re_1}$ with $\bar{a}_ix = \overline{xe_2r_ie_2}$ by (ii). Hence $x' = (\Sigma\, y_i\bar{a}_i)x = \bar{b}x$, where $\bar{b} = \Sigma\, y_i\bar{a}_i$. Since $\bar{b} \neq 0$, $x = \bar{b}^{-1}x'$. Define a $g \in \mathrm{Hom}_R(I',I)$ by setting $g(x'z) = xz$; $z \in R$. Let f be any element in $\mathrm{Hom}_R(I,I')$. Then $gf \in \mathrm{End}_R(I)$. Hence there exists an $\bar{a}$ in $\overline{e_1Re_1}$ such that $gf(x) = \bar{a}x$ by (ii). Therefore, $f(x) = g^{-1}(\bar{a}x) = \overline{ba}x$ and f is extendible to an element in $\mathrm{End}_R(e_1R)$. We know similarly that $\mathrm{End}_R(I') = \bar{b}^{-1}\mathrm{End}_R(I)\bar{b} = \{\bar{c} \in \overline{e_1Re_1} \mid cI' \subset I'\}$. Hence I' satisfies (ii). Let $h \in \mathrm{Hom}_R(I',R)$ and $R = \Sigma_{i=1}^m \oplus e_iR$. Let $\pi_i : R \to e_iR$ be the projection. If $\pi_ih \neq 0$, $e_iR \approx e_1R$ by (i). Assume that $\pi_ih = h_i \neq 0$ for i = 1, 2, ..., t and $h_j = 0$ for j > t. Since $e_1R \approx e_iR$ for $i \le t$, there exist $c_i \in e_1Re_i$ and $d_i \in e_iRe_1$ such that $c_id_i = e_1$ and $d_ic_i = e_i$ by (3.1.9). Using d_i and c_i, we know that any element in $\mathrm{Hom}_R(I',h_i(I'))$ is extendible to an element in $\mathrm{Hom}_R(e_1R,e_iR)$ for $i \le t$. Take $p_i \in R$ such that $p_ix' = h_i(x')$. Then $h(x') = \Sigma\, h_i(x') = (\Sigma\, p_i)x'$. Hence R is right self-mini-injective by (10.8.2).

10.9.8 PROPOSITION Let K be a field and R a K-algebra with finite dimension. If R is mini-injective as a right R-module, then R is right QF-2.

Proof: Let I_1 be a minimal right ideal in e_1R, where e_1 is primitive. We assume that $I_1 \approx \overline{e_2'R}$. Since $I_1 \subset e_iJ^{k_1}$ and $e_1J^{k_1+1} = 0$, each element in $\overline{e_1Re_1}$ gives us an element in $\mathrm{Hom}_R(I_i,e_1R)$ [= $\mathrm{Hom}_R(I_1,S(e_1R))$] via left-sided multiplication, and $\mathrm{Hom}_R(I_1,e_1R) = e_1Re_1$ as R-modules by the assumption. Put $I_1 = x\bar{e}_2'R$ and consider an automorphism f of I_1 by setting f(x) = xa for $a \in \overline{e_2'Re_2'}$. Then f is extendible to an element in $\mathrm{Hom}_R(e_1R,e_1R)$ by the assumption. Hence xa = bx for some b in $\overline{e_1Re_1}$. This relation gives us a K-monomorphism of $\overline{e_2'Re_2'}$ to $\overline{e_1Re_1}$ Hence $[e_1Re_1 : K] \ge [e_2'Re_2' : K]$. Repeating this argument, we obtain a chain of primitive idempotents $e_1, e_2', \ldots, e_i', \ldots$ such that a minimal right ideal I_i in $e_i'R$ is isomorphic to $\overline{e_{i+1}'R}$ and $[\overline{e_i'Re_i'} : K] \ge [\overline{e_{i+1}'Re_{i+1}'} : K]$. We may assume that $e_i'R \approx e_j'R$ for some $i < j$. Then $I_{i-1} \approx \overline{e_i'R} \approx \overline{e_j'R} \approx I_{j-1}$. Hence $e_{i-1}'R \approx e_{j-1}'R$ by (10.9.3). Therefore, $e_1R \approx e_k'R$ for some k. Accordingly, $[\overline{e_1Re_1} : K] = [\overline{e_2'Re_2'} : K] = [\overline{e_k'Re_k'} : K]$. Hence $\mathrm{Hom}_R(I_1,e_1R) = \overline{e_2'Re_2'}$. Let S be a minimal right ideal in e_1R. Then there exists an a in $\overline{e_1Re_1}$ such that $aI_1 = S$ by the assumption. However, since $\mathrm{Hom}_R(I_1,e_1R) = \overline{e_2'Re_2'}$ as above, there exists a b in $\overline{e_2'Re_2'}$ such that $ax = xb$. Hence $S = aI_1 = axR = xbR \subset I_1$. Therefore, $S(e_1R)$ is simple.

10.10 QF-RINGS AND LIFTING (EXTENDING) PROPERTY

We shall give some characterizations of QF-rings involving the extending property of projectives.

Let R be a right artinian ring and let $R = \Sigma_{i=1}^{n}\Sigma_{j=1}^{\rho(i)} \oplus e_{ij}R$ be as in Sec. 10.3.

10.10.1 THEOREM Let R be left and right artinian. Then the following conditions are equivalent:

1. R is a QF-ring.
2. Every right (and left) R-projective has the extending property of direct decompositions of the socle.
3. Every right R-projective has the extending property of direct decompositions of the socle and $r_R(J) \subset \ell_R(J)$.
4. Every right R-injective E has the lifting property of direct decompositions modulo radical and R is right QF-2.
5. R is right and left QF-2 and mini-injective as a right R-module.
6. R is mini-injective as a left and right R-module.
7. R is uni-injective as a right R-module and right QF-2.

Proof: (1) ⇒ (2)-(7), (2) ⇒ (1), and (5) ⇒ (1). They are clear from (10.9.1), (10.9.3), (10.4.3), (10.7.3), and (5.7.3).

(3) ⇒ (1). It is sufficient to show that R is left QF-2, since R is right QF-2 and R-mini-injective by (10.9.1). We take the unique minimal right ideal x_1R in e_1R. We may assume that $x_1 \in e_1Re_2'$ as in the proof of (10.9.8). Since $r_R(J) \supset \ell_R(J)$, $Jx = 0$. Hence Rx_1 is semisimple. On the other hand, since $Rx_1 = Re_1x$, Rx_1 is a minimal left ideal in Re_2'. Let Rx_2 be another minimal left ideal in Re_2' and $x_2 \in e_3'Re_2'$. Then $S(e_1R) = x_1R \approx \overline{e_2'R} \approx x_2R = S(e_3'R)$ since $r_R(J) \subset \ell_R(J)$ by the assumption. Hence $e_1R \approx e_3'R$ by (10.9.3). Noting that x_1R is minimal, we obtain an isomorphism f : $x_1R \to x_2R$ with $f(x_1) = x_2$. f is extendible to an element $y \in \mathrm{Hom}_R(e_1R,e_3'R)$ by (9.6.6). Hence $x_2 = yx_1$, so $Rx_2 = Rx_1$. The above correspondence $e_1 \to e_2'$ gives a permutation of the set $\{e_{i1}\}_{i=1}^{n}$ by (10.9.3). Hence R is left QF-2.

(4) ⇒ (1). We know by (10.4.3) that there exists the representative set $\{e_{i1}R/e_{i1}A_{ij}\}_{i=1,j=1}^{p,\ n(i)}$ of indecomposable injectives. Since R is artinian, n(i) = 1 for all i by (9.3.1), and $e_{i1}R$ is uniform by the assumption. Hence $E(e_{i1}R) \approx e_{j1}R/e_{j1}A_{j1}$ for some j. We consider a diagram

$$\begin{array}{ccccc} 0 \rightarrow & e_iR & \rightarrow & E(e_iR) \approx & e_jR/e_jA_j \\ & \downarrow \varphi & \swarrow h & & \\ & e_iR/e_iA_i & & & \end{array}$$

where $e_k = e_{k1}$, $A_k = A_{k1}$, and φ is the natural epimorphism. Since e_iR/e_iA_i is injective, we obtain an epimorphism $h : e_jR/e_jA_j \rightarrow e_iR/e_iA_i$. Hence i = j and $e_iA_i = 0$. Since n(i) = 1 for all i, p = n. Therefore, R = $\Sigma_{i=1}^{n}\Sigma_{j=1}^{\rho(i)} \oplus e_{ij}R$ is self-injective as a right R-module.

(6) ⇒ (1). We assume that R is self-mini-injective. Let xR be a minimal right ideal in e_1R, where e_i is primitive. Then $xR = xe_2'R$ and $x \in e_1Re_2'$. Since Jx = 0 by (10.9.3), Rx is minimal in Re_2' as in the proof of (3) ⇒ (1). Therefore, for any element b in $\overline{e_1Re_1}$, there exists an a in $\overline{e_2'Re_2'}$ such that bx = xa as in the proof of (10.9.8), for R is left mini-injective. Again using the same argument, we know that $xR = S(e_1R)$. Hence R is QF-2. Therefore, R is a QF-ring by (10.9.3) and (5.7.6).

(7) ⇒ (1). We shall show that R is self-injective as a right R-module. We can use the standard argument (4.4.2). Let I be a right ideal in R and $f \in \mathrm{Hom}_R(I,R)$. We can find a maximal member among the set of all extensions of f by Zorn's lemma, say $(I_0, f_0 : I_0 \rightarrow R)$. We assume that $I_0 \neq R$. Then there exists a primitive idempotent e such that $e \notin I_0$. Put $K = eR \cap I_0$ and $I_1 = I_0 + eR$. We take an extension f_1 of $f_0 \mid K$ from the assumption. We put $g(x) = f_0(x_1) + f_1(er)$, where $x = x_1 + er$, $x_1 \in I_0$, and $r \in R$. Then $g \in \mathrm{Hom}_R(I_1,R)$, which contradicts the assumption on I_0. Hence $I_0 = R$ and R is self-injective.

10.10.2 THEOREM Let R be a right artinian ring. Then the following conditions are equivalent:

1. R is a QF-ring.
2. R is right QF-2 and QF-2* and a minimal injective cogenerator is max. quasi-projective.
3. Every injective E has the lifting property of direct decomposition modulo radical and $\ell_R(J) \subset r_R(J)$.

Proof: (1) ⇒ (2). Since R is an injective cogenerator and a projective module as a right R-module by (5.6.2), we obtain (2).

(2) ⇒ (1). Let e be a primitive idempotent. R being QF-2 and QF-2*, E(eR) is hollow. Hence E(eR) = eR by (2) and (10.5.7). Therefore, R is self-injective.

(1) ⇒ (3). Since every injective module is projective by (5.7.7) and $\ell_R(J) = r_R(J)$ by (5.7.3), we have (3).

(3) ⇒ (1). Let $R = \Sigma_{i=1}^{n}\Sigma_{j=1}^{\rho(i)} \oplus e_{ij}R$ be as in the beginning of this section. Put $e_{i1} = e_i$. We have a nonisomorphic representative set of indecomposable injectives $\{e_1R/e_1A_1, \ldots, e_nR/e_nA_n\}$ as in the proof of (4) ⇒ (1) of (10.10.1). We shall show that $e_iA_i = 0$ for all i. We assume that $e_iA_i \neq 0$. Let I be a minimal right ideal contained in e_iA_i. Then there exists a monomorphism g of I to an injective module $e_{i'}R/e_{i'}A_{i'}$ for some i'. Hence there exists an element r in $e_{i'}R$ such that $g(x) = rx + e_{i'}A_{i'}$. We may assume that $r \in e_{i'}Re_i$. The fact that $I \subset \ell_R(J) \subset r_R(J)$ implies that $r \notin J$, so $e_{i'} = e_i$. The left-sided multiplication by r in e_iR/e_iJ gives us an element in $\mathrm{End}_R(e_iR/e_iJ)$. Since e_iR/e_iA_i is max. quasi-projective, there exists an element $r' \in \mathrm{End}_R(e_iR) = e_iRe_i$ such that $r - r' \in e_iJe_i$ and $r'e_iA_i \subset e_iA_i$. Hence we can replace r by r'. We consider a composition of R-homomorphisms

$$f : e_iR \xrightarrow{r'} e_iR \xrightarrow{\nu} e_iR/e_iA_i$$

where ν is the natural epimorphism. Then $\ker f = \{y \in e_iR \mid r'y \in e_iA_i\}$. Since r' is invertible in e_iRe_i, $\ker f = r'^{-1}e_iA_i \supset e_iA_i \supset I$. However, $0 \neq g(I) = r'I + e_iA_i = f(I) = 0$, which is a contradiction. Hence $e_iA_i = 0$. Therefore, R is self-injective.

10.10.3 COROLLARY Let R be right artinian. Then R is a QF-ring if and only if every injective E and every projective P have the lifting and extending property of direct decompositions of $\bar{E}$ and S(P), respectively. Furthermore, if $\ell_R(J) = r_R(J)$ (e.g., $J^2 = 0$ or R is commutative), R is a QF-ring if and only if every injective E (or projective P) has the lifting (or the extending) property of direct decompositions of $\bar{E}$ [or S(P)].

Proof: This is clear from (10.10.2) and (10.9.2).

As is well known, R is a QF-ring if and only if R is self-injective as a right R-module. However, R is actually quasi-injective as a right R-module in this case, so R is self-injective by Baer's criterion. Hence the concept dual to the above is the following: A (minimal) injective cogenerator is quasi-projective. We have the following.

10.10.4 COROLLARY Let R be as in (10.10.3). Then R is a QF-ring if and only if the minimal injective cogenerator is quasi-projective.

Proof: We assume that the minimal injective cogenerator is quasi-projective. Then every injective is quasi-projective by (5.4.10) [cf. (5.4.8)]. Put $E = E(R)$. Then $E \approx \Sigma_{i=1}^{k} \oplus e_iR/e_iA_i$, where the e_i are primitive idempotents, the A_i are right ideals, and $e_jRe_iA_i \subset e_jA_j$ from (5.4.11). We may assume that $R \subset \Sigma_{i=1}^{k} \oplus e_iR/e_iA_i$ as right R-modules and R is basic. We note from (10.5.3) that the set $\{e_i\}$ contains the set of all nonisomorphic primitive idempotents. Let $1 = \Sigma\, \bar{a}_i$; $a_i = e_ia_i \in e_iR$, where $\bar{a}_i = a_i + e_iA_i$. Then $J = J(R) \subset \Sigma_{i=1}^{k} \oplus \bar{a}_iJ$. We shall show that $a_1Je_i\ell_R(J) = 0$ for all i. Since $\ell_R(J) = \Sigma_K \oplus e_i\ell_R(J)$; $K \subset \{1,2,\ldots,k\}$, $\ell_R(J) \subset r_R(J)$. We note that $e_jRe_i = \mathrm{Hom}_R(e_iR, e_jR)$ and each element in e_jRe_i induces an element in $\mathrm{Hom}_R(e_i/e_iA_i, e_jR/e_jA_j)$ from the above. Now $\bar{a}_1Je_1\ell_R(J) = \bar{e}_1[a_1Je_1\ell_R(J)] = (e_1a_1Je_1)[\bar{e}_1\ell_R(J)]$, where we note that $e_1a_1Je_1 \subset J(\mathrm{End}_R(e_1R))$ and each element in $J(\mathrm{End}_R(e_1R))$ induces an element in $J(\mathrm{End}_R(e_1R/e_1A_1))$. Hence $(e_1a_1Je_1)[\bar{e}_1\ell_R(J)] = 0$ for e_1R/e_1A_1 is uniform. Next we consider $e_1a_1Je_2\ell_R(J)$. If $e_2R \not\approx e_1R$, then we have $e_1a_1Je_2\ell_R(J) = 0$ from the above. We assume that $e_2R \approx e_1R$.

Case 1. $e_2\ell_R(J) \subset e_2A_2$. Then $e_1A_1 \supset e_1Re_2\ell_R(J) \supset e_1a_1Je_2\ell_R(J)$. Hence $\bar{a}_1Je_2\ell_R(J) = 0$.

Case 2. $e_2\ell_R(J) \not\subset e_2A_2$ and $e_1a_1Je_2\ell_R(J) \not\subset e_1A_1$. Then e_1R/e_1A_1 and e_2R/e_2A_2 contain the isomorphic simple module $\bar{e}_2\ell_R(J)$, which is a contradiction. Hence, if $e_2\ell_R(J) \not\subset e_2A_2$, $e_1a_1e_2\ell_R(J) \subset e_1A_1$. Therefore, $\bar{a}_1Je_2\ell_R(J) = 0$, so $\ell_R(J) \subset r_R(J)$. Since a quasi-projective is max. quasi-projective, we have (10.10.4) by (10.5.2) and (10.10.2).

Finally, we take an algebra. Let K be a field and let R be a K-algebra of finite dimension. In this case we note that we have the duality functor $\mathrm{Hom}_K(-,K) = (-)^*$. Then every right R-injective E has the lifting property of direct decompositions of $\bar{E}$ if and only if every left R-projective has the extending property of direct decompositions of the socle; i.e., R is left mini-injective, so R is a QF-algebra by (5.7.9). We note that we may restrict ourselves to the case where every module is f.g. by Sec. 5.6.

10.10.5 THEOREM Let R be an algebra over a field K with [R:K] finite. Then the following conditions are equivalent:

1. R is a QF-ring.

2. A minimal injective cogenerator is max. quasi-projective.
3. R is a right self-mini-injective ring.

Proof: (1) ⇒ (3). This is clear by (10.10.1).

(3) ⇒ (2). Since R* is an injective cogenerator as a left R-module, we obtain (2) for the left R-modules.

(2) ⇒ (1). We may assume that R is basic and we use the same notation as in the proof of (10.10.4). Since R is an algebra of finite dimension, every indecomposable injective is f.g. and isomorphic to $(Re_i)^*$ for some e_i. We denote $(Re_i)^*$ by E_i. Then $S_i = \mathrm{End}_R(E_i)$ is anti-isomorphic to e_iRe_i. Let N_i be a maximal submodule of E_i and put $\tilde{S}_i = \{x \in S_i \mid x(N_i) \subset N_i\}$. Then $\mathrm{End}_R(E_i/N_i) = \tilde{S}_i/\tilde{S}_i \cap J(S_i)$ by (2). Hence $\tilde{S}_i/\tilde{S}_i \cap J(S_i) \subset S_i/J(S_i)$, and $\mathrm{End}_R(E_i/N_i)$ is anti-isomorphic to a K-subfield of $\overline{e_iRe_i}$. Put $E_i/N_i \approx \overline{e_{i'}R}$. Then $[\overline{e_{i'}Re_{i'}} : K] \leq [\overline{e_iRe_i} : K]$ from the above. We know, from (10.5.7) and the same argument as in the proof of (10.9.8), that the mapping $i \to i'$ gives us a permutation of $\{1,2,\ldots,n\}$, where $\Sigma_{i=1}^{n} \oplus (Re_i)^*$ is a minimal injective cogenerator. Hence $[\overline{e_{i'}Re_{i'}} : K] = [\overline{e_iRe_i} : K]$, so N_i is the unique maximal submodule of E_i by (10.5.7). Therefore, R is right QF-2*. Accordingly, every injective E has the lifting property of direct decompositions of $\bar{E}$ by (10.5.2). Then we obtain a nonisomorphic representative set of indecomposable injectives $\{e_1R/e_1A_1, \ldots, e_nR/e_nA_n\}$ by (10.4.3) and (9.3.1). Hence $\{(e_iR/e_iA_i)^*\}$ is a nonisomorphic representative set of indecomposable and projective left R-modules. Therefore, $\Sigma_{i=1}^{n} \oplus (e_iR/e_iA_i)^* \approx R$ as left R-modules. Accordingly, $[R:K] = \Sigma_{i=1}^{n}[(e_iR/e_iA_i)^* : K] = \Sigma_{i=1}^{n}[e_iR/e_iA_i : K]$. Hence $e_iA_i = 0$ for all i, so R is self-injective.

Examples Let K be a field.

1. Put

$$R = \begin{bmatrix} K & 0 & 0 \\ K & K & 0 \\ K & 0 & K \end{bmatrix}$$

Then $\mathrm{Hom}_R(S(e_{22}R), S(e_{33}R))$ is not extendible to $\mathrm{Hom}_R(e_{22}R, e_{33}R)$. Hence R is right generalized uniserial, but R does not have the extending property of simple modules as a right R-module.

2. We shall give an example showing that artinian and right self-mini-injective rings are not right QF-2 in general. Let x be an indeterminate and Q a field. Put $L = Q(x)$ and $K = Q(x^2)$. Then we have an isomorphism σ of L onto K and $[L : K] = 2$. Let $R = L1 \oplus Lu$ be a left vector space over

L. We put $(Lu)^2 = 0$ and $ua = \sigma(a)u$ for $a \in L$. Then R is a ring and $[R : K] = 2$ as a left L-module and $[R : L] = 3$ as a right L-module. Hence R is a left and right artinian ring. J = Lu contains minimal right ideals Ku and xKu. Let I be a minimal right ideal in J. Then I = aL; a = bu ($b \in L$) and $End_R(Ku) = K$. Therefore, R is self right mini-injective. We note that $End_R(J)$ as a left R-module properly contains the set of the right-sided multiplications of R, and R is left QF-2. Furthermore, R satisfies the conditions in (10.9.3) as a left R-module. However, R is not left mini-injective.

In case of QF-rings, right artinian and right self-injective rings satisfy the same condition for the left side. However, this fact is not true for the self-mini-injective rings from this example.

3. Let K and L be as in Example 2. Put

$$R = \begin{bmatrix} L & L \\ 0 & K \end{bmatrix}$$

Then R is right weakly mini-injective. However, R is not right QF-2, and hence not right mini-injective. $e_{22}R$ is weakly uni-injective, but not mini-injective [cf. (10.9.5)].

4. Put

$$R = \left\{ \begin{bmatrix} a & b & c \\ o & d & e \\ o & o & a \end{bmatrix} \;\middle|\; \text{a-e are in } K \right\}$$

Then R is weakly mini-injective but not weakly uni-injective, for $f : e_{11}R \to e_{11}J^2$ is not extendible.

5. Put

$$R = \begin{bmatrix} K & uK + vK & K \\ 0 & K & K \\ 0 & 0 & K \end{bmatrix}$$

and $e_{12}(uk_1 + vk_2)e_{23}k_3 = e_{13}(k_1k_3 + k_2k_3)$ for $k_i \in K$. Then $e_{11}R$ is mini-injective. On the other hand, $e_{11}R$ contains two isomorphic uniform submodules (0,uK,K) and (0,vK,K). The above isomorphism is not extendible to an element in $Hom_R(e_{11}R, e_{11}R)$. Hence $e_{11}R$ is not uni-injective.

Notes and References

NOTES

Chapter 1. This chapter is an exposition of the fundamentals of noncommutative ring theory. The reader may consult [6,8,9,150,239].

Chapter 2. The content of this chapter duplicates the first chapters in Mitchell's book [159]. See [65] for (2.2.4), and Stenström's book [244] for (2.2.5). For general references the reader may consult [3,6,8,15,31, 65,73-75,87-89,91,159,194,195,202].

Chapter 3. General references are [6,8,15,26,31,45,58,91,125,195]. The concepts of small submodules and essential submodules are due to Bass [20] and Johnson [133], respectively. Hollow modules appeared in [62]. Minimal epimorphisms and maximal monomorphisms are frequently used in [11] and [12]. As far as we know, the concept of summable matrices first appeared in [13].

Chapter 4. (4.1.2) is due to Cartan [30]. See [7,30,79,175] for related results in Sec. 4.2. To the best of our knowledge, the notion of a cofinitely generated module was defined after 1965. (4.3.6) is due to Auslander and Goldman [242]. The notion of the projective cover is due to [20]. Refer to [31] for the homological material in Sec. 4.3. (4.4.2) is due to Baer [17], and (4.4.7) is due to Eckmann and Schopf [46]. (4.5.9) is due to Hopkins [121]. The proof of (4.5.18) follows Asano's book [241]. (4.5.21) is due to Harada and Sai [118]. The content of Sec. 4.6 is due to Nicholson [179]. (4.7.20) and (4.7.21) are due to Matlis [156]. The notion of a small hull of a ring and (4.7.41) are due to Harada [101] and [102]. General references are [6,8,15,26,31,45,58,91,120,125,195]; [7,30, 79,175] for Sec. 4.2; [11,20,33,36,37,72,77,83,119,172,181,185,230] for Sec. 4.3; [18,22,27,41,44,57,139,156,165,174,200,201,209,237,238] for Sec. 4.4; [7,121,157,176,177] for Sec. 4.5; [84,179] for Sec. 4.6. The reader may consult [171,200,201,209,240] for commutative rings.

Chapter 5. The description of Sec. 5.1 is due to [21]. Morita [163] gave the characterization of generators in $\underline{M}_R$ [cf. (5.2.1)]. (5.2.2) is essentially due to Gabriel and Popescu [75]. The trace ideal in $\underline{M}_R$ is defined in [242]. (5.3.5) is due to [163]. Harada [83] and Auslander [11] independently found (5.4.1). (5.4.2) is due to Johnson and Wong [135], and (5.4.4) is due to Faith and Utumi [59] and its proof is given by Chase (see [56]). (5.4.7) is given by [85]. The contents of (5.4.10) on quasi-projective modules and (5.4.11) are due to Jans and Wu [130]. Almost every result in the first half of Sec. 5.5 is given in [93] and [114]. (5.5.1) is due to [135] and the origins of this section are Faith [55] and Rosenberg and Zelinsky [200]. Some part of the proof of (5.5.12) follows Ikeda and Nakayama [122]. The reader may consult Nagata's book [171] for the last half of Sec. 5.5 (commutative noetherian rings). (5.5.32) and (5.5.36) are due to [93], and (5.5.34) is due to [156]. The content of Sec. 5.6 is due to [163] and Azumaya [14]. The fundamental results in Sec. 5.7 are due to Nakayama [173] and Azumaya and Nakayama's book [15]. (5.7.7) is due to Faith and Walker [60] and (5.7.9) is due to [110] (cf. Kiričenko [147]). (5.8.1) is given by the author and (5.5.2) is due to Vinsonhaler [221,222] (cf. Masaike [155]). The definitions in Sec. 5.8 are due to Thrall [215]. General references for Chapter 5 are [6,8,15,45,91,125,163,195]; [24] for Sec. 5.2; [4,21,71,143,146,190,206,207] for Sec. 5.3; [1,2,28,55,56,69,86, 128,135,148,160] for Sec. 5.4; [28,55,56,59,93,114,122,135,156] for Sec. 5.5; [14,35,158,167,169,183,184,186,211,219,236] for Sec. 5.6; [39,50,134, 164,170,173,214] for Sec. 5.7; [38,40,67,129,149,151,155,192,213,214,215, 221,222] for Sec. 5.8.

Chapter 6. The concept of (semi-) perfect modules is due to Mares [154] and Bass [20]. (6.1.1) is due to Ware and Zelmanowitz [225] (cf. Harada and Kanbara [116]). (6.1.3) is due to [20], (6.1.12) is due to [116], and (6.1.14) is due to [154]. We can find (6.1.7) in [95]. Almost every result in Secs. 6.2-6.4 is given by Harada and Sai [118]. See Ehresmann [47] and Kelly [248]. The reader may consult Azumaya [13] for the idea in Sec. 6.3. (6.3.6) is given by [89]. The content of Sec. 6.5 follows Harada and Kanbara [116]. (6.6.1) is due to [98]. General references for Chapter 6 are [6,15,91,96,195,243,244,245]; [20,25,95,141,154,161,168,204,212] for Sec. 6.1; see [12,47,199] for Sec. 6.2 and [60,116,174,177,193] for Sec. 6.4. The reader may consult [123] for Sec. 6.5.

Chapter 7. The content of Secs. 7.1-7.3 is due to [90,92,116,118]. (7.1.2) is due to [13]. The exposition of Sec. 7.4 follows [100] and that of Sec. 7.5 follows [96]. Results in Sec. 7.6 are given in [20]. The content of Sec. 7.7 is taken from [92] and [96]. The reader may consult [98] and [123] for Sec. 7.8. General references for Chapter 7 are [6,15,58,91,95,96]; [5,10,13,32,42,43,54,60,90,92,94,118,202,227,231,232] for Sec. 7.1; [5,17, 19,44,90,92,94,138] for Sec. 7.2; [53,90,92,94,137] for Sec. 7.3; [100,153, 194,203,209,224,225] for Sec. 7.4; [20,25,33,44,61,140,161,166,198,204,205, 212,229] for Sec. 7.6; [29,63,76,78,210,220] for Sec. 7.7.

Chapter 8. The content of Sec. 8.1 is due to [43] and that of Sec. 8.2 is due to [90] and Kanbara [138]. Almost every result in Secs. 8.3 and 8.4 is given in [97]. The results from (8.4.1) through (8.4.8) are due to Crawly and Jonnson [43]. (8.4.9) is due to Yamagata [235], and (8.4.11) is due to [115]. (8.4.13) is due to Yamagata [233] and [235]. (8.5.1) is due to [179], and (8.5.5) and (8.5.10) were obtained from [115,180,234]. (8.5.20) is due to Kado [136], and (8.6.1) and (8.6.2) are due to [43]. General references for Chapter 8 are [6,96,97,115]; [54,66,138,179,226, 228,235] for Sec. 8.1; [98,123] for Sec. 8.7.

Chapters 9 and 10. Almost everything is new. We shall cite related papers. General references for Chap. 9 are [107-109,115,117,187,188,216-218,246,249, 250,251,252,253]; [182] for Sec. 9.1; [34] and [85] for Sec. 9.3; [51,70,80, 82,104,145,247,255] for Sec. 9.5; [10,126,127] for Secs. 9.6 and 9.7. General references for Chap. 10 are [105,106,189]; [62,64,77,78,103,152,197] for Sec. 10.1; [38,40,67] for Sec. 10.2; [112] for Sec. 10.5; [68,149,151, 192,197] for Sec. 10.6; [16,111] for Sec. 10.8; [16,110,147,173,213] for Sec. 10.10.

REFERENCES

1. Ahsan, J. Submodules of quasi-projective modules. Comm. Algebra, 975-988 (1981).
2. Alamelu, S. On quasi-injectives over noetherian rings. J. Indian Math. Soc. *39*, 121-130 (1975).
3. Albu, T. On commutative Grothendieck categories having a noetherian cogenerator. Arch. Math. *34*, 210-219 (1980).
4. Anderson, F. W. Endomorphism rings of projective modules. Math. Z. *111*, 322-332 (1969).
5. ———, and Fuller, K. R. Modules with decompositions that complement direct summands. J. Algebra *22*, 241-253 (1972).

6. ———, and ———. *Rings and Categories of Modules*. New York: Springer-Verlag, 1973.

7. Artin, E. Zur Theorie der hyperkomplexen Zahlen. Abh. Math. Sem. Univ. Hamburg *5*, 251-260 (1927).

8. ———, Nesbitt, C. J., and Thrall, R. M. *Rings with Minimum Condition*. Ann Arbor: Univ. of Michigan Press, 1944.

9. Asano, K., and Nagao, H. *Group Theory* (in Japanese). Tokyo: Iwanami, 1965.

10. Asashiba, H., and Sumioka, T. On Krull-Schmidt's theorem and the indecomposability of amalgamated sums. (To appear.)

11. Auslander, M. On regular group rings. Proc. Amer. Math. Soc. *8*, 658-664 (1957).

12. ———. Representation theory of artinian algebras, I, II. Comm. Algebra, 177-268 (1974); 269-310 (1974).

13. Azumaya, G. Corrections and supplementations to my paper concerning Krull-Remak-Schmidt's theorem. Nagoya Math. J. *1*, 117-124 (1950).

14. ———. A duality theory for injective modules. Amer. J. Math. *81*, 249-278 (1959).

15. ———, and Nakayama, T. *Algebra, II* (in Japanese). Tokyo: Iwanami, 1954.

16. ———, Mbuntum, F., and Varadarajan, K. On M-projective and M-injective modules. Pacific J. Math. *59*, 9-16 (1957).

17. Baer, R. Abelian groups that are direct summands of every containing abelian group. Bull. Amer. Math. Soc. *46*, 800-806 (1940).

18. ———. Direct decompositions. Trans. Amer. Math. Soc. *62*, 62-98 (1947).

19. ———. Direct decompositions into infinitely many summands. Trans. Amer. Math. Soc. *64*, 519-551 (1948).

20. Bass, H. Finitistic dimension and a homological generalization of semi-primary rings. Trans. Amer. Math. Soc. *95*, 466-488 (1960).

21. ———. The Morita theorems. Lecture notes, Univ. of Oregon, 1962.

22. Beck, I. Σ-Injective module. J. Algebra *21*, 232-249 (1972).

23. ———. Projective and free modules. Math. Z. *129*, 231-234 (1972).

24. Beckwith, H. B. On a trace for modules. Nagoya Math. J. *40*, 125-131 (1970).

25. Björk, J. E. Rings satisfying a minimum condition on principal ideals. J. reine angew. Math. *236*, 112-119 (1969).

26. Bourbaki, N. Algèbre. *Modules et anneaux semi-simples*. Paris: Hermann, 1958, Chap. VIII.

27. Caillear, A. Une caractérisation des modules Σ-injectifs. C. R. Acad. Sci. Paris *A269*, 997-999 (1969).

28. ———, and Renault, G. Anneau associé à une somme directe infinie de modules quasi-injectifs. Arch. Math. *21*, 561-566 (1970).

29. Camillo, V. P., and Zelmanovitz, J. M. Dimension modules. Pacific J. Math. *91*, 249-261 (1980).

30. Cartan, E. Sur les systèmes de nombres complexes. C. R. Acad. Sci. Paris *A124*, 1217-1220 (1897).

31. ———, and Eilenberg, S. *Homological Algebra.* Princeton, N.J.: Princeton Univ. Press, 1956.

32. Chamard, J. Y. Modules riches en coirréducible et sous-modules compléments. C. R. Acad. Sci. Paris *A264*, 987-988 (1967).

33. Chase, S. U. Direct products of modules. Trans. Amer. Math. Soc. *97*, 457-473 (1960).

34. ———. A generalization of the triangular matrices. Nagoya Math. J. *18*, 13-25 (1961).

35. Cheatham, T., and Cumbie, R. Test modules. Proc. Amer. Math. Soc. *49*, 311-314 (1975).

36. ———, and Stone, D. R. Flat and projective character modules. Proc. Amer. Math. Soc. *81*, 175-177 (1981).

37. Choudhury, D. P., and Tewari, K. Tensor purities, cyclic quasi-projectives and cocyclic copurity. Comm. Algebra, 1559-1572 (1979).

38. Colby, R. R., and Rutter, E. A., Jr. Semiprimary QF-3 rings. Nagoya Math. J. *32*, 371-386 (1968).

39. ———, and ———. Generalization of QF-3 algebras. Trans. Amer. Math. Soc. *153*, 371-386 (1971).

40. ———, and ———. Rings which have flat injective modules. J. Algebra *35*, 239-252 (1975).

41. ———, and ———. Exactness and double dual. Proc. Amer. Math. Soc. *82*, 521-526 (1981).

42. Conlon, S. B. An extension of the Krull-Schmidt theorem. Bull. Austral. Math. Soc. *1*, 109-114 (1969).

43. Crawly, P., and Jonnson, B. Refinements for infinite direct decomposition of algebraic systems. Pacific J. Math. *14*, 797-855 (1964).

44. Cunningham, R. S., and Rutter, E. A., Jr. Perfect modules. Math. Z. *140*, 105-110 (1974).

45. Curtis, C. W., and Reiner, I. *Representation Theory of Finite Groups and Associative Algebras.* New York: Interscience, 1962.

46. Eckmann, B., and Schopf, A. Über injektive Moduln. Arch. Math. *4*, 75-78 (1953).

47. Ehresmann, C. *Catégories et structures.* Paris: Dunod, 1965.

48. Eilenberg, S., and MacLane, S. Natural isomorphisms in group theory. Proc. Nat. Acad. Sci. U.S. *A28*, 537-543 (1942).

49. ———, and ———. General theory of natural equivalences. Trans. Amer. Math. Soc. *58*, 231-294 (1945).

50. ———, and Nakayama, T. On the dimension of modules and algebras, II. Nagoya Math. J. *9*, 1-16 (1955).

51. Eisenbund, D., and Griffith, P. The structure of serial rings. Pacific J. Math. *36*, 109-121 (1971).

52. Elliger, S. Zu dem Satz von Krull-Remak-Schmidt-Azumaya. Math. Z. *115*, 227-230 (1970).

53. Elliger, S. Interdirekte Summen von Moduln. J. Algebra *18*, 271-303 (1971).

54. Evans, E. G., Jr. Krull-Schmidt and cancellation over local rings. Pacific J. Math. *46*, 115-121 (1973).

55. Faith, C. Rings with ascending condition on annihilators. Nagoya Math. J. *46*, 179-191 (1966).

56. ———. *Lectures on Injective Modules and Quotient Rings*. Lecture Notes in Mathematics 49. New York: Springer-Verlag, 1967.

57. ———. When are proper cyclics injectives? Pacific J. Math. *45*, 97-112 (1973).

58. ———. *Algebra: Rings, Modules and Categories, I, II*. New York: Springer-Verlag, 1973, 1976.

59. ———, and Utumi, Y. Quasi-injective modules and their endomorphism rings. Arch. Math. *15*, 166-174 (1964).

60. ———, and Walker, E. A. Direct sum representations of injective modules. J. Algebra *5*, 202-221 (1967).

61. Fisher, J. W. Endomorphism of modules. Notices Amer. Math. Soc. *18* (1971).

62. Fleury, P. A note on dualizing Goldie dimension. Canad. Math. Bull. *17*, 511-517 (1974).

63. ———. Hollow modules and local endomorphism rings. Pacific J. Math. *53*, 379-385 (1975).

64. ———. On a class of idealizers. J. Algebra *58*, 342-349 (1979).

65. Freyd, P. *Abelian Categories*. New York: Harper & Row, 1964.

66. Fuchus, L. On quasi-injective modules. Ann. Sc. Norm. Sup. Pisa *23*, 541-546 (1969).

67. Fuller, K. R. The structure of QF-3 rings. Trans. Amer. Math. Soc. *95*, 343-354 (1968).

68. ———. On indecomposable injectives over artinian rings. Pacific J. Math. *29*, 115-135 (1969).

69. ———. On direct representations of quasi-injectives and quasi-projectives. Arch. Math. *20*, 459-502 (1969).

70. ———. On generalized uniserial rings and decompositions that complement direct summands. Math. Ann. *200*, 175-178 (1973).

71. ———. Density and equivalence. J. Algebra *29*, 528-550 (1974).

72. ———, and Shutters, W. A. Projective modules over noncommutative local rings. Tôhoku Math. J. *27*, 303-311 (1975).

73. Gabriel, P. Des catégories abéliennes. Bull Soc. Math. France *90*, 323-488 (1962).

74. ———, and Oberst, U. Spektral Kategorien und regulare Rings in von Neumann Sinn. Math. Z. *92*, 389-395 (1966).

75. ———, and Popescu, N. Caractérisation des catégories abéliennes avec générateurs et limites inductives exactes. C. R. Acad. Sci. Paris *A258*, 4188-4190 (1964).

76. Gentile, E. R. Singular submodule and injective hull. Indag. Math. *24*, 426-433 (1968).

77. Goel, S. C., Jain, S. K., and Singh, S. Rings whose cyclic modules are injective or projective. Proc. Amer. Math. Soc. *53*, 16-18 (1975).

78. Goldie, A. W. Torsion-free modules and rings. J. Algebra *1*, 268-287 (1964).

79. ———. Rings with maximal condition. Lecture notes. Yale Univ., 1964.

80. Griffith, P. On the decomposition of modules and generalized left uniserial rings. Math. Ann. *184*, 300-308 (1970).

81. Grothendieck, A. Sur quelques points d'algebre homologique. Tôhoku Math. J. *9*, 119-221 (1957).

82. Haack, J. K. Self duality and serial rings. J. Algebra *59*, 435-463 (1979).

83. Harada, M. Note on the dimension of modules and algebras. J. Inst. Polytech., Osaka City Univ. *7*, 17-27 (1956).

84. ———. Note on raising idempotent. J. Inst. Polytech., Osaka City Univ. *10*, 63-65 (1959).

85. ———. Note on quasi-injective modules. Osaka J. Math. *2*, 351-356 (1965).

86. ———. Hereditary semiprimary rings and triangular matrix rings. Nagoya Math. J. *27*, 463-484 (1966).

87. ———. On special types of hereditary abelian categories. Osaka J. Math. *4*, 243-255 (1967).

88. ———. On semiprimary abelian categories. Osaka J. Math. *5*, 189-198 (1968)

89. ———. On semisimple abelian categories. Osaka J. Math. *7*, 89-95 (1970).

90. ———. On categories of indecomposable modules, II. Osaka J. Math. *8*, 309-321 (1971).

91. ———. *Introduction to Ring Theory* (in Japanese). Tokyo: Kyoritu, 1971.

92. ———. Supplementary remarks on categories of indecomposable modules. Osaka J. Math. *9*, 49-55 (1972).

93. ———. On quasi-injective modules with a chain condition over a commutative ring. Osaka J. Math. *9*, 421-426 (1972).

94. ———. Note on categories of indecomposable modules. Publ. Dept. Math. Lyon *9*, 11-25 (1972).

95. ———. Perfect categories, I, II, III, IV. Osaka J. Math. *10*, 329-341, 343-355, 357-367, 585-596 (1973).

96. ———. Application of factor categories to completely indecomposable modules. Publ. Dept. Math. Lyon *11*, 19-104 (1974).

97. ———. On the exchange property in a direct sum of indecomposable modules. Osaka J. Math. *12*, 719-736 (1975).

98. ———. A ring-theoretical proof in a factor category of indecomposable modules. J. Math. Soc. Japan *28*, 160-167 (1976).

99. ———. A note on relative T-nilpotency. Osaka J. Math. *13*, 431-438 (1976).

100. ———. Small submodules in a projective module and semi-T-nilpotent sets. Osaka J. Math. *14*, 355-364 (1977).

101. ———. On small ring homomorphisms. Osaka J. Math. *15*, 365-370 (1978).

102. ———. On the small hull of a commutative ring. Osaka J. Math *15*, 679-682 (1978).

103. ———. A note on hollow modules. Rev. Union Mat. Argentina *28*, 186-194 (1978).

104. ———. Non-small modules and non-cosmall modules. In *Ring Theory: Proceedings of the 1978 Antwerp Conference*, F. Van Oystaeyen, ed. New York: Marcel Dekker, 1979, pp. 669-689.

105. ———. Factor rings of a hereditary and QF-3 ring. Osaka J. Math. *17*, 1-8 (1980).

106. ———. On one-sided QF-2 rings, I, II. Osaka J. Math. *17*, 421-431, 433-438 (1980).

107. ———. On lifting property of direct sums of hollow modules. Osaka J. Math. *17*, 783-791 (1980).

108. ———. Modules with the lifting property. Osaka J. Math. *19*, 189-201 (1982).

109. ———. Modules with the extending property. Osaka J. Math. *19*, 203-215 (1982).

110. ———. A characterization of QF-algebras. Osaka J. Math. *20*, 1-4 (1983).

111. ———. On self mini-injective rings. Osaka J. Math *19*, 587-597 (1982).

112. ———. Max. quasi-projective modules. (To appear.)

113. ———. Uniserial rings and lifting properties. Osaka J. Math. *19*, 217-229 (1982).

114. ———, and Ishii, T. On endomorphism rings of noetherian quasi-injective modules. Osaka J. Math. *9*, 217-223 (1972).

115. ———, and ———. On perfect rings and the exchange property. Osaka J. Math. *12*, 483-491 (1975).

116. ———, and Kanbara, H. On categories of projective modules. Osaka J. Math. *8*, 471-483 (1971).

117. ———, and Oshiro, K. On extending property of direct sums of uniform modules. Osaka J. Math. *18*, 767-785 (1981).

118. ———, and Sai, Y. On categories of indecomposable modules, I. Osaka J. Math. *7*, 343-344 (1970).

119. Hiremath, V. A. Finitely projective modules over a Dedekind domain. J. Austral. Math. Soc. Ser. A *26*, 330-336 (1978).

120. Hochschild, G. Cohomology groups of an associative algebra. Ann. Math. *46*, 58-67 (1945).

121. Hopkins, C. Rings with minimal conditions for left ideals. Ann. Math. *40*, 712-730 (1939).

122. Ikeda, M., and Nakayama, T. On some characteristic properties of quasi-Frobenius and regular rings. Proc. Amer. Math. Soc. *5*, 15-19 (1954).

123. Ishii, T. On locally direct summands of modules. Osaka J. Math. *12*, 409-413 (1972).

124. Jacobson, N. The theory of rings. Amer. Math. Soc. Surveys *2* (1943).

125. ———. Structure of rings. Amer. Math. Soc. Colloq. Publ. *37* (1956).

126. Jain, S. K., and Mohamed, S. Rings whose cyclic modules are continuous. J. Indian Math. Soc. *42*, 197-202 (1978).

127. ———, and Mueller, B. Semiperfect rings whose proper modules are continuous. Arch. Math. *37*, 140-143 (1981).

128. ———, Singh, S., and Symonds, G. Rings whose proper cyclic modules are quasi-injective. Pacific J. Math. *67*, 461-472 (1976).

129. Jans, J. P. Projective injective modules. Pacific J. Math. *9*, 1103-1108 (1959).

130. ———, and Wu, L. E. On quasi-projectives. Illinois J. Math. *11*, 439-448 (1967).

131. Jategaonkar, A. V. Morita duality and noetherian rings. J. Algebra *69*, 358-371 (1981).

132. Jeremy, L. Modules et anneaux quasi-continus. Canad. Math. Bull. *17*, 217-228 (1974).

133. Johnson, R. E. The extended centralizer of a ring over a module. Proc. Amer. Math. Soc. *2*, 891-895 (1951).

134. ———, and Wong, E. T. Self-injective rings. Canad. Math. Bull. *2*, 167-173 (1959).

135. ———, and ———. Quasi-injective modules and irreducible rings. J. London Math. Soc. *36*, 260-268 (1961).

136. Kado, J. Note on the exchange property. Math. J. Okayama Univ. *18*, 153-157 (1975-76).

137. Kahlon, U. S. Problem of Krull-Remak-Schmidt-Azumaya-Matlis. J. Indian Math. Soc. *35*, 255-261 (1971).

138. Kanbara, H. Note on Krull-Remak-Schmidt-Azumaya's theorem. Osaka J. Math. *9*, 409-413 (1972).

139. Kaplansky, I. Modules over Dedekind rings and valuation rings. Trans. Amer. Math. Soc. *72*, 327-340 (1952).

140. ———. Projective modules. Ann. Math. *68*, 372-377 (1958).

141. Kasch, F., and Mares, E. Eine Kennzeichnung semi-perfekter Moduln. Nagoya Math. J. *27*, 525-529 (1966).

142. Kato, T. Dominant modules. J. Algebra *14*, 341-349 (1970).

143. ———. Structure of dominant modules. J. Algebra *39*, 563-570 (1976).

144. ———, and Ohtake, K. Morita contexts and equivalences. J. Algebra *61*, 360-366 (1979).

145. Kawada, Y. A generalization of Morita's theorem concerning generalized uniserial algebras. Proc. Japan Acad. *34*, 404-406 (1958).

146. ———. On dominant modules and dominant rings. J. Algebra *56*, 409-435 (1979).

147. Kiricěnko, V. V. On quasi-Frobenius rings and Gorenstein order. Proc. Steklov Inst. Math. *148*, 171-177 (1980).

148. Koehler, A. Quasi-projective and quasi-injective modules. Pacific J. Math. *36*, 713-720 (1971).

149. Krause, G. On rings whose injective indecomposable modules are finitely generated. Tôhoku Math. J. *22*, 333-346 (1970).

150. Kurosh, A. G. *Theory of Groups*. New York: Chelsea, 1955.

151. Kurshan, R. P. Rings whose cyclic injective indecomposable modules have finitely generated socles. J. Algebra *15*, 376-386 (1970).

152. Leonard, W. W. Small modules. Proc. Amer. Math. Soc. *17*, 527-531 (1966).

153. Liebert, W. Radical of some endomorphism rings. J. reine angew. Math. *261/263*, 166-170 (1973).

154. Mares, E. Semiperfect modules. Math. Z. *83*, 347-463 (1963).

155. Masaike, K. QF-3 rings with semiprimary quotient ring. (To appear.)

156. Matlis, E. Injective modules over noetherian rings. Pacific J. Math. *8*, 511-528 (1958).

157. Miller, R. W., and Teply, M. The descending chain condition relative to a torsion theory. Pacific J. Math. *83*, 207-219 (1979).

158. ———, and Turnidge, D. R. Morita duality for endomorphism rings. Proc. Amer. Math. Soc. *31*, 91-94 (1972).

159. Mitchell, B. *Theory of Categories*. New York: Academic, 1965.

160. Miyashita, T. On quasi-injective modules, a generalization of the study of completely reducible modules. J. Fac. Sci. Hokkaiko Univ. *18*, 158-187 (1965).

161. Mohamed, S., and Singh, S. Generalizations of decomposition theorems known over perfect rings. J. Austral. Math. Soc. Ser. A, 496-510 (1977).

162. Monk, G. S. A characterization of exchange rings. Proc. Amer. Math. Soc. *35*, 349-353 (1972).

163. Morita, K. Duality for modules and its application to the theory of rings with minimum condition. Sci. Rep. Tokyo Kyoiku Daigaku *6*, 83-142 (1958).

164. ———, and Tachikawa, H. Character modules, submodules of a free module and quasi-Frobenius rings. Math. Z. *65*, 414-428 (1956).

165. ———, Kawaka, T., and Tachikawa, H. On injective modules. Math. Z. *68*, 217-226 (1957).

166. Motose, K. On the Sexauer-Warnock theorem. J. Fac. Sci. Shinshu Univ. *5*, 33-34 (1970).

167. Mueller, B. J. On Morita duality. Canad. J. Math. *21*, 1338-1347 (1969).

168. ———. Semi-perfect rings. Illinois J. Math. *14*, 464-467 (1970).

169. ———. Linear compactness and Morita duality. J. Algebra *16*, 60-66 (1970).

170. ———. The structure of quasi-Frobenius rings. Canad. J. Math. *26*, 1141-1151 (1974).

171. Nagata, M. *Local Rings*. Tracts in Pure and Applied Mathematics 13. New York: Wiley-Interscience, 1962.

172. Nakano, T. A near semisimple ring. Comm. Math. Univ. Sanci. Pauli *7*, 27-33 (1959).

173. Nakayama, T. On Frobenius algebras, I, II. Ann. Math. *40*, 611-633 (1941); *42*, 1-21 (1941).

174. Năstăsescu, C. Anneaux de quotients et modules Σ(Δ)-injectifs. Inst. Mat. Bucuresti, Preprint Ser. Math. *78* (1979).

175. ———. Conditions de finitude pour les modules, I, II. Rev. Roum. Math. Pure Appl. *24*(5), 745-758 (1979); Inst. Mat. Bucuresti, Preprint Ser. Math. *63* (1979).

176. ———. Théorème de Hopkins pour les catégories de Grothendieck. Inst. Mat. Bucuresti, Preprint Ser. Math. *26* (1980).

177. ———. Quelques observations sur les modules Σ-injectifs. Inst. Mat. Bucuresti, Preprint Ser. Math. *57* (1980).

178. ———, and Popescu, N. Anneaux semi-artiniens. Bull. Math. France *96*, 357-368 (1968).

179. Nicholson, W. K. Lifting idempotents and exchange rings. Trans. Amer. Math. Soc. *229*, 269-278 (1977).

180. ———. On semiperfect modules. Canad. Math. J. *18*, 77-80 (1975).

181. Oberst, L., and Schneider, H. J. Die Struktur von projektiven Moduln. Invent. Math. *13*, 295-304 (1971).

182. Okado, M. Remarks on the lifting property of simple modules. (To appear.)

183. Onodera, T. Über Kogeneratoren. Arch. Math. *19*, 402-410 (1968).

184. ———. Codominant dimensions and Morita equivalences. Hokkaido Math. J. *6*, 169-182 (1977).

185. ———. On modules which are flat over their endomorphism rings. Hokkaido Math. J. *7*, 179-182 (1978).

186. ———. A note on locally compact modules. Kokkaido Math. J. *8*, 121-125 (1979).

187. Oshiro, K. Continuous modules and quasi-continuous modules. (To appear.)

188. ———. Semi-perfect modules and quasi-perfect modules. (To appear.)

189. ———. Lifting modules, extending modules and their applications, to QF-rings. (To appear.)

190. Osima, M. Note on basic rings. Math. J. Ikayama Univ. *2*, 103-110 (1952-53).

191. Osofsky, B. L. A generalization of quasi-Frobenius rings. J. Algebra *4*, 373-387 (1966).

192. ———. Noncommutative rings whose cyclic modules have cyclic injective hulls. Pacific J. Math. *25*, 331-340 (1968).

193. Papp, Z. On algebraically closed modules. Publ. Math. Debrecen *6*, 311-327 (1959).

194. Pareigis, B. *Categories and Functors*. New York: Academic, 1970.

195. Patterson, E. M. On the radical of rings of row-finite matrices. Proc. Roy. Soc. Edinburgh *A66*, 42-46 (1962).

196. Popescu, N. *Abelian Categories with Applications to Rings and Modules*. London: Academic, 1973.

197. Rayer, M. Small and co-small modules. Ph.D. dissertation, Indiana Univ., 1971.

198. Rengswamy, K. M. Modules with finite spanning dimension. Canad. Math. Bull. *20*, 255-262 (1977).

199. Rjabuhin, J. M. Radical in categories. Bull. Akad. Stiince RSS Moldoven *6*, 58-74 (1964).

200. Rosenberg, A., and Zelinsky, D. Finiteness of the injective hull. Math. Z. *70*, 372-380 (1959).

201. ———, and ———. Annihilators. Port. Math. *20*, 53-65 (1961).

202. Sai, Y. On regular categories. Osaka J. Math. *7*, 301-306 (1970).

203. Salver, R. The Jacobson radical of row-finite matrices. J. Algebra *12*, 345-359 (1969).

204. Sandomierski, F. L. On semiperfect and perfect rings. Proc. Amer. Math. Soc. *21*, 205-207 (1969).

205. ———. Modules over the endomorphism ring of a finitely generated module. Proc. Amer. Math. Soc. *31*, 27-31 (1972).

206. Sato, M. Fuller's theorem on equivalences. J. Algebra *52*, 274-284 (1978).

207. ———. On equivalence between module subcategories. J. Algebra *59*, 412-420 (1979).

208. Sexauer, N. F., and Warnock, J. E. The radical of the row-finite matrices over an arbitrary ring. Trans. Amer. Math. Soc. *139*, 287-295 (1965).

209. Sharpe, D. W., and Vamos, P. *Injective Modules*. Cambridge: Cambridge Univ. Press, 1972.

210. Shock, R. C. The ring of endomorphisms of a finite dimension module. Israel J. Math. *11*, 309-314 (1972).

211. Sugano, K. A note on Azumaya's theorem. Osaka J. Math. *4*, 157-160 (1967).

212. Szeto, G. The structure of semiperfect rings. Comm. Algebra, 219-229 (1977).

213. Tachikawa, H. On left QF-3 rings. Pacific J. Math. *32*, 255-268 (1970).

214. ———. *Quasi-Frobenius Rings and Generalizations*. Lecture Notes in Mathematics 351. New York: Springer-Verlag, 1973.

215. Thrall, R. M. Some generalizations of quasi-Frobenius algebra. Trans. Amer. Math. Soc. *64*, 173-183 (1948).

216. Utumi, Y. On continuous regular rings and semi-simple self-injective rings. Canad. J. Math. *12*, 597-605 (1960).

217. ———. On continuous regular rings. Canad. Math. Bull. *4*, 63-69 (1961).

218. ———. On continuous rings and self-injective rings. Trans. Amer. Math. Soc. *118*, 158-173 (1965).

219. Vamos, P. Semi-local noetherian PI-rings. Bull. London Math. Soc. *9*, 251-256 (1977).

220. Varadarajan, K. Dual Goldie dimension. Comm. Algebra, 565-610 (1979).

221. Vinsonhaler, C. I. Orders in QF-3 rings. J. Algebra *14*, 83-90 (1971).

222. ———. Supplement to the paper "Orders in QF-3 rings." J. Algebra *17*, 149-151 (1971).

223. Walker, C. L., and Warfield, R. B., Jr. Unique decomposition and isomorphic refinement theorems in additive categories. J. Pure Appl. Algebra *7*, 347-359 (1976).

224. Ware, R. Endomorphism rings of projective modules. Trans. Amer. Math. Soc. *155*, 233-256 (1971).

225. ———, and Zelmanowitz, J. The radical of the endomorphism ring of a projective module. Proc. Amer. Math. Soc. *26*, 15-20 (1970).

226. Warfield, R. B., Jr. Decompositions of injective modules. Pacific J. Math. *31*, 263-276 (1969).

227. ———. A Krull-Schmidt theorem for infinite sums of modules. Proc. Amer. Math. Soc. *22*, 260-465 (1969).

228. ———. Exchange rings and decompositions of modules. Math. Ann. *199*, 31-36 (1972).

229. Weidenfeld, M., and Weidenfeld, G. Idéaux d'une catégorie préadditive, application aux catégories semi-parfaits. C. R. Acad. Paris *A270*, 1569-1571 (1970).

230. Whitehead, J. M. Projective modules and their trace ideals. Comm. Algebra, 1843-1901 (1980).

231. Yamagata, K. Nonsingular and Matlis's problem. Sci. Rep. Tokyo Kyoiku Daigaku *11*, 114-120 (1972).

232. ———. A note on a problem of Matlis. Proc. Japan Acad. Sci. *49*, 145-147 (1973).

233. ———. The exchange property and direct sums of indecomposable injective modules. Pacific J. Math. *55*, 301-317 (1974).

234. ———. On projective modules with the exchange property. Sci. Rep. Tokyo Kyoiku Daigaku Sec. A, 149-158 (1974).

235. ———. On rings of finite representation type and modules with the finite exchange property. Sci. Rep. Tokyo Kyoiku Daigaku *13*, 1-6 (1975).

236. ———. On Morita duality for additive group valued functors. Comm. Algebra, 367-392 (1979).

237. Zimmermann, W. Über die aufsteignende Kettenbedingung für Annullatoren. Arch. Math. *27*, 261-266 (1976).

238. ———. Rein injektive direkte Summen von Moduln. Comm. Algebra, 1083-1117 (1977).

239. Zassenhaus, H. *The Theory of Groups*. New York: Chelsea, 1949.

240. Zariski, O., and Samuel, P. *Commutative Algebra, I*. Princeton, N.J.: D. Van Nostrand, 1958.

Additional References

241. Asano, K. *Rings and Ideals* (in Japanese). Tokyo: Kyoritu, 1949.

242. Auslander, M., and Goldman, O. Maximal orders. Trans. Amer. Math. Soc. *97*, 1-24 (1960).

243. Beck, I. On modules whose endomorphism ring is local. Israel J. Math. *29*, 393-407 (1978).

244. ———. An independence structure on indecomposable modules. Ark. Mat. *16*, 171-178 (1978).

245. ———. A note on functors that vanish on indecomposable modules. Math. Scand. *50*, 8-18 (1982).

246. Bouhy, T., and Mohamed, S. Continuous modules. Albanian J. Sci. Eng. *2*, 107-112 (1977).

247. Eisenbund, D., and Griffith, P. Serial rings. J. Algebra *17*, 389-400 (1971).

248. Kelly, M. On the radical of a category. J. Austral. Math. Soc. *4*, 299-307 (1964).

249. Mohamed, S., and Mueller, B. J. *Decomposition of Dual Continuous Modules*. Lecture Notes in Mathematics 700. New York: Springer-Verlag, 1979, pp. 87-94.

250. Mueller, B. J., and Rizvi, S. T. On the decomposition of continuous modules. Canad. Math. Bull. *25*, 296-301 (1982).

251. ———. On the existence of continuous hulls. Comm. Algebra *10*, 1819-1838 (1982).

252. ———. On injective and quasi-continuous modules. (To appear.)

253. Rizvi, S. T. Contributions to the theory of continuous modules. Ph.D. thesis, McMaster Univ., Hamilton, Ontario, 1980.

254. Stenström, B. *Rings of Quotients*. Die Grundlehren Mathematischen Wissenschaften 217. New York: Springer-Verlag, 1975.

255. Warfield, R. B., Jr. Serial rings and finitely presented modules. J. Algebra *37*, 187-222 (1975).

Notation

$\underline{(1.1.1)}$	number of formula or defining phrase
(1.1.1)	number of proposition or theorem

Let S and S' be sets.

$S \supset S'$	S properly contains S' or S = S'
$S \supsetneq S'$	S properly contains S' (sometimes $S \supset S'$ means also $S \supsetneq S'$ if there is no confusion)
$\lvert S \rvert$	cardinal of S
$\underline{S}$	set of subsets of S

Let M and $\{M_n\}$ be modules and let R and T be rings.

$\bar{M}$	factor module of M
$\bar{m}$ $(m \in M)$	class of m in $\bar{M}$
ΠM_n	product of $\{M_n\}$
$M^{(I)}$	direct sum of $\lvert I \rvert$ copies of M
$\lVert M \rVert$	composition length of M
$(m_1, m_2, \ldots, m_n)$ or $<m_1, m_2, \ldots, m_n>$	submodule of M generated by $\{m_i\}$
c.ind.	completely indecomposable
f.c.	finitely contained
f.g.	finitely generated
M_R	right R-module
${}_TM_R$	left T- and right R-bimodule
E(M)	injective hull of M
J(M)	Jacobson radical of M
(P, ν)	projective cover
S(M)	socle of M
Z(M)	singular submodule of M

$\ell_M(X)$ $(X \subseteq R)$	$\{m \in M \mid mX = 0\}$
$r_R(A)$ $(A \subseteq M)$	$\{r \in R \mid Ar = 0\}$
$M_1 \precsim M_2$	there exists an epimorphism of M_2 onto M_1
$M_1 \precsim^* M_2$	there exists a monomorphism of M_1 into M_2
$M_1 \subset_e M_2$	M_1 is essential in M_2

Let g be a homomorphism of M_1 to M_2

$g\|M_1'$ $(M_1' \subseteq M_1)$	restriction of g to M_1'
$M_1(g)$	$\{m_1 + g(m_1) \mid m \in M_1\} \subseteq M_1 \oplus M_2$
1_M	identity mapping of M
1_R	identity element in a ring R

Let R be a commutative ring and P a prime.

M[P]	P^∞-divisible submodule of M
M[[P]]	P-primary component of M
M(t)	torsion submodule of M

Let e and f be idempotents.

$e \perp f$	$ef = fe = 0$
$e < f$	$ef = fe = e$

Categories

$\underline{A}$, $\underline{M}$, $\underline{C}(M)$	category
$\underline{J}'$, $\underline{C}$	ideal of category
$\bar{\underline{A}}$, $\bar{\underline{C}}(M)$	factor category

Index